Cluster Algebras, Hall Algebras and Representation Theory

CONTEMPORARY
MATHEMATICS

845

Cluster Algebras, Hall Algebras and Representation Theory

AMS Special Session
Cluster Algebras, Hall Algebras and Representation Theory
April 20–21, 2024
University of Wisconsin-Milwaukee, Milwaukee, WI

Xueqing Chen
Yiqiang Li
Editors

AMERICAN MATHEMATICAL SOCIETY

2020 *Mathematics Subject Classification.* Primary 13-XX, 16-XX, 17-XX, 20-XX, 55-XX.

Library of Congress Cataloging-in-Publication Data

Names: AMS Special Session Cluster Algebras, Hall Algebras, and Representation Theory (2024 : Milwaukee, Wis.) author | Chen, Xueqing, 1971- editor | Li, Yiqiang editor

Title: Cluster algebras, Hall algebras, and representation theory : AMS Special Session Cluster Algebras, Hall Algebras, and Representation Theory, April 20-21, 2024, University of Wisconsin-Milwaukee, WI / [edited by] Xueqing Chen, Yiqiang Li.

Description: Providence, Rhode Island : American Mathematical Society, [2026] | Series: Contemporary mathematics, 0271-4132 ; volume 845 | Includes bibliographical references.

Identifiers: LCCN 2026017174 | ISBN 9781470478711 paperback | ISBN 9781470486099 ebook

Subjects: LCSH: Cluster algebras | Representations of groups | LCGFT: Conference papers and proceedings | Conference papers and proceedings

Classification: LCC QA251.3 .A47 2024

LC record available at https://lccn.loc.gov/2026017174

Contemporary Mathematics ISSN: 0271-4132 (print); ISSN: 1098-3627 (online)

DOI: https://doi.org/10.1090/conm/845

∞ The paper used in this book is acid-free and falls within the guidelines
established to ensure permanence and durability.
Visit the AMS home page at `https://www.ams.org/`

10 9 8 7 6 5 4 3 2 1 31 30 29 28 27 26

CONTEMPORARY MATHEMATICS

845

Cluster Algebras, Hall Algebras and Representation Theory

AMS Special Session
Cluster Algebras, Hall Algebras and Representation Theory
April 20–21, 2024
University of Wisconsin-Milwaukee, Milwaukee, WI

Xueqing Chen
Yiqiang Li
Editors

AMERICAN MATHEMATICAL SOCIETY

2020 *Mathematics Subject Classification.* Primary 13-XX, 16-XX, 17-XX, 20-XX, 55-XX.

Library of Congress Cataloging-in-Publication Data

Names: AMS Special Session Cluster Algebras, Hall Algebras, and Representation Theory (2024 : Milwaukee, Wis.) author | Chen, Xueqing, 1971- editor | Li, Yiqiang editor

Title: Cluster algebras, Hall algebras, and representation theory : AMS Special Session Cluster Algebras, Hall Algebras, and Representation Theory, April 20-21, 2024, University of Wisconsin-Milwaukee, WI / [edited by] Xueqing Chen, Yiqiang Li.

Description: Providence, Rhode Island : American Mathematical Society, [2026] | Series: Contemporary mathematics, 0271-4132 ; volume 845 | Includes bibliographical references.

Identifiers: LCCN 2026017174 | ISBN 9781470478711 paperback | ISBN 9781470486099 ebook

Subjects: LCSH: Cluster algebras | Representations of groups | LCGFT: Conference papers and proceedings | Conference papers and proceedings

Classification: LCC QA251.3 .A47 2024

LC record available at https://lccn.loc.gov/2026017174

Contemporary Mathematics ISSN: 0271-4132 (print); ISSN: 1098-3627 (online)

DOI: https://doi.org/10.1090/conm/845

∞ The paper used in this book is acid-free and falls within the guidelines
established to ensure permanence and durability.
Visit the AMS home page at `https://www.ams.org/`

10 9 8 7 6 5 4 3 2 1 31 30 29 28 27 26

Contents

Preface

Cluster algebras are a class of commutative algebras invented by Fomin and Zelevinsky at the beginning of the century. Since their inception, they have flourished into an active and fruitful field, even earning a designated code in MathSciNet.

Hall algebras date back to the early twentieth century in the study of finite abelian p-groups. In the late twentieth century, Ringel significantly generalized Hall algebras and established the first connection to the theory of quantum groups, a subfield of representation theory inspired by quantum physics that aims to quantize various structures in representation theory.

These three subjects–cluster algebras, Hall algebras, and quantum groups–are intimately related and have significantly influenced each other. For example, Fomin and Zelevinsky's motivation to study cluster algebras stemmed from a desire to understand Lusztig's canonical basis, which arose from the geometric interpretation of Ringel's construction of Hall algebras and can only be seen in quantum groups.

These three subjects have explosive developments in recent years. In this volume, we present some of the recent progress in these areas and their intersections. Some contributions are expository and they summarize and survey recent exciting breakthroughs and topics in the three areas. Some contributions are research articles. They contain original ideas, reveal new insights, connections, and results, enriching our understanding of the areas. We hope that the volume will add to the literature a resource that experts for the three areas and beyond will be referencing in their research.

The idea of compiling this volume is germinated during the AMS Spring Central Sectional Meeting Special Session on Cluster Algebras, Hall algebras and Representation Theory held from April 20-21, 2024, at University of Wisconsin-Milwaukee, WI. After the meeting, we invited several speakers and experts in the field to contribute to this volume. Speakers from the AMS Special Session contribute most of the work in the volume. We also include a work by Mr. Adhish Rele, a participant of the Special Session, which is well aligned with the theme of the volume.

This volume includes a work from Jiepeng Fang, Yixin Lan, and Jie Xiao on the interactions of canonical bases, when the quantum parameter is specialized at 1, and semicanonical bases. They show that the associated colored graphs coincide and that the transition matrix between the two bases is uni-upper triangular.

The work by Ruoyu Guo provides recent progress in representation theory of finite dimensional algebras related to the finitistic dimension conjecture. The latter conjecture is an important homological conjecture and has implications in many other unsolved problems such as the Nakayama conjecture and Gorenstein symmetry conjecture. In the work, three new homological invariants are introducted in the general context of Noetherian rings to upper bound the big finitistic dimension. The new invariants are shown to improve upon the previous upper bound called "delooping level." Applications to tilting theory and contravariantly finite subcategories are also presented.

Fan Li, Jiongkai Pan and Shijie Zhu provide an expository survey on some recent development of representation theory with applications in topological data analysis. They first surveyed $\mathbb{R}$-representations and $\mathbb{S}^1$-representations which serve as technical foundations of modern persistence theory. They then survey applications of poset representations including rank decompositions and homological invariants. They also included a result on morphisms between thin modules over locally finite posets. Some open problems are presented.

Marko and Zubkov's work provides an explicit description of the induced supermodules $H^0_-(\lambda)$ for the queer supergroup $\mathbf{Q}(2)$, characters of simple supermodules $L_-(\lambda)$, simple composition series of $H^0_-(\lambda)$ and blocks of $\mathbf{Q}(2)$ over a field of positive characteristic $p > 2$.

Fan Qin's work present a construction of a so-called triangular bases from cluster algebras, unifying several existing constructions including those related to the coordinate rings of algebraic groups, double Bott-Samelson cells, and braid varieties.

We also include a work on the behavior of Hall algebras associated with quivers with loops from Adhish Rele. The edge contraction is a very basic operation in graph theory. In Li's work, it is shown that Hall algebras behave well under such an operation. His work requires that there is no other arrow joining the incident vertices of the contracted arrow. Adhish Rele's work extends to the general setting that the quiver is allowed to have loops. He shows that the Hall algebra associated with a contracted quiver is still a subalgebra of the Hall algebra of the original quiver.

Sharma, Durai and Nam's work addresses the theory of radical Weyl algebras. His results led to the interesting observation that there are uncountably non-isomorphic infinite dimensional simple Lie algebras.

Finally, we are sincerely grateful to Michael Haggett, Bridget Manzella and Christine Thivierge of the AMS editorial staff for their invaluable support and professional help in bringing this volume to publication.

List of Participants

Pablo Boixeda Alvarez
Yale University

Elie Casbi
Northeastern University

Hunter Dinkins
Massachusetts Institute of Technology

Tucker Jerome Ervin
University of Alabama

Ruoyu Guo
Brandeis University

Peter Koroteev
University at Buffalo

Vasily Krylov
Massachusetts Institute of Technology

Michael K Lau
Universite Laval

Kyungyong Lee
University of Alabama

Fang Li
Zhejiang University

Zongzhu Lin
Kansas State University

Shiping Liu
University of Sherbrooke

Raymond Maresca
Brandeis University

František Marko
Pennsylvania State University

Ki-Bong Nam
University of Wisconsin-Whitewater

Fan Qin
Beijing Normal University

Jie Ren
State University of New York at Buffalo

Daniele Rosso
Indiana University Northwest

Dylan Spence
University of Wisconsin-Whitewater

Gordan Glisa Todorov
Northeastern University

Jie Xiao
Beijing Normal University

Contemporary Mathematics
Volume **845**, 2026
https://doi.org/10.1090/conm/845/16954

The correspondence between the canonical and semicanonical bases

Jiepeng Fang, Yixin Lan, and Jie Xiao

Abstract. Given any symmetric Cartan datum, Lusztig has provided a pair of inductive key lemmas to construct the perverse sheaves over the corresponding quiver and the functions of irreducible components over the corresponding preprojective algebra respectively. In the present article, we prove that these two inductive algorithms of Lusztig coincide. Consequently we can define two $\mathbb{Z}_2 \times I$-colored graphs and prove that they are isomorhic. This result finishes the proof of the statement that Lusztig's functions f_Z of irreducible components form a basis of the enveloping algebra. We can also deduces its crystal structure from those Lusztig's functions. As an application, we prove that the transition matrix between the canonical basis (at $v = 1$) and the semicanonical basis is upper triangular with diagonal entries 1.

1. Introduction

A symmetric Cartan datum $(I, (-,-))$ is given by a finite set I and a symmetric bilinear form $(-,-) : \mathbb{Z}^I \times \mathbb{Z}^I \to \mathbb{Z}$. The associated generalized Cartan matrix $(a_{i,j})_{i,j\in I}$ satisfies $a_{i,i} = 2$ and $a_{i,j} = (i,j)$ for $i \neq j$. Let $\mathfrak{g}$ be the Kac-Moody Lie algebra associated to the generalized Cartan matrix $(a_{i,j})_{i,j\in I}$, let $\mathbf{U}(\mathfrak{g})$ be the enveloping algebra of $\mathfrak{g}$ and let $\mathbf{U}_v(\mathfrak{g})$ be the quantized enveloping algebra. The (quantized) enveloping algebra admits triangular decomposition.

Given a symmetric Cartan datum, we can associate a finite graph $\mathbf{\Gamma}$ without loops such that the set of vertices is I and there are exactly $|a_{i,j}|$ edges connecting i and j for $i \neq j$. Choose an orientation Ω, we get a quiver $Q = (I, \Omega)$. Lusztig considered the moduli space $\mathbf{E}_{\mathbf{V},\Omega}$ of representations of the quiver Q and categorified the positive (or negative) part $\mathbf{U}_v^+(\mathfrak{g})$ of the quantized enveloping algebra via perverse sheaves on $\mathbf{E}_{\mathbf{V},\Omega}$ in [**9**].

More precisely, in [**9**] Lusztig considered the flag variety $\tilde{\mathcal{F}}_{\underline{\nu},\Omega}$ for each flag type $\underline{\nu}$ of $\mathbf{V}$, which is smooth, and defined a proper map $\pi_{\underline{\nu},\Omega} : \tilde{\mathcal{F}}_{\underline{\nu},\Omega} \to \mathbf{E}_{\mathbf{V},\Omega}$. Let $\bar{\mathbb{Q}}_{l,\tilde{\mathcal{F}}_{\underline{\nu},\Omega}}$ be the constant sheaf on $\tilde{\mathcal{F}}_{\underline{\nu},\Omega}$, then $L_{\underline{\nu}} = (\pi_{\underline{\nu},\Omega})_! \bar{\mathbb{Q}}_{l,\tilde{\mathcal{F}}_{\underline{\nu},\Omega}}$ is semisimple by [**3**]. Let $\mathcal{Q}_{\mathbf{V},\Omega}$ be the category consisting of direct sums of shifts of direct summands of such

2020 *Mathematics Subject Classification.* Primary 16G20, 17B37.

J. Fang is supported by the New Cornerstone Science Foundation through the New Cornerstone Investigator Program awarded to Professor Xuhua He. Y. Lan is supported by the National Natural Science Foundation of China [Grant No. 1288201]. J. Xiao is supported by National Natural Science Foundation of China [Grant No. 12031007] and National Natural Science Foundation of China [Grant No. 12471030].

$L_{\underline{v}}$ and let $\mathcal{K}_{\mathbf{V},\Omega}$ be the $\mathbb{Z}[v,v^{-1}]$-module spanned by $\{[L]|L \in \mathcal{Q}_{\mathbf{V},\Omega}\}$ modulo the following relations $[X \oplus Y] = [X]+[Y], [X[1]] = v[X]$. The induction and restriction functors introduced by Lusztig make $\mathcal{K}_{\Omega} = \bigoplus_{\mathbf{V}} \mathcal{K}_{\mathbf{V},\Omega}$ a bialgebra, which is isomorphic to the integeral form ${}_{\mathbb{Z}}\mathbf{U}_v^+(\mathfrak{g})$. The main result of Lusztig's categorification is that the set $\mathcal{P}_{\mathbf{V},\Omega}$ of simple obejects in $\mathcal{Q}_{\mathbf{V},\Omega}$ form a basis of $\mathbf{U}_v^+(\mathfrak{g})_{|\mathbf{V}|}$, which is called the canonical basis. The canonical basis has some remarable properties, like integrality and positivity.

When constructing the canonical basis, Lusztig provided a key lemma (See details in [**9**] or Lemma 3.4 in the present article) to prove that the image of each direct summand of $L_{\underline{v}}$ in $\mathcal{Q}_{\mathbf{V},\Omega}$ is contained in ${}_{\mathbb{Z}}\mathbf{U}_v^+(\mathfrak{g})$. After a more detailed analysis in [**10**], the key lemma for Lusztig's sheaves induces the Kashiwara's operators (in the sense of [**6**]) in a categorification level and implies that Lusztig's sheaves has a crystal structure isomorphic to $B(\infty)$. Actually, with the key lemma, we can define a $\mathbb{Z}_2 \times I$-colored graph $\mathcal{G}_1$ for Lusztig's sheaves, which essentially realizes the crystal graph of $B(\infty)$.

Let $Q = (I, \Omega)$ be the acyclic quiver associated with the given Cartan datum, we let $\Pi(\bar{Q}) = \mathbb{C}\bar{Q}/\mathcal{I}$ be the preprojective algebra of Q. More precisely, let $\bar{Q} = (I, \Omega \cup \bar{\Omega})$ be the double quiver of Q, then $\Pi(\bar{Q})$ is the path algbera of $\bar{Q}$ modulo the addmissible relation $\mathcal{I}$ generated by

$$\sum_{h \in \Omega \cup \bar{\Omega}, h''=i} x_h x_{\bar{h}} - \sum_{h \in \Omega \cup \bar{\Omega}, h'=i} x_{\bar{h}} x_h, i \in I.$$

Lusztig [**11**] introduced the moduli space $\Lambda_{\mathbf{V}}$ of nilpotent representations of $\Pi(\bar{Q})$. By considering the convolution product of the constructible functions on $\Lambda_{\mathbf{V}}$, he defined an algebra $\mathcal{M}$, which is isomorphic to the positive part $\mathbf{U}^+(\mathfrak{g})$. For each irreducible component Z of $\Lambda_{\mathbf{V}}$, Lusztig has constructed some function f_Z. With the result $|Irr\Lambda_{\mathbf{V}}| = \dim_{\mathbb{Q}} \mathbf{U}^+(\mathfrak{g})_{|\mathbf{V}|}$ in [**7**], Lusztig's functions f_Z form a basis of $\mathbf{U}^+(\mathfrak{g})$, which is called the semicanonical basis by Lusztig.

When constructing the functions f_Z, Lusztig provided a key lemma for constructible functions. (See details in [**11**] or Lemma 5.4 in the present article) We have an important observation that the inductive construction of Lusztig's functions f_Z coincides with the inductive algorithm of Lusztig's perverse sheaves. Indeed, we can define a $\mathbb{Z}_2 \times I$-colored graph $\mathcal{G}_2$ for certain equivalent classes of Lusztig's functions f_Z and prove the following main theorem (See details in Section 7): There is a natural isomorphism of $\mathbb{Z}_2 \times I$-colored graph $\Phi : \mathcal{G}_1 \to \mathcal{G}_2$.

Notice that the set of equivalent classes of Lusztig's functions $f_Z \in \mathcal{M}_{\mathbf{V}}$ can be identified with the set of irreducible components of $\Lambda_{\mathbf{V}}$, so we can deduce the crystal structure of $\bigcup_{\mathbf{V}} Irr\Lambda_{\mathbf{V}}$ directly inside Lusztig's construction. As a corollary, using the fact that $|\mathcal{P}_{\mathbf{V},\Omega}| = \dim_{\mathbb{Q}(v)} \mathbf{U}_v^+(\mathfrak{g}) = \dim_{\mathbb{Q}} \mathbf{U}^+(\mathfrak{g})$, we can deduce that Lusztig's functions f_Z form a basis of $\mathbf{U}^+(\mathfrak{g})$ without assuming the main theorem in [**7**] holds.

Taking the classical limit $v \to 1$, the canonical basis becomes a basis of $\mathbf{U}^+(\mathfrak{g})$, still called by the canonical basis. Then the isomorphism Φ of $\mathbb{Z}_2 \times I$-colored graphs gives the correspondence between the canonical and semicanonical bases of $\mathbf{U}^+(\mathfrak{g})$. Notice that there are natural constructible vector bundles ψ_{Ω} and $\tilde{\psi}_{\Omega}$, under the correspondence given by Φ, our second main result (Theorem 8.6) is that the transition matrix between the canonical basis and the semicanonical basis is upper triangular with all diagonal entries equal to 1. Notice that Geiss, Leclerc and Schroer have provided an example in [**4**] to show that these two bases are not equal. As a byproduct, we also obtain a construction of monomial basis of $\mathbf{U}^+(\mathfrak{g})$ depending on an order of I.

In Section 2, we define the convolution algebras $\mathcal{M}_\Omega$ and $\mathcal{M}$ of the quiver and its preprojective algebra respectively and prove that the pushforward of the natural projection ψ_Ω induces an isomorphism of algebras from $\mathcal{M}$ to $\mathcal{M}_\Omega$. In Section 3, we recall the geometric realization of $\mathbf{U}_v^+(\mathfrak{g})$ by Lusztig [**9**] and the key inductive lemmas for perverse sheaves. In Section 4, we define the $\mathbb{Z}_2 \times I$-colored graph for Lusztig's perverse sheaves and study the commutative relations of arrows in the graph. In Section 5, we recall the construction of Lusztig's functions f_Z in [**11**] and the key inductive lemmas for those f_Z. In Section 6, we define the $\mathbb{Z}_2 \times I$-colored graph for equivalent classes of Lusztig's functions and study the commutative relations of arrows in the graph. In Section 7, we prove our first main result that the $\mathbb{Z}_2 \times I$-colored graphs defined in Section 4 and Section 6 are isomorphic and deduce the results in [**7**]. In Section 8, with an order of I, we define an order of vertices of $\mathbb{Z}_2 \times I$-colored graphs and construct a monomial basis. We also prove the second main theorem: The transition matrix between the canonical basis and the semicanonical basis is upper triangular with all diagonal entries equal to 1.

2. The convolution algebras on $\mathbf{E}_{\mathbf{V},\Omega}$ and $\Lambda_{\mathbf{V}}$

2.1. The variety $\mathbf{E}_{\mathbf{V},\Omega}$ and $\Lambda_{\mathbf{V}}$. Given a symmetric Cartan datum $(I,(-,-))$, let $\mathbf{\Gamma}$ be the finite graph without loops associated to $(I,(-,-))$. Let I be its set of vertices and H be the set of pairs consisting of egdes with orientation. More precisely, to give an egde with orientation is equivalent to give $h', h'' \in I$ and we adapt the notation $h' \xrightarrow{h} h''$. Let . $h \mapsto \bar{h}$ be the involution of H such that $\bar{h}' = h'', \bar{h}'' = h'$ and $\bar{h} \neq h$. An orientation of the graph Γ is a subset $\Omega \subset H$ such that $\Omega \cap \bar{\Omega} = \emptyset$ and $\Omega \cup \bar{\Omega} = H$.

Let $k = \bar{k}$ be an algebraic closed field. Given $v \in \mathbb{N}^I$, a subset $\tilde{H} \subseteq H$ and a graded k-vector space $\mathbf{V}$ with dimension vector $|\mathbf{V}| = v$, we let

$$\mathbf{E}_{\mathbf{V}} = \bigoplus_{h \in H} \mathrm{Hom}(\mathbf{V}_{h'}, \mathbf{V}_{h''})$$

$$\mathbf{E}_{\mathbf{V},\tilde{H}} = \bigoplus_{h \in \tilde{H}} \mathrm{Hom}(\mathbf{V}_{h'}, \mathbf{V}_{h''})$$

In particular, for an orientation Ω, we let

$$\mathbf{E}_{\mathbf{V},\Omega} = \bigoplus_{h \in \Omega} \mathrm{Hom}(\mathbf{V}_{h'}, \mathbf{V}_{h''})$$

The algebraic group $G_{\mathbf{V}} = \prod_{i \in I} \mathbf{GL}(\mathbf{V}_i)$ acts on $\mathbf{E}_{\mathbf{V}}, \mathbf{E}_{\mathbf{V},\tilde{H}}$ and $\mathbf{E}_{\mathbf{V},\Omega}$ by $(g \cdot x)_h = g_{h''} x_h g_{h'}^{-1}$.

We fix a function $\epsilon : H \to k^*$ such that $\epsilon(h) + \epsilon(\bar{h}) = 0$ for all h. The moment map $\mu : \mathbf{E}_{\mathbf{V}} \to \mathbf{gl}_{\mathbf{V}} = \bigoplus_{i \in I} \mathbf{End}(\mathbf{V}_i)$ is defined by $(\mu(x))_i = \sum_{h \in H, h''=i} \epsilon(h) x_h x_{\bar{h}}$. We let $\Lambda_{\mathbf{V}}$ be the closed subvariety of $\mathbf{E}_{\mathbf{V}}$ consisting of nilponent elements x such that $\mu(x) = 0$. It is easy to see that $\Lambda_{\mathbf{V}}$ is $G_{\mathbf{V}}$-stable.

2.2. Convolution algebra arising from $\mathbf{E}_{\mathbf{V},\Omega}$. If $f: X \to Y$ is a morphism of varieties and g is a constructible function on Y, then following [**12**], we define $f^*(g)$ to be the constructible function such that $f^*(g)(x) = g(f(x))$. If h is a constructible function on X, then we define $f_!(h)$ to be the constructible function such that $f_!(h)(y) = \sum_{c \in \mathbb{Q}} \chi(h^{-1}(c) \cap f^{-1}(y))$. Here $\chi(X)$ means the Euler characteristic for the constructible set X.

Assume that $k = \mathbb{C}$ and we fix an orientation Ω. Given $v' + v'' = v \in \mathbb{N}^I$ and graded vector spaces $\mathbf{V}, \mathbf{V}', \mathbf{V}''$ with dimension vectors v, v', v'' respectively, let $\mathbf{E}'_\Omega$ be the set consisting of $(x, \tilde{\mathbf{W}}, \rho_1, \rho_2)$ where $x \in \mathbf{E}_{\mathbf{V},\Omega}, \tilde{\mathbf{W}}$ is a x-stable subspace of $\mathbf{V}$ with dimension vector v'' and $\rho_1 : \mathbf{V}/\tilde{\mathbf{W}} \simeq \mathbf{V}', \rho_2 : \tilde{\mathbf{W}} \simeq \mathbf{V}''$ are linear isomorphisms. We also let $\mathbf{E}''_\Omega$ be the set consisting of $(x, \tilde{\mathbf{W}})$ as above. Here we say $\tilde{\mathbf{W}}$ is x-stable if and only if $x_h(\tilde{\mathbf{W}}_{h'}) \subset \tilde{\mathbf{W}}_{h''}$ for any $h \in \Omega$.

Lusztig has introduced the following diagram in [**9**]:

$$\mathbf{E}_{\mathbf{V}',\Omega} \times \mathbf{E}_{\mathbf{V}'',\Omega} \xleftarrow{p_1} \mathbf{E}'_\Omega \xrightarrow{p_2} \mathbf{E}''_\Omega \xrightarrow{p_3} \mathbf{E}_{\mathbf{V},\Omega}$$

where $p_1(x, \tilde{\mathbf{W}}, \rho_1, \rho_2) = (\rho_{1,*}(\bar{x}|_{\mathbf{V}/\tilde{\mathbf{W}}}), \rho_{2,*}(x|_{\tilde{\mathbf{W}}}))$, $p_2(x, \tilde{\mathbf{W}}, \rho_1, \rho_2) = (x, \tilde{\mathbf{W}})$ and $p_3(x, \tilde{\mathbf{W}}) = x$. Here $x|_{\tilde{\mathbf{W}}}$ is the restriction of x on the subspace $\tilde{\mathbf{W}}$ and $\bar{x}|_{\mathbf{V}/\tilde{\mathbf{W}}}$ is the natural induced linear map of x on the quotient space $\mathbf{V}/\tilde{\mathbf{W}}$. $\rho_{1,*}(\bar{x}|_{\mathbf{V}/\tilde{\mathbf{W}}}) = \rho_1 \circ (\bar{x}|_{\mathbf{V}/\tilde{\mathbf{W}}}) \circ \rho_1^{-1}$ is the element in $\mathbf{E}_{\mathbf{V}'}$ induced by $\bar{x}|_{\mathbf{V}/\tilde{\mathbf{W}}}$ and $\rho_{2,*}(x|_{\tilde{\mathbf{W}}}) = \rho_2 \circ (x|_{\tilde{\mathbf{W}}}) \circ \rho_2^{-1}$ is the element in $\mathbf{E}_{\mathbf{V}''}$ induced by $x|_{\tilde{\mathbf{W}}}$.

Notice that p_1 is smooth with connected fibers, p_2 is a principle $G_{\mathbf{V}'} \times G_{\mathbf{V}''}$-bundle and p_3 is proper. (See details in [**9**].)

Now let $\tilde{M}(\mathbf{E}_{\mathbf{V},\Omega})$ be the $\mathbb{Q}$-space consisting of all $G_{\mathbf{V}}$-invariant constructible functions $f : \mathbf{E}_{\mathbf{V},\Omega} \to \mathbb{Q}$. We define a bilinear operator $* : \tilde{M}(\mathbf{E}_{\mathbf{V}',\Omega}) \times \tilde{M}(\mathbf{E}_{\mathbf{V}'',\Omega}) \to \tilde{M}(\mathbf{E}_{\mathbf{V},\Omega})$ as the following: Given $f_1 \in \tilde{M}(\mathbf{E}_{\mathbf{V}',\Omega})$, $f_2 \in \tilde{M}(\mathbf{E}_{\mathbf{V}'',\Omega})$, $f_1 * f_2 = (p_3)_! f''$ where f'' is the unique constructible function on $\mathbf{E}''_\Omega$ such that $p_2^*(f'') = p_1^*(f_1 \otimes f_2)$. Lusztig has proved in [**9**] that $*$ makes $\tilde{\mathcal{M}}_\Omega = \bigoplus_{\mathbf{V}} \tilde{M}(\mathbf{E}_{\mathbf{V},\Omega})$ be an associative algebra. This is a constructible function version of Ringle-Hall algebra in [**13**].

Denote the constant function (taking value 1) on $\mathbf{E}_{\mathbf{V},\Omega}$ with $|\mathbf{V}| = mi$ by $\mathbf{1}_{i^m} = \mathbf{1}_{i(m)}$ and let $\mathcal{M}_\Omega$ be the subalgebra of $\tilde{\mathcal{M}}_\Omega$ generated by those $\mathbf{1}_{i^m}$. We denote $\mathcal{M}_\Omega \cap \tilde{M}(\mathbf{E}_{\mathbf{V},\Omega})$ by $\mathcal{M}_{\mathbf{V},\Omega}$. Then following Section 10.19 of [**9**] or [**13**], we can see that $\mathcal{M}_\Omega$ is isomorphic to the enveloping algebra $\mathbf{U}^+ = \mathbf{U}^+(\mathfrak{g})$ of the Kac-Moody Lie algebra $\mathfrak{g}$ associated with the Cartan datum which corresponds to the given graph $\mathbf{\Gamma}$.

2.3. Convolution algebra arising from $\Lambda_{\mathbf{V}}$. Given $v' + v'' = v \in \mathbb{N}^I$ and graded vector spaces $\mathbf{V}, \mathbf{V}', \mathbf{V}''$ with dimension vectors v, v', v'' respectively, we let Λ' be the set consisting of $(x, \tilde{\mathbf{W}}, \rho_1, \rho_2)$ where $x \in \Lambda_{\mathbf{V}}$, $\tilde{\mathbf{W}}$ is a x-stable subspace of $\mathbf{V}$ with dimension v'' and $\rho_1 : \mathbf{V}/\tilde{\mathbf{W}} \simeq \mathbf{V}', \rho_2 : \tilde{\mathbf{W}} \simeq \mathbf{V}''$ are linear isomorphisms. We also let Λ'' be the set consisting of $(x, \tilde{\mathbf{W}})$ as above. Then there is a diagram in [**9**]:

$$\Lambda_{\mathbf{V}'} \times \Lambda_{\mathbf{V}''} \xleftarrow{p} \Lambda' \xrightarrow{r} \Lambda'' \xrightarrow{q} \Lambda_{\mathbf{V}}$$

$p(x, \tilde{\mathbf{W}}, \rho_1, \rho_2) = (\rho_{1,*}(\bar{x}|_{\mathbf{V}/\tilde{\mathbf{W}}}), \rho_{2,*}(x|_{\tilde{\mathbf{W}}}))$, $r(x, \tilde{\mathbf{W}}, \rho_1, \rho_2) = (x, \tilde{\mathbf{W}})$ and $q(x, \tilde{\mathbf{W}}) = x$. Notice that r is a principle $G_{\mathbf{V}'} \times G_{\mathbf{V}''}$-bundle and q is proper.

Now let $\tilde{M}(\Lambda_{\mathbf{V}})$ be the $\mathbb{Q}$-space consisting of all $G_{\mathbf{V}}$-invariant constructible functions $f : \Lambda_{\mathbf{V}} \to \mathbb{Q}$. We define a bilinear operator $* : \tilde{M}(\Lambda_{\mathbf{V}'}) \times \tilde{M}(\Lambda_{\mathbf{V}''}) \to \tilde{M}(\Lambda_{\mathbf{V}})$ as the following:

$$f_1 * f_2 = q_! f''$$

where $f_1 \in \tilde{M}(\Lambda_{\mathbf{V}'}), f_2 \in \tilde{M}(\Lambda_{\mathbf{V}''})$ and f'' is the unique constructible function on Λ'' such that $r^*(f'') = p^*(f_1 \otimes f_2)$. Then $*$ makes $\tilde{\mathcal{M}} = \bigoplus_{\mathbf{V}} \tilde{M}(\Lambda_{\mathbf{V}})$ be an associative algebra. Denote the constant function (taking value 1) on $\Lambda_{\mathbf{V}}$ with $|\mathbf{V}| = mi$ by $\mathbf{1}_{i^m} = \mathbf{1}_{i^{(m)}}$ and let $\mathcal{M}$ be the $\mathbb{Q}$-subalgebra of $\tilde{\mathcal{M}}$ generated by those $\mathbf{1}_{i^m}$. We denote $\mathcal{M} \cap \tilde{M}(\Lambda_{\mathbf{V}})$ by $\mathcal{M}_{\mathbf{V}}$ and denote the integral form of $\mathcal{M}$ by ${}_{\mathbb{Z}}\mathcal{M}$. More precisely, ${}_{\mathbb{Z}}\mathcal{M} = \mathrm{span}_{\mathbb{Z}}\{\mathbf{1}_{i_1^{m_1}} * \mathbf{1}_{i_2^{m_2}} \cdots * \mathbf{1}_{i_l^{m_l}} \mid l \in \mathbb{N}, i_j \in I, m_j \in \mathbb{N}$ for any $1 \leq j \leq l\}$. Lusztig has proved in [**9**] that $\mathcal{M}$ is canonically isomorphic to $\mathbf{U}^+(\mathfrak{g})$.

2.4. The flag varieties and the map ψ_Ω. For any $\underline{v} = (v^1, v^2, \cdots, v^m)$ with each $v^l \in \mathbb{N}^I$, let $\tilde{\mathcal{F}}_{\underline{v}}$ be the variety consisting of (x, f) where $x \in \Lambda_{\mathbf{V}}$ and $f = (0 = \mathbf{V}^m \subset \mathbf{V}^{m-1} \cdots \subset \mathbf{V}^1 \subset \mathbf{V}^0 = \mathbf{V})$ is an x-stable flag of type $\underline{v}$. More precisely, $\dim(\mathbf{V}^{l-1}/\mathbf{V}^l) = v^l$ for any l. There is a proper morphism $\pi_{\underline{v}} : \tilde{\mathcal{F}}_{\underline{v}} \to \Lambda_{\mathbf{V}}$. Denote the constant function on $\tilde{\mathcal{F}}_{\underline{v}}$ by $\mathbf{1}_{\tilde{\mathcal{F}}_{\underline{v}}}$.

Then we have the following result:

LEMMA 2.1. *Let $\underline{v} = (v^1, v^2, \cdots, v^l)$ be a flag type such that $v^k = m_k i_k$ for $1 \leq k \leq l$, then $(\pi_{\underline{v}})_!(\mathbf{1}_{\tilde{\mathcal{F}}_{\underline{v}}}) = \mathbf{1}_{i_1^{m_1}} * \mathbf{1}_{i_2^{m_2}} \cdots * \mathbf{1}_{i_l^{m_l}}$.*

PROOF. We prove the lemma by induction on l. When $l = 1$, the lemma holds trivially. Now we assume $\mathbf{1}_{i_1^{m_1}} * \mathbf{1}_{i_2^{m_2}} \cdots * \mathbf{1}_{i_{l-1}^{m_{l-1}}} = (\pi_{\underline{v}'})_!(\mathbf{1}_{\tilde{\mathcal{F}}_{\underline{v}'}})$ where $\underline{v}' = (v^1, v^2, \cdots v^{l-1})$. Now we fix a decomposition $\mathbf{V} = \mathbf{V}' \oplus \mathbf{V}''$ with $|\mathbf{V}''| = v^l$ and $|\mathbf{V}'| + |\mathbf{V}''| = |\mathbf{V}|$. Let F be the subvariety of $\Lambda_{\mathbf{V}}$ consisting of $x \in \Lambda_{\mathbf{V}}$ such that $\mathbf{V}'$ is x-stable and $\tilde{\mathcal{F}}_{v,0}$ be the subvariety of $\tilde{\mathcal{F}}_v$ consisting of (x, f) such that the $(l-1) - th$ component $\mathbf{V}^{l-1}$ is exactly $\mathbf{V}'$. Then the restriction of $\pi_{\underline{v}}$ on $\tilde{\mathcal{F}}_{\underline{v},0}$ gives a morphism $u : \tilde{\mathcal{F}}_{\underline{v},0} \to F$ and it induces a proper map $\tilde{u} : G_{\mathbf{V}} \times_{Q_{\mathbf{V}}} \tilde{\mathcal{F}}_{\underline{v},0} \to G_{\mathbf{V}} \times_{Q_{\mathbf{V}}} F = \Lambda''$. Here $Q_{\mathbf{V}}$ is the stablizer of $\mathbf{V}'$ in $G_{\mathbf{V}}$. Then we can check by definition that

$$r^*(\tilde{u})_!(\mathbf{1}_{G_{\mathbf{V}} \times_{Q_{\mathbf{V}}} \tilde{\mathcal{F}}_{\underline{v},0}}) = p^*((\pi_{\underline{v}'})_!(\mathbf{1}_{\tilde{\mathcal{F}}_{\underline{v}'}}) \otimes \mathbf{1}_{i_l^{m_l}}).$$

Notice that $(r\tilde{u})_!((\mathbf{1}_{G_{\mathbf{V}} \times_{Q_{\mathbf{V}}} \tilde{\mathcal{F}}_{\underline{v},0}}) = (\pi_{\underline{v}})_!(\mathbf{1}_{\tilde{\mathcal{F}}_{\underline{v}}})$, we are done. □

REMARK 2.2. By the above lemma, it is easy to see that

$$(\pi_{\underline{v}'})_!(\mathbf{1}_{\tilde{\mathcal{F}}_{\underline{v}'}}) * (\pi_{\underline{v}''})_!(\mathbf{1}_{\tilde{\mathcal{F}}_{\underline{v}''}}) = (\pi_{\underline{v}'\underline{v}''})_!(\mathbf{1}_{\tilde{\mathcal{F}}_{\underline{v}'\underline{v}''}}).$$

This is an analogy of Lemma 3.2 in [**9**].

Similarly, we can define $\tilde{\mathcal{F}}_{\underline{v},\Omega}$ to be the smooth variety consisting of (x, f) where $x \in \mathbf{E}_{\mathbf{V},\Omega}$ and $f = (0 = \mathbf{V}^m \subset \mathbf{V}^{m-1} \cdots \subset \mathbf{V}^1 \subset \mathbf{V}^0 = \mathbf{V})$ is an x-satble flag of type $\underline{v}$. More precisely, $\mathbf{dim}(\mathbf{V}^{l-1}/\mathbf{V}^l) = v^l$ for any l. There is a proper morphism $\pi_{\underline{v},\Omega} : \tilde{\mathcal{F}}_{\underline{v}} \to E_{\mathbf{V},\Omega}$. Denote the constant function on $\tilde{\mathcal{F}}_{\underline{v},\Omega}$ by $\mathbf{1}_{\tilde{\mathcal{F}}_{\underline{v},\Omega}}$. Then by the same argument in Lemma 2.1, the following equation holds in in $\mathcal{M}_\Omega$ for a flag type $\underline{v} = (v^1, v^2, \cdots, v^l)$ satisfying $v^k = m_k i_k$ for $1 \leq k \leq l$:

$$(\pi_{\underline{v},\Omega})_!(\mathbf{1}_{\tilde{\mathcal{F}}_{\underline{v},\Omega}}) = \mathbf{1}_{i_1^{m_1}} * \mathbf{1}_{i_2^{m_2}} \cdots * \mathbf{1}_{i_l^{m_l}}.$$

Fix an orientation Ω, there is a natural map $\psi_\Omega : \mu^{-1}(0) \to \mathbf{E}_{\mathbf{V},\Omega}$ defined by forgetting the components in $\bar{\Omega}$:

$$\psi_\Omega((x_h)_{h\in H}) = ((x_h)_{h\in\Omega}).$$

Then ψ_Ω induces a natural morphism $\tilde{\psi}_\Omega : \tilde{\mathcal{F}}_{\underline{v}} \to \tilde{\mathcal{F}}_{\underline{v},\Omega}$, which is defined by forgetting the components of $x \in \Lambda_{\mathbf{V}}$ in $\bar{\Omega}$:

$$\tilde{\psi}_\Omega((x_h)_{h\in H}, f) = ((x_h)_{h\in\Omega}, f).$$

PROPOSITION 2.3. *Each fiber of the map ψ_Ω and $\tilde{\psi}_\Omega$ is isomorphic to an affine space. Moreover, the following diagram is commutative:*

$$\begin{array}{ccc} \tilde{\mathcal{F}}_{\underline{v}} & \xrightarrow{\tilde{\psi}_\Omega} & \tilde{\mathcal{F}}_{\underline{v},\Omega} \\ \downarrow{\scriptstyle \pi_{\underline{v}}} & & \downarrow{\scriptstyle \pi_{\underline{v},\Omega}} \\ \mu^{-1}(0) & \xrightarrow{\psi_\Omega} & \mathbf{E}_{\mathbf{V},\Omega} \end{array}$$

PROOF. To give a fiber of ψ_Ω at $(x_h)_{h\in\Omega}$ is equivalent to give the solution $(x_h)_{h\in\bar{\Omega}}$ of $\mu(x) = 0$, which is a family of homogenous linear equations for fixed $(x_h)_{h\in\Omega}$, hence each fiber of ψ_Ω is an affine space.

Notice that if $x \in \mu^{-1}(0)$ fixes a flag, then x is automatically nilpotent. We can see that for a fixed $(x_h)_{h\in\Omega} \in \mathbf{E}_{\mathbf{V}}$ and an $(x_h)_{h\in\Omega}$-stable flag $f = (0 = \mathbf{V}^m \subset \mathbf{V}^{m-1} \cdots \subset \mathbf{V}^1 \subset \mathbf{V}^0 = \mathbf{V})$, to give a fiber of $\tilde{\psi}_\Omega$ at (x, f) is equivalent to give the solution of $\mu(x) = 0$ and $x_h(\mathbf{V}^l_{h'}) \subseteq \mathbf{V}^l_{h''}$ for any $0 \leq l \leq m$ and $h \in \bar{\Omega}$, which is also a family of homogenous linear equations. Hence each fiber of $\tilde{\psi}_\Omega$ is an affine space.

The diagram commutes by definition. □

After extensions by zero, the constructible functions on $\Lambda_{\mathbf{V}}$ can be naturally regarded as constructible functions on $\mu^{-1}(0)$. Then $(\psi_\Omega)_! : \mathcal{M} \to \mathcal{M}_\Omega$ is well-defined.

COROLLARY 2.4. *The map ψ_Ω induces a surjective homomorphism of algebras $(\psi_\Omega)_! : \mathcal{M} \to \mathcal{M}_\Omega$.*

PROOF. We only need to prove that $(\psi_\Omega)_!((\pi_{\underline{v}})_!(\mathbf{1}_{\tilde{\mathcal{F}}_{\underline{v}}})) = (\pi_{\underline{v},\Omega})_!(\mathbf{1}_{\tilde{\mathcal{F}}_{\underline{v},\Omega}})$.

Since each fiber of $\tilde{\psi}_\Omega$ is isomorphic to an affine space , whose Euler characteristic equals to 1, $(\tilde{\psi}_\Omega)_!(\mathbf{1}_{\tilde{\mathcal{F}}_{\underline{v}}}) = \mathbf{1}_{\tilde{\mathcal{F}}_{\underline{v},\Omega}}$. Since $\psi_\Omega \pi_{\underline{v}} = \pi_{\underline{v},\Omega}\tilde{\psi}_\Omega$, we are done. □

Lusztig has proved that the generators $\mathbf{1}_{i^m}$ satisfy the Serre relation in $\mathcal{M}$ (see details in[**9**] Lemma 12.11), hence there is a surjetive algebra morphism $\Psi : \mathbf{U}^+ \to \mathcal{M}$ given by $\Psi(e_i^{(m)}) = \mathbf{1}_{i^m}$. Here $e_i^{(m)}$ is the m-th divided power of the Chevalley generator e_i. Combining this result with the corollary above, we can see that the following theorem holds:

THEOREM 2.5. *The map $(\psi_\Omega)_! : \mathcal{M} \to \mathcal{M}_\Omega$ is indeed an isomorphism of algebras.*

PROOF. The composition $\mathbf{U}^+ \xrightarrow{\Psi} \mathcal{M} \xrightarrow{(\psi_\Omega)_!} \mathcal{M}_\Omega \cong \mathbf{U}^+$ is surjective, sending $e_i^{(p)}$ to itself. Hence it is an isomorphism and so does $(\psi_\Omega)_!$. □

Given $v' + v'' = v \in \mathbb{N}^I$ and graded vector spaces $\mathbf{V}, \mathbf{V}', \mathbf{V}''$ with dimension vectors v, v', v'' respectively, assume $\underline{v}' = ((v')^1, (v')^2, \cdots, (v')^{m'})$ is a flag type of $\mathbf{V}'$ and $\underline{v}'' = ((v'')^1, (v'')^2, \cdots, (v'')^{m''})$ is a flag type of $\mathbf{V}''$. Let $\tilde{\mathcal{F}}'_{\underline{v}'\underline{v}''}$ be the set consisting of (x, f, ρ_1, ρ_2) where $x \in \Lambda_{\mathbf{V}}$, $f = (0 = \mathbf{V}^{m'+m''} \subset \mathbf{V}^{m'+m''-1} \cdots \subset \mathbf{V}^1 \subset \mathbf{V}^0 = \mathbf{V})$ is an x-stable flag of type $\underline{v}'\underline{v}''$ and $\rho_1 : \mathbf{V}/\mathbf{V}^{m''} \cong \mathbf{V}', \rho_2 : \mathbf{V}^{m''} \cong \mathbf{V}''$ are linear isomorphisms. Similarly, let $\tilde{\mathcal{F}}'_{\underline{v}'\underline{v}'',\Omega}$ be the set consisting of (x, f, ρ_1, ρ_2) where $x \in \mathbf{E}_{\mathbf{V},\Omega}$, $f = (0 = \mathbf{V}^{m'+m''} \subset \mathbf{V}^{m'+m''-1} \cdots \subset \mathbf{V}^1 \subset \mathbf{V}^0 = \mathbf{V})$ is an x-stable flag of type $\underline{v}'\underline{v}''$ and $\rho_1 : \mathbf{V}/\mathbf{V}^{m''} \cong \mathbf{V}', \rho_2 : \mathbf{V}^{m''} \cong \mathbf{V}''$ are linear isomorphisms. There are natural morphisms defined as the following:

$$\tilde{\mathcal{F}}'_{\underline{v}'\underline{v}''} \xrightarrow{\tilde{r}} \tilde{\mathcal{F}}_{\underline{v}'\underline{v}''} : (x, f, \rho_1, \rho_2) \mapsto (x, f),$$

$$\tilde{\mathcal{F}}_{\underline{v}'\underline{v}''} \to \Lambda'' : (x, f) \mapsto (x, \mathbf{V}^{m''}),$$

$$\tilde{\mathcal{F}}'_{\underline{v}'\underline{v}''} \to \Lambda' : (x, f, \rho_1, \rho_2) \mapsto (x, \mathbf{V}^{m''}, \rho_1, \rho_2),$$

and

$$\tilde{\mathcal{F}}'_{\underline{v}'\underline{v}''} \xrightarrow{\tilde{p}} \tilde{\mathcal{F}}_{\underline{v}'} \times \tilde{\mathcal{F}}_{\underline{v}''} : (x, f, \rho_1, \rho_2) \mapsto ((\rho_{1,*}(\bar{x}|_{\mathbf{V}/\mathbf{V}^{m''}}), f_1)(, \rho_{2,*}(x|_{\mathbf{V}^{m''}}), f_2)),$$

where f_1 is the flag of $\mathbf{V}' \cong \mathbf{V}/\mathbf{V}^{m''}$ induced from f and f_2 is the flag of $\mathbf{V}'' \cong \mathbf{V}^{m''}$ induced from f.

Similarly, we can set:

$$\tilde{\mathcal{F}}'_{\underline{v}'\underline{v}'',\Omega} \xrightarrow{\tilde{p}_2} \tilde{\mathcal{F}}_{\underline{v}'\underline{v}'',\Omega} : (x, f, \rho_1, \rho_2) \mapsto (x, f),$$

$$\tilde{\mathcal{F}}_{\underline{v}'\underline{v}'',\Omega} \to \mathbf{E}''_\Omega : (x, f) \mapsto (x, \mathbf{V}^{m''}),$$

$$\tilde{\mathcal{F}}'_{\underline{v}'\underline{v}'',\Omega} \to \mathbf{E}'_\Omega : (x, f, \rho_1, \rho_2) \mapsto (x, \mathbf{V}^{m''}, \rho_1, \rho_2),$$

and

$$\tilde{\mathcal{F}}'_{\underline{v}'\underline{v}'',\Omega} \xrightarrow{\tilde{p}_1} \tilde{\mathcal{F}}_{\underline{v}',\Omega} \times \tilde{\mathcal{F}}_{\underline{v}'',\Omega} : (x, f, \rho_1, \rho_2) \mapsto (\rho_{1,*}(\bar{x}|_{\mathbf{V}/\mathbf{V}^{m''}}), f_1)(, \rho_{2,*}(x|_{\mathbf{V}^{m''}}), f_2).$$

Then we have the following commutative diagram:

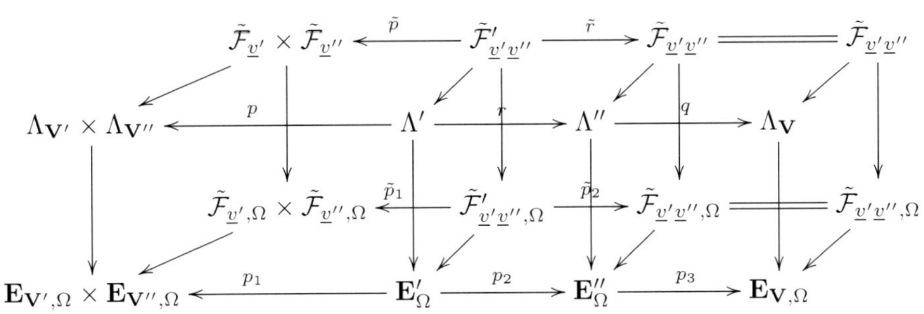

where all morphisms from up to down are naturally induced by ψ_Ω, whose fibers are affine spaces, and the other outward morphisms are the naturally projection given by forgetting flags. The commutative diagram above gives another explaination of Corollary 2.4 and suggests that the two different construction of $\mathbf{U}^+(\mathfrak{g})$ have more detailed relations.

3. Lusztig's sheaves and the quantized enveloping algebra

3.1. Induction functor and restriction functor. Consider the following diagram defined in Section 2.2:

$$\mathbf{E}_{\mathbf{V}',\Omega}\times\mathbf{E}_{\mathbf{V}'',\Omega}\xleftarrow{p_1}\mathbf{E}'_{\Omega}\xrightarrow{p_2}\mathbf{E}''_{\Omega}\xrightarrow{p_3}\mathbf{E}_{\mathbf{V},\Omega}$$

Let d_1 be the dimension of the fibers of p_1 and d_2 be the dimension of the fibers of p_2.

We fix a decomposition $\mathbf{T}\oplus\mathbf{W}=\mathbf{V}$ of graded vector space, let F_Ω be the closed subvariety of $E_{\mathbf{V}}$ consisting of x such that $\mathbf{W}$ is x-stable. Let $Q_{\mathbf{V}}$ be the stablizer of $\mathbf{W}$ in $G_{\mathbf{V}}$ and let d_3 be the dimension of $G_{\mathbf{V}}/Q_{\mathbf{V}}$. Consider the following diagram

$$\mathbf{E}_{\mathbf{T},\Omega}\times\mathbf{E}_{\mathbf{W},\Omega}\xleftarrow{\kappa_\Omega}F_\Omega\xrightarrow{\iota_\Omega}\mathbf{E}_{\mathbf{V},\Omega}$$

where ι_Ω is the natural embedding and $\kappa_\Omega(x)=(x',x'')\in\mathbf{E}_{\mathbf{T},\Omega}\times\mathbf{E}_{\mathbf{W},\Omega}$ is induced by $x\in F$. Notice that κ_Ω is a vector bundle.

Since for any $\underline{v}=(v^1,v^2,\cdots,v^m)$ with each $v^l\in\mathbb{N}^I$, the flag variety $\tilde{\mathcal{F}}_{\underline{v},\Omega}$ is smooth and the map $\pi_{\underline{v},\Omega}$ is proper, we can see that $L_{\underline{v}}=(\pi_{\underline{v},\Omega})_!\bar{\mathbb{Q}}_l$ is a semisimple perverse sheaf on $\mathbf{E}_{\mathbf{V},\Omega}$ by [**3**]. Here $\bar{\mathbb{Q}}_l$ is the constant sheaf on $\tilde{\mathcal{F}}_{\underline{v},\Omega}$.

Let $\mathcal{D}^b_{G_{\mathbf{V}}}(\mathbf{E}_{\mathbf{V},\Omega})$ be the $G_{\mathbf{V}}$-equivariant derived category of constructible sheaves on $E_{\mathbf{V},\Omega}$, and $\mathcal{Q}_{\mathbf{V},\Omega}$ be the subcategory consisting of direct sums of shifts of direct summands of such $L_{\underline{v}}$. We say L is a Lusztig's sheaf if it is a simple object in $\mathcal{Q}_{\mathbf{V},\Omega}$.

Let $\mathcal{K}_{\mathbf{V},\Omega}$ be the Grothendieck group of $\mathcal{Q}_{\mathbf{V},\Omega}$ and $\mathcal{K}_\Omega=\bigoplus_{\mathbf{V}}\mathcal{K}_{\mathbf{V},\Omega}$. More precisely, $\mathcal{K}_{\mathbf{V},\Omega}$ is a $\mathbb{Z}[v,v^{-1}]$-module spanned by $\{[L]|L\in\mathcal{Q}_{\mathbf{V},\Omega}\}$ modulo the following relations:

$$[X\oplus Y]=[X]+[Y],$$
$$[X[1]]=v[X].$$

Let $\mathcal{P}_{\mathbf{V},\Omega}$ be the set of simple perverse sheaves in $\mathcal{Q}_{\mathbf{V},\Omega}$ and denote the union of $\mathcal{P}_{\mathbf{V},\Omega}$ by $\mathcal{P}_\Omega$, then the images $\{[L]|L\in\mathcal{P}_\Omega\}$ of $\mathcal{P}_\Omega$ in $\mathcal{K}_\Omega$ indeed form a $\mathbb{Z}[v,v^{-1}]$-basis of $\mathcal{K}_\Omega$, which is called the canonical basis. (See details in [**9**].) In particular, We denote the image in $\mathcal{K}_\Omega$ of constant sheaf $\bar{\mathbb{Q}}_l$ on $\mathbf{E}_{\mathbf{V},\Omega}$ with $|\mathbf{V}|=pi$ by $E_i^{(p)}$.

The Verdier duality functor $\mathbf{D}:\mathcal{D}^b_{G_{\mathbf{V}}}(\mathbf{E}_{\mathbf{V},\Omega})\to\mathcal{D}^b_{G_{\mathbf{V}}}(\mathbf{E}_{\mathbf{V},\Omega})$ induces a $\mathbb{Z}$-linear bar involution $^-:\mathcal{K}_\Omega\to\mathcal{K}_\Omega$ such that $\overline{[L]}=[L]$ for any $L\in\mathcal{P}_\Omega$. (See details in [**10**].)

Lusztig has introduced the induction functor in [**9**]

$$\mathbf{Ind}^{\mathbf{V}}_{\mathbf{V}',\mathbf{V}''}=(p_3)_!(p_2)_\flat(p_1)^*:\mathcal{D}^b_{G_{\mathbf{V}'}}(\mathbf{E}_{\mathbf{V}',\Omega})\times\mathcal{D}^b_{G_{\mathbf{V}''}}(\mathbf{E}_{\mathbf{V}'',\Omega})\to\mathcal{D}^b_{G_{\mathbf{V}}}(\mathbf{E}_{\mathbf{V},\Omega})$$

and the restriction functor

$$\mathbf{Res}^{\mathbf{V}}_{\mathbf{T},\mathbf{W}}=(\kappa_\Omega)_!(\iota_\Omega)^*:\mathcal{D}^b_{G_{\mathbf{V}}}(\mathbf{E}_{\mathbf{V},\Omega})\to\mathcal{D}^b_{G_{\mathbf{V}'}\times G_{\mathbf{V}''}}(\mathbf{E}_{\mathbf{V}',\Omega}\times\mathbf{E}_{\mathbf{V}'',\Omega})$$

Lusztig has also proved the following results (See details in [**9**] Section 3 and Section 4.):

$$\mathbf{Ind}^{\mathbf{V}}_{\mathbf{V}',\mathbf{V}''}(L_{\underline{v}'}\boxtimes L_{\underline{v}''})=L_{\underline{v}'\underline{v}''},$$
$$\mathbf{Res}^{\mathbf{V}}_{\mathbf{T},\mathbf{W}}L_{\underline{v}}=\bigoplus_{\underline{\tau}+\underline{\omega}=\underline{v}}L_{\underline{\tau}}\boxtimes L_{\underline{\omega}}[-2M(\underline{\tau},\underline{\omega})],$$

here $M(\underline{\tau},\underline{\omega})=\sum_{h\in\Omega,l'<l}\tau^{l'}_{h'}\omega^l_{h''}+\sum_{i\in I,l<l'}\tau^{l'}_i\omega^l_i$.

Set $\mathbf{ind}^{\mathbf{V}}_{\mathbf{V}',\mathbf{V}''} = \mathbf{Ind}^{\mathbf{V}}_{\mathbf{V}',\mathbf{V}''}[d_1 - d_2]$, then we can define a $\mathbb{Z}[v, v^{-1}]$-bilinear operator $* : \mathcal{K}_\Omega \otimes \mathcal{K}_\Omega \to \mathcal{K}_\Omega$ as the following: Given $A \in \mathcal{Q}_{\mathbf{V}',\Omega}, B \in \mathcal{Q}_{\mathbf{V}'',\Omega}$, we set

$$[A] * [B] = [\mathbf{ind}^{\mathbf{V}}_{\mathbf{V}',\mathbf{V}''}(A \boxtimes B)].$$

Then $(\mathcal{K}_\Omega, *)$ is an associative algebra. Lusztig has proved that there is an isomorphism of algebras from $(\mathcal{K}_\Omega, *)$ to ${}_{\mathbb{Z}}U_q^+(\mathfrak{g})$, sending $E_i^{(p)}$ to the p-th divided power $e_i^{(p)}$ of the Chevalley generator e_i in ${}_{\mathbb{Z}}U_q^+(\mathfrak{g})$. (See detials in [**9**] Section 10.)

Set $\mathbf{res}^{\mathbf{V}}_{\mathbf{T},\mathbf{W}} = \mathbf{Res}^{\mathbf{V}}_{\mathbf{T},\mathbf{W}}[d_1 - d_2 - 2d_3]$, then we can define a $\mathbb{Z}[v, v^{-1}]$-linear operator $\bar{\Delta} : \mathcal{K}_\Omega \to \mathcal{K}_\Omega \otimes \mathcal{K}_\Omega$ as the following: Given $A \in \mathcal{Q}_{\mathbf{V},\Omega}$, we set

$$\bar{\Delta}([A]) = [\bigoplus_{\mathbf{T},\mathbf{W}} \mathbf{res}^{\mathbf{V}}_{\mathbf{T},\mathbf{W}}(A)],$$

then $\bar{\Delta}$ gives a coalgebra structure of $\mathcal{K}_\Omega$.

Similarly, we can also define $\Delta : \mathcal{K}_\Omega \to \mathcal{K}_\Omega \otimes \mathcal{K}_\Omega$ as the following: Given $A \in \mathcal{Q}_{\mathbf{V},\Omega}$, we set

$$\Delta([A]) = [\bigoplus_{\mathbf{T},\mathbf{W}} \mathbf{D}(\mathbf{res}^{\mathbf{V}}_{\mathbf{T},\mathbf{W}}(\mathbf{D}A))],$$

then $\bar{\Delta} = (^- \otimes ^-) \circ \Delta \circ ^-$ and Δ also gives a coalgebra structure of $\mathcal{K}_\Omega$. The coproduct Δ indeed realizes the coproduct of ${}_{\mathbb{Z}}U_q^+(\mathfrak{g})$. (See detials in [**9**] Section 10.)

3.2. Fourier–Deligne transformation. Assume k is an algebraic closure of a finite field $\mathbb{F}_q$ and fix a nontrivial character $\mathbb{F}_q \to \bar{\mathbb{Q}}_l^*$. This character defines an Artin-Schreier local system of rank 1 on k.

For two orientations Ω, Ω', we define

$$T : \mathbf{E}_{\mathbf{V},\Omega\cup\Omega'} \to k$$

by $T(x) = \sum_{h \in \Omega \backslash \Omega'} tr(x_h x_{\bar{h}})$. Then the inverse image of the Artin-Schreier local system under T is a well-defined $G_{\mathbf{V}}$-equivariant local system of rank 1 on $\mathbf{E}_{\mathbf{V},\Omega\cup\Omega'}$ and we denote it by $\mathcal{L}_T$.

Consider the following diagram:

$$\mathbf{E}_{\mathbf{V},\Omega} \xleftarrow{\delta} \mathbf{E}_{\mathbf{V},\Omega\cup\Omega'} \xrightarrow{\delta'} \mathbf{E}_{\mathbf{V},\Omega'}$$

here δ, δ' are the forgetting maps defined by:

$$\delta((x_h)_{h\in\Omega\cup\Omega'}) = ((x_h)_{h\in\Omega}),$$
$$\delta'((x_h)_{h\in\Omega\cup\Omega'}) = ((x_h)_{h\in\Omega'}).$$

Then following [**9**] we can define a functor

$$\mathcal{F}_{\Omega,\Omega'} : \mathcal{D}^b_{G_{\mathbf{V}}}(E_{\mathbf{V},\Omega}) \to \mathcal{D}^b_{G_{\mathbf{V}}}(E_{\mathbf{V},\Omega'}), \mathcal{F}_{\Omega,\Omega'}(L) = \delta'_!(\delta^*(L) \otimes \mathcal{L}_T)[D],$$

here $D = \sum_{h\in\Omega\backslash\Omega'} \dim \mathbf{V}_{h'} \dim \mathbf{V}_{h''}$.

The functor $\mathcal{F}_{\Omega,\Omega'}$ is called the Fourier-Deligne transformation for quivers by Lusztig. Moreover, he has proved the following proposition:

PROPOSITION 3.1 ([**9**, Theorem 5.4]). *With the notations above, we have:*

$$\mathcal{F}_{\Omega,\Omega'}(\mathbf{Ind}^{\mathbf{V}}_{\mathbf{V}',\mathbf{V}''} L_1 \boxtimes L_2) \cong \mathbf{Ind}^{\mathbf{V}}_{\mathbf{V}',\mathbf{V}''}(\mathcal{F}_{\Omega,\Omega'}(L_1) \boxtimes \mathcal{F}_{\Omega,\Omega'}(L_2))[C],$$

where $C = \sum_{h\in\Omega\backslash\Omega'} (v''_{h'} v'_{h''} - v'_{h'} v''_{h''})$.

With the proposition above, we can easily get the following two results:

COROLLARY 3.2 ([**9**, Theorem 10.14]). *The functor $\mathcal{F}_{\Omega,\Omega'}$ induces an isomorphism between algebras $\mathcal{K}_\Omega$ and $\mathcal{K}_{\Omega'}$.*

COROLLARY 3.3 ([**9**, Corollary 5.6]). *The functor $\mathcal{F}_{\Omega,\Omega'}$ indeuces a bijection $\mathcal{P}_\Omega$ and $\mathcal{P}_{\Omega'}$ and defiens a equivalence of categories $\mathcal{Q}_{\mathbf{V},\Omega} \cong \mathcal{Q}_{\mathbf{V},\Omega'}$. Moreover, if we denote the bijection between the sets $\mathcal{P}_\Omega$ and $\mathcal{P}_{\Omega'}$ by $\eta_{\Omega,\Omega'}$, then $\eta_{\Omega',\Omega''}\eta_{\Omega,\Omega'} = \eta_{\Omega,\Omega''}$.*

With the two corollaries above, we can denote $\mathcal{K}_\Omega$ by $\mathcal{K}$ and $\mathcal{P}_{\mathbf{V},\Omega}$ by $\mathcal{P}_\mathbf{V}$ respectively if there is no ambiguity.

3.3. Bilinear pairing. Following [**5**] and [**10**], we choose a smooth irreducible variety Γ with free $G_\mathbf{V}$-action such that the $\bar{\mathbb{Q}}_l$-cohomology of Γ vanishes in degree $1, 2 \cdots m$ for a large enough m, then $G_\mathbf{V}$ acts freely on $\Gamma \times \mathbf{E}_{\mathbf{V},\Omega}$. Let us consider the diagram

$$\mathbf{E}_{\mathbf{V},\Omega} \xleftarrow{s} \Gamma \times \mathbf{E}_{\mathbf{V},\Omega} \xrightarrow{t} G_\mathbf{V}\backslash\Gamma \times \mathbf{E}_{\mathbf{V},\Omega}$$

here $G_\mathbf{V}\backslash\Gamma \times \mathbf{E}_{\mathbf{V},\Omega}$ is the quotient space.

Let $u : G_\mathbf{V}\backslash\Gamma \times \mathbf{E}_{\mathbf{V},\Omega} \to \{pt\}$ and $d = dim(G_\mathbf{V}\backslash\Gamma)$, then for any $A, B \in \mathcal{P}_{\mathbf{V},\Omega}$ and $j \in \mathbb{Z}$, we denote the $\bar{\mathbb{Q}}_l$-vector space $\mathcal{H}^{j+2d}(u_!(t_\flat s^* A \otimes t_\flat s^* B))$ by $\mathbf{D}_j(\mathbf{V}, A, B)$. This vector space does not depend on the choice of Γ and m. (See details in [**8**] or [**5**].).

Lusztig introduced a symmetric $\mathbb{Z}[v, v^{-1}]$-bilinear pairing $(-,-) : \mathcal{K}_{\mathbf{V},\Omega} \times \mathcal{K}_{\mathbf{V},\Omega} \to \mathbb{Z}((v))$ by:

$$([L],[K]) = \sum_j \dim \mathbf{D}_j(\mathbf{V}, L, K)v^j$$

Here $L, K \in \mathcal{P}_{\mathbf{V},\Omega}$ are simple perverse sheaves. In particular, $(E_i, E_i) = (1-v^{-2})^{-1}$.

Set $(\mathcal{K}_{\mathbf{V},\Omega}, \mathcal{K}_{\mathbf{V}',\Omega}) = 0$ for $|\mathbf{V}| \neq |\mathbf{V}'|$, then $(-,-) : \mathcal{K}_\Omega \times \mathcal{K}_\Omega \to \mathbb{Z}((v))$ is a well-defined symmetric $\mathbb{Z}[v, v^{-1}]$-bilinear pairing. Moreover, the pairing $(-,-)$ satisfies the following property: (See details in [**10**] Chapter 10 and Chapter 13.)

(1)$(-,-)$ doesn't depend on the choice of Ω;

(2) $([L],[L']) \in v^{-1}\mathbb{Z}[[v^{-1}]] \cap \mathbb{Q}(v)$ if $[L] \neq [L']$ for $L, L' \in \mathcal{P}_{\mathbf{V},\Omega}$;

(3) $([L],[L]) \in 1 + v^{-1}\mathbb{Z}[[v^{-1}]] \cap \mathbb{Q}(v)$ for $L \in \mathcal{P}_{\mathbf{V},\Omega}$.

3.4. Analysis at sink. Fix $i \in I$ and an orientation Ω such that i is a sink, we define $\mathbf{E}_{\mathbf{V},i,p}$ to be the subset of $\mathbf{E}_{\mathbf{V},\Omega}$ consisting of x such that $\mathrm{codim}_{\mathbf{V}_i}(\mathrm{Im} \bigoplus_{h\in\Omega, h''=i} x_h) = p$. Then $\mathbf{E}_\mathbf{V}$ has a partition $\mathbf{E}_{\mathbf{V},\Omega} = \bigcup_p \mathbf{E}_{\mathbf{V},i,p}$. and the union $\mathbf{E}_{\mathbf{V},i,\geq p} = \bigcup_{p' \geq p} \mathbf{E}_{\mathbf{V},i,p'}$ is a closed subset of $\mathbf{E}_{\mathbf{V},\Omega}$.

Given $L \in \mathcal{P}_{\mathbf{V},\Omega}$, there exists a unique integer t such that $supp(L) \subseteq \mathbf{E}_{\mathbf{V},i,\geq t}$ but $supp(L) \nsubseteq \mathbf{E}_{\mathbf{V},i,\geq t+1}$ and we write $t_i(L) = t$. Notice that $t_i(L) \leq v_i$.

The following lemma is proved by Lusztig, which is called the key lemma in [**10**]:

LEMMA 3.4 ([**9**, Lemma 6.4]). *With the notation above, fix $0 \leq t \leq v_i$. Assume $|\mathbf{T}| = |\mathbf{V}'| = ti$ and let $d = t(v_i - t)$.*

(1) Let $L \in \mathcal{P}_{\mathbf{V},\Omega}$ be such that $t_i(L) = t$, then $\mathbf{Res}^\mathbf{V}_{\mathbf{T},\mathbf{W}} L \in \mathcal{Q}_{\mathbf{W},\Omega}$ is a direct sum of finitely many summands of the form $K'[f']$ for various $K' \in \mathcal{P}_{\mathbf{W},\Omega}$ and $f' \in \mathbb{Z}$. Moreover, exactly one of these summands, denoted by $K[f]$, satisfies $t_i(K) = 0$ and $f = d$ and the others satisfy $t_i(K') > 0$.

(2) Let $K \in \mathcal{P}_{\mathbf{V}'',\Omega}$ be such that $t_i(K)=0$, then $\mathbf{Ind}^{\mathbf{V}}_{\mathbf{V}',\mathbf{V}''}(\bar{\mathbb{Q}}_l \boxtimes K)$ is a direct sum of finitely many summands of the form $L'[g']$ for variuos $L' \in \mathcal{P}_{\mathbf{V},\Omega}$ and $g' \in \mathbb{Z}$. Moreover, exactly one of these summands, denoted by $L[g]$, satisfies $t_i(L)=t$ and $g=d$ and the others satisfy $t_i(L')>t$.

(3) There is a bijectiion

$$\pi_{i,t} : \{K \in \mathcal{P}_{\mathbf{V}'',\Omega} | t_i(K)=0\} \to \{L \in \mathcal{P}_{\mathbf{V},\Omega} | t_i(L)=t\}$$

induced by the decompositions of the direct sums above.

If $|\mathbf{V}'| = ri$, we denote $\mathbf{V}'$ by $\mathbf{V}'_r$ and $\mathbf{V}''$ by $\mathbf{V}''_r$. For an orientation Ω' and $L \in \mathcal{P}_{\mathbf{V},\Omega'}$, we define $s_i(L)$ to be the largest integer r satisfying that there exists $L' \in \mathcal{P}_{\mathbf{V}''_r,\Omega'}$ such that L is isomorphic to a shift of a direct summand of $\mathbf{Ind}^{\mathbf{V}}_{\mathbf{V}'_r,\mathbf{V}''_r}(\bar{\mathbb{Q}}_l \boxtimes L')$. Notice that the definition of $s_i(L)$ does not depend on the choice of Ω' by Proposition 3.1.

PROPOSITION 3.5 ([**9**, Proposition 6.6]). *With the notations above,*

(1) There exist $L'_{r'} \in \mathcal{P}_{\mathbf{V}''_{r'},\Omega'}$ for $r' > s_i(L)$ and $L''_{r'} \in \mathcal{P}_{\mathbf{V}''_{r'},\Omega'}$ for $r' \geq s_i(L)$ such that

$$L \oplus \bigoplus_{r' > s_i(L)} \mathbf{Ind}^{\mathbf{V}}_{\mathbf{V}'_{r'},\mathbf{V}''_{r'}}(\bar{\mathbb{Q}}_l \boxtimes L'_{r'}) \cong \bigoplus_{r' \geq s_i(L)} \mathbf{Ind}^{\mathbf{V}}_{\mathbf{V}'_{r'},\mathbf{V}''_{r'}}(\bar{\mathbb{Q}}_l \boxtimes L''_{r'}).$$

(2) $s_i(L) = t_i(L)$ if i is a sink in Ω'.

We sketch a proof which is given by Lusztig in [**9**] as follows:

PROOF. Without loss of generality, we assume that i is a sink in Ω'. By Lemma 3.4, we can see that $s_i(L) \geq t_i(L)$. Assume that $L' \in \mathcal{P}_{\mathbf{V}''_r,\Omega'}$ is a shift of a direct summand of $L_{\underline{\nu}}$ such that L is isomorphic to a shift of a direct summand of $\mathbf{Ind}^{\mathbf{V}}_{\mathbf{V}'_r,\mathbf{V}''_r}(\bar{\mathbb{Q}}_l \boxtimes L')$, then L is a shift of a direct summand of $L_{\underline{\nu}'}$ for $\underline{\nu}' = (ri, \underline{\nu})$. Notice that $supp(L_{\underline{\nu}'}) \subset E_{\mathbf{V},\Omega',\geq r}$, we can see that $t_i(L) \geq r$. Hence $t_i(L) \geq s_i(L)$. (2) is proved. By desending induction on $s_i(L) = t_i(L)$, (1) can be proved easily. □

3.5. Analysis at source. Fix $i \in I$ and an orientation Ω such that i is a source, we define $\mathbf{E}^p_{\mathbf{V},i}$ to be the subset of $\mathbf{E}_{\mathbf{V},\Omega}$ consisting of x such that $\dim(\mathrm{Ker} \bigoplus_{h \in \Omega, h'=i} x_h) = p$. Then $\mathbf{E}_{\mathbf{V}}$ has a partition $\mathbf{E}_{\mathbf{V},\Omega} = \bigcup_p \mathbf{E}^p_{\mathbf{V},i}$. and the union $\mathbf{E}^{\geq p}_{\mathbf{V},i} = \bigcup_{p' \geq p} \mathbf{E}^{p'}_{\mathbf{V},i}$ is a closed subset.

Given $L \in \mathcal{P}_{\mathbf{V},\Omega}$, there exists a unique integer t such that $supp(L) \subseteq \mathbf{E}^{\geq t}_{\mathbf{V},i}$ but $supp(L) \nsubseteq \mathbf{E}^{\geq t+1}_{\mathbf{V},i}$ and we write $t^*_i(L) = t$. Notice that $t^*_i(L) \leq \nu_i$.

By a similar arguments, we can prove the following result, which is dual to Lemma 3.4:

LEMMA 3.6. *With the notation above, fix $0 \leq t \leq \nu_i$. Assume $|\mathbf{W}| = |\mathbf{V}''| = ti$ and let $d = t(\nu_i - t)$.*

(1) Let $L \in \mathcal{P}_{\mathbf{V},\Omega}$ be such that $t^*_i(L) = t$, then $\mathbf{Res}^{\mathbf{V}}_{\mathbf{T},\mathbf{W}} L \in \mathcal{Q}_{\mathbf{T},\Omega}$ is a direct sum of finitely many summands of the form $K'[f']$ for various $K' \in \mathcal{P}_{\mathbf{T},\Omega}$ and $f' \in \mathbb{Z}$. Moreover, exactly one of these summands, denoted by $K[f]$, satisfies $t^*_i(K) = 0$ and $f = d$ and the others satisfy $t^*_i(K') > 0$.

(2) Let $K \in \mathcal{P}_{\mathbf{V}',\Omega}$ be such that $t^*_i(K) = 0$, then $\mathbf{Ind}^{\mathbf{V}}_{\mathbf{V}',\mathbf{V}''}(K \boxtimes \bar{\mathbb{Q}}_l)$ is a direct sum of finitely many summands of the form $L'[g']$ for variuos $L' \in \mathcal{P}_{\mathbf{V},\Omega}$ and $g' \in \mathbb{Z}$.

Moreover, exactly one of these summands, denoted by $L[g]$, satisfies $t_i^*(L)=t$ and $g=d$ and the others satisfy $t_i^*(L')>t$.

(3) There is a bijectiion

$$\pi_{i,t}^*:\{K\in\mathcal{P}_{\mathbf{V}',\Omega}|t_i^*(K)=0\}\to\{L\in\mathcal{P}_{\mathbf{V},\Omega}|t_i^*(L)=t\}$$

induced by the decompositions of the direct sums above.

In this section, if $|\mathbf{V}''|=ri$, we denote $\mathbf{V}'$ by $\mathbf{V}'_r$ and $\mathbf{V}''$ by $\mathbf{V}''_r$. For an orientation Ω' and $L\in\mathcal{P}_{\mathbf{V},\Omega'}$, we define $s_i^*(L)$ to be the largest integer r satisfying that there exists $L'\in\mathcal{P}_{\mathbf{V}'_r,\Omega'}$ such that L is isomorphic to a direct summand of $\mathbf{Ind}^{\mathbf{V}}_{\mathbf{V}'_r,\mathbf{V}''_r}(L'\boxtimes\bar{\mathbb{Q}}_l)$.

With the notations above, we have the following proposition, which is dual to Proposition 3.5:

PROPOSITION 3.7. *(1) There exist $L'_{r'}\in\mathcal{P}_{\mathbf{V}''_{r'},\Omega'}$ for $r'>s_i^*(L)$ and $L''_{r'}\in\mathcal{P}_{\mathbf{V}''_{r'},\Omega'}$ for $r'\geq s_i^*(L)$ such that*

$$L\oplus\bigoplus_{r'>s_i^*(L)}\mathbf{Ind}^{\mathbf{V}}_{\mathbf{V}'_{r'},\mathbf{V}''_{r'}}(L'_{r'}\boxtimes\bar{\mathbb{Q}}_l)\cong\bigoplus_{r'\geq s_i^*(L)}\mathbf{Ind}^{\mathbf{V}}_{\mathbf{V}'_{r'},\mathbf{V}''_{r'}}(L''_{r'}\boxtimes\bar{\mathbb{Q}}_l).$$

(2) $s_i^*(L)=t_i^*(L)$ if i is a source in Ω'.

4. An $\mathbb{Z}_2\times I$-colored graph $\mathcal{G}_1$ of $\mathcal{K}$

4.1. Kashiwara's operator in rank one case. We fix an $i\in I$. For any $\mathbf{V}$, we choose a decompositions $\mathbf{V}=\mathbf{T}\oplus\mathbf{V}''$ such that $|\mathbf{T}|=i$. Then $\mathbf{E}_{\mathbf{T},\Omega}\times\mathbf{E}_{\mathbf{V}'',\Omega}\cong\mathbf{E}_{\mathbf{V}'',\Omega}$, hence $\mathbf{res}^{\mathbf{V}}_{\mathbf{T},\mathbf{V}''}:\mathcal{D}^b_{G_{\mathbf{V}}}(\mathbf{E}_{\mathbf{V},\Omega})\to\mathcal{D}^b_{G_{\mathbf{V}''}}(\mathbf{E}_{\mathbf{V}'',\Omega})$. Similarly, we choose $\mathbf{V}=\mathbf{V}'\oplus\mathbf{W}$ such that $|\mathbf{W}|=i$, then $\mathbf{res}^{\mathbf{V}}_{\mathbf{V}',\mathbf{W}}:\mathcal{D}^b_{G_{\mathbf{V}}}(\mathbf{E}_{\mathbf{V},\Omega})\to\mathcal{D}^b_{G_{\mathbf{V}'}}(\mathbf{E}_{\mathbf{V}',\Omega})$

DEFINITION 4.1. We define ${}_i\mathcal{R}$ to be the functor $\bigoplus_{\mathbf{V}}\mathbf{D}\circ\mathbf{res}^{\mathbf{V}}_{\mathbf{T},\mathbf{V}''}\circ\mathbf{D}$ and $\mathcal{R}_i$ to be the functor $\bigoplus_{\mathbf{V}}\mathbf{D}\circ\mathbf{res}^{\mathbf{V}}_{\mathbf{V}',\mathbf{W}}\circ\mathbf{D}$, then ${}_i\mathcal{R}$ and $\mathcal{R}_i$ induces operators ${}_ir$ and $r_i:\mathcal{K}\to\mathcal{K}$ defined by:

$$\begin{aligned}{}_ir([A])&=[{}_i\mathcal{R}(A)],\\ r_i([A])&=[\mathcal{R}_i(A)]\end{aligned}$$

for any $A\in\mathcal{Q}_{\mathbf{V},\Omega}$.

For $n\in\mathbb{N}$, we define operators $\theta_i^{(n)}$ and $\theta_i^{*(n)}:\mathcal{K}\to\mathcal{K}$ by setting:

$$\begin{aligned}\theta_i^{(n)}([A])&=E_i^{(n)}*[A],\\ \theta_i^{*(n)}([A])&=[A]*E_i^{(n)}\end{aligned}$$

for any $A\in\mathcal{Q}_{\mathbf{V},\Omega}$. In particular, we denote $\theta_i^{(n)}$ and $\theta_i^{*(n)}$ by θ_i and θ_i^* respectively if $n=1$.

The following relation is easy to know and it implies that the operators ${}_ir,r_i$ coincide with the derivative operators of $U_q^+(\mathfrak{g})$(See details in [**17**].):

LEMMA 4.2. *The operators satisfies the following relation:*

$${}_ir\theta_i^{(n)}=v^{2n}\theta_i^{(n)}{}_ir+v^{n-1}\theta_i^{(n-1)}\ \textit{for } n\geq 1.$$

Notice that for $N > \dim \mathbf{V}_i$, $({}_ir)^N([A]) = 0$ for any $A \in \mathcal{Q}_{\mathbf{V},\Omega}$, hence the operator $\Pi_t = \sum_{s\geq 0}(-1)^s v^{(s-1)s/2}\theta_i^{(s)}({}_ir)^{(s+t)} : \mathcal{K} \to \mathcal{K}$ is well-defined and satisfies the following properties:

LEMMA 4.3 ([**10**, Lemma 16.1.2]).
(1) ${}_ir\Pi_t = 0$ *for any* $t \geq 0$*;*
(2) *Notice that for any* $x \in \mathcal{K}$, $\sum_{t\geq 0} v^{-(t-1)t/2}\theta_i^{(t)}\Pi_t(x)$ *is a finite sum, and*

$$\sum_{t\geq 0} v^{-(t-1)t/2}\theta_i^{(t)}\Pi_t = \mathbf{Id}_{\mathcal{K}};$$

(3) *Given* $x \in \mathcal{K}$*, there is a unique collection of* $\{x_N \in \mathcal{K}|\ N \geq 0\}$ *such that* $x = \sum_{N\geq 0}\theta_i^{(N)}x_N$ *and* ${}_ir(x_N) = 0$*. Indeed,* $x_N = v^{-(N-1)N/2}\Pi_N(x)$*. More precisely, if we denote* $\mathrm{Ker}\,{}_ir$ *by* $\mathcal{K}(0)$ *and set* $\mathcal{K}(n) = \theta_i^{(n)}\mathcal{K}(0)$*,* $\mathcal{K} = \bigoplus_{n\geq 0}\mathcal{K}(n)$*.*

With the lemma above, we can introduce the Kashiwara's operators of $\mathcal{K}$ following [**10**].

DEFINITION 4.4. We define linear operators $\tilde{\phi}_i, \tilde{\epsilon}_i : \mathcal{K} \to \mathcal{K}$ as the following:

$$\begin{aligned}\tilde{\phi}_i(\theta_i^{(n)}y) =&\theta_i^{(n+1)}y,\\ \tilde{\epsilon}_i(\theta_i^{(n)}y) =&\theta_i^{(n-1)}y \text{ if } n > 0,\\ \tilde{\epsilon}_i(\theta_i^{(n)}y) =&0 \text{ if } n = 0,\end{aligned}$$

for $n \in \mathbb{N}$ and $y \in \mathcal{K}(0)$.

REMARK 4.5. If we identify $\mathcal{K}$ with ${}_{\mathbb{Z}}U_q^+(\mathfrak{g})$, then the operators $\tilde{\phi}_i, \tilde{\epsilon}_i$ coincide with the Kashiwara's operators $\tilde{f}_i, \tilde{e}_i$ on ${}_{\mathbb{Z}}U_q^+(\mathfrak{g})$ in [**6**].

Assume i is a sink in Ω, let

$$\mathcal{B} = \{[L]|L \in \mathcal{P}_\Omega\}$$

and

$$\mathcal{B}_{i,n} = \{[L] \in \mathcal{P}_\Omega|t_i(L) = n\}, n \geq 0, i \in I,$$

then $\pi_{i,n}$ induces a bijection between the sets $\mathcal{B}_{i,0}$ and $\mathcal{B}_{i,n}$. Moreover, by Lemma 3.4 and Proposition 3.5 we can see that the basis $\mathcal{B}$ is apated in the sense of Lusztig in [**10**] Section 16.3.1. More precisely, the following results hold.

PROPOSITION 4.6 ([**10**, Section 17.3.2]).
(1) *The set* $\bigcup_{N\geq n}\mathcal{B}_{i,n}$ *form a basis of* $\theta_i^{(n)}\mathcal{K}$*;*
(2) *For* $[L_0] \in \mathcal{B}_{i,0}$, $\theta_i^{(n)}[L_0] - \pi_{i,n}([L_0]) \in \theta_i^{(n+1)}\mathcal{K}$*.*

Let

$$\mathcal{L} = \{x \in \mathcal{K}|(x,x) \in \mathbb{Z}[[v^{-1}]] \cap \mathbb{Q}(v)\}$$

and set

$$\begin{aligned}T_{i,0} &= \{x \in \mathcal{K}(0)|(x,x) \in 1 + v^{-1}\mathbb{Z}[v,v^{-1}]\},\\ T_{i,n} &= \theta_i^{(n)}T_{i,0}, n > 0.\end{aligned}$$

Let $\mathcal{B}_i(n) = \mathcal{B} \cap (T_{i,n} + v^{-1}\mathcal{L})$, then $\mathcal{B} = \bigcup_{N \geq 0} \mathcal{B}_i(N)$. The following proposition shows that the operators $\tilde{\phi}_i, \tilde{\epsilon}_i$ can be described by the bijections $\pi_{i,n}$ modulo $v^{-1}\mathcal{L}$:

PROPOSITION 4.7 ([**10**, Proposition16.2.5, Proposition 16.2.8 and Proposition 16.3.5]). *We have the following results:*

(1) *$\mathcal{L}$ is a $\mathbb{Z}[v^{-1}]$-module with the basis $\mathcal{B}$.*
(2) *$\mathcal{L}$ is preserved by $\tilde{\phi}_i, \tilde{\epsilon}_i$.*
(3) *$\mathcal{B}_{i,n} = \mathcal{B}_i(n)$.*
(4) *Let $[L_0] \in \mathcal{B}_{i,0}$ and $[L] = \pi_{i,n}([L_0])$, then*

$$\tilde{\phi}_i([L]) = \pi_{i,n+1}([L_0]) \mod v^{-1}\mathcal{L}$$

and

$$\tilde{\epsilon}_i([L]) = \begin{cases} \pi_{i,n-1}([L_0]) \mod v^{-1}\mathcal{L} & n > 0 \\ 0 \mod v^{-1}\mathcal{L} & n = 0 \end{cases}$$

If we replace $\theta_i^{(n)}$ and ${}_i r$ by $\theta_i^{*(n)}$ and r_i respectively and let $\mathcal{K}(0)^* = \operatorname{Ker} r_i, \mathcal{K}(n)^* = \theta_i^{*(n)}\mathcal{K}(0)^*$, we can introduce linear operators $\tilde{\phi}_i^*$ and $\tilde{\epsilon}_i^*$ in a similar way.

Assume i is a source in Ω, let $\mathcal{B}_{i,n}^* = \{[L] | t_i^*(L) = n\}$ for $n \in \mathbb{N}$, then $\pi_{i,n}^*$ induces a bijection between the sets $\mathcal{B}_{i,0}^*$ and $\mathcal{B}_{i,n}^*$. The following result is dual to Proposition 4.6:

PROPOSITION 4.8.
(1) *The set $\bigcup_{N \geq n} \mathcal{B}_{i,n}^*$ form a basis of $\theta_i^{*(n)}\mathcal{K}$;*
(2) *For $[L_0] \in \mathcal{B}_{i,0}^*$, $\theta_i^{*(n)}[L_0] - \pi_{i,n}^*([L_0]) \in \theta_i^{*(n+1)}\mathcal{K}$.*

Define $T_{i,0}^* = \{x \in \mathcal{K}(0)^* | (x,x) \in 1 + v^{-1}\mathbb{Z}[v, v^{-1}]\}$ and $T_{i,n}^* = \theta_i^{*(n)} T_{i,0}^*$ for $n > 0$. Let $\mathcal{B}_i^*(n) = \mathcal{B} \cap (T_{i,n}^* + v^{-1}\mathcal{L})$, then $\mathcal{B} = \bigcup_{N \geq 0} \mathcal{B}_i^*(N)$. The following result is dual to Proposition 4.7:

PROPOSITION 4.9. (1) *For any $n \in \mathbb{N}$, $\mathcal{B}_{i,n}^* = \mathcal{B}_i^*(n)$.*
(2) *Let $[L_0] \in \mathcal{B}_{i,0}^*$ and $[L] = \pi_{i,n}^*([L_0])$, then*

$$\tilde{\phi}_i^*([L]) = \pi_{i,n+1}^*([L_0]) \mod v^{-1}\mathcal{L}$$

and

$$\tilde{\epsilon}_i^*([L]) = \begin{cases} \pi_{i,n-1}^*([L_0]) \mod v^{-1}\mathcal{L} & n > 0 \\ 0 \mod v^{-1}\mathcal{L} & n = 0 \end{cases}$$

4.2. The graph $\mathcal{G}_1$. For each $i \in I$, we fix orientations Ω_i and Ω^i such that i is a sink in Ω_i but a source in Ω^i.

For a given $L \in \mathcal{P}_\Omega$ and $i \in I$, we consider $L' = \mathcal{F}_{\Omega,\Omega_i}(L)$. We assume $p = t_i(L')$, then there exists a unique L'' such that $\pi_{i,p}(L'') = L'$. Let $K' = \pi_{i,p+1}(L'')$, then $K = \mathcal{F}_{\Omega_i,\Omega}(K') \in \mathcal{P}_\Omega$ is a well-defined simple perverse sheaf, which does not depend on the choice of Ω_i. We associate an arrow $[L] \xrightarrow{i_+} [K]$ and adapt the notation $[K] = i_+([L])$ or $K = i_+(L)$.

Dually, if we consider $L' = \mathcal{F}_{\Omega,\Omega^i}(L)$ and assume $p = t_i^*(L')$, let $K' = \pi^*_{i,p+1}(\pi^*_{i,p})^{-1}(L')$ and $K = \mathcal{F}_{\Omega^i,\Omega}(K') \in \mathcal{P}_\Omega$. We associate an arrow $[L] \xrightarrow{i^+} [K]$ and write $[K] = i^+([L])$ or $K = i^+(L)$.

DEFINITION 4.10. We define an $I \times \mathbb{Z}_2$-colored graph $\mathcal{G}_1 = (\mathcal{V}_1, \mathcal{E}_1)$ as follows:

(1) The set of vertices is $\mathcal{V}_1 = \{[L] | L \in \mathcal{P}_\Omega\}$.

(2) The set of arrows is $\mathcal{E}_1 = \{[L] \xrightarrow{i^+} [K], [L] \xrightarrow{i_+} [K] |\ i \in I, K, L \in \mathcal{P}_\Omega\}$.

We also define i_- and i^- to be the inverse of i_+ and i^+ respectively.

By Corollary 3.3, we can see that $\mathcal{G}_1 = (\mathcal{V}_1, \mathcal{E}_1)$ is a well-defined $I \times \mathbb{Z}_2$-colored graph, which is independent of the choice of Ω.

4.3. The commutative relation between i^+ and j_+. In this and the next sections, we fix $i \neq j \in I$ and choose orientations Ω_i, Ω^i and Ω_j such that i is a sink in Ω_i but a source in Ω^i and j is a sink in Ω_j. Given $K \in \mathcal{P}_{\mathbf{V},\Omega}$, we shorten $t_i(\mathcal{F}_{\Omega,\Omega_i}(K)), t_i^*(\mathcal{F}_{\Omega,\Omega^i}(K))$ and $t_j(\mathcal{F}_{\Omega,\Omega_j}(K))$ by $t_i(K), t_i^*(K)$ and $t_j(K)$ in this and the next sections respectively.

LEMMA 4.11. *Assume $[L_0] \in \mathcal{B}^*_{i,0}$ and $[L] = \pi^*_{i,c}([L_0])$ for some $c \geq 0$, then $t_j(L_0) = t_j(L)$.*

PROOF. Assume $t_j(L_0) = d$. By Proposition 4.7, $\mathcal{B}_{j,d} = \mathcal{B}_j(d)$ and there exists $x' \in T_{j,0}$ and $z' \in \mathcal{L}$ such that $[L_0] = \theta_j^{(d)} x' + v^{-1}z'$. Similarly, there exists $x'' \in T^*_{i,0}$ and $z'' \in \mathcal{L}$ such that $[L_0] = x'' + v^{-1}z''$. Then we can see that

$$x'' - E_j^{(d)} * x' \in v^{-1}\mathcal{L}.$$

Notice that $[L] = (\tilde{\phi}_i^*)^c[L_0] \bmod v^{-1}\mathcal{L}$, we can see that $[L] = \theta_i^{*(c)} x'' + v^{-1}z$ for some $z \in \mathcal{L}$. Hence $[L] = E_j^{(d)} * x' * E_i^{(c)} \bmod v^{-1}\mathcal{L}$. Notice that $x' * E_i^{(c)} \in T_{j,0}$,

$$[L] \in \mathcal{B}_j(d) = \mathcal{B}_{j,d},\ t_j(L) = d = t_j(L_0).$$

□

COROLLARY 4.12. *If $[L'] = i^+([L])$, then $t_j(L) = t_j(L')$.*

PROOF. Assume $t_i^*(L) = c$. Let L_0 be the unique simple pervese sheaf such that $t_i^*(L_0) = 0$ and $\pi^*_{i,c}(L_0) = L$, then $\pi^*_{i,c+1}(L_0) = L'$. Then $t_j(L) = t_j(L_0) = t_j(L')$. □

LEMMA 4.13. *Assume $[K] \in \mathcal{B}_{j,0} \cap \mathcal{B}^*_{i,0}$, then $\pi_{j,m}\pi^*_{i,n}([K]) = \pi^*_{i,n}\pi_{j,m}([K])$ for $m, n \geq 0$.*

PROOF. Take $x' \in T^*_{i,0}$ and $x'' \in T_{j,0}$ such that $[K] = x' + v^{-1}z'$ and $[K] = x'' + v^{-1}z''$ with $z', z'' \in \mathcal{L}$ as in the proof of Lemma 4.11, then we can see that $K_1 = \pi^*_{i,n}\pi_{j,m}(K)$ is the unique simple perverse sheaf such that $[K_1] = E_j^{(m)} * x'' * E_i^{(n)} \bmod v^{-1}\mathcal{L}$. Similarly, $K_2 = \pi_{j,m}\pi^*_{i,n}(K)$ is the unique simple perverse sheaf such that $[K_2] = E_j^{(m)} * x' * E_i^{(n)} \bmod v^{-1}\mathcal{L}$. Notice that $x' = x'' \bmod v^{-1}\mathcal{L}$, we can see that $[K_1] = [K_2] \bmod v^{-1}\mathcal{L}$. By Proposition 4.7, the canonical basis induces a basis of $\mathcal{L}/v^{-1}\mathcal{L}$, we can see that $[K_1] = [K_2]$. □

COROLLARY 4.14. *For any $K \in \mathcal{P}_{\mathbf{V},\Omega}$, $i^+j_+([K]) = j_+i^+([K])$.*

PROOF. Assume $t_j(K) = m$ and $t_i^*(K) = n$, after applying Lemma 3.4 and Lemma 3.6, we can take $K_0 = (\pi_{j,m})^{-1}(\pi_{i,n}^*)^{-1}(K)$. Then K_0 is the unique simple perverse sheaf such that $\pi_{j,m}\pi_{i,n}^*(K_0) = \pi_{i,n}^*\pi_{j,m}(K_0) = K$. Then by Lemma 4.13,

$$\begin{aligned} i^+j_+([K]) =& i^+j_+\pi_{j,m}\pi_{i,n}^*([K_0]) = i^+\pi_{j,m+1}\pi_{i,n}^*([K_0]) \\ =& i^+\pi_{i,n}^*\pi_{j,m+1}([K_0]) = \pi_{i,n+1}^*\pi_{j,m+1}([K_0]) \end{aligned}$$

and

$$\begin{aligned} j_+i^+([K]) =& j_+i^+\pi_{i,n}^*\pi_{j,m}([K_0]) = j_+\pi_{i,n+1}^*\pi_{j,m}([K_0]) \\ =& j_+\pi_{j,m}\pi_{i,n+1}^*([K_0]) = \pi_{j,m+1}\pi_{i,n+1}^*([K_0]). \end{aligned}$$

Then by Lemma 4.13, $i^+j_+([K]) = j_+i^+([K])$. □

4.4. The commutative relation between i^+ and i_+. In this section we study the commutative relation between i^+ and i_+.

LEMMA 4.15. *Assume $L_0 \in \mathcal{P}_{\mathbf{V}',\Omega}$ and $t_i^*(L_0) = 0, t_i(L_0) = d$. Let $L = \pi_{i,c}^*(L_0)$ for some $c \in \mathbb{N}$, then*

$$t_i(L) = \begin{cases} d & c + (|\mathbf{V}'|, i) \leq d \\ c + (|\mathbf{V}'|, i) & \textit{otherwise} \end{cases}$$

PROOF. Take $x' \in T_{i,0}$ and $z \in \mathcal{L}$ such that $[L_0] = E_i^{(d)} * x' + v^{-1}z$ as in the proof of Lemma 4.11, then we can see that $[L] = E_i^{(d)} * x' * E_i^{(c)} \mod v^{-1}\mathcal{L}$. We denote $E_i^{(d)} * x' * E_i^{(c)}$ by x. Then by Lemma 4.3, $x = \sum_{N\geq 0} \theta_i^{(N)} x_N$ with

$$x_N = v^{-(N-1)N/2}\Pi_N(x) = \sum_{s\geq 0}(-1)^s v^{(s-1)s/2-(N-1)N/2} E_i^{(s)} * ({}_ir)^{(s+N)}(x).$$

Using the fact ${}_ir(xy) = {}_ir(x)y + v^{(|\mathbf{V}|,i)}x{}_ir(y)$ for any homogenous elements $x \in \mathcal{K}_{\mathbf{V}}, y \in \mathcal{K}_{\mathbf{V}'}$, we easily get:

$$({}_ir)^m(xy) = \sum_{t\geq 0}^{m} v^{(|\mathbf{V}|-ti,(m-t)i)+t(m-t)} \frac{[m]_v!}{[t]_v![m-t]_v!}({}_ir)^t(x) * ({}_ir)^{m-t}(y).$$

Take $x = E_i^{(d)} * x'$ and $y = E_i^{(c)}$, we obtain equation (1):

$$\begin{aligned} ({}_ir)^m(E_i^{(d)} * x' * E_i^{(c)}) =& \sum_{t\geq 0}^{m} v^{(|\mathbf{V}'|-ti,(m-t)i)+t(m-t)} \frac{[m]_v!}{[t]_v![m-t]_v!} \\ & *({}_ir)^t(E_i^{(d)} * x') * ({}_ir)^{m-t}(E_i^{(c)}). \end{aligned}$$

Notice that by definition of ${}_ir$:

$$({}_ir)^{(a)}E_i^{(b)} = \begin{cases} v^{(2b-1-a)a/2}E_i^{(b-a)} & a \geq b \\ 0 & \text{otherwise} \end{cases}$$

so equation (2)

$$({}_ir)^t(E_i^{(d)} * x') = \begin{cases} v^{(2d-1-t)t/2}E_i^{(d-t)} * x' & d \geq t \\ 0 & \text{otherwise} \end{cases}$$

and equation (3)

$$({}_ir)^{m-t}(E_i^{(c)}) = \begin{cases} v^{(2c-1-m+t)(m-t)/2}E_i^{(c-m+t)} & c \geq m-t \\ 0 & \text{otherwise} \end{cases}$$

We apply the equations (1),(2) and (3) to x_N and assume $c + (|\mathbf{V}'|, i) \leq d$. Notice that

$$\frac{[a+b]_v!}{[a]_v![b]_v!} \in v^{ab}\mathbb{Z}[[v^{-1}]]$$

for $a, b \in \mathbb{N}$, we can see that $x_N = 0 \bmod v^{-1}\mathcal{L}$ for $N \neq d$ but $x_d = x' * E_i^{(c)} \bmod v^{-1}\mathcal{L}$. Hence $x \in T_{i,d}$ and $t_i(L) = d$.

Otherwise, we assume that $k = c + (|\mathbf{V}'|, i) - d > 0$. Applying the equations (1),(2) and (3) to x_N, we can see that $x_{d+k} = x' * E_i^{(c-k)} \bmod v^{-1}\mathcal{L}$ and $x_N = 0 \bmod v^{-1}\mathcal{L}$ for $N \neq d+k$. Hence $x \in T_{i,d+k}$ and $t_i(L) = d+k = c + (|\mathbf{V}'|, i)$. □

By the similar argument, we have the following result dual to Lemma 4.15:

LEMMA 4.16. *Assume $L_0 \in \mathcal{P}_{\mathbf{V}',\Omega}$ and $t_i(L_0) = 0, t_i^*(L_0) = c$. Let $L = \pi_{i,d}(L_0)$ for some $d \in \mathbb{N}$, then :*

$$t_i^*(L) = \begin{cases} c & d + (|\mathbf{V}'|, i) \leq c \\ d + (|\mathbf{V}'|, i) & \textit{otherwise} \end{cases}$$

The following two lemmas give the commutative relation of i_+ and i^+ case by case:

LEMMA 4.17. *Assume $L_0 \in \mathcal{P}_{\mathbf{V}',\Omega}$ and $t_i^*(L_0) = 0, t_i(L_0) = d > 0$. Take $c \in \mathbb{N}_{>0}$ such that $c+(|\mathbf{V}'|, i) \leq d$ and set $L = \pi_{i,c}^*(L_0)$. Let K_0 be the unique simple perverse sheaf such that $i_+(K_0) = L_0$. Then $t_i^*(K_0) = 0$, hence $\pi_{i,c}^*(K_0)$ is well-defined. Moreover, $i_+\pi_{i,c}^*(K_0) = L$. Or equivalently, $i_+(i^+)^c([K_0]) = (i^+)^c i_+([K_0])$. In this case, $t_i^*(i_-(L)) = t_i^*(L) = c$.*

PROOF. After applying Lemma 3.4, we can take $L_0' \in \mathcal{P}_{\mathbf{V}'',\Omega}$ to be the unique simple perverse sheaf such that $\pi_{i,d}(L_0') = L_0$ and $t_i(L_0') = 0$. We assume $t_i^*(L_0') = c'$. Then by Lemma 4.16, $0 = t_i^*(L_0) = max\{d + (|\mathbf{V}''|, i), c'\}$, hence $c' = 0$ and $d + (|\mathbf{V}''|, i) \leq 0$. Then $t_i^*(K_0) = max\{d - 1 + (|\mathbf{V}''|, i), c'\} = 0$ and $\pi_{i,c}^*(K_0)$ can be defined.

From the proof of Lemma 4.15, we can see that $[L] = E_i^{(d)} * x' * E_i^{(c)} \bmod v^{-1}\mathcal{L}$ for some $x' \in T_{i,0}$ and $[L_0] = E_i^{(d)} * x' \bmod v^{-1}\mathcal{L}$. By definition, $[K_0] = E_i^{(d-1)} * x' \bmod v^{-1}\mathcal{L}$. Then $[\pi_{i,c}^*(K_0)] = E_i^{(d-1)} * x' * E_i^{(c)} = \tilde{\epsilon}_i([L]) \bmod v^{-1}\mathcal{L}$. Hence $i_+\pi_{i,c}^*(K_0) = L$ and $i_+(i^+)^c([K_0]) = (i^+)^c i_+([K_0])$. □

LEMMA 4.18. *Assume $L_0 \in \mathcal{P}_{\mathbf{V}',\Omega}$ and $t_i^*(L_0) = 0, t_i(L_0) = d > 0$. Take $c \in \mathbb{N}_{>0}$ such that $c + (|\mathbf{V}'|, i) > d$ and set $L = \pi_{i,c}^*(L_0)$. Then*

$$i_-([L]) = i^-([L]),$$

or equivalently,

$$i_-(i^+)^c([L_0]) = (i^+)^{c-1}([L_0]).$$

Moreover, $t_i^*(i_-(L)) = t_i^*(L) - 1 = c - 1$.

PROOF. Let $k = c + (|\mathbf{V}'|, i) - d > 0$, then by the proof of Lemma 4.15, we can see that

$$[L] = E_i^{(d+k)} * x' * E_i^{(c-k)} = E_i^{(d)} * x' * E_i^{(c)} \bmod v^{-1}\mathcal{L}.$$

Then by the definition of i^-, i_- and Proposition 4.7 and Proposition 4.9,

$$[i_-(L)] = E_i^{(d+k-1)} * x' * E_i^{(c-k)} \bmod v^{-1}\mathcal{L}$$

and

$$[i^-(L)] = x'' * E_i^{c-1} = E_i^{(d)} * x' * E_i^{(c-1)} \bmod v^{-1}\mathcal{L},$$

where $x'' \in T^*_{i,0}$. If $k - 1 > 0$, then the proof of Lemma 4.15 implies that

$$E_i^{(d+k-1)} * x' * E_i^{(c-k)} = E_i^{(d)} * x' * E_i^{(c-1)} \bmod v^{-1}\mathcal{L}.$$

On the other hand, if $k - 1 = 0$, the statement

$$E_i^{(d+k-1)} * x' * E_i^{(c-k)} = E_i^{(d)} * x' * E_i^{(c-1)} \bmod v^{-1}\mathcal{L}$$

trivially holds. Since we always have $[i^-(L)] = [i_-(L)] \bmod v^{-1}\mathcal{L}$, we can see that $i_-(L) = i^-(L)$. $\square$

5. The key lemmas for $\mathcal{M}$

In this section, we will introduce some lemmas about $\mathcal{M}$ parallel to Lemma 3.4 and Lemma 3.6.

5.1. The key lemma for the left mutiplication. Let $\Lambda_{\mathbf{V},i,p}$ be the subset of $\Lambda_{\mathbf{V}}$ defined by:

$$\Lambda_{\mathbf{V},i,p} = \{x \in \Lambda_{\mathbf{V}} | \mathrm{codim}_{\mathbf{V}_i}(\mathrm{Im} \bigoplus_{h \in H, h''=i} x_h) = p\}$$

and $\Lambda_{\mathbf{V},i,\geq p} = \bigcup_{p' \geq p} \Lambda_{\mathbf{V},i,p'}$. Each $\Lambda_{\mathbf{V},i,p}$ is locally closed in $\Lambda_{\mathbf{V}}$. Since $\bigcup_p \Lambda_{\mathbf{V},i,p} = \Lambda_{\mathbf{V}}$, there is a unique p such that $Z \cap \Lambda_{\mathbf{V},i,p}$ is dense in Z for each irreducible component Z of $\Lambda_{\mathbf{V}}$. In this case, we say that Z generically belongs to $\Lambda_{\mathbf{V},i,p}$ and write $t_i(Z) = p$.

We assume that $|\mathbf{V}'| = pi$ and denote $qr : \Lambda' \to \Lambda_{\mathbf{V}}$ by q'. Then $p^{-1}(\Lambda_{\mathbf{V}'',i,0}) = (q'^{-1})(\Lambda_{\mathbf{V},i,p})$. We denote $p^{-1}(\Lambda_{\mathbf{V}'',i,0})$ by $\Lambda'_{i,p}$ and denote $(q^{-1})(\Lambda_{\mathbf{V},i,p})$ by $\Lambda''_{i,p}$. Then the following diagram is commutative:

$$\begin{array}{ccccccc}
\Lambda_{\mathbf{V}'',i,0} & \xleftarrow{p} & \Lambda'_{i,p} & \xrightarrow{r} & \Lambda''_{i,p} & \xrightarrow{q} & \Lambda_{\mathbf{V},i,p} \\
\downarrow{\scriptstyle i_1} & & \downarrow{\scriptstyle i_2} & & \downarrow{\scriptstyle i_3} & & \downarrow{\scriptstyle i_4} \\
\Lambda_{\mathbf{V}''} & \xleftarrow{p} & \Lambda' & \xrightarrow{r} & \Lambda'' & \xrightarrow{q} & \Lambda_{\mathbf{V}}
\end{array}$$

here i_1, i_2, i_3 and i_4 are the natural embeddings.

The following lemma is from [**15**] and we give a sketch of proof here:

LEMMA 5.1 ([**15**] 4.17). *(1) $q' : \Lambda'_{i,p} \to \Lambda_{\mathbf{V},i,p}$ is a principle $G_{\mathbf{V}''} \times G_{\mathbf{V}'}$ -bundle.*

(2) $p : \Lambda'_{i,p} \to \Lambda_{\mathbf{V}'',i,0}$ is a smooth map whose fibers are connected of dimension $(\sum_{i \in I} v_i^2) - p(v'', i)$

PROOF. By definition, $r : \Lambda'_{i,p} \to \Lambda''_{i,p}$ is a principle bundle. Notice that when $x \in \Lambda_{\mathbf{V},i,p}$, the x-stable subspace with dimension $|\mathbf{W}|$ must be $\sum_{h\in H,h''=i} \mathrm{Im}(x_h)$, we get the proof of (1).

The fiber of p at x'' is the set consisting of $(x, \widetilde{\mathbf{W}}, \rho_1, \rho_2)$ such that $(\widetilde{\mathbf{W}}, \rho_1, \rho_2)$ can be chosen arbitarily and x is an extension of x''. $(\widetilde{\mathbf{W}}, \rho_1, \rho_2)$ is given by a point in $G_{\mathbf{V}}/U$, here U is the unipotent radical of the parabolic subgroup of type $(|\mathbf{V}'|, |\mathbf{V}''|)$. To give an extension of x'' is equivalent to give $(y_h)_{h\in H,h'=i}$ such that the following composition vanishes

$$\mathbf{V}'_i \xrightarrow{\bigoplus_{h\in H,h'=i} y_h} \bigoplus_{i-j} \mathbf{V}''_j \xrightarrow{\sum_{h\in H,h'=i} \epsilon(h)x'_{\bar{h}}} \mathbf{V}''_i$$

By assumption, the rank of the second linear map in the composition is equal to $\dim \mathbf{V}''_i$. By linear algebra and direct calculation, we get the proof. □

COROLLARY 5.2. *$q'p^{-1}$ induces a bijection $\eta_{i,p} : Irr\Lambda_{\mathbf{V}'',i,0} \to Irr\Lambda_{\mathbf{V},i,p}$ from the set of irreducible components of $\Lambda_{\mathbf{V}'',i,0}$ to the set of irreducible components $\Lambda_{\mathbf{V},i,p}$. The map $\eta_{i,p}$ also induces a bijection (still denoted by $\eta_{i,p}$) from $\{Z \in Irr\Lambda_{\mathbf{V}''} | t_i(Z) = 0\}$ to $\{Z \in Irr\Lambda_{\mathbf{V}} | t_i(Z) = p\}$ defined by:*

$$\eta_{i,p}(\bar{Z}) = \overline{\eta_{i,p}(Z)}$$

for $Z \in Irr\Lambda_{\mathbf{V}'',i,0}$. Here $\bar{Z}$ and $\overline{\eta_{i,p}(Z)}$ are the closure of Z and $\eta_{i,p}(Z)$ respectively.

DEFINITION 5.3. We say a constructible function f generically takes value c on an irreducible component Z if $f^{-1}(c) \cap Z$ is dense in Z. In this case, we adapt the notation $\rho_Z(f) = c$. Given an irreducible component Z, if f generically takes value 1 on Z but generically takes value 0 on the other irreducible components, we say f has property $\mathcal{P}(Z)$.

If a constructible function $f \in \mathcal{M}$ has property $\mathcal{P}(Z)$, we say f is a Lusztig's function in $\mathcal{M}$ of the irreducible component Z. Lusztig has proved the following lemma which implies the existence of such functions and we sketch the proof. (See details in [**11**])

LEMMA 5.4 ([**11**]). *For each irreducible component Z of $\Lambda_{\mathbf{V}}$, there exists an $f \in \mathcal{M}_{\mathbf{V}}$ with property $\mathcal{P}(Z)$.*

PROOF. We prove by inducton on $|\mathbf{V}| = v$. Asuume that for any $Z' \in Irr\Lambda_{\mathbf{V}'}$ with $|\mathbf{V}'| < v$, we have construct some $f_{Z'} \in \mathcal{M}$ which has property $\mathcal{P}(Z')$. For a given $Z \in Irr\Lambda_{\mathbf{V}}$, by [**7**] Corollary 1.9, there exists $i \in I$ such that $t_i(Z) = p > 0$. Then $Z_1 = Z \cap \Lambda_{\mathbf{V},i,p}$ is an irreducible componnet of $\Lambda_{\mathbf{V},i,p}$. Let Z_2 be the irreducible component of $\Lambda_{\mathbf{V}'',i,0}$ corresponding to Z_1 given by Corollary 5.2, then the closure Z'' of Z_2 is an irreducible component of $\Lambda_{\mathbf{V}''}$ with $t_i(Z'') = 0$. By induction hypothesis, there exists $f_{Z''} \in \mathcal{M}$ which has property $\mathcal{P}(Z'')$. Then $\tilde{f} = \mathbf{1}_{i^p} * f_{Z''} \in \mathcal{M}$. Notice that $\tilde{f}(x) = 0$ for $x \in \Lambda_{\mathbf{V}} \backslash \Lambda_{\mathbf{V},i,\geq p}$ and the restriction of $\tilde{f}$ on $\Lambda_{\mathbf{V},i,p}$ is uniquely determined by the restriction of f on $\Lambda_{\mathbf{V}'',i,0}$. Now we argue by descending induction on p. If $p = \dim \mathbf{V}_i$, then $\tilde{f} = \mathbf{1}_{i^p} * f_{Z''}$ has property $\mathcal{P}(Z)$ by construction. If for all $t_i(Z'') = p' > p$, we can construct $f_{Z''} \in \mathcal{M}$ which has property $\mathcal{P}(Z'')$. Then the constructible function

$$f = \tilde{f} - \sum_{Z'' \in Irr\Lambda_{\mathbf{V}}, t_i(Z'') > p} \rho_{Z''}(\tilde{f}) f_{Z''}$$

is what we desired. □

REMARK 5.5. Since there maybe exist more than one $i \in I$ such that $t_i(Z) > 0$ for a given irreducible component Z, the construction of f depends on the choices of $f_{Z''}$ and the choice of a sequence of irreducible components $\underline{Z} = (Z_0, Z_1, Z_2, \cdots, Z_s)$, $Z_l \in Irr\Lambda_{\mathbf{V}_l}$ for $0 \leq l \leq s$ and a sequence $\underline{i} = (i_1, i_2, \cdots, i_s)$, $i_l \in I$ for $1 \leq l \leq s$ satisfying the following conditions:

(1) $Z_0 = Z$

(2) $|\mathbf{V}_s| = qj$ for some $q > 0, j \in I$ and Z_s is the unique irreducible component of $\Lambda_{\mathbf{V}_s}$.

(3) For any $1 \leq l \leq s$, $t_{i_l}(Z_{l-1}) = p_l > 0$ and Z_l is the corresponding irreducible component of Z_{l-1} given by Corollary 5.2. More precisely, $\eta_{i_l, p_l}(Z_l) = Z_{l-1}$.

If a pair $\underline{p} = (\underline{Z}, \underline{i})$ of sequences satisfies the three conditions above, we say $\underline{p}$ is a left-addmissible path of Z. Given a left-addmissible path $\underline{p}$, we can construct a constructible function f having property $\mathcal{P}(Z)$. We denote this constructible function (still not unique) by $f_{\underline{p}}$.

From the proof above, we also get the following result:

LEMMA 5.6. *If Z' is an irreducible component of $\Lambda_{\mathbf{V}''}$ with $t_i(Z') = 0$ and $Z = \eta_{i,p}(Z')$, and we choose a constructible function $f_{Z'}$ having property $\mathcal{P}(Z')$, then*

$$\mathbf{1}_{i^p} * f_{Z'} = f_Z + \sum_{Z'' \in Irr\Lambda_{\mathbf{V}}, t_i(Z'') > p} c_{Z''} f_{Z''},$$

where $f_Z, f_{Z''} \in \mathcal{M}_V$ are some constructible functions having property $\mathcal{P}(Z), \mathcal{P}(Z'')$ respectively and $c_{Z''} \in \mathbb{Z}$.

We call the lemma above the key lemma of Lusztig's functions for left mutiplication.

5.2. The key lemma for the right multiplication. Let $\Lambda^p_{\mathbf{V},i}$ be the subset of $\Lambda_{\mathbf{V}}$ defined by

$$\Lambda^p_{\mathbf{V},i} = \{x \in \Lambda_{\mathbf{V}} | \dim(\mathrm{Ker} \bigoplus_{h \in H, h' = i} x_h) = p\}$$

and $\Lambda^{\geq p}_{\mathbf{V},i} = \bigcup_{p' \geq p} \Lambda^{p'}_{\mathbf{V},i}$. Since $\bigcup_p \Lambda^p_{\mathbf{V},i} = \Lambda_{\mathbf{V}}$, for each irreducible component Z, there is a unique p such that $\Lambda^p_{\mathbf{V},i} \cap Z$ is dense in Z and we write $t^*_i(Z) = p$.

If we asuume $|\mathbf{V}''| = pi$, then $p^{-1}(\Lambda^0_{\mathbf{V}',i}) = (q')^{-1}(\Lambda^p_{\mathbf{V},i})$ and we denote this space by Λ'^p_i and let $\Lambda''^p_i = r(\Lambda'^p_i)$. The following diagram is commutative:

$$\begin{array}{ccccccc}
\Lambda^0_{\mathbf{V}',i} & \xleftarrow{p} & \Lambda'^p_i & \xrightarrow{r} & \Lambda''^p_i & \xrightarrow{q} & \Lambda^p_{\mathbf{V},i} \\
\downarrow{\scriptstyle j_1} & & \downarrow{\scriptstyle j_2} & & \downarrow{\scriptstyle j_3} & & \downarrow{\scriptstyle j_4} \\
\Lambda_{\mathbf{V}'} & \xleftarrow{p} & \Lambda' & \xrightarrow{r} & \Lambda'' & \xrightarrow{q} & \Lambda_{\mathbf{V}}
\end{array}$$

here j_1, j_2, j_3 and j_4 are the natural embeddings. Then by the same argument, we get the following results dual to those in Section 4.1.

LEMMA 5.7. *(1) $q' : \Lambda_i^{'p} \to \Lambda_{\mathbf{V},i}^p$ is a principle $G_{\mathbf{V}'} \times G_{\mathbf{V}''}$-bundle.*

(2) $p : \Lambda_i^{'p} \to \Lambda_{\mathbf{V}',i}^0$ is a smooth map whose fibers are connected of dimension $(\sum_{i\in I} v_i^2) - p(v', i) + p(v_i' - p)$.

The map $q'p^{-1}$ also induces a bijection of irreducible components parallel to Corollary 5.2.

COROLLARY 5.8. *We have a bijection $\eta_{i,p}^* : Irr\Lambda_{\mathbf{V}',i}^0 \to Irr\Lambda_{\mathbf{V},i}^p$ between the sets of irreducible components induced by $q^{'}p^{-1}$. The map $\eta_{i,p}^*$ also induces a bijection from $\{Z \in \Lambda_{\mathbf{V}'}|t_i^*(Z) = 0\}$ to $\{Z \in Irr\Lambda_{\mathbf{V}}|t_i^*(Z) = p\}$ defined by:*

$$\eta_{i,p}^*(\bar{Z}) = \overline{\eta_{i,p}^*(Z)}$$

for $Z \in Irr\Lambda_{\mathbf{V}',i}^0$. Here $\bar{Z}$ and $\overline{\eta_{i,p}^*(Z)}$ are the closure of Z and $\eta_{i,p}^*(Z)$ respectively.

We also has the following key lemma of Lusztig's functions for right multiplication:

LEMMA 5.9. *If Z' is an irreducible component of $\Lambda_{\mathbf{V}'}$ with $t_i^*(Z') = 0$ and $Z = \eta_{i,p}^*(Z')$, and we choose a constructible function $f_{Z'}$ having property $\mathcal{P}(Z')$, then*

$$f_{Z'} * \mathbf{1}_{i^p} = f_Z + \sum_{Z'' \in Irr\Lambda_{\mathbf{V}}, t_i^*(Z'')>p} c'_{Z''} f_{Z''},$$

where $f_Z, f_{Z''} \in \mathcal{M}_V$ are some constructible functions having property $\mathcal{P}(Z), \mathcal{P}(Z'')$ respectively and $c'_{Z''} \in \mathbb{Z}$.

DEFINITION 5.10. Assume that $\underline{Z} = (Z_0, Z_1, Z_2, \cdots, Z_s), Z_l \in Irr\Lambda_{\mathbf{V}_l}$ for $0 \leq l \leq s$ is a sequence of irreducible components and $\underline{i} = (i_1, i_2, \cdots, i_s), i_l \in I$ for $1 \leq l \leq s$ is a sequence of vertex. We say $\underline{p} = (\underline{Z}, \underline{i})$ is a right-admissible path of Z if the following three conditions holds:

(1) $Z_0 = Z$

(2) $|\mathbf{V}_s| = qj$ for some $q > 0, j \in I$ and Z_s is the unique irreducible component of $\Lambda_{\mathbf{V}_s}$.

(3) For any $1 \leq l \leq s$, $t_{i_l}^*(Z_{l-1}) = p_l > 0$ and Z_l is the corresponding irreducible component of Z_{l-1} given by corollary 5.8. More precisely, $\eta_{i_l,p_l}^*(Z_l) = Z_{l-1}$.

Given a right-admissible path $\underline{p}$ of Z, we can construct a constructible function f having property $\mathcal{P}(Z)$ (not unique) by right mutiplication as in Lemma 5.4. We denote this constructible function by $f_{\underline{p}}$.

5.3. The key lemmas for the coproduct. Fix a decomposition $\mathbf{T} \oplus \mathbf{W} = \mathbf{V}$ of graded vector space, we can define $Res_{\mathbf{T},\mathbf{W}}^{\mathbf{V}} : \tilde{M}(\Lambda_{\mathbf{V}}) \to \tilde{M}(\Lambda_{\mathbf{T}}) \times \tilde{M}(\Lambda_{\mathbf{W}})$ by:

$$Res_{\mathbf{T},\mathbf{W}}^{\mathbf{V}}(f)(x', x'') = f(x' \oplus x'')$$

Geiss, Leclerc and Schroer has proved in [4] that

$$Res_{\mathbf{T},\mathbf{W}}^{\mathbf{V}}(\pi_{\underline{v}})_!(\mathbf{1}_{\tilde{\mathcal{F}}_{\underline{v}}}) = \sum_{\underline{v}', \underline{v}''} (\pi_{\underline{v}'})_!(\mathbf{1}_{\tilde{\mathcal{F}}_{\underline{v}'}}) \otimes (\pi_{\underline{v}''})_!(\mathbf{1}_{\tilde{\mathcal{F}}_{\underline{v}''}})$$

here $\underline{v}', \underline{v}''$ run over all flag types such that $\underline{v}'$ is a flag type of $\mathbf{T}$, $\underline{v}''$ is a flag type of $\mathbf{W}$ and $\underline{v}' + \underline{v}'' = \underline{v}$. Hence $Res^{\mathbf{V}}_{\mathbf{T},\mathbf{W}} : \mathcal{M}_{\mathbf{V}} \to \mathcal{M}_{\mathbf{T}} \otimes \mathcal{M}_{\mathbf{W}}$. Consider all graded space $\mathbf{V}$, Res gives a coproduct $\Delta : \mathcal{M} \to \mathcal{M} \otimes \mathcal{M}$.

Now we assume $|\mathbf{T}| = pi$ and consider the following diagram:

$$\Lambda_{\mathbf{W}} \cong \Lambda_{\mathbf{T}} \times \Lambda_{\mathbf{W}} \xleftarrow{\kappa} F \xrightarrow{\iota} \Lambda_{\mathbf{V}}$$

here F is the closed subset of $\Lambda_{\mathbf{V}}$ consisting of x such that $\mathbf{W}$ is x-stable. ι is the natural embbeding and $\kappa(x) = x|_{\mathbf{W}}$. κ admits a section $i : \Lambda_{\mathbf{W}} \to F$, sending $x'' \in \Lambda_{\mathbf{W}}$ to the direct sum of x'' and the semisimple module supported on $i \in I$. Then by definition, we can see that $Res^{\mathbf{V}}_{\mathbf{T},\mathbf{W}}(f) = i^*\iota^*(f)$.

Notice that $\kappa^{-1}(\Lambda_{\mathbf{W},i,0}) = \iota^{-1}(\Lambda_{\mathbf{V},i,p})$, we denote this set by $F_{i,p}$. We also denote $q^{-1}(\Lambda_{\mathbf{V},i,p})$ by $\Lambda''_{i,p}$. Then we can prove the following proposition.

Proposition 5.11. *(1) Each fiber of $\kappa : F_{i,p} \to \Lambda_{\mathbf{W},i,0}$ is an affine space of dimension $p(\sum\limits_{j\in I, j-i} v''_j - v''_i)$. Here we adapt the notation $j - i$ if there exists $h \in H$ connecting i and j.*

(2) $q : \Lambda''_{i,p} \to \Lambda_{\mathbf{V},i,p}$ is an isomorphism.

(3) Let $\mathbf{G}$ be the variety consisting of $\{\tilde{\mathbf{W}} \subset \mathbf{V} \mid |\mathbf{W}| = |\tilde{\mathbf{W}}|\}$. Consider the projection $\rho : \Lambda''_{i,p} \to \mathbf{G}, \rho(x, \tilde{\mathbf{W}}) = \tilde{\mathbf{W}}$, then $F_{i,p}$ is naturally identified with the fiber of ρ at $\mathbf{W}$. There is a natural embedding $j : F_{i,p} \to \Lambda''_{i,p}$ such that $qj = \iota$.

$$\begin{array}{ccccc}
\Lambda_{\mathbf{W},i,0} & \xleftarrow{\kappa} & F_{i,p} & \xrightarrow{\iota} & \Lambda_{\mathbf{V},i,p} \\
\downarrow{\scriptstyle i_1} & & \downarrow{\scriptstyle i_2} & & \downarrow{\scriptstyle i_3} \\
\Lambda_{\mathbf{W}} & \xleftarrow{\kappa} & F & \xrightarrow{\iota} & \Lambda_{\mathbf{V}}
\end{array}$$

Proof. (1) A fiber at $x' \in \Lambda_{\mathbf{W},i,0}$ can be identified with the closed subset consisting of $y \in \mathrm{Hom}(\mathbf{T}_i, \bigoplus\limits_{i-j} \mathbf{W}_j)$ such that the composition

$$\mathbf{T}_i \xrightarrow{\bigoplus\limits_{h\in H, h'=i} y_h} \bigoplus_{i-j} \mathbf{W}_j \xrightarrow{\sum\limits_{h\in H, h'=i} \epsilon(h) x'_{\bar{h}}} \mathbf{W}_i$$

is zero. Since $x' \in \Lambda_{\mathbf{W},i,0}$, the second map is surjective. Hence by linear algebra, we can see that the fiber is an affine space with dimension $p(\sum\limits_{j\in I, j-i} v''_j - v''_i)$.

(2) Notice that when $x \in \Lambda_{\mathbf{V},i,p}$, the x-stable subspace with dimension $|\mathbf{W}|$ must be $\mathrm{Im}(\bigoplus\limits_{h\in H, h''=i} x_h)$, then (2) is trivial.

(3) Check by definition. ☐

Then we can see that

Corollary 5.12. *The map $\kappa\iota^{-1}$ induces a bijection $\epsilon_{i,p} : Irr\Lambda_{\mathbf{V},i,p} \to Irr\Lambda_{\mathbf{W},i,0}$. And the map $\epsilon_{i,p}$ also induces a bijection from $\{Z \in Irr\Lambda_{\mathbf{V}} | t_i(Z) = p\}$ to $\{Z \in Irr\Lambda_{\mathbf{W}} | t_i(Z) = 0\}$, which is the inverse of $\eta_{i,p}$.*

Proof. By Proposition 5.11, we can see that $\epsilon_{i,p} : Irr\Lambda_{\mathbf{V},i,p} \to Irr\Lambda_{\mathbf{W},i,0}$ is a bijection. Notice that both $\eta_{i,p}(Z')$ and $\epsilon^{-1}_{i,p}(Z')$ are exactly the unique irreducible component containing those x' which is isomorphic to a direct sum of some $x \in$

$Z' \cap \Lambda_{\mathbf{W},i,0}$ and p simple representations at i, we can see that $\epsilon_{i,p}$ is the inverse of $\eta_{i,p}$. □

As a corollary, we have the following key lemma of Lusztig's functions for coproduct:

LEMMA 5.13. *With the noation above, assume that Z is an irreducible component of $\Lambda_{\mathbf{V}}$ with $t_i(Z) = p$ and $Z' = \epsilon_{i,p}(Z)$. Let $\underline{p} = (\underline{Z}, \underline{i})$ be a left-admissible path of Z such that $i_1 = i$ and $Z_1 = Z'$, then*

$$Res^{\mathbf{V}}_{\mathbf{T},\mathbf{W}}(f_{\underline{p}}) = f_{Z'} + \sum_{Z'' \in Irr\Lambda_{\mathbf{W}}, t_i(Z'')>0} d_{Z''} f_{Z''},$$

where $f_{Z'}, f_{Z''} \in \mathcal{M}_V$ are some constructible functions having property $\mathcal{P}(Z')$, $\mathcal{P}(Z'')$ respectively and $d_{Z''} \in \mathbb{Z}$.

$$\begin{array}{ccccc} \Lambda_{\mathbf{W},i,0} & \xrightarrow{i} & F_{i,p} & \xrightarrow{\iota} & \Lambda_{\mathbf{V},i,p} \\ \downarrow{\scriptstyle i_1} & & \downarrow{\scriptstyle i_2} & & \downarrow{\scriptstyle i_3} \\ \Lambda_{\mathbf{W}} & \xrightarrow{i} & F & \xrightarrow{\iota} & \Lambda_{\mathbf{V}} \end{array}$$

PROOF. Notice that if $(Z_0, Z_1, \cdots, Z_s)$ and $(i_1, i_2, \cdots, i_s)$ form a left-addmissible path of Z, then $(Z_1, Z_2, \cdots, Z_s)$ and $(i_2, i_3, \cdots, i_s)$ form a left-admissible path of Z'. We denote this left-addmissible path by $\underline{p}'$. Notice that $Res^{\mathbf{V}}_{\mathbf{T},\mathbf{W}}(f_{\underline{p}})|_{\Lambda_{\mathbf{W},i,0}}$ is determined by $f_{\underline{p}}|_{\Lambda_{\mathbf{V},i,p}}$. From the construction of $f_{\underline{p}}$, we can see that $Res^{\mathbf{V}}_{\mathbf{T},\mathbf{W}}(f_{\underline{p}})|_{\Lambda_{\mathbf{W},i,0}} = f_{\underline{p}'}|_{\Lambda_{\mathbf{W},i,0}}$ genericly takes value 1 on Z' and genericly takes value 0 on the other Z'' with $t_i(Z'') = 0$. Take $d_{Z''} = \rho_{Z''}(Res^{\mathbf{V}}_{\mathbf{T},\mathbf{W}}(f_Z))$ for those Z'' with $t_i(Z'') > 0$ and choose a $f_{Z''} \in \mathcal{M}$ having property $\mathcal{P}(Z'')$ for each Z'' with $t_i(Z'') > 0$. Let $f_{Z'} = Res^{\mathbf{V}}_{\mathbf{T},\mathbf{W}}(f_Z) - \sum_{Z'' \in Irr\Lambda_{\mathbf{W}}, t_i(Z'')>0} d_{Z''} f_{Z''} \in \mathcal{M}$, then $f_{Z'} \in \mathcal{M}$ has property $\mathcal{P}(Z')$ and we get the proof. □

Dually, we assume $|\mathbf{W}| = pi$ and consider the following diagram:

$$\Lambda_{\mathbf{T}} \cong \Lambda_{\mathbf{T}} \times \Lambda_{\mathbf{W}} \xleftarrow{\kappa} F \xrightarrow{\iota} \Lambda_{\mathbf{V}}$$

Notice that $\kappa^{-1}(\Lambda^0_{\mathbf{T},i}) = \iota^{-1}(\Lambda^p_{\mathbf{V},i})$ and we denote this space by F^p_i, then we have the folliwng result dual to Proposition 5.11:

PROPOSITION 5.14. *(1) Each fiber of $\kappa : F^p_i \to \Lambda^0_{\mathbf{T},i}$ is an affine space of dimension $p(\sum_{j \in I, j - i} v'_j - v'_i)$.*

(2) $q : \Lambda''^p_i \to \Lambda^p_{\mathbf{V},i}$ is an isomorphism.

(3) Consider $\rho : \Lambda''^p_i \to \mathbf{G}, \rho(x, \widetilde{\mathbf{W}}) = \widetilde{\mathbf{W}}$, then F^p_i is naturally identifed with the fiber of ρ at $\mathbf{W}$ and there is a natural embedding $j : F^p_i \to \Lambda''^p_i$ such that $qj = \iota$.

With the above proposition, we can easily obtain the following corollary dual to Corollary 5.12:

COROLLARY 5.15. *There is a natural bijection $\epsilon^*_{i,p} : Irr\Lambda^p_{\mathbf{V},i} \to Irr\Lambda^0_{\mathbf{T},i}$ indeuced by $\kappa\iota^{-1}$. And the map $\epsilon^*_{i,p}$ also induces a bijection from $\{Z \in Irr\Lambda_{\mathbf{V}} | t^*_i(Z) = p\}$ to $\{Z \in Irr\Lambda_{\mathbf{T}} | t^*_i(Z) = 0\}$, which is the inverse of $\eta^*_{i,p}$.*

The following lemma is the key lemma of Lusztig's functions for coproduct, which is dual to Lemma 5.13:

LEMMA 5.16. *With the notations above, assume that Z is an irreducible component of $\Lambda_{\mathbf{V}}$ with $t_i^*(Z) = p$ and $Z' = \epsilon_{i,p}^*(Z)$. Let $\underline{p} = (\underline{Z}, \underline{i})$ be a right-addmissible path of Z such that $Z_1 = Z'$ and $i_1 = i$, then*

$$Res_{\mathbf{T},\mathbf{W}}^{\mathbf{V}}(f_{\underline{p}}) = f_{Z'} + \sum_{Z'' \in Irr\Lambda_{\mathbf{W}}, t_i^*(Z'')>0} d'_{Z''} f_{Z''},$$

where $f_{Z'}, f_{Z''} \in \mathcal{M}_V$ are some constructible functions having property $\mathcal{P}(Z')$, $\mathcal{P}(Z'')$ respectively and $d'_{Z''} \in \mathbb{Z}$.

6. An $\mathbb{Z}_2 \times I$-colored graph $\mathcal{G}_2$ of $\mathcal{M}$

6.1. The graph $\mathcal{G}_2$.

DEFINITION 6.1. Given two constructible functions f, g, if on each irreducible componnet Z, f and g generically take the same value, we say f generically equals to g. It defines an equivalent relation. Given $f \in \mathcal{M}$, we denote the equivalent class of f by $[f]$.

Given an irreducible component $Z \in Irr\Lambda_{\mathbf{V}'}$, we let $[f_Z]$ be the equivalent class consisting of those $f_Z \in \mathcal{M}$ with property $\mathcal{P}(Z)$. Then by Lemma 5.6, $\eta_{i,p}$ induces a bijection (still denoted by $\eta_{i,p}$) between the set of the equivalent classes $\{[f]|f$ has property $\mathcal{P}(Z')$ for some $t_i(Z') = 0\}$ and the set of the equivalent classes $\{[f]|f$ has property $\mathcal{P}(Z)$ for some $t_i(Z) = p\}$ as the following: Let $f \in [f]$ with property $\mathcal{P}(Z')$ with $t_i(Z') = 0$, then

$$1_{i^p} * f = f_Z + \sum_{Z'' \in Irr\Lambda_{\mathbf{V}}, t_i(Z'')>p} c_{Z''} f_{Z''}$$

for some f_Z with property $\mathcal{P}(Z)$. Even though f_Z is not unique, the equivalent class $[f_Z]$ is unique and consists of elements having property $\mathcal{P}(\eta_{i,p}(Z'))$.

Similarly, $\eta_{i,p}^*$ induces a bijection (still denoted by $\eta_{i,p}^*$) from the set of the equivalent classes $\{[f]|f$ has property $\mathcal{P}(Z')$ for some $t_i^*(Z') = 0\}$ to the set of the equivalent classes $\{[f]|f$ has property $\mathcal{P}(Z)$ for some $t_i^*(Z) = p\}$.

Given an equivalent class $[f]$ such that $f \in \mathcal{M}$ has property $\mathcal{P}(Z)$ for $t_i(Z) = n$, we define $[g] = i_+([f]) = \eta_{i,n+1}\eta_{i,n}^{-1}([f])$ and associate an arrow $[f] \xrightarrow{i_+} [g]$. Similarly, if f has property $\mathcal{P}(Z)$ for $t_i^*(Z) = n$, we define $[g] = i^+([f]) = \eta_{i,n+1}^*(\eta_{i,n}^*)^{-1}([f])$ and associate an arrow $[f] \xrightarrow{i^+} [g]$.

Dually, if $[f]$ is an equivalent class such that $f \in \mathcal{M}$ has property $\mathcal{P}(Z)$ for $t_i(Z) = n > 0$, we define $i_-([f]) = \eta_{i,n-1}\eta_{i,n}^{-1}([f])$. If $[f]$ is an equivalent class such that $f \in \mathcal{M}$ such that $f \in \mathcal{M}$ has property $\mathcal{P}(Z)$ for $t_i^*(Z) = n > 0$, we define $i^-([f]) = \eta_{i,n-1}^*(\eta_{i,n}^*)^{-1}([f])$. Then by construction, we can see that i_-, i^- are the inverse of i_+, i^+ respectively.

DEFINITION 6.2. We define a $\mathbb{Z}_2 \times I$-colored graph $\mathcal{G}_2 = (\mathcal{V}_2, \mathcal{E}_2)$ as the following:

(1) The set of vertices is $\mathcal{V}_2 = \{[f]|f \in \mathcal{M}_{\mathbf{V}}$ has property $\mathcal{P}(Z)$ for some $\mathbf{V}$ and irreducible component $Z \subset \Lambda_{\mathbf{V}}\}$.

(2) The set of arrows is $\mathcal{E}_2 = \{[f] \xrightarrow{i_+} [g], [f] \xrightarrow{i^+} [g] | i \in I, [f], [g] \in \mathcal{V}_2\}$

REMARK 6.3. Notice that $\mathcal{V}_2$ is naturally bijective to the set $\bigcup\limits_{\mathbf{V}} Irr\Lambda_{\mathbf{V}}$. If $[f]\in\mathcal{V}_2$ such that $f\in[f]$ has proerty $\mathcal{P}(Z)$, we adapt the notation $t_i([f])=t_i(Z)$ and $t_i^*([f])=t_i^*(Z)$.

6.2. The commutative relation between i^+ and j_+. In this section, we prove the lemmas parallel to those in Section 4.3

LEMMA 6.4. *Assume $[f_0]\in\mathcal{V}_2$ such that $t_i^*([f_0])=0$ and $[f]=\eta^*_{i,c}([f_0])$, then $t_j([f_0])=t_j([f])$.*

PROOF. Assume $t_j([f_0])=d$ and f_0 has property $\mathcal{P}(Z')$, then $t_j(Z')=d$ and $t_i^*(Z')=0$. Consider the diagram for $|\mathbf{V}''|=ci$.

$$\begin{array}{ccccccc}
\Lambda^0_{\mathbf{V}',i} & \xleftarrow{p} & \Lambda_i'^{c} & \xrightarrow{r} & \Lambda_i''^{c} & \xrightarrow{q} & \Lambda^c_{\mathbf{V},i}\\
\downarrow{\scriptstyle j_1} & & \downarrow{\scriptstyle j_2} & & \downarrow{\scriptstyle j_3} & & \downarrow{\scriptstyle j_4}\\
\Lambda_{\mathbf{V}'} & \xleftarrow{p} & \Lambda' & \xrightarrow{r} & \Lambda'' & \xrightarrow{q} & \Lambda_{\mathbf{V}}
\end{array}$$

and the composition

$$\mathbf{V}_i' \xrightarrow{\bigoplus\limits_{h\in H,h'=i} x_h'} \bigoplus_{i-j}\mathbf{V}_j' \xrightarrow{\sum\limits_{h\in H,h'=i}\epsilon(h)y_{\bar{h}}} \mathbf{V}_i''$$

Notice that for $x'\in\Lambda^0_{\mathbf{V}',i}\cap\Lambda_{\mathbf{V}',j,d}$ and $x\in q'p^{-1}(x')$, the number:

$$\mathrm{codim}_{\mathbf{V}_j}(\sum_{h\in H,h''=j}\mathrm{Im}x_h)=\mathrm{codim}_{\mathbf{V}_j}(\sum_{h\in H,h''=j}\mathrm{Im}x_h')=d$$

does not depend on the choice of those $y_{\bar{h}}$, we can see that $t_j([f_0])=t_j([f])$. □

By similar argument as in the proof of Corollary 4.12, we obtain the following corollary.

COROLLARY 6.5. *If $[f]\in\mathcal{V}_2$ and $[g]=i^+([f])$, then $t_j([g])=t_j([f])$.*

LEMMA 6.6. *Assume $[f_0]\in\mathcal{V}_2$ such that $t_i^*([f_0])=t_j([f_0])=0$, then*

$$\eta_{j,d}\eta^*_{i,c}([f_0])=\eta^*_{i,c}\eta_{j,d}([f_0])$$

for any $c,d\in\mathbb{N}$.

PROOF. By Lemma 6.4, we can see that $t_j(\eta^*_{i,c}([f_0]))=0$, hence $\eta_{j,d}\eta^*_{i,c}([f_0])$ is well-defined. By the similar argument, $\eta^*_{i,c}\eta_{j,d}([f_0])$ is also well-defined. Let $[f_1]=\eta_{j,d}\eta^*_{i,c}([f_0]),[f_2]=\eta^*_{i,c}\eta_{j,d}([f_0])$, then $t_j([f_1])=d=t_j([f_2])$ and $t_i^*([f_1])=c=t_i^*([f_2])$.

Given any $x\in\Lambda^c_{\mathbf{V},i}\cap\Lambda_{\mathbf{V},j,d}$, there exist unique x-stable subspaces $\mathbf{W},\mathbf{T}$ such that $|\mathbf{W}|+dj=|\mathbf{V}|,|\mathbf{T}|=ci$ respectively. Since $\mathbf{T}\subset\mathbf{W}$, we can define $x|_{\mathbf{W}/\mathbf{T}}$. It determines an orbit $\mathcal{O}(x)\subset\Lambda_{\mathbf{V}''}$ with $|\mathbf{V}''|+dj+ci=|\mathbf{V}|$. We define $\tilde{f}:\Lambda^c_{\mathbf{V},i}\cap\Lambda_{\mathbf{V},j,d}$ by $\tilde{f}(x)=f_0(x')$ if $x'\in\mathcal{O}(x)$. Then by the construction of f_1 in Lemma 5.4, Lemma 5.6 and Lemma 5.9, we can see that $f_1|_{\Lambda^c_{\mathbf{V},i}\cap\Lambda_{\mathbf{V},j,d}}$ generically equals to $g|_{\Lambda^c_{\mathbf{V},i}\cap\Lambda_{\mathbf{V},j,d}}$ with $g=\mathbf{1}_{j^{(d)}}*f_0*\mathbf{1}_{i^{(c)}}$. By the similar argument, we can see that $f_2|_{\Lambda^c_{\mathbf{V},i}\cap\Lambda_{\mathbf{V},j,d}}$ generically equals to $g|_{\Lambda^c_{\mathbf{V},i}\cap\Lambda_{\mathbf{V},j,d}}$, too. Hence

$$\eta_{j,d}\eta^*_{i,c}([f_0])=\eta^*_{i,c}\eta_{j,d}([f_0]).$$

□

By the similar argument as in Corollary 4.14, we have the following corollary:

COROLLARY 6.7. *Given* $[f] \in \mathcal{V}_2$, $j_+ i^+([f]) = i^+ j_+([f])$.

6.3. The commutative relation between i^+ and i_+. In this subsection, we prove the lemmas parallel to those in Section 4.4.

LEMMA 6.8. *Assume* $[f_0] \in \mathcal{V}_2$ *with* $f_0 \in \mathcal{M}_{\mathbf{V}'}$ *and* $t_i^*([f_0]) = 0, t_i([f_0]) = d$. *Let* $[f] = \eta_{i,c}^*([f_0])$, *then*

$$t_i([f]) = \begin{cases} d & c + (|\mathbf{V}'|, i) \leq d \\ c + (|\mathbf{V}'|, i) & \textit{otherwise} \end{cases}$$

PROOF. Assume that f_0 has property $\mathcal{P}(Z')$ and f has property $\mathcal{P}(Z)$. Consider the diagram for $|\mathbf{V}''| = ci$.

$$\begin{array}{ccccccc} \Lambda^0_{\mathbf{V}',i} & \xleftarrow{p} & \Lambda'^c_i & \xrightarrow{r} & \Lambda''^c_i & \xrightarrow{q} & \Lambda^c_{\mathbf{V},i} \\ \downarrow{\scriptstyle j_1} & & \downarrow{\scriptstyle j_2} & & \downarrow{\scriptstyle j_3} & & \downarrow{\scriptstyle j_4} \\ \Lambda_{\mathbf{V}'} & \xleftarrow{p} & \Lambda' & \xrightarrow{r} & \Lambda'' & \xrightarrow{q} & \Lambda_{\mathbf{V}} \end{array}$$

and the composition

$$\mathbf{V}'_i \xrightarrow{\bigoplus_{h\in H,h'=i} x'_h} \bigoplus_{i-j} \mathbf{V}'_j \xrightarrow{\begin{pmatrix} \sum_{h\in H,h'=i} \epsilon(h) x'_{\bar{h}} \\ \sum_{h\in H,h'=i} \epsilon(h) y_{\bar{h}} \end{pmatrix}} \mathbf{V}'_i \oplus \mathbf{V}''_i$$

The fiber of p at x' is the set consisting of $(x, \widetilde{\mathbf{W}}, \rho_1, \rho_2)$ such that $(\widetilde{\mathbf{W}}, \rho_1, \rho_2)$ can be chosen arbitarily and x is an extension of x'. $(\widetilde{\mathbf{W}}, \rho_1, \rho_2)$ is given by a point in $G_{\mathbf{V}}/U$, here U is the unipotent radical of the parabolic subgroup of type $(|\mathbf{V}'|, |\mathbf{V}''|)$. To give an extension of x' is equivalent to give $(y_{\bar{h}})_{h\in H,h'=i}$ such that the above composition vanishes.

For any $x' \in \Lambda^0_{\mathbf{V}',i} \cap \Lambda_{\mathbf{V}',i,d}$, $\bigoplus_{h\in H,h'=i} x'_h$ is injective and $\dim(\mathrm{Im} \bigoplus_{h\in H,h'=i} \epsilon(h) x'_{\bar{h}}) = v'_i - d$. Fix bases of $\mathbf{V}'$ and $\mathbf{V}''$, we denote the matrix of $\bigoplus_{h\in H,h'=i} x'_h$ by X', $\bigoplus_{h\in H,h'=i} \epsilon(h) x'_{\bar{h}}$ by $\bar{X}'$ and $\bigoplus_{h\in H,h'=i} \epsilon(h) y_{\bar{h}}$ by $\bar{Y}$ respectively. Then each raw of $\bar{X}', \bar{Y}$ is contained in $\mathbf{K} = \{z | zX' = 0\}$. Since $\bigoplus_{h\in H,h'=i} x'_h$ is injective , $\mathrm{rank}(X') = v'_i$, we can see that

$$\dim(\mathbf{K}) = \sum_{i-j} v'_j - v'_i = -(|\mathbf{V}'|, i) + v'_i.$$

Hence

$$\mathrm{rank}(\begin{pmatrix} \bar{X}' \\ \bar{Y} \end{pmatrix}) \leq -(|\mathbf{V}'|, i) + v'_i.$$

If $c + (|\mathbf{V}'|, i) \leq d$, then $v_i - d \leq \sum_{i-j} v'_j - v'_i$. Let $\mathcal{O}_1 = f_0^{-1}(1) \cap Z' \cap \Lambda^0_{\mathbf{V}',i} \cap \Lambda_{\mathbf{V}',i,d}$, then $\mathcal{O}_1$ is dense in Z'. Let $\mathcal{O}_2$ be the set consisting of x such that x is an extension of $x' \in \mathcal{O}_1$ and $\mathrm{rank}(\begin{pmatrix} \bar{X}' \\ \bar{Y} \end{pmatrix}) = \mathrm{rank}(\bar{X}') + \mathrm{rank}(\bar{Y})$. Notice that on each fiber $p^{-1}(x')$, the condition that $\mathrm{rank}(\begin{pmatrix}\bar{X}'\\ \bar{Y}\end{pmatrix}) = \mathrm{rank}(\bar{X}') + \mathrm{rank}(\bar{Y})$ is an open condition, we can see that $\mathcal{O}_2$ is dense in Z. On the other hand, $f_0 * \mathbf{1}_{i^{(c)}}|_{\mathcal{O}_2} = 1$ by the construction of $\mathcal{O}_2$ and $\mathcal{O}_2 \subset \Lambda_{\mathbf{V},i,d}$. Hence $t_i([f]) = t_i(Z) = d$.

If $c + (|\mathbf{V}'|, i) > d$, then on each fiber $p^{-1}(x')$, the condition that $\mathrm{rank}(\begin{pmatrix}\bar{X}'\\ \bar{Y}\end{pmatrix}) = \dim(\mathbf{K})$ becomes an open condition. If we let $\mathcal{O}_3$ be the set consisting of x such that x is an extension of $x' \in \mathcal{O}_1$ and $\mathrm{rank}(\begin{pmatrix}\bar{X}'\\ \bar{Y}\end{pmatrix}) = \dim(\mathbf{K})$, we can see that $f_0 * \mathbf{1}_{i^{(c)}}|_{\mathcal{O}_3} = 1$ and $\mathcal{O}_3 \subset \Lambda_{\mathbf{V},i,c+(|\mathbf{V}'|,i)}$. Hence $t_i([f]) = c + (|\mathbf{V}'|, i)$. □

Replace t_i, t_i^* by t_i^*, t_i respectively, we have the following dual result:

LEMMA 6.9. *Assume $[f_0] \in \mathcal{V}_2$ with $f_0 \in \mathcal{M}_{\mathbf{V}'}$ and $t_i([f_0]) = 0, t_i^*([f_0]) = c$. Let $[f] = \eta_{i,d}([f_0])$, then*

$$t_i^*([f]) = \begin{cases} c & d + (|\mathbf{V}'|, i) \le c \\ d + (|\mathbf{V}'|, i) & \textit{otherwise} \end{cases}$$

The following lemma is parallel to Lemma 4.15:

LEMMA 6.10. *Assume $[f_0] \in \mathcal{V}_2$ with $f_0 \in \mathcal{M}_{\mathbf{V}'}$ and $t_i^*([f_0]) = 0, t_i(f_0) = d > 0$. Take $c \in \mathbb{N}_{>0}$ such that $c + (|\mathbf{V}'|, i) \le d$ and set $[f] = \eta_{i,c}^*([f_0])$. Let $[g_0]$ be the unique equivalent class such that $i_+([g_0]) = [f_0]$. Then $t_i^*([g_0]) = 0$, hence $\eta_{i,c}^*([g_0])$ is well-defined. Moreover,*

$$i_+\eta_{i,c}^*([g_0]) = [f].$$

Or equivalently,

$$i_+(i^+)^c([g_0]) = (i^+)^c i_+(g_0).$$

In this case,

$$t_i^*(i_-([f])) = t_i^*([f]) = c.$$

PROOF. We assume that f_0 has property $\mathcal{P}(Z')$ and f has property $\mathcal{P}(Z)$. By the similar argument in Lemma 4.15, we can prove that $t_i^*([g_0]) = 0$, hence $[g] = \eta_{i,c}^*([g_0])$ is well-defined.

Let $[h_0]$ be the unique equivalent class such that $\eta_{i,d}([h_0]) = [f_0]$ with $h_0 \in \mathcal{M}_{\tilde{\mathbf{V}}}$ having property $\mathcal{P}(Z_0)$. Then by Lemma 6.9, we can see that $t_i^*([h_0]) = 0$, otherwise $t_i^*([f_0]) > 0$. Hence $[h] = \eta_{i,c}^*([h_0])$ is well-defined. Moreover, with the assumption $c + (|\mathbf{V}'|, i) \le d$, we can see that $t_i([h]) = 0$ for the similar reason, hence $\eta_{i,d}([h])$ is defined.

We claim that $\eta_{i,d}([h]) = \eta_{i,c}^*([f_0]) = [f]$. We assume that h has property $\tilde{Z}$. Then from the proof of Lemma 6.8 and Lemma 6.9, we can see that Z contains a dense subset $\mathcal{O}_1 \subset \Lambda_{\mathbf{V},i,d} \cap \Lambda_{\mathbf{V},i}^c \cap Z$ such that $\mathbf{1}_{i^{(d)}} * h_0 * \mathbf{1}_{i^{(c)}}|_{\mathcal{O}_1} = 1$. Similarly, $\tilde{Z}$ contains a dense subset $\mathcal{O}_2 \subset \Lambda_{\mathbf{V},i,d} \cap \Lambda_{\mathbf{V},i}^c \cap \tilde{Z}$ such that $\mathbf{1}_{i^{(d)}} * h_0 * \mathbf{1}_{i^{(c)}}|_{\mathcal{O}_2} = 1$. Notice that there exists a unique irreducible componet of $\Lambda_{\mathbf{V},i,d} \cap \Lambda_{\mathbf{V},i}^c$ such that $\mathbf{1}_{i^{(d)}} * h_0 * \mathbf{1}_{i^{(c)}}$ generically takes value 1 on it. We can see that $Z = \tilde{Z}$ and

$$\eta_{i,d}([h]) = \eta_{i,c}^*([f_0]) = [f].$$

Replace $[f], [f_0]$ by $[g], [g_0]$ respectively, for the same reason we can see that

$$\eta_{i,d-1}([h]) = \eta_{i,c}^*([g_0]) = [g],$$

hence

$$i_+([g]) = i_+\eta_{i,d-1}([h]) = \eta_{i,d}([h]) = [f].$$

□

The following lemma is parallel to Lemma 4.16:

LEMMA 6.11. *Assume* $[f_0] \in \mathcal{V}_2$ *with* $f_0 \in \mathcal{M}_{\mathbf{V}'}$ *and* $t_i^*([f_0]) = 0, t_i([f_0]) = d$. *Take* $c \in \mathbb{N}_{>0}$ *such that* $c + (|\mathbf{V}'|, i) > d$ *and set* $[f] = \eta_{i,c}^*([f_0])$. *Then*

$$i_-([f]) = i^-([f]).$$

Moreover,

$$t_i^*(i_-([f])) = t_i^*([f]) - 1 = c - 1.$$

PROOF. Let $k = c + (|\mathbf{V}'|, i) - d > 0$. Let $[h_0]$ be the unique equivalent class such that $\eta_{i,d}([h_0]) = [f_0]$ with $h_0 \in \mathcal{M}_{\tilde{\mathbf{V}}}$ having property $\mathcal{P}(Z_0)$ and set $[h] = \eta_{i,c-k}^*([h_0])$. We claim that $\eta_{i,d+k}([h]) = [f]$.

Indeed, if we assume f has proerty $\mathcal{P}(Z_1)$ and $f' \in \eta_{i,d+k}([h])$ has property $\mathcal{P}(Z_2)$, then by the proof of Lemma 6.8, we can see that Z_1 contains a dense subset $\mathcal{O}_1$ consisting of x such that x is an extension of x' given by the following composition

$$\mathbf{V}_i' \xrightarrow{\bigoplus_{h\in H,h'=i} x_h'} \bigoplus_{i-j} \mathbf{V}_j' \xrightarrow{\begin{pmatrix} \sum_{h\in H,h'=i} \epsilon(h)x_{\bar{h}}' \\ \sum_{h\in H,h'=i} \epsilon(h)y_{\bar{h}} \end{pmatrix}} \mathbf{V}_i' \oplus \mathbf{V}_i''$$

with $\operatorname{rank}(\begin{pmatrix} \bar{X}' \\ \bar{Y} \end{pmatrix}) = \sum_{i-j} v_j - v_i + c$.

Similarly, Z_2 contains a dense subset $\mathcal{O}_2$ consisting of x such that x is an extension of x'' given by the following composition

$$\mathbf{V}_i' \oplus \mathbf{V}_i'' \xrightarrow{(\bigoplus_{h\in H,h'=i} y_h, \bigoplus_{h\in H,h'=i} x_h'')} \bigoplus_{i-j} \mathbf{V}_j'' \xrightarrow{\sum_{h\in H,h'=i} \epsilon(h)x_{\bar{h}}''} \mathbf{V}_i''$$

with $\operatorname{rank}((Y, X'')) = v_i - c$. Here (Y, X'') is the matrix of $(\bigoplus_{h\in H,h'=i} y_h, \bigoplus_{h\in H,h'=i} x_h'')$ under fixed bases.

Consider the composition $\mathbf{V}_i \xrightarrow{\bigoplus_{h\in H,h'=i} x_h} \bigoplus_{i-j} \mathbf{V}_j \xrightarrow{\sum_{h\in H,h'=i} \epsilon(h)x_{\bar{h}}} \mathbf{V}_i$ and let $X, \bar{X}$ be the matrix of $\bigoplus_{h\in H,h'=i} x_h, \bigoplus_{h\in H,h'=i} x_{\bar{h}}$ under fixed bases respectively. Notice that for a fixed X_0 with $\operatorname{rank}(X_0) = v_i - c$, then any $\bar{X}$ such that $\bar{X}X_0 = 0$ has rank $\leq \sum_{i-j} v_j - v_i + c$, so $\{\bar{X} | \bar{X}X_0 = 0, \operatorname{rank}(\bar{X}) = \sum_{i-j} v_j - v_i + c\}$ is a dense subset of $\{\bar{X} | \bar{X}X_0 = 0\}$. Dually, for fixed $\bar{X}_0$ with rank $\sum_{i-j} v_j - v_i + c$, $\{X | \bar{X}_0 X = 0, \operatorname{rank}(X) = \sum_{i-j} v_j - v_i + c\}$ is a dense subset of $\{X | \bar{X}_0 X = 0\}$. We can see that $\mathcal{O}_1 \cap \mathcal{O}_2$ is noempty and dense in $\mathcal{O}_1, \mathcal{O}_2$, so $Z_1 = Z_2$ and $\eta_{i,d+k}([h]) = [f]$.

By Lemma 6.10,

$$\eta_{i,d}\eta_{i,c-k}^*([h_0]) = \eta_{i,c-k}^*\eta_{i,d}([h_0]).$$

We denote this equivalent class by $[g]$. By the claim, we can see that $[f] = (i^+)^k([g]) = (i_+)^k([g])$. Replace k by $k-1$, we also have $(i^+)^{k-1}([g]) = (i_+)^{k-1}([g])$, hence $i_-([f]) = i^-([f])$. and $t_i^*(i_-([f])) = t_i^*([f]) - 1 = c - 1$. $\square$

7. The isomorphism of the graphs

In this section, we will construct an isomorphism between the colored graph $\mathcal{G}_1$ and $\mathcal{G}_2$ by induction. We will denote $t_i(\mathcal{F}_{\Omega,\Omega_i}(L))$ and $t_i^*(\mathcal{F}_{\Omega,\Omega^i}(L))$ by $t_i(L)$ and $t_i^*(L)$ respectively in this section, if there is no ambiguity.

Let L_0 be the unique simple perverse sheaf on $\mathbf{E}_{\mathbf{V}_0}$ with $\mathbf{V}_0 = 0$. By Lemma 7.2 in [**9**], for any $L \in \mathcal{V}_1$, there exists a path $[L_0] \xrightarrow{i_{1+}} [L_1] \xrightarrow{i_{2+}} \cdots \xrightarrow{i_{l+}} [L_l] = [L]$ in $\mathcal{G}_1$ and we write $h(L) = l$. Notice that $h(L) = \sum\limits_{i \in I} \dim \mathbf{V}_i$ for $L \in \mathcal{P}_{\mathbf{V}}$, it does not depend on the choice of path.

Similarly, let $\mathbf{1}_0$ be the constant function on $\Lambda_{\mathbf{V}_0}$ with $\mathbf{V}_0 = 0$. By Corollary 1.6 in [**11**], for any $[f] \in \mathcal{V}_2$, there exists a path $[\mathbf{1}_0] \xrightarrow{i_{1+}} [f_1] \xrightarrow{i_{2+}} \cdots \xrightarrow{i_{l+}} [f_l] = [f]$ in $\mathcal{G}_2$ and we write $h([f]) = l$. Notice that $h([f]) = \sum\limits_{i \in I} \dim \mathbf{V}_i$ for $f \in \mathcal{M}_{\mathbf{V}}$, it does not depend on the choice of path.

Let $\mathcal{V}_1^{\leq m}$ be the subset of $\mathcal{V}_1$ consisting of those $[L]$ with $h(L) \leq m$ and let $\mathcal{E}_1^{\leq m}$ be the subset of $\mathcal{E}_1$ consisting of arrows between vertex in $\mathcal{V}_1^{\leq m}$:

$$\mathcal{V}_1^{\leq m} = \{[L] \in \mathcal{V}_1 | h(L) \leq m\}$$

$$\mathcal{E}_1^{\leq m} = \{[L] \xrightarrow{i^+} [K], [L] \xrightarrow{i_+} [K] | i \in I, [K], [L] \in \mathcal{V}_1^{\leq m}\}$$

Then we can see that $\mathcal{G}_1^{\leq m} = (\mathcal{V}_1^{\leq m}, \mathcal{E}_1^{\leq m})$ is a full subgraph of $\mathcal{G}_1$.

Similarly, let $\mathcal{V}_2^{\leq m}$ be the subset of $\mathcal{V}_2$ consisting of those $[f]$ with $h([f]) \leq m$ and let $\mathcal{E}_2^{\leq m}$ be the subset of $\mathcal{E}_2$ consisting of arrows between vertices in $\mathcal{V}_2^{\leq m}$:

$$\mathcal{V}_2^{\leq m} = \{[f] \in \mathcal{V}_2 | h([f]) \leq m\}$$

$$\mathcal{E}_2^{\leq m} = \{[f] \xrightarrow{i^+} [g], [f] \xrightarrow{i_+} [g] | i \in I, [f], [g] \in \mathcal{V}_2^{\leq m}\}$$

Then $\mathcal{G}_2^{\leq m} = (\mathcal{V}_2^{\leq m}, \mathcal{E}_2^{\leq m})$ is a full subgraph of $\mathcal{G}_2$.

For any fixed $L \in \mathcal{V}_1$, we choose a path $[L_0] \xrightarrow{i_{1+}} [L_1] \xrightarrow{i_{2+}} \cdots \xrightarrow{i_{l+}} [L_l] = [L]$ in $\mathcal{G}_1$ and set $\Phi([L]) = i_{l+} i_{(l-1)+} \cdots i_{1+}([\mathbf{1}_0])$.

THEOREM 7.1. *$\Phi : \mathcal{G}_1 \to \mathcal{G}_2$ is well-defined and is an isomorphism of $\mathbb{Z}_2 \times I$-colored graphs. More precisely, $\Phi([L])$ does not depend on the choice of the path $[L_0] \xrightarrow{i_{1+}} [L_1] \xrightarrow{i_{2+}} \cdots \xrightarrow{i_{l+}} [L_l] = [L]$ for any $[L] \in \mathcal{V}_1$. $\Phi : \mathcal{V}_1 \to \mathcal{V}_2$ is bijective and Φ commutes with i_+ and i^+ for any $i \in I$. Moreover, for any $i \in I$ and $L \in \mathcal{V}_1$, $t_i(L) = t_i(\Phi([L])), t_i^*(L) = t_i^*(\Phi([L]))$.*

PROOF. We only need to show the restriction $\Phi : \mathcal{G}_1^{\leq m} \to \mathcal{G}_2^{\leq m}$ is an isomorphism of colored graphs for any m. When $m = 1$, the theorem trivially holds. Now we assume that $\Phi([L])$ is well-defined for all L with $h(L) \leq n-1$ and $\Phi : \mathcal{G}_1^{\leq n-1} \to \mathcal{G}_2^{\leq n-1}$ is an isomorphism of $\mathbb{Z}_2 \times I$-colored graphs which preserves t_i and t_i^*. We need to show the statement also holds for n under the inductive assumption. We divide our proof into four parts: Firstly, we prove that $\Phi([L])$ does not depend on the choice of path for $h(L) = n$, then $\Phi : \mathcal{V}_1^{\leq n} \to \mathcal{V}_2^{\leq n}$ is well-defined. Next we prove that $\Phi : \mathcal{V}_1^{\leq n} \to \mathcal{V}_2^{\leq n}$ is bijective. Then we prove that Φ

commutes with arrows in $\mathcal{E}_1^{\leq n}$ and $\mathcal{E}_2^{\leq n}$. Finally, we prove that $t_i(L) = t_i(\Phi([L]))$ and $t_i^*(L) = t_i^*(\Phi([L]))$ for any L with $h(L) = n$.

Step one: We prove that $\Phi([L])$ is well-defined for $h(L) = n$. Fix a simple perverse sheaf L with $h(L) = n$ and choose different paths

$$[L_0] \xrightarrow{i_{1+}} [L_1] \xrightarrow{i_{2+}} \cdots \xrightarrow{i_{n+}} [L_n] = [L]$$

$$[L_0] \xrightarrow{i'_{1+}} [L'_1] \xrightarrow{i'_{2+}} \cdots \xrightarrow{i'_{n+}} [L'_n] = [L],$$

we need to show that

$$i_{l+}i_{(l-1)+}\cdots i_{1+}([\mathbf{1}_0]) = i'_{l+}i'_{(l-1)+}\cdots i'_{1+}([\mathbf{1}_0]).$$

We denote the left hand side by $[f_1]$ and the right hand side by $[f_2]$. By the induction hypothesis,

$$[g_1] = \Phi([L_{n-1}]) = i_{(l-1)+}\cdots i_{1+}([\mathbf{1}_0])$$

and

$$[g_2] = \Phi([L'_{n-1}]) = i'_{(l-1)+}\cdots i'_{1+}([\mathbf{1}_0])$$

are well-defined and $[f_1] = i_{n+}([g_1]), [f_2] = i'_{n+}([g_2])$.

CASE 1. If $i_n = i'_n = i$, notice that $[L'_{n-1}] = i_-([L]) = [L_{n-1}]$ and $[g_1] = [g_2] = \Phi(i_-([L]))$ is well-defined, then $[f_1] = [g_1] = i_{n+}(\Phi(i_-([L])))$ in this case.

CASE 2. If $i_n \neq i'_n$ and there exists $j \in I\backslash\{i_n, i'_n\}$ such that $t_j^*([L]) > 0$. Consider $[K] = j^-([L])$, then $[K] \in \mathcal{V}_1^{\leq n-1}$. By the induction hypothesis, $\Phi(K)$ is well-defined. Since $t_j^*(L) > 0$, then by Corollary 4.12, $t_j^*(L_{n-1}) = t_j^*(i_{n-}(L)) > 0$. Hence we can take $K_1 = j^-(L_{n-1}) \in \mathcal{V}_1^{\leq n-2}$. Similarly, we can take $K_2 = j^-(L'_{n-1}) \in \mathcal{V}_1^{\leq n-2}$. By the induction hypothesis, $[h] = \Phi([K]), [h_1] = \Phi([K_1])$ and $[h_2] = \Phi([K_2])$ are well-defined and $[g_1] = j^+([h_1]), [g_2] = j^+([h_2])$. Since $j \neq i_n, i'_n$, by Corollary 4.14 j^+ commutes with i_{n+} and i'_{n+} in $\mathcal{G}_1$, so $K = i_{n+}(K_1) = i'_{n+}(K_2)$. By the indcution hypothesis, $[h] = i_{n+}([h_1]) = i'_{n+}([h_2])$. By Corollary 6.7 j^+ also commutes with i_{n+} and i'_{n+} in $\mathcal{G}_2$,

$$[f_1] = i_{n+}([g_1]) = i_{n+}j^+([h_1]) = j^+i_{n+}([h_1]) = j^+([h]),$$

$$[f_2] = i'_{n+}([g_2]) = i'_{n+}j^+([h_2]) = j^+i'_{n+}([h_2]) = j^+([h]).$$

Hence in this case $[f_1] = [f_2]$.

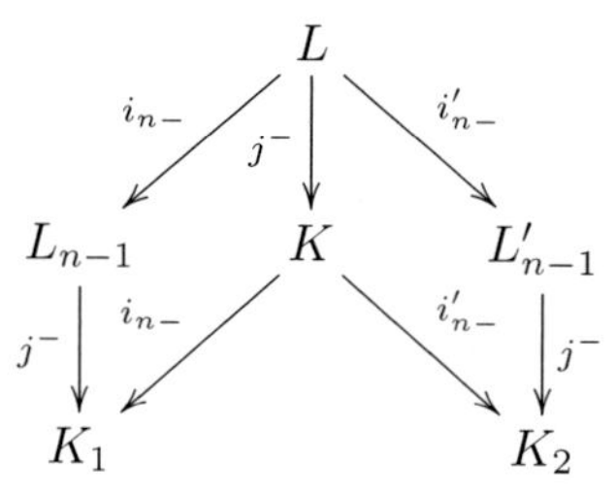

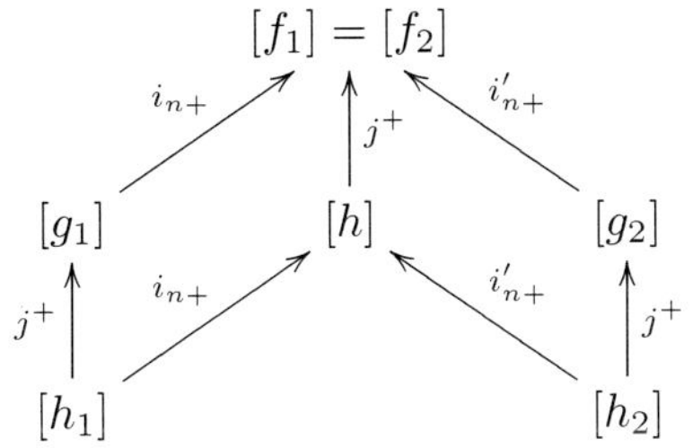

CASE 3. If $i_n \neq i'_n$ and there only exists $j = i_n$ or $i'_n \in I$ such that $t^*_j(L) = c > 0$ (the other $t^*_j(L) = 0$). Without loss of generality, we can assume $j = i_n$. Consier $K = (j^-)^c(L) \in \mathcal{V}_1^{\leq n-1}$, then $\Phi([K]) = [h]$ is well-defined. Notice that by Lemma 4.17 and Lemma 4.18,

$$\begin{cases} (j^-)^c j_-(L) = j_-(j^-)^c(L) & \text{if } t^*_j(L_{n-1}) = c \\ (j^-)^{c-1} j_-(L) = (j^-)^c(L) & \text{if } t^*_j(L_{n-1}) = c-1 \end{cases}$$

If the conidtions in Lemma 4.17 holds and $t^*_j(L_{n-1}) = c$, we can take $K_1 = j_-(K), K_2 = i'_{n-}(K)$ and set $[h_1] = \Phi([K_1]), [h_2] = \Phi([K_2])$. Then

$$[f_1] = j_+([g_1]) = j_+\Phi([L_{n-1}]) = j_+\Phi((j^+)^c([K_1])) = j_+(j^+)^c([h_1]).$$

Notice that in this case, the conditions in Lemma 6.10 holds, so in $\mathcal{G}_2$

$$j_+(j^+)^c([h_1]) = (j^+)^c j_+([h_1]).$$

Hence $[f_1] = (j^+)^c([h])$ holds. On the other hand, since $j \neq i'_n$, j^+ commutes with i'_{n+}. By the same argument as in case (2), we can prove $[f_2] = (j^+)^c([h])$. Then $[f_1] = [f_2]$ in this case.

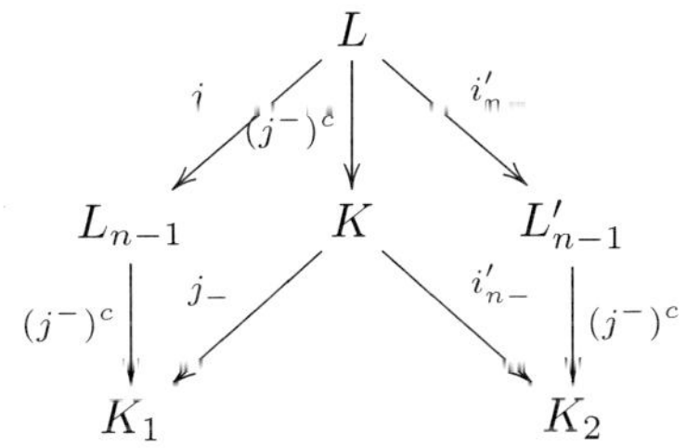

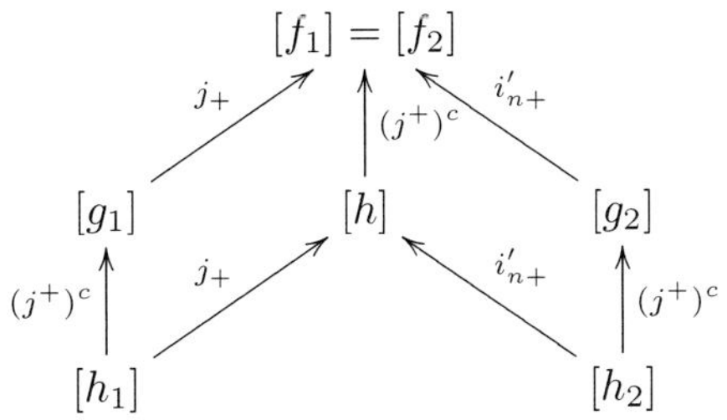

If the conditions in Lemma 4.18 holds and $t^*_j(L_{n-1}) = c-1$, we take $K_1 = K$ and take K_2 as before. Then $(j^+)^{c-1}(K) = L_{n-1}$. Hence

$$[f_1] = j_+([g_1]) = j_+\Phi(L_{n-1}) = j_+\Phi((j^+)^{c-1}(K_1)).$$

Notice that in this case the conditions in Lemma 6.11 holds and in $\mathcal{G}_2$,

$$j_+(j^+)^{c-1}([h]) = (j^+)^c([h]).$$

Hence $[f_1] = (j^+)^c([h])$. On the other hand, $[f_2] = (j^+)^c([h])$ for the same reason as in the case (2). Hence $[f_1] = [f_2]$ in this case.

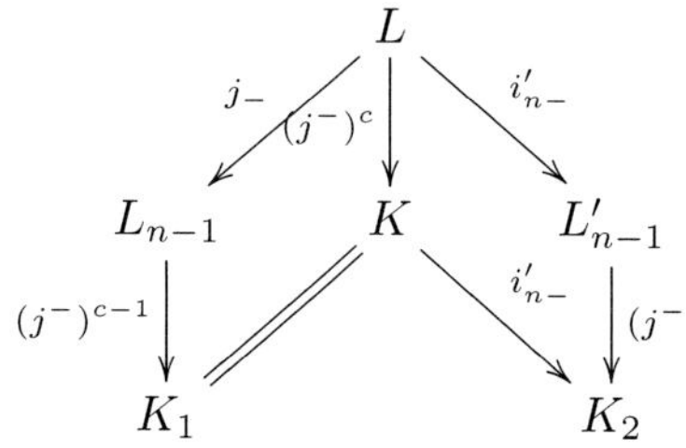

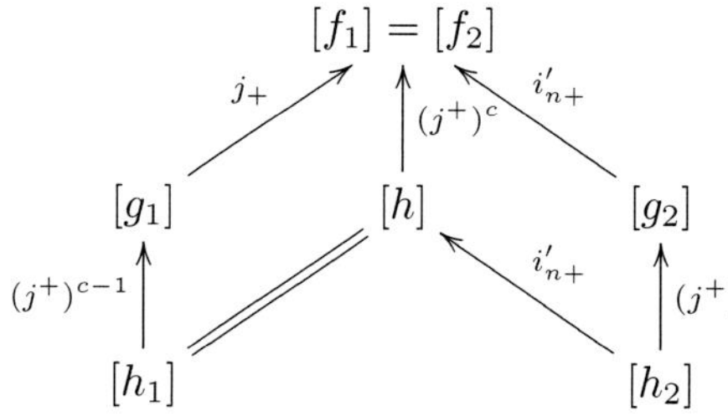

Since in any cases we always have $[f_1] = [f_2]$, $\Phi([L])$ is independent of the choice of the path $[L_0] \xrightarrow{i_{1+}} [L_1] \xrightarrow{i_{2+}} \cdots \xrightarrow{i_{n+}} [L_n] = [L]$ and $\Phi : \mathcal{V}_1^{\leq n} \to \mathcal{V}_2^{\leq n}$ is well-defined.

Step two: We prove that $\Phi : \mathcal{V}_1^{\leq n} \to \mathcal{V}_2^{\leq n}$ is bijective. Indeed, we can construct a well-defined map $\Psi : \mathcal{G}_2^{\leq n} \to \mathcal{G}_1^{\leq n}$ in a similar way.

If $\Phi : \mathcal{V}_1^{\leq n} \to \mathcal{V}_2^{\leq n}$ is not injective, we can find simple perverse sheaves $L \neq K$ and their paths $[L_0] \xrightarrow{i_{1+}} [L_1] \xrightarrow{i_{2+}} \cdots \xrightarrow{i_{n+}} [L_n] = [L]$ and $[L_0] \xrightarrow{j_{1+}} [K_1] \xrightarrow{j_{2+}} \cdots \xrightarrow{j_{n+}} [K_n] = [K]$ such that $[f] = i_{n+}i_{(n-1)+}\cdots i_{1+}([\mathbf{1}_0]) = j_{n+}j_{(n-1)+}\cdots j_{1+}([\mathbf{1}_0])$. Then by construction of Ψ, we can choose paths $[\mathbf{1}_0] \xrightarrow{i_{1+}} [f_1] \xrightarrow{i_{2+}} \cdots \xrightarrow{i_{n+}} [f]$ and $[\mathbf{1}_0] \xrightarrow{j_{1+}} [g_1] \xrightarrow{j_{2+}} \cdots \xrightarrow{j_{n+}} [f]$ for $[f]$. Then $\Psi([f]) = [K] = [L]$, which is a contradiction.

For the same reason, Ψ is injective and inverse to Φ. Hence $\Phi : \mathcal{V}_1^{\leq n} \to \mathcal{V}_2^{\leq n}$ is bijective.

Step three: Now we show that $\Phi : \mathcal{G}_1^{\leq n} \to \mathcal{G}_2^{\leq n}$ commutes with arrows i_+ and i^+ for $i \in I$. It suffices to show Φ commutes with arrows $[K] \xrightarrow{i_+} [L]$ and $[K] \xrightarrow{i^+} [L]$ with $i \in I, h(L) = n$.

For any $[L] = i_+([K])$ with $h(L) = n$, we need to prove $\Phi([L]) = i_+(\Phi([K]))$. We choose a path $[L_0] \xrightarrow{j_{1+}} [K_1] \xrightarrow{j_{2+}} \cdots \xrightarrow{j_{(n-1)+}} [K_{n-1}] = [K]$ of K, then $[L_0] \xrightarrow{j_{1+}} [K_1] \xrightarrow{j_{2+}} \cdots \xrightarrow{j_{(n-1)+}} [K_{n-1}] \xrightarrow{i_+} [L]$ is a path of L. Since Φ does not depend on the choice of path, we can see that $\Phi([L]) = i_+j_{(n-1)+}\cdots j_{1+}[\mathbf{1}_0] = i_+(\Phi([K]))$. Hence Φ commutes with $i_+, i \in I$.

For $[L] = i^+([K])$, we prove $\Phi([L]) = i^+(\Phi([K]))$ case by case. Take $j \in I$ such that $t_j(L) > 0$.

case(1): If $j \neq i$, then i^+ commutes with j_+. Let $K' = j_-(K)$, then $j_+(\Phi([K'])) = \Phi([K])$. Hence

$$i^+\Phi([K]) = i^+j_+(\Phi([K'])) = j_+i^+(\Phi([K'])).$$

Notice that $h(K') = n-2$, $i^+(\Phi([K'])) = \Phi(i^+([K']))$ by the induction hypothesis, so

$$i^+\Phi([K]) = j_+\Phi(i^+([K'])) = \Phi(j_+i^+([K'])) = \Phi([L]).$$

case(2): If $j = i$ and $t_i^*(L) = c > 0$, we consider $K_0 = (i^-)^c(L)$, then $K_0 = (i^-)^{c-1}(K)$. It suffices to show that $\Phi([L]) = (i^+)^c(\Phi([K_0]))$. Take $K' = i_-(L)$.

If the conditions in Lemma 4.17 and Lemma 6.10 holds for i and L, then in this case $t_i^*(K') = c$. Let $K'' = (i^-)^c(K')$, then by Lemma 4.17, we can see that $i_+(K') = i_+(i^+)^c(K'')$ and $K_0 = i_+(K'')$. By a similar argument as in the case $i \neq j$,

$$(i^+)^c(\Phi([K_0])) = (i^+)^c i_+\Phi([K'']) = i_+(i^+)^c\Phi([K'']) = i_+\Phi([K']) = \Phi([L]).$$

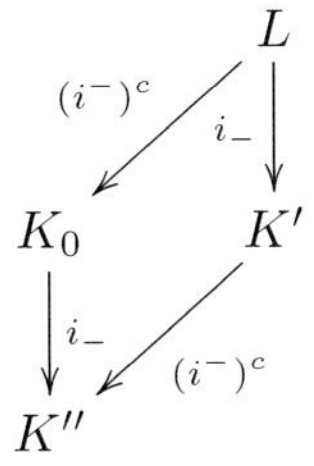

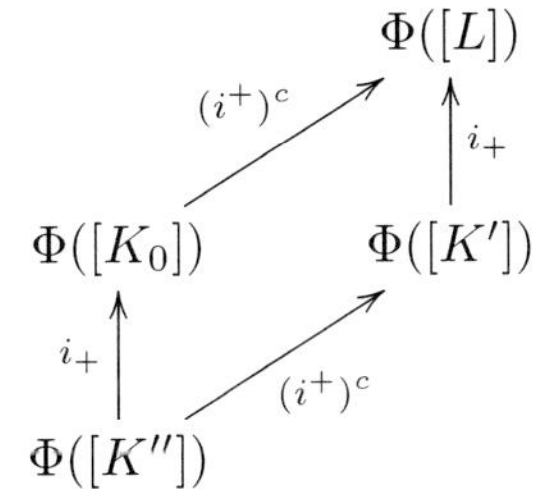

If the conditions in Lemma 4.18 and Lemma 6.11 holds for i and L, then $t_i^*(K') = c-1$ and $K_0 = (i^-)^{c-1}(K')$. In this case $K' = K, i^+(K) = i_+(K) = L$, then $i^+\Phi([K]) = i_+\Phi([K']) = \Phi([L])$. Hence Φ commutes with $i^+, i \in I$ in any case.

Step four: We prove that $t_i(L) = t_i(\Phi([L]))$ and $t_i^*(L) = t_i^*(\Phi([L]))$ for $h(L) = n$.

Notice that for any $i, j \in I$ and $L \in \mathcal{P}_{\mathbf{V}}$, $t_i(j^+(L))$ is uniquely determined by $t_i(L)$ and the dimension vector $|\mathbf{V}|$ by Lemma 4.11 and Lemma 4.15, and $t_i(j^+(\Phi([L])))$ is also determined by $t_i(\Phi([L]))$ and the dimension vector $|\mathbf{V}|$ in the same way by Lemma 6.4 and Lemma 6.8. Since Φ commutes with j^+, Φ preserves the value of t_i. Replace j^+ by j_+, we can see that Φ also preserves the value of t_i^*. □

As an corollary, we have the following proposition as in [**11**]:

PROPOSITION 7.2. *For any irreducible component Z, there exists a unique $f_Z \in \mathcal{M}$ such that f_Z has property $\mathcal{P}(Z)$. Moreover, the set $\{f_Z | Z$ is an irreducible component of $\Lambda_{\mathbf{V}}$ for some $\mathbf{V}\}$ is a $\mathbb{Q}$-basis of $\mathcal{M}$ and a $\mathbb{Z}$-basis of ${}_{\mathbb{Z}}\mathcal{M}$. We call $\{f_Z | Z \in \Lambda_{\mathbf{V}}$ for some $\Lambda_{\mathbf{V}}\}$ the semicanonical basis of $\mathcal{M}$.*

PROOF. By Lemma 5.4, we can choose an f_Z for each Z and consider the $\mathbb{Q}$-vector space $\mathbf{S} = \mathrm{span}\{f_Z | f_Z$ is chosen for $Z\}$, then it is easy to see that these f_Z are linear independent and form a basis of $\mathbf{S}$. Let $\mathbf{S}_{\mathbf{V}} = \mathbf{S} \cap \mathcal{M}_{\mathbf{V}}$. For any $\mathbf{V}$, we can see that $\dim \mathbf{S_V} = |\{f_Z|Z \in Irr\Lambda_\mathbf{V}\}| = |\{[f] \in \mathcal{V}_2|f \in \mathcal{M}_\mathbf{V}\}| = |\mathcal{P}_\mathbf{V}| = \dim \mathbf{U}_q^+(\mathfrak{g})$. On the other hand, by Theorem 2.5, $\dim \mathcal{M}_\mathbf{V} = \dim \mathcal{M}_{\mathbf{V},\Omega} = \dim \mathbf{U}^+$. Hence $\mathcal{S}_\mathbf{V} = \mathcal{M}_\mathbf{V}$ and those chosen f_Z form a $\mathbb{Q}$-basis of $\mathcal{M}$. By the construction of f_Z, it is also a $\mathbb{Z}$-basis of ${}_\mathbb{Z}\mathcal{M}$. To see the uniqueness of f_Z, we assume that one can choose another $f'_Z \in \mathcal{P}$ having property $\mathcal{P}_Z$. Express f'_Z as a linear combination of the basis $\{f_Z|Z \in Irr\Lambda_\mathbf{V}\}$ by $f'_Z = \sum_{Z'} c_{Z'} f_{Z'}$, then the right hand side generically takes value $c_{Z'}$ on Z'. It implies that $c_Z = 1$ and $c_{Z'} = 0$ for the other Z', hence $f'_Z = f_Z$ is unique. □

REMARK 7.3. We can naturally identify the set $\mathcal{V}_2$ with the set $\bigcup_\mathbf{V} Irr\Lambda_\mathbf{V}$. Then the arrows i_+, i_-, i^+, i^- in $\mathcal{G}_2$ coincide with the abstract crystal operator $\tilde{f}_i, \tilde{e}_i, \tilde{f}_i^*, \tilde{e}_i^*$ on $\bigcup_\mathbf{V} Irr\Lambda_\mathbf{V}$ respectively. The Theorem 7.1 is equivalent to the following result in [**7**] : $\bigcup_\mathbf{V} Irr\Lambda_\mathbf{V}$ has an abstract crystal structure and is isomorphic to $B(\infty)$. Our proof does not rely on the abstract crystal theory (the embedding theorem and the criterion of $B(\infty)$, in particular) and shows that $\{f_Z|Z \in Irr\Lambda_\mathbf{V}\}$ is a basis of $\mathcal{M}$ without assuming the results in [**7**].

REMARK 7.4. If we admit that $\{f_Z|Z \in Irr\Lambda_\mathbf{V}\}$ is a basis of $\mathbf{U}^+$, then it is a basis of canonical type in the sense of [**1**]. It is also proved in [**1**] that there is an abstract crystal basis structure arising from a basis of canonical type, which is isomorphic to $B(\infty)$.

By the similar argument as Proposition 3.5, we can prove that the semicanonical basis $\{f_Z|Z \in \Lambda_\mathbf{V}$ for some $\Lambda_\mathbf{V}\}$ is adapted:

PROPOSITION 7.5 ([**11**, Section 2.9]).
(1) *$\{f_Z|t_i(Z) \geq d, Z \in \Lambda_\mathbf{V}$ for some $\Lambda_\mathbf{V}\}$ form a $\mathbb{Q}$-basis of $\mathbf{1}_{i^{(d)}} * \mathcal{M}$ and a $\mathbb{Z}$-basis of $\mathbf{1}_{i^{(d)}} * {}_\mathbb{Z}\mathcal{M}$.*
(2) *If Z' is an irreducible component such that $t_i(Z') = 0$ and $f_Z = \eta_{i,d}(f_{Z'})$, then $f_Z - \mathbf{1}_{i^{(d)}} * f_{Z'} \in \mathbf{1}_{i^{(d+1)}} * \mathcal{M}$.*
(3) *$\{f_Z|t_i^*(Z) \geq d, Z \in \Lambda_\mathbf{V}$ for some $\Lambda_\mathbf{V}\}$ form a $\mathbb{Q}$-basis of $\mathcal{M} * \mathbf{1}_{i^{(d)}}$ and a $\mathbb{Z}$-basis of ${}_\mathbb{Z}\mathcal{M} * \mathbf{1}_{i^{(d)}}$.*
(4) *If Z' is an irreducible component such that $t_i^*(Z') = 0$ and $f_Z = \eta_{i,d}^*(f_{Z'})$, then $f_Z - f_{Z'} * \mathbf{1}_{i^{(d)}} \in \mathcal{M} * \mathbf{1}_{i^{(d+1)}}$.*

8. Comparision of the canonical basis and the semicanonical basis

8.1. An order in the colored graphs. Assume $|I| = n$ and fix an order $(i_1 \prec i_2 \prec \cdots \prec i_n)$ of I, we define an order of $I \times \mathbb{N}^+$ as the following: $(i_{n_2}, m_2) \prec (i_{n_1}, m_1)$ if and only if $i_{n_1} \prec i_{n_2}$ or $i_{n_1} = i_{n_2}$ and $m_1 < m_2$.

Let $\mathcal{S}$ be the set of sequneces of $I \times \mathbb{N}^+$. Given

$$\underline{s}_1 = ((i_{n_1}, m_1), (i_{n_2}, m_2), \cdots, (i_{n_k}, m_k)) \in \mathcal{S}$$

and

$$\underline{s}_2 = ((i_{n'_1}, m'_1), (i_{n'_2}, m_2), \cdots, (i_{n'_l}, m'_l)) \in \mathcal{S}$$

such that $\sum_{1 \leq s \leq k} m_s i_{n_s} = \sum_{1 \leq s \leq l} m'_s i_{n'_s}$, we say $\underline{s}_2 \prec \underline{s}_1$ if there exists $r \in \mathbb{N}$ such that $(i_{n_t}, m_t) = (i_{n'_t}, m'_t)$ for $1 \leq t < r$ and $(i_{n'_r}, m'_r) \prec (i_{n_r}, m_r)$. Let $\mathcal{S}_\mathbf{V}$ be the subset of $\mathcal{S}$ consisting of the sequences $\underline{s} = ((i_{n_1}, m_1), (i_{n_2}, m_2), \cdots, (i_{n_k}, m_k))$ such that $\sum_{1 \le s \le k} m_s i_{n_s} = |\mathbf{V}|$, then $(\mathcal{S}_\mathbf{V}, \prec)$ becomes a partially ordered set.

Now we can inductively define a map $\underline{s}^\prec = \underline{s} : \mathcal{V}_1 \to \mathcal{S}$ as the following:

(1) If L is the constant sheaf $\bar{\mathbb{Q}}_l$ on $\mathbf{E}_{\mathbf{V},\Omega}$ with $|\mathbf{V}| = pi$, we define $\underline{s}([L]) = (i, p)$. In particular, If L is the constant sheaf on $\mathbf{E}_{\mathbf{V},\Omega}$ with $\mathbf{V} = 0$, we define $\underline{s}([L]) = \emptyset$.

(2) For the other $L \in \mathcal{P}_\mathbf{V}$, there exists a unique $r \in \mathbb{N}$ such that $t_{i_r}(\mathcal{F}_{\Omega,\Omega_r}(L)) = n > 0$ but $t_{i_s}(\mathcal{F}_{\Omega,\Omega_s}(L)) = 0$ for any $r < s$. Then applying Lemma 3.4, we get a simple perverse sheaf K with $t_{i_r}(\mathcal{F}_{\Omega,\Omega_r}(K)) = 0$, we define $\underline{s}([L]) = ((i_r, n), \underline{s}([K]))$.

By construction, we can see that $\underline{s}$ is injective. Now we can define an order of $\mathcal{P}_\mathbf{V}$ as the following: $L \prec L' \in \mathcal{P}_\mathbf{V}$ if and only if $\underline{s}([L]) \prec \underline{s}([L'])$.

Notice that $\mathcal{G}_1$ is invariant under the Fourier Deligne transform, we can choose an orientation Ω for any fixed order $\prec$ such that $(I, \prec)$ is adapted to Ω. More precisely, we can choose Ω such that i_l is a sink in the subquiver consisting of vertices $\{i_l\} \cup \{i_j \in I | i_l \prec i_j\}$ for any l.

We can also define a map $\underline{s}' : \mathcal{V}_2 \to \mathcal{S}$ in a similar way, then $\mathcal{V}_2$ also becomes a partially ordered set. It is easy to see that $\underline{s}(L) = \underline{s}'(\Phi([L]))$ and Φ preserves the order $\prec$.

8.2. Monomial bases. Given $\underline{s} = ((i_{n_1}, m_1), (i_{n_2}, m_2), \cdots, (i_{n_k}, m_k)) \in \mathcal{S}$, we define

$$m_{\underline{s}} = \mathbf{1}_{i_{n_1}^{m_1}} * \mathbf{1}_{i_{n_2}^{m_2}} * \cdots * \mathbf{1}_{i_{n_k}^{m_k}} \in \mathcal{M}$$

and

$$[M_{\underline{s}}] = E_{i_{n_1}}^{(m_1)} * E_{i_{n_2}}^{(m_2)} * \cdots * E_{i_{n_k}}^{(m_k)} \in \mathcal{K}.$$

LEMMA 8.1. *Given $Z \in Irr\Lambda_\mathbf{V}$ and $f_Z \in \mathcal{M}$, then*

$$m_{\underline{s}'(f_Z)} = f_Z + \sum_{\underline{s}'(f_{Z'}) \succ \underline{s}'(f_Z)} c_{Z,Z'} f_{Z'}.$$

PROOF. We prove the lemma by induction on the length k of the sequence

$$\underline{s}'(f_Z) = ((i_{n_1}, m_1), (i_{n_2}, m_2), \cdots, (i_{n_k}, m_k)).$$

(1) When $k = 1$, then $f_Z = \mathbf{1}_{i_{n_1}^{(m_1)}} = m_{\underline{s}'(f_Z)}$. The lemma trivially holds.

(2) For $k > 1$, by Lemma 3.4, we can find $\tilde{Z} \in \Lambda_{\mathbf{V}'}$ with $|\mathbf{V}| = |\mathbf{V}'| + m_1 i_{n_1}$ such that

$$\mathbf{1}_{i_{n_1}^{m_1}} * f_{\tilde{Z}} = f_Z + \sum_{Z'' \in Irr\Lambda_\mathbf{V}, t_{i_{n_1}}(Z'') > m_1} c_{Z''} f_{Z''}.$$

By the induction hypothesis,

$$m_{\underline{s}'(f_{\tilde{Z}})} = f_{\tilde{Z}} + \sum_{\underline{s}'(f_{\tilde{Z}'}) \succ \underline{s}'(f_{\tilde{Z}})} c_{\tilde{Z},\tilde{Z}'} f_{\tilde{Z}'}.$$

Notice that $\underline{s}'(f_Z) = ((i_{n_1}, m_1), \underline{s}'(f_{\tilde{Z}}))$,

$$\begin{aligned} m_{\underline{s}'(f_Z)} =& \mathbf{1}_{i_{n_1}^{m_1}} * m_{\underline{s}'(f_{\tilde{Z}})} \\ =& \mathbf{1}_{i_{n_1}^{m_1}} * (f_{\tilde{Z}} + \sum_{\underline{s}'(f_{\tilde{Z}'}) \succ \underline{s}'(f_{\tilde{Z}})} c_{\tilde{Z},\tilde{Z}'} f_{\tilde{Z}'}) \\ =& f_Z + \sum_{Z'' \in Irr\Lambda_\mathbf{V}, t_{i_{n_1}}(Z'') > m_1} c_{Z''} f_{Z''} + \sum_{\underline{s}'(f_{\tilde{Z}'}) \succ \underline{s}'(f_{\tilde{Z}})} c_{\tilde{Z},\tilde{Z}'} \mathbf{1}_{i_{n_1}^{m_1}} * f_{\tilde{Z}'} \end{aligned}$$

Notice that if $t_{i_{n_1}}(Z'') = m > m_1$, then $\underline{s}'(f_{Z''})$ starts with (i_{n_1}, m) or (i_r, m') such that $r < n_1$. So we have the following fact $(\sharp)$

$$(\sharp)\text{: } \underline{s}'(f_{Z''}) \succ \underline{s}'(f_Z) \text{ for those } Z'' \text{ with } t_{i_{n_1}}(Z'') = m > m_1.$$

Now it suffices to show that $\mathbf{1}_{i_{n_1}^{m_1}} * f_{\tilde{Z}'} = \sum_{\underline{s}'(f_{Z'''}) \succ \underline{s}'(f_Z)} d_{Z'''} f_{Z'''}$ for those $\tilde{Z}'$ such that $\underline{s}'(f_{\tilde{Z}'}) \succ \underline{s}'(f_{\tilde{Z}})$. For those $f_{\tilde{Z}'}$ with $t_{i_{n_1}}(\tilde{Z}') > 0$, $f_{\tilde{Z}'}$ belongs to the $\mathbb{Q}$-space $\mathbf{1}_{i_{n_1}^1} * \mathcal{M}$ and $\mathbf{1}_{i_{n_1}^{m_1}} * f_{\tilde{Z}'}$ belongs to the $\mathbb{Q}$-space $\mathbf{1}_{i_{n_1}^{(m_1+1)}} * \mathcal{M}$. Since $\{f_Z | t_{i_{n_1}} \geq m_1 + 1\}$ forms a basis of the $\mathbb{Q}$-space $\mathbf{1}_{i_{n_1}^{(m_1+1)}} * \mathcal{M}$ by Proposition 7.5, $\mathbf{1}_{i_{n_1}^{m_1}} * f_{\tilde{Z}'} = \sum_{t_{i_{n_1}}(Z''') > m_1} d_{Z'''} f_{Z'''}$. Since $t_{i_{n_1}}(Z''') = m > m_1$, $\underline{s}'(f_{Z'''}) \succ \underline{s}'(f_Z)$ by the fact $(\sharp)$.

On the other hand, for those $f_{\tilde{Z}'}$ with $t_{i_{n_1}}(\tilde{Z}') = 0$,

$$\mathbf{1}_{i_{n_1}^{m_1}} * f_{\tilde{Z}'} = f_{Z'} + \sum_{Z'' \in Irr\Lambda_{\mathbf{V}}, t_{i_{n_1}}(Z'') > m_1} c_{Z''} f_{Z''},$$

where Z' is an irreducible component with $t_{i_{n_1}} = m_1$. Since $t_{i_{n_1}}(Z'') > m_1$, we can see that $\underline{s}'(f_{Z''}) \succ \underline{s}'(f_Z)$ by $(\sharp)$. Now we prove that $\underline{s}'(f_{Z'}) \succ \underline{s}'(f_Z)$. If $t_{i_r}(Z') > 0$ for some $r < n_1$, then $\underline{s}'(f'_Z)$ starts with (i_r, m') with $r < n_1$, hence $\underline{s}'(f_{Z'}) \succ \underline{s}'(f_Z)$. Otherwise, $\underline{s}'(f_{Z'}) = ((i_{n_1}, m_1), \underline{s}'(f_{\tilde{Z}'}))$. Since $\underline{s}'(f_{\tilde{Z}'}) \succ \underline{s}'(f_{\tilde{Z}})$,

$$\underline{s}'(f_{Z'}) = ((i_{n_1}, m_1), \underline{s}'(f_{\tilde{Z}'})) \succ ((i_{n_1}, m_1), \underline{s}'(f_{\tilde{Z}})) = \underline{s}'(f_Z).$$

In a conclusion, $m_{\underline{s}'(f_Z)} = f_Z + \sum_{\underline{s}'(f_{Z'}) \succ \underline{s}'(f_Z)} c_{Z,Z'} f_{Z'}$. □

By basic linear algebra, we obtain the following result:

COROLLARY 8.2. *The set* $\mathbf{M}'_{\mathbf{V}} = \{m_{\underline{s}'(f_Z)} | f_Z \in \mathcal{V}_2 \text{ for } Z \in Irr\Lambda_{\mathbf{V}}\}$ *is a* $\mathbb{Q}$-*basis of* $\mathcal{M}_{\mathbf{V}}$ *and a* $\mathbb{Z}$-*basis of* ${}_{\mathbb{Z}}\mathcal{M}_{\mathbf{V}}$. *Moreover, the transition matrix between the semicanonical basis* $\mathbf{B}_{2,\mathbf{V}} = \{f_Z | Z \in Irr\Lambda_{\mathbf{V}}\}$ *and* $\mathbf{M}'_{\mathbf{V}}$ *is upper triangular (with respect to the order of* $\mathcal{S}$ *and* $\mathcal{V}_2$*) and the diagonal entries of this matrix are all equal to* 1.

PROOF. With the lemma above, we can easily see that $\{m_{\underline{s}'(f_Z)} | f_Z \in \mathcal{V}_2$ for $Z \in Irr\Lambda_{\mathbf{V}}\}$ and $\{f_Z | f_Z \in \mathcal{V}_2$ for $Z \in Irr\Lambda_{\mathbf{V}}\}$ can be linear expressed by each other. Hence $\mathbf{M}'_{\mathbf{V}} = \{m_{\underline{s}'(f_Z)} | f_Z \in \mathcal{V}_2$ for $Z \in Irr\Lambda_{\mathbf{V}}\}$ is a $\mathbb{Q}$-basis of $\mathcal{M}_{\mathbf{V}}$. Notice that $c_{Z,Z'} = \rho_{Z'}(m_{\underline{s}'(f_Z)}) \in \mathbb{Z}$, we can see that it is also a $\mathbb{Z}$-basis of ${}_{\mathbb{Z}}\mathcal{M}_{\mathbf{V}}$. □

Since $\{[L] | L \in \mathcal{P}_{\mathbf{V}}\}$ shares the similar adapted property of $\{f_Z | Z \in Irr\Lambda_{\mathbf{V}}\}$, we can also prove the following results in a similar way:

LEMMA 8.3. *Given* $L \in \mathcal{P}_{\mathbf{V}}$, *then the following equation holds in* $\mathcal{K}$:

$$[M_{\underline{s}(L)}] = [L] + \sum_{\underline{s}([L'])\succ\underline{s}([L])} c_{L,L'}[L'].$$

COROLLARY 8.4. *The set* $\mathbf{M}_{\mathbf{V}} = \{m_{\underline{s}(L)} | L \in \mathcal{V}_1\}$ *forms a* $\mathbb{Z}[v, v^{-1}]$-*basis of* $\mathcal{K}_{\mathbf{V}}$. *Moreover, the transition matrix between the canonical basis* $\mathbf{B}_{1,\mathbf{V}} = \{[L] | L \in \mathcal{P}_{\mathbf{V}}\}$ *and* $\mathbf{M}_{\mathbf{V}}$ *is upper triangular (with respect to the order of* $\mathcal{S}$ *and* $\mathcal{V}_1$*) and the diagonal entries of this matrix are all equal to* 1.

REMARK 8.5. For any simple perverse sheaf $L \in \mathcal{P}_{\mathbf{V},\Omega}$, we can define a linear map $\rho_L : \mathcal{K} \to \mathbb{Z}[v, v^{-1}]$ as the following: $\rho_L([L']) = \sum_{k\in\mathbb{Z}} \dim \mathrm{Hom}_{\mathcal{Q}_{\mathbf{V}}}(L[k], L')v^k$ for any $L' \in \mathcal{Q}_{\mathbf{V}}$ and $\rho_L([L']) = 0$ for $L' \in \mathcal{Q}_{\mathbf{V}'}$ with $|\mathbf{V}| \neq |\mathbf{V}'|$. Then $c_{L,L'} = \rho_{L'}(M_{\underline{s}(L)})$.

8.3. The transition matrix between the canonical basis and the semicanonical basis. Recall that there are isomorphisms between algebras ($\sigma : \mathcal{K} \to {}_{\mathbb{Z}}\mathbf{U}_q(\mathfrak{g})^+$ and $\varkappa : {}_{\mathbb{Z}}\mathcal{M} \to {}_{\mathbb{Z}}\mathbf{U}(\mathfrak{g})^+$. Let $\varsigma : \mathcal{K} \to {}_{\mathbb{Z}}\mathbf{U}(\mathfrak{g})^+$ be the composition of σ and taking the classical limits. Then $\varsigma(\mathbf{B}_1)$ is a basis of ${}_{\mathbb{Z}}\mathbf{U}(\mathfrak{g})^+$, we still denote the basis by $\mathbf{B}_1$. Similarly, $\varkappa(\mathbf{B}_2)$ is a basis of ${}_{\mathbb{Z}}\mathbf{U}(\mathfrak{g})^+$, we still denote it by $\mathbf{B}_2$.

$$\mathcal{K} \xrightarrow{\varsigma} {}_{\mathbb{Z}}\mathbf{U}(\mathfrak{g})^+ \xleftarrow{\varkappa} {}_{\mathbb{Z}}\mathcal{M}$$

THEOREM 8.6. *The transition matrix between the bases* $\mathbf{B}_1 = \bigcup_{\mathbf{V}} \mathbf{B}_{1,\mathbf{V}}$ *and* $\mathbf{B}_2 = \bigcup_{\mathbf{V}} \mathbf{B}_{2,\mathbf{V}}$ *of* ${}_{\mathbb{Z}}\mathbf{U}(\mathfrak{g})^+$ *is upper triangular (with respect to the order of* $\mathcal{S}$*) and the diagonal entries of this matrix are all equal to* 1. *More precisely, assume that* $L \in \mathcal{P}_{\mathbf{V}}$ *and* $f_Z = \Phi([L])$, *then*

$$\varsigma([L]) = \varkappa(f_Z) + \sum_{\underline{s}'(f_{Z'}) \succ \underline{s}'(f_Z)} c_{Z'}\varkappa(f_{Z'})$$

with $c_{z'} \in \mathbb{Z}$.

PROOF. Notice that $\underline{s}(L) = \underline{s}'(\Phi(L))$, $\varkappa(m_{\underline{s}'(\Phi(L))}) = \varsigma([M_{\underline{s}(L)}])$, hence $\varkappa(\mathbf{M}') = \varsigma(\mathbf{M}) = \tilde{\mathbf{M}}$ and $\tilde{\mathbf{M}}$ is a basis of $\mathbf{U}(\mathfrak{g})^+$.Let $P_i, i = 1, 2$ be the transition matrix P_i between $\tilde{\mathbf{M}}$ and $\mathbf{B}_i, i = 1, 2$ respectively. By Lemma 8.3, let $P_1 = (a_{\underline{s}(L),[L']})_{[L],[L']\in\mathcal{V}_1}$, we can see that

$$a_{\underline{s}(L),[L']} = \begin{cases} \rho_{L'}([M_{\underline{s}(L)}])|_{v=1} & \underline{s}(L) \prec \underline{s}(L') \\ 1 & L = L' \\ 0 & \text{otherwise} \end{cases}$$

Similarly, let $P_2 = (b_{\underline{s}'(f_Z),f_{Z'}})_{f_Z,f_{Z'}\in\mathcal{V}_2}$,

$$b_{\underline{s}'(f_Z),f_{Z'}} = \begin{cases} \rho_{Z'}(m_{\underline{s}'(f_Z)}) & \underline{s}'(f_Z) \prec \underline{s}'(f_{Z'}) \\ 1 & f_Z = f_{Z'} \\ 0 & \text{otherwise} \end{cases}$$

Since Φ preserves the order of $\mathcal{V}_i$, we can see that the transition matrix $P_2P_1^{-1}$ between the bases $\mathbf{B}_1$ and $\mathbf{B}_2$ is upper triangular (with respect to the order of $\mathcal{V}_1$ and $\mathcal{V}_2$) and with diagonal entries equal to 1. □

REMARK 8.7. We can see that the theorem does not depend on the choice of the order of I. More precisely, if we choose another order $\tilde{\prec}$ of I, then $\tilde{\prec}$ induces an order of $\mathcal{S}$. We can also define $\underline{s}^{\tilde{\prec}} : \mathcal{V}_1 \to \mathcal{S}$ and $\underline{s}'^{\tilde{\prec}} : \mathcal{V}_2 \to \mathcal{S}$ in a similar way. Then with the notation above, we still have $\varsigma([L]) = \varkappa(f_Z) + \sum_{\underline{s}'^{\tilde{\prec}}(f_{Z'}) \tilde{\succ} \underline{s}'^{\tilde{\prec}}(f_Z)} c_{Z'}\varkappa(f_{Z'})$. Moreover, the transition matrix $P_2P_1^{-1}$ does not depend on the choice of $\prec$ (up to a permutation).

DEFINITION 8.8. Given $L, K \in \mathcal{P}_{\mathbf{V}}$, we say $L \preceq K$ if and only if for any order $\prec$ of I and the induced map $\underline{s}^{\prec} : \mathcal{V}_1 \to \mathcal{S}$, $\underline{s}^{\prec}([L]) \prec \underline{s}^{\prec}([K])$. Similarly, given $f_Z, f_{Z'} \in \mathcal{V}_2$, we say $f_Z \preceq' f_{Z'}$ if and only if for any order $\prec$ of I and the induced map $\underline{s}'^{\prec} : \mathcal{V}_2 \to \mathcal{S}$, $\underline{s}'^{\prec}(f_Z) \prec \underline{s}'^{\prec}(f_{Z'})$. Then $(\mathcal{V}_1, \preceq)$ and $(\mathcal{V}_2, \preceq')$ are partially ordered sets and $L \preceq K$ if and only if $\Phi([L]) \preceq' \Phi([K])$.

COROLLARY 8.9. *The transition matrix between the bases* $\mathbf{B}_1$ *(with the order* $\preceq$*) and* $\mathbf{B}_2$ *(with the order* $\preceq'$*) of* ${}_{\mathbb{Z}}\mathbf{U}(\mathfrak{g})^+$ *is upper triangular and the diagonal entries of this matrix are all equal to* 1. *More precisely, assume that* $L \in \mathcal{P}_{\mathbf{V}}$ *and* $f_Z = \Phi([L])$, *then*

$$\varsigma([L]) = \varkappa(f_Z) + \sum_{f_Z \preceq' f_{Z'}} c_{Z'} \varkappa(f_{Z'}).$$

REMARK 8.10. Since both $\mathbf{B}_1$ and $\mathbf{B}_2$ are basis of canonical type, Baumann in [**1**] has proved the similar result as the following: The transition matrix between two bases of canonical type is upper triangular and the diagonal entries of this matrix are all equal to 1.

REMARK 8.11. If the graph $\mathbf{\Gamma}$ is of type A, D, E, by [**14**] or [**2**], we can see that the transition matrix between the PBW basis and the canonical basis is upper triangular with diagonal entries equal to 1. Hence, we can see that the transition matrix between the PBW basis and the semicanonical basis is upper triangular with diagonal entries equal to 1 in this case, which includes the result of [**16**] in type A.

With Proposition 4.8 and Proposition 7.5, we have the following result:

PROPOSITION 8.12 ([**11**, Section 2.9]). *Fix* $\underline{d} = (d_i)_{i \in I}$ *with* $d_i \in \mathbb{N}$, *then the following submodules of* ${}_{\mathbb{Z}}\mathbf{U}(\mathfrak{g})^+$ *are equal:*

(1) $\sum_{i \in I} {}_{\mathbb{Z}}\mathbf{U}(\mathfrak{g})^+ * e_i^{(d_i)}$, *here* $e_i^{(d_i)}$ *is the divided power in* ${}_{\mathbb{Z}}\mathbf{U}(\mathfrak{g})^+$.
(2) *The* $\mathbb{Z}$*-span of* $\mathbf{B}_{1,\underline{d}} = \{\varsigma([L]) | t_i^*(\mathcal{F}_{\Omega,\Omega^i}(L)) \geq d_i$ *for some* $i \in I\}$.
(3) *The* $\mathbb{Z}$*-span of* $\mathbf{B}_{2,\underline{d}} = \{\varkappa(f_Z) | t_i^*(Z) \geq d_i$ *for some* $i \in I\}$.

Given $\lambda \in \mathfrak{h}^*$ such that $\lambda(h_i) \leq 0$ for any i, ($\{h_i, i \in I\}$ is a basis of $\mathfrak{h}$) let $d_i = 1 - \lambda(h_i)$, then $V(\lambda) \cong \mathbf{U}(\mathfrak{g})^+ / \sum_{i \in I} \mathbf{U}(\mathfrak{g})^+ * e_i^{(d_i)}$ admits a $\mathbf{U}(\mathfrak{g})$-module structure such that e_i acts by left multiplication. Let $\pi : \mathbf{U}^+(\mathfrak{g}) \to V(\lambda)$ be the obvious quotient map, then we have the following proposition.

PROPOSITION 8.13 ([**11**],[**9**]). *With the notations above, set* $\mathbf{B}_1(\lambda) = \{\pi(\varsigma([L])) | [L] \in \mathbf{B}_1, [L] \notin \mathbf{B}_{1,\underline{d}}\}$ *and* $\mathbf{B}_2(\lambda) = \{\pi(\varkappa(f_Z)) | \kappa(f_Z) \in \mathbf{B}_2, \kappa(f_Z) \notin \mathbf{B}_{2,\underline{d}}\}$. *Then* $\mathbf{B}_1(\lambda)$ *and* $\mathbf{B}_2(\lambda)$ *are* $\mathbb{Q}$*-basis of* $V(\lambda)$.

DEFINITION 8.14. The order $\preceq$ of $\mathcal{V}_1$ naturally induced a partially order of $\mathbf{B}_1(\lambda)$ as the following: Given L, K such that $t_i^* \geq d_i$ for some i, we say $\pi(\chi([L])) \preceq \pi(\chi([K]))$ if and only if $L \preceq K$. We still denote the partially order of $\mathbf{B}_1(\lambda)$ by $\preceq$. Similarly, we can define a partially order $\preceq'$ on $\mathbf{B}_2(\lambda)$.

Since Φ preserves t_i^*, we can see that, for $\Phi([L]) = f_Z$, $\varsigma([L]) \in \mathbf{B}_{1,\underline{d}}$ if and ony if $\varkappa(f_Z) \in \mathbf{B}_{2,\underline{d}}$. By Corollary 8.9 and Proposition 8.13, we have the following corollary:

COROLLARY 8.15. *The transition matrix between the bases* $\mathbf{B}_1(\lambda)$ *(with the order* $\preceq$*) and* $\mathbf{B}_2(\lambda)$ *(with the order* $\preceq'$*) of* $V(\lambda)$ *is upper triangular and the diagonal entries of this matrix are all equal to* 1.

Acknowledgments

We are grateful to George Lusztig. After he saw our first version of the present article on arxiv, he informed us the preprint [**1**] by Pierre Baumann, in which similar results have been obtained. (See Remark 7.4 and Remark 8.10.) We are also grateful to the referees for their suggestions.

References

[1] P. Baumann, *The canonical basis and the quantum frobenius morphism*, `arXiv:1201.0303`, 2011.

[2] Pierre Baumann and Joel Kamnitzer, *Preprojective algebras and MV polytopes*, Represent. Theory **16** (2012), 152–188, DOI 10.1090/S1088-4165-2012-00413-7. MR2892443

[3] A. A. Beĭlinson, J. Bernstein, and P. Deligne, *Faisceaux pervers* (French), Analysis and topology on singular spaces, I (Luminy, 1981), Astérisque, vol. 100, Soc. Math. France, Paris, 1982, pp. 5–171. MR751966

[4] Christof Geiss, Bernard Leclerc, and Jan Schröer, *Semicanonical bases and preprojective algebras* (English, with English and French summaries), Ann. Sci. École Norm. Sup. (4) **38** (2005), no. 2, 193–253, DOI 10.1016/j.ansens.2004.12.001. MR2144987

[5] I. Grojnowski and G. Lusztig, *A comparison of bases of quantized enveloping algebras*, Linear algebraic groups and their representations (Los Angeles, CA, 1992), Contemp. Math., vol. 153, Amer. Math. Soc., Providence, RI, 1993, pp. 11–19, DOI 10.1090/conm/153/01304. MR1247495

[6] M. Kashiwara, *On crystal bases of the Q-analogue of universal enveloping algebras*, Duke Math. J. **63** (1991), no. 2, 465–516, DOI 10.1215/S0012-7094-91-06321-0. MR1115118

[7] Masaki Kashiwara and Yoshihisa Saito, *Geometric construction of crystal bases*, Duke Math. J. **89** (1997), no. 1, 9–36, DOI 10.1215/S0012-7094-97-08902-X. MR1458969

[8] George Lusztig, *Cuspidal local systems and graded Hecke algebras. I*, Inst. Hautes Études Sci. Publ. Math. **67** (1988), 145–202. MR972345

[9] G. Lusztig, *Quivers, perverse sheaves, and quantized enveloping algebras*, J. Amer. Math. Soc. **4** (1991), no. 2, 365–421, DOI 10.2307/2939279. MR1088333

[10] George Lusztig, *Introduction to quantum groups*, Progress in Mathematics, vol. 110, Birkhäuser Boston, Inc., Boston, MA, 1993. MR1227098

[11] G. Lusztig, *Semicanonical bases arising from enveloping algebras*, Adv. Math. **151** (2000), no. 2, 129–139, DOI 10.1006/aima.1999.1873. MR1758244

[12] R. D. MacPherson, *Chern classes for singular algebraic varieties*, Ann. of Math. (2) **100** (1974), 423–432, DOI 10.2307/1971080. MR361141

[13] Claus Michael Ringel, *Hall algebras and quantum groups*, Invent. Math. **101** (1990), no. 3, 583–591, DOI 10.1007/BF01231516. MR1062796

[14] Claus Michael Ringel, *The Hall algebra approach to quantum groups*, XI Latin American School of Mathematics (Spanish) (Mexico City, 1993), Aportaciones Mat. Comun., vol. 15, Soc. Mat. Mexicana, México, 1995, pp. 85–114. MR1360930

[15] Olivier Schiffmann, *Lectures on canonical and crystal bases of Hall algebras* (English, with English and French summaries), Geometric methods in representation theory. II, Sémin. Congr., vol. 24, Soc. Math. France, Paris, 2012, pp. 143–259. MR3202708

[16] HongBo Yin and ShunHua Zhang, *The transition matrix between PBW basis and semicanonical basis of $U^+(\mathfrak{sl}_n(\mathbb{C}))$*, Sci. China Math. **57** (2014), no. 7, 1427–1434, DOI 10.1007/s11425-014-4804-4. MR3213879

[17] M. Zhao, *Derivation functors and Lusztig's induction functors*, `arXiv:2202.00430`, 2022.

Department of Mathematics and New Cornerstone Science Laboratory,The University of Hong Kong, Pokfulam, Hong Kong, Hong Kong SAR, People's Republic of China

Email address: fangjp@hku.hk

Academy of Mathematics and Systems Science, Chinese Academy of Sciences, Beijing 100190, People's Republic of China

Email address: lanyixin@amss.ac.cn

School of mathematical seciences, Beijing Normal University, Beijing 100875, People's Republic of China

Email address: jxiao@bnu.edu.cn

Contemporary Mathematics
Volume **845**, 2026
https://doi.org/10.1090/conm/845/16955

Homological invariants of left and right serial quiver algebras

Ruoyu Guo

Abstract. We investigate the relationship between the delooping level (dell) and the finitistic dimension of left and right serial quiver algebras. These 2-syzygy finite algebras have finite delooping level, and it can be calculated with an easy and finite algorithm. When the algebra is right serial, its right finitistic dimension is equal to its left delooping level. When the algebra is left serial, the above equality only holds under certain conditions. We provide examples to demonstrate this and include discussions on the sub-derived (sub-ddell) and derived delooping level (ddell). Both sub-ddell and ddell are improvements of the delooping level. We motivate their definitions and showcase how they can behave better than the delooping level in certain situations throughout the paper.

1. Introduction and Definitions

We present some new results on the finitistic dimension conjecture over finite dimensional algebras. This important homological conjecture in representation theory is the sufficient condition for numerous other conjectures including the Nakayama conjecture, Gorenstein symmetry conjecture, and Anslander-Reiten conjecture, to name a few. The representational approaches covered in this paper rely on studying the properties of homological invariants and creating new ones that are upper or lower bounds of the finitistic dimension. They are the delooping level [**5**], sub-derived delooping level, and the derived delooping level [**9**]. We study their behavior in monomial algebras and specialize to left and right serial quiver algebras. As subclasses of monomial algebras, left and right serial quiver algebras enjoy a lot of good properties such as 2-syzygy finiteness and having a tractable syzygy structure. We can also calculate their little and big finitistic dimensions with the margin of error at most one [**12**]. From the class of monomial algebras we find one of the first examples where the big and the little finitistic dimensions differ [**13**]. Various bounds of the finitistic dimension of monomial algebras are studied using different methods [**6**, **7**, **15**, **22**]. Despite the large amount of work on monomial algebras, their finitistic dimensions still need to be calculated on a case-by-case basis. Our main result shows we can calculate the right finitistic dimension through the left delooping level if the quiver algebra is right serial or left serial with an additional condition. This adds to the list of algebras whose big finitistic dimension is described by the delooping level of the opposite algebra. The main results are a series of equalities when Λ is a right serial algebra or left serial under an additional condition

$$\operatorname{findim}\Lambda = \operatorname{Findim}\Lambda = \operatorname{dell}\Lambda^{\mathrm{op}} = \operatorname{ddell}\Lambda^{\mathrm{op}} < \infty,$$

where all the homological dimensions use **right** modules by default. When the equality does not hold in the left serial case, we provide a representation-finite algebra in Example 4.7 such that $\operatorname{findim}\Lambda = \operatorname{Findim}\Lambda = \operatorname{ddell}\Lambda^{\mathrm{op}} < \operatorname{dell}\Lambda$. We also recover known results on Nakayama algebras and monomial algebras of acyclic quivers using our proposed method along the way. While $\operatorname{Findim}\Lambda = \operatorname{ddell}\Lambda^{\mathrm{op}}$ holds in several examples in which $\operatorname{Findim}\Lambda < \operatorname{dell}\Lambda^{\mathrm{op}}$ such as in [**9**, Example 3.8] and Example 4.7, $\operatorname{Findim}\Lambda = \operatorname{ddell}\Lambda^{\mathrm{op}}$ may not be true in general considering the example [**2**, Example 4.22]. There is still much to study to what extent the derived delooping level can describe the big finitistic dimension.

For the rest of the paper, let Λ be a finite dimensional algebra over an algebraically closed field $\mathbb{K}$. Let $\operatorname{mod}\Lambda$ and $\operatorname{Mod}\Lambda$ be the category of finitely generated **right** Λ-modules and the category of all **right** Λ-modules. The (right) little and big finitistic dimension conjectures state that for a finite dimensional algebra Λ, the little finitistic dimension $\operatorname{findim}\Lambda$ and the big finitistic dimension $\operatorname{Findim}\Lambda$, respectively defined as

$$\operatorname{findim}\Lambda = \sup\{\operatorname{pd}M \mid M \in \operatorname{mod}\Lambda, \operatorname{pd}M < \infty\}$$

and

$$\operatorname{Findim}\Lambda = \sup\{\operatorname{pd}M \mid M \in \operatorname{Mod}\Lambda, \operatorname{pd}M < \infty\},$$

are finite, where $\operatorname{pd}M$ is the projective dimension of M. Note that the original formulation of the finitistic dimension conjecture contains another statement $\operatorname{findim}\Lambda = \operatorname{Findim}\Lambda$. This is proved to be false in the case of monomial algebras [**13**] and is not the focus of this paper.

A quiver $Q = (Q_0, Q_1, s, t)$ is a directed graph with four pieces of information, where Q_0 is the vertex set, Q_1 is the arrow set, and $s, t : Q_1 \to Q_0$ are the starting and terminal vertices of an arrow in Q_1. We frequently consider paths in the quiver, so for convenience, we extend the domain of s and t to include all paths in Q in the natural way. For each quiver Q, we can associate with it a path algebra whose $\mathbb{K}$-basis is the set of all paths in Q and multiplication is path concatenation. Introductions to quiver representations can be found in [**1, 20**]. For a quiver algebra $\Lambda = \mathbb{K}Q/I$ where I is an admissible ideal, we denote by P_v and S_v the indecomposable projective and simple modules whose top is supported on the vertex v, respectively.

Quiver algebras provide a wealth of examples for studying this conjecture, and they are very general in the finite dimensional algebra case in the following sense. Every finite dimensional algebra over a field $\mathbb{K}$ is Morita equivalent to a basic finite dimensional algebra, which is then isomorphic to some quiver algebra $\mathbb{K}Q/I$ subject to relations I when $\mathbb{K}$ is algebraically closed. Since the finitistic dimension is invariant under Morita equivalence and field extensions [**16**], it suffices to assume $\mathbb{K}$ is algebraically closed and study the finitistic dimensions of quiver algebras.

The organization of the paper is as follows. In Section 2, we recall two invariants related to the delooping level called the sub-derived delooping level and derived delooping level, focusing on the motivation of their definition. The derived delooping level is especially better in terms of its properties and as an upper bound. In Section 3, we introduce the technique that we use for proving the main theorems and recover some known results along the way. Section 4 contains the main result on left and right serial algebras and an illuminating representation-finite algebra example.

Acknowledgments. The author is grateful for his advisor Kiyoshi Igusa for helpful discussions on the paper.

2. Variants of the Delooping Level

We first recall the definition of the delooping level $\operatorname{dell}\Lambda$ for an algebra Λ. For two modules M and N, let $M \overset{\oplus}{\hookrightarrow} N$ mean M is a **direct summand** of N. Since we do not need to consider projective summands when calculating the projective dimension, modules considered hereafter have their projective summands omitted unless stated otherwise.

DEFINITION 2.1 ([**5**]). Let $\mho = \operatorname{Tr}\Omega\operatorname{Tr}$ be the left adjoint of the syzygy functor Ω in $\underline{\operatorname{mod}}\,\Lambda$. Define the delooping level of a Λ-module M as

$$\operatorname{dell} M = \inf\{n \in \mathbb{N} \mid \Omega^n M \text{ is a direct summand of } \Omega^{n+1}N \text{ for some module } N\},$$

and we can show that

$$\operatorname{dell} M = \inf\{n \in \mathbb{N} \mid \Omega^n M \overset{\oplus}{\hookrightarrow} \Omega^{n+1}\mho^{n+1}\Omega^n M\}. \tag{2.1}$$

Define the **delooping level** of an algebra Λ as

$$\operatorname{dell}\Lambda = \sup\{\operatorname{dell} S \mid S \text{ is a simple } \Lambda\text{-module}\}.$$

If M is isomorphic to the direct summand of $\Omega^k(N)$ for some k and some module N, then we say M is a **k-syzygy** and that M is **k-deloopable**. If $M \overset{\oplus}{\hookrightarrow} \Omega^i N_i$ for some N_i and every $i \in \mathbb{N}$, then we say M is **infinitely deloopable**.

It is proved in [**5**] that $\operatorname{Findim}\Lambda \leq \operatorname{dell}\Lambda^{\mathrm{op}}$. The delooping level does not need to be finite a priori, but it is often small and in fact equal to the finitistic dimension in many cases. The cases where we know $\operatorname{Findim}\Lambda = \operatorname{dell}\Lambda^{\mathrm{op}}$ include

- algebras with finite global dimension
- Gorenstein algebras [**5**]
- Nakayama algebras [**19, 21**]
- radical square zero algebras [**4**]
- truncated path algebras [**2**]

and we add to this list right serial quiver algebras and a subclass of left serial quiver algebras in Section 4. However, the equality is not true for monomial algebras in general. If Λ is monomial, the difference $\operatorname{dell}\Lambda^{\mathrm{op}} - \operatorname{Findim}\Lambda$ is finite but can be arbitrarily large.

Despite many promising results about the delooping level, there is an example in [**17**] where $\operatorname{dell}\Lambda = \infty$ for a finite dimensional algebra Λ, where Λ is a quiver algebra with only two vertices. Moreover, that same example shows that the set of finitely generated modules with finite delooping level is not closed under extension, submodules, or quotients. This means the delooping level does not necessarily behave well under exact sequences, so the delooping level of an arbitrary module has to be calculated or estimated independently using the definition (2.1) without notable properties. However, if we know the delooping level of a module is finite, (2.1) provides a finite algorithm for finding what it is.

The φ-dimension and ψ-dimension [**14**] are two other related invariants that have been widely used for over two decades. They satisfy $\operatorname{findim}\Lambda \leq \varphi\dim\Lambda \leq$

$\psi \dim \Lambda$. In contrast to the delooping level, the ψ-dimension and projective dimension interact very well under short exact sequences in the sense that if $0 \to M_1 \to M_2 \to M_3 \to 0$ is a short exact sequence and $\operatorname{pd} M_3 < \infty$, then $\operatorname{pd} M_3 \leq \psi(M_1 \oplus M_2)+1$. This property has allowed many applications for the ψ-dimension, one of the strongest of which shows findim $\Lambda < \infty$ for Λ with representation dimension no more than 3. Other applications include the Igusa-Todorov algebras [**23**] and the more general LIT algebras [**3**]. Due to the difference in delooping level's properties, these results cannot yet be replicated for the delooping level. For example, it is still an open question if Findim $\Lambda < \infty$ for representation dimension 3 algebras.

The invariants in the rest of the section were introduced in [**9**], but we hope to present their motivations more cohesively and pose future questions regarding their applications. In order to strengthen the properties of the delooping level, we first consider if there are any implications for submodules of a module with finite delooping level. This question leads to the definition of the sub-derived delooping level sub-ddell Λ and a better upper bound for Findim Λ^{op}.

DEFINITION 2.2 ([**9**]). The **sub-derived delooping levels** of a module M and of an algebra Λ are

- $\operatorname{sub-ddell} M = \inf\{\operatorname{dell} N \mid M \hookrightarrow N\}$
- $\operatorname{sub-ddell} \Lambda = \sup\{\operatorname{sub-ddell} S \mid S \text{ is a simple } \Lambda\text{-module}\}$

THEOREM 2.3. *For any algebra* Λ,

$$\operatorname{Findim} \Lambda^{\mathrm{op}} \leq \operatorname{sub-ddell} \Lambda \leq \operatorname{dell} \Lambda.$$

The sub-derived delooping level can be strictly better than the delooping level as an upper bound in Example 2.4 below and can also give no improvement in [**9**, Example 3.8], where $\infty = \operatorname{sub-ddell} \Lambda = \operatorname{dell} \Lambda > \operatorname{Findim} \Lambda^{\mathrm{op}} = 1$. The next example is one of the smallest possible with $\operatorname{dell} \Lambda > \operatorname{Findim} \Lambda^{\mathrm{op}} = \operatorname{sub-ddell} \Lambda$.

EXAMPLE 2.4. Let $\Lambda = \mathbb{K}Q/I$ be the quiver algebra of the following quiver Q

$$\circlearrowleft 1 \rightleftarrows 2 \circlearrowright$$

with the following indecomposable projectives and injectives

$$\text{Projectives: } \begin{smallmatrix} & 1 & \\ 1 & & 2 \\ 2 & & 1 \\ 1 & & \end{smallmatrix}, \quad \begin{smallmatrix} & 2 & \\ 2 & & 1 \\ 1 & & 1 \end{smallmatrix}; \qquad \text{Injectives: } \begin{smallmatrix} 2 & & 2 & \\ 1 & & & 2 \\ & 1 & & \end{smallmatrix}, \quad \begin{smallmatrix} 1 & & & \\ 1 & & & 1 \\ & 2 & & 1 \\ & & 2 & \end{smallmatrix}.$$

The module $\begin{smallmatrix} 2 \\ 1 \end{smallmatrix}$ is infinitely deloopable, since it is the direct summand of its own syzygy. Therefore, we omit the summand $\begin{smallmatrix} 2 \\ 1 \end{smallmatrix}$ when calculating the delooping level of S_2. We know $\operatorname{dell} S_2 > 0$ since S_2 is not a submodule of a projective module. We find $\Omega S_2 = \begin{smallmatrix} 2 \\ 1 \end{smallmatrix} \oplus \begin{smallmatrix} 1 \\ 1 \end{smallmatrix}$ and will show that $\begin{smallmatrix} 1 \\ 1 \end{smallmatrix}$ is not 2-deloopable. It suffices to show $\begin{smallmatrix} 1 \\ 1 \end{smallmatrix}$ is not a direct summand of $\Omega^2 \mho^2 \Omega S_2 = \Omega^2 \mho^2 \left(\begin{smallmatrix} 2 \\ 1 \end{smallmatrix} \oplus \begin{smallmatrix} 1 \\ 1 \end{smallmatrix} \right)$. Using the notation that modules in $\operatorname{mod} \Lambda^{\mathrm{op}}$ are underlined, we can calculate the following

- $\Omega^2 \operatorname{Tr} \begin{pmatrix} 2 \\ 1 \end{pmatrix} = \underline{S_2} \oplus \begin{matrix} 1 \\ \underline{1} \end{matrix}$,
- $\Omega^2 \operatorname{Tr} \underline{S_2} = \begin{matrix} 2 \\ 1 \end{matrix} \oplus S_1$,
- $\Omega^2 \operatorname{Tr} \begin{pmatrix} 1 \\ \underline{1} \end{pmatrix} = \begin{matrix} 2 \\ 1 \end{matrix} \oplus \begin{matrix} 2 \\ 1 \end{matrix}$,
- $\Omega^2 \operatorname{Tr} \begin{pmatrix} 1 \\ 1 \end{pmatrix} = \begin{matrix} 2 \\ \underline{1} \end{matrix} \oplus \begin{matrix} 2 \\ \underline{1} \end{matrix}$,
- $\Omega^2 \operatorname{Tr} \begin{pmatrix} 2 \\ \underline{1} \end{pmatrix} = \begin{matrix} 2 \\ 1 \end{matrix} \oplus S_1$.

Therefore, $\Omega^2 \mho^2 \Omega S_2 = \begin{pmatrix} 2 \\ 1 \end{pmatrix}^5 \oplus (S_1)^3$. Since it does not contain $\begin{matrix} 1 \\ 1 \end{matrix}$ as a summand, $\operatorname{dell} S_2 > 1$.

Since $\Omega^2 S_2 = \begin{pmatrix} 2 \\ 1 \end{pmatrix}^3 \oplus S_1$, the module $\begin{matrix} 2 \\ 1 \end{matrix}$ is infinitely deloopable, and S_1 is a direct summand of $\Omega^3 \begin{pmatrix} 2 \\ 1 \end{pmatrix}$, we find $\operatorname{dell} S_2 = \operatorname{dell} \Lambda = 2$.

On the other hand, S_2 is a submodule of $\begin{smallmatrix} & 2 & \\ 2 & & 1 \end{smallmatrix}$ whose syzygy is $(S_1)^2$ and infinitely deloopable, so

$$\text{sub-ddell}\, S_2 = \text{sub-ddell}\, \Lambda = 1 = \operatorname{Findim} \Lambda^{\mathrm{op}} < \operatorname{dell} \Lambda = 2.$$

In a similar vein, we ask if there are any interesting properties for quotient modules M of some module C_0 of finite delooping level. It turns out that it is not sufficient to only consider C_0 and the quotient M. We need to consider the kernel of the projection from C_0 to M and moreover all exact sequences that end in M. This motivates the definition of the derived delooping level.

DEFINITION 2.5 ([**9**]). Let M be a Λ-module.

- The k-**delooping level** of M is

$$k\text{-dell}\, M = \inf\{n \in \mathbb{N} \mid \Omega^n M \overset{\oplus}{\hookrightarrow} \Omega^{n+k} N \text{ for some } N\}$$

- The **derived delooping level** of M is

$$\begin{aligned} \operatorname{ddell} M = \inf\{m \in \mathbb{N} \mid \exists n \leq m & \text{ and an exact sequence in } \operatorname{mod} \Lambda \text{ of the form} \\ & 0 \to C_n \to C_{n-1} \to \cdots \to C_1 \to C_0 \to M \to 0, \\ & \text{where } (i+1)\text{-dell}\, C_i \leq m - i,\ i = 0, 1, \ldots, n\}, \end{aligned}$$

- The **derived delooping level** of Λ is

$$\operatorname{ddell} \Lambda = \sup\{\operatorname{ddell} S \mid S \text{ is a simple } \Lambda\text{-module}\}.$$

THEOREM 2.6. *For any algebra* Λ,

$$\operatorname{Findim} \Lambda^{\mathrm{op}} \leq \operatorname{ddell} \Lambda \leq \operatorname{dell} \Lambda.$$

The derived delooping level is a better upper bound compared to the delooping level in the same Example 2.4.

EXAMPLE 2.7 (Example 2.4 Revisited). Let Λ be the same as in Example 2.4. The short exact sequence

$$0 \to S_1 \to \begin{smallmatrix} 2 \\ 1 \end{smallmatrix} \to S_2 \to 0$$

shows $\operatorname{ddell} S_2 = 1$. Therefore in this case,

$$1 = \operatorname{Findim} \Lambda^{\mathrm{op}} = \operatorname{ddell} \Lambda = \text{sub-ddell}\, \Lambda < \operatorname{dell} \Lambda = 2.$$

Note that if there exists a counterexample Λ of the finitistic dimension conjecture, the derived delooping level of some simple Λ^{op}-module S must be infinite, so by Definition 2.5, there must be a large number of right Λ^{op}-modules with infinite k-delooping level for some k so that $(i+1)$-$\operatorname{dell} C_i = \infty$ for some i in **every** exact sequence of the form $0 \to C_n \to C_{n-1} \to \cdots \to C_1 \to C_0 \to S \to 0$. Currently, the only example with $\operatorname{dell} \Lambda = \infty$ in **[17]** features a local algebra, and it has one module with infinite delooping level. Designing algebras which have more modules of infinite delooping level or k-delooping levels is crucial to advancing the progress or finding counterexamples of the finitistic dimension conjecture.

QUESTION 2.8. Can we design a finite dimensional algebra that has enough modules with infinite k-delooping level to make $\operatorname{ddell} S = \infty$ for some simple module S?

Moreover, the property of having finite derived delooping level is preserved under extensions and submodules, so the class of modules with finite derived delooping level forms a torsion-free class in $\operatorname{mod} \Lambda$.

THEOREM 2.9 (**[9]**). *Let $0 \to A \to B \to C \to 0$ be a short exact sequence in* $\operatorname{Mod} \Lambda$. *Then*

- *If* $\operatorname{ddell} A < \infty$ *and* $\operatorname{ddell} C < \infty$, *then* $\operatorname{ddell} B \leq \operatorname{ddell} A + \operatorname{ddell} C + 1$.
- *If* $\operatorname{ddell} B < \infty$, *then* $\operatorname{ddell} A \leq \operatorname{ddell} B + 1$.

Another direction of future interest is to use the derived delooping level to replicate the results for representation dimension 3 algebras and the more general Igusa-Todorov and LIT algebras mentioned in the introduction. It is still unknown whether the big finitistic dimension of these algebras are finite.

3. Monomial Algebras

We discuss some preliminary results that we will need for the rest of the paper and use them to first provide a different proof of $\operatorname{dell} \Lambda = \operatorname{Findim} \Lambda^{\mathrm{op}}$ for monomial algebras whose quiver is acyclic in this section. A quiver is **acyclic** if there is no oriented cycle in the quiver. We use the idea of the proof to introduce new theorems on two special classes of algebras called left and right serial quiver algebras in Section 4.

DEFINITION 3.1. A module is **uniserial** if its submodules are totally ordered. Alternatively, a module is uniserial if and only if it has a unique composition series. A quiver algebra $\Lambda = \mathbb{K}Q/I$ is **left** (resp. **right**) **serial** if each indecomposable left (resp. right) projective Λ-module is uniserial.

Let $\Lambda = \mathbb{K}Q/I$ be a monomial algebra, where I is an admissible ideal generated by monomials. There are $|Q_0|$ **trivial paths** in Q, one for each vertex, denoted by e_v for vertex v. There exists a canonical set of **minimal relations** for the ideal I

consisting of only monomials, and we use that set of minimal relations by default and call it B_I. All paths in B_I are called **zero paths**, and note the difference between trivial and zero paths. We call B_Q the set of all nonzero paths in Q including the trivial paths, so B_Q is a $\mathbb{K}$-basis of the quiver algebra Λ. When we consider the opposite algebra Λ^{op}, its corresponding set of minimal relations and $\mathbb{K}$-basis are $B_{\tilde{I}}$ and $B_{Q^{\mathrm{op}}}$, respectively.

It is easy to understand the second and higher syzygies of modules over a monomial algebra Λ since all second syzygies are direct sums of $p\Lambda$ where p is a path of length greater than or equal to 1 in Q [**12**]. Finding the syzygy of modules of the form $p\Lambda$ is also straightforward. The process amounts to finding nonzero paths to concatenate with p to become a minimal relation in I. This has been pointed out in different languages in several papers including the original proof that $\operatorname{Findim}\Lambda < \infty$ for monomial algebras Λ [**7**] and in later works such as the proof that $\operatorname{dell}\Lambda = \operatorname{Findim}\Lambda^{\mathrm{op}}$ for truncated path algebras Λ where the authors name the similar concept "right complementary" path [**2**] as opposed to "right completion" in our next Definition 3.2.

DEFINITION 3.2. Let $Q = (Q_0, Q_1, s, t)$ be a quiver and $\Lambda = \mathbb{K}Q/I$ be a monomial algebra. For each nonzero path α in Q, we say β is a **right completion** of α if $\alpha\beta \in I$, $\beta \notin I$, and for all factorizations $\beta = \beta_1\beta_2$ with β_2 nontrivial, $\alpha\beta_1 \notin I$.

Alternatively, Definition 3.2 says β is a right completion of α if β is a minimal nonzero path to make $\alpha\beta$ zero. Note that there may be more than one right completions for a path. We will consider sequences of right completions $\alpha_0\alpha_1\cdots\alpha_n$ where α_{i+1} is a right completion of α_i for $i = 0, 1, \ldots, n-1$. We say a sequence $\alpha_0\alpha_1\cdots\alpha_n$ of right completions is **right maximal** if α_n has no right completion, and is **left maximal** if α_0 is not a right completion of any nonzero path. If a sequence of right completions is both left and right maximal, we say it is **maximal**. The **length** of a sequence of right completions is the number of right completions in the sequence, so the sequence $\alpha_0\alpha_1\cdots\alpha_n$ has length n. Note that the length is not necessarily the number of minimal relations in the sequence since there may be multiple ways to factor the whole path $\alpha_0\alpha_1\cdots\alpha_n$ into right completions. By definition, maximal sequences of right completions have finite length, and for modules with infinite projective dimension, we say they are associated with sequences of right completions of infinite length. We prove two important lemmas related to right completions.

LEMMA 3.3. *Suppose $\Lambda = \mathbb{K}Q/I$ is a monomial algebra. Let μ, ν be nonzero paths of length at least 1. Then the right Λ-module $\nu\Lambda$ is a direct summand of $\Omega(\mu\Lambda)$ if and only if ν is a right completion of μ.*

PROOF. Consider the short exact sequence

$$0 \to \Omega(\mu\Lambda) \to P_{t(\mu)} \to \mu\Lambda \to 0.$$

As $\mathbb{K}$-vector spaces, $\mu\Lambda$ has the basis $B_{\mu\Lambda} = \{\omega \in B_Q \mid \mu\omega \notin I\}$, and $P_{t(\mu)}$ has the basis $B_{P_{t(\mu)}} = \{\omega \in B_Q \mid s(\omega) = t(\mu)\}$. It is clear that $P_{t(\mu)}$ maps onto $\mu\Lambda$, and the $\mathbb{K}$-basis of the kernel $\Omega(\mu\Lambda)$ is

$$B_{\Omega(\mu\Lambda)} = \{\nu \in B_Q \mid s(\nu) = t(\mu), \mu\nu \in I\}. \tag{3.1}$$

From (3.1), we note that the generators of $\Omega(\mu\Lambda)$ as a Λ-module are those $\omega r \in B_{P_{t(\mu)}}$ such that $\mu\omega r \in I$ but $\mu\omega \notin I$ for some path ω and arrow r. Therefore, we can rewrite

(3.2) $B_{\Omega(\mu\Lambda)} = \{\omega r x \in B_{P_{t(\mu)}} \mid x \text{ is a path}, \mu\omega \notin I, \mu\omega r \in I \text{ for some arrow } r\},$

so as a Λ-module $\Omega(\mu\Lambda)$ is exactly the direct sum of all $(\omega r)\Lambda$ where ωr is a right completion of μ. □

In Lemma 3.3, if $\mu\Lambda$ is already projective, then by definition μ has no right completions, so indeed $\nu\Lambda$ is zero. On the other hand, if μ has no right completion, then $\mu\Lambda$ is projective. Since $\operatorname{dell}\Lambda$ bounds the finitistic dimension of the opposite algebra, we need to study how a sequence of right completions and its reverse are related to each other. This is discussed in Lemma 3.4. To avoid confusion about which algebra we are working over, we default to using right finitistic dimensions $\operatorname{findim}\Lambda$ and $\operatorname{Findim}\Lambda$, where paths are denoted with lowercase letters in Q, and left delooping level $\operatorname{dell}\Lambda^{\mathrm{op}}$, where every path has a tilde above the letter in Q^{op}.

Suppose we have a monomial algebra $\Lambda = \mathbb{K}Q/I$. Given a sequence of right completions in Q^{op} with $n \geq 1$

$$\tilde{\mu} = \widetilde{p_0}\widetilde{p_1}\cdots\widetilde{p_n}, \tag{3.3}$$

there is a unique way to factor each path $\widetilde{p_i}$ into $\widetilde{x_i}\widetilde{y_i}$ such that $\widetilde{y_i}\widetilde{x_{i+1}}\widetilde{y_{i+1}}$ for $i = 0, 1, \ldots, n-1$ and $\widetilde{y_{n-1}}\widetilde{p_n}$ are minimal relations in $\tilde{I}$. The paths $\widetilde{y_i}$ are never trivial, and $\widetilde{x_i}$ can be the trivial path. We rewrite the factored version of (3.3) as

$$\widetilde{x_0}\widetilde{y_0}\widetilde{x_1}\widetilde{y_1}\cdots\widetilde{x_{n-1}}\widetilde{y_{n-1}}\widetilde{p_n}. \tag{3.4}$$

The reverse of (3.4) in Q is

$$p_n y_{n-1} x_{n-1} \cdots y_1 x_1 y_0 x_0. \tag{3.5}$$

Note that even if the sequence in (3.3) is maximal, the reversed sequence (3.5) may not be left maximal or right maximal due to the possibility of multiple arrows going in and out of each vertex. However, we know for sure that the Λ-module $M = \Lambda/p_n\Lambda$ has projective dimension at least $n+1$ by identifying all possible minimal relations that can occur in (3.5).

LEMMA 3.4. *Using the notation in* (3.3), (3.4), *and* (3.5), *there are at least* n *minimal relations in* (3.5), *and* $\operatorname{pd}_\Lambda(\Lambda/p_n\Lambda) \geq n+1$.

PROOF. It is clear that the numbers of minimal relations in (3.4) and (3.5) are the same and that there are at least n minimal relations in (3.5), which are $p_n y_{n-1}$, $y_{n-1}x_{n-1}y_{n-2}, \ldots, y_1 x_1 y_0$. There may be more minimal relations in (3.4). If there are additional minimal relations in (3.4), they cannot start with any arrow in $\widetilde{x_i}$ for $i = 0, 1, \ldots, n-1$. Without the loss of generality, suppose $\widetilde{x_0}$ is nontrivial and for a contradiction that there is a minimal relation starting with some arrow of $\widetilde{x_0}$. The terminal arrow of the relation cannot be in or before $\widetilde{y_0}$ since $\widetilde{x_0}\widetilde{y_0}$ is nonzero. The terminal arrow cannot be the last arrow of or after $\widetilde{y_1}$ since $\widetilde{y_0}\widetilde{x_1}\widetilde{y_1}$ is a minimal relation. Lastly, if the terminal arrow is in $\widetilde{x_1}$ or $\widetilde{y_1}$ except for the last arrow, then $\widetilde{x_1}\widetilde{y_1}$ is not a right completion of $\widetilde{x_0}\widetilde{y_0}$ and should instead be shorter. Therefore, there is no minimal relation in (3.4) starting with any arrow in $\widetilde{x_i}$. On the other hand, there may be relations whose starting arrow is an arrow in $\widetilde{y_i}$ and terminal arrow before the end of $\widetilde{y_{i+2}}$, but this does not affect the results in the rest of the proof.

Let $N \in \operatorname{mod}\Lambda$ be the non-projective summand of $\Lambda/p_n\Lambda$. If $n \leq 3$, then by applying Lemma 3.3 with the minimal relations in B_I, we get $\Omega_\Lambda N = p_n\Lambda$,

$y_{n-1}x_{n-1}\Lambda \overset{\oplus}{\hookrightarrow} \Omega^2_\Lambda N$, $y_{n-2}\Lambda \overset{\oplus}{\hookrightarrow} \Omega^3_\Lambda N$, and $x_{n-2}y_{n-3}\Lambda \overset{\oplus}{\hookrightarrow} \Omega^4_\Lambda N$ whenever applicable. For $n \geq 4$, since we showed in the previous paragraph that all other minimal relations in (3.5) must have their starting and terminal arrows in y_{i+2} and y_i for some i, we can describe the higher syzygies of N as $\Omega^j_\Lambda N = x_{n-j+2}y_{n-j+1}\Lambda$ for $j = 5, \ldots, n+1$. Note that $\Omega^j_\Lambda N$ is not projective for $j < n+1$ since their generator is a path that has a right completion, so $\operatorname{pd}_\Lambda N \geq n+1$. □

It is already shown that if Λ is a Nakayama algebra [**19**, **21**] or a monomial algebra whose underlying quiver is acyclic (for example in [**8**, Proposition 2.3] since the algebra has finite global dimension), then

(3.6)
$$\begin{aligned}\operatorname{ddell}\Lambda = \operatorname{ddell}\Lambda^{\mathrm{op}} = \operatorname{dell}\Lambda = \operatorname{dell}\Lambda^{\mathrm{op}} &= \operatorname{findim}\Lambda \\ &= \operatorname{findim}\Lambda^{\mathrm{op}} = \operatorname{Findim}\Lambda = \operatorname{Findim}\Lambda^{\mathrm{op}}.\end{aligned}$$

However, we present a different proof for the case of monomial algebras of acyclic quivers that can be generalized to more cases in the next section by demonstrating how the delooping level of a simple module in one algebra corresponds to the projective dimension of modules generated by paths in the opposite algebra.

PROPOSITION 3.5. *If Q is an acyclic quiver and $\Lambda = \mathbb{K}Q/I$ is a monomial algebra, then*

$$\operatorname{gldim}\Lambda = \operatorname{Findim}\Lambda = \operatorname{findim}\Lambda = \operatorname{dell}\Lambda^{\mathrm{op}} < \infty.$$

PROOF. Since Q is an acyclic quiver, $\operatorname{gldim}\Lambda < \infty$, so the statement is automatically true, but we present a proof that can be generalized to other classes of monomial algebras. We will prove $\operatorname{dell}\Lambda^{\mathrm{op}} \leq \operatorname{findim}\Lambda$ since $\operatorname{findim}\Lambda \leq \operatorname{Findim}\Lambda \leq \operatorname{dell}\Lambda^{\mathrm{op}}$. We can assume $\operatorname{dell}\Lambda^{\mathrm{op}} \geq 1$ since it is known that $\operatorname{Findim}\Lambda = 0$ if and only if $\operatorname{dell}\Lambda^{\mathrm{op}} = 0$, for example in [**18**].

Consider every simple module $S = S_{\tilde{v}} \in \operatorname{mod}\Lambda^{\mathrm{op}}$ supported on some vertex $\tilde{v}$. There exist sequences of right completions corresponding to $S_{\tilde{v}}$ all of the form

$$\widetilde{p_0}\widetilde{p_1}\widetilde{p_2}\cdots\widetilde{p_{n-1}}\widetilde{p_n}, \tag{3.7}$$

where $\widetilde{p_0}$ is an arrow starting at $\tilde{v}$, $\widetilde{p_{i+1}}$ is a right completion of $\widetilde{p_i}$ for $i = 0, 1, \ldots, n-1$, and $\widetilde{p_n}\Lambda^{\mathrm{op}}$ is projective so that (3.7) is right maximal. Iterating over all simple Λ^{op} modules and all such sequences of right completions corresponding to them, we pick any longest sequence and write it in the form (3.7). This implies $\operatorname{dell}\Lambda^{\mathrm{op}} \leq n+1$. Note that this does not imply $\operatorname{dell}\Lambda^{\mathrm{op}} = n+1$ since earlier syzygies of $S_{\tilde{v}}$ could be more deloopable making its delooping level smaller, but we show later this does not happen and the equality must hold.

To describe the minimal relations in $B_{\tilde{I}}$ that occur in (3.7), we factor each path $\widetilde{p_i}$ into a concatenation of two paths as in (3.4). The sequence (3.7) becomes

$$\widetilde{y_0}\widetilde{x_1}\widetilde{y_1}\cdots\widetilde{x_{n-1}}\widetilde{y_{n-1}}\widetilde{p_n}, \tag{3.8}$$

where $\widetilde{x_0}$ is always trivial and is therefore omitted because the syzygy of $S_{\tilde{v}}$ is the direct sum of all $\widetilde{y_0}\Lambda^{\mathrm{op}}$ with $\widetilde{y_0}$ an arrow starting from $\tilde{v}$.

We reverse (3.8) to obtain a sequence of right completions in Q

$$p_n y_{n-1} x_{n-1} \cdots y_1 x_1 y_0. \tag{3.9}$$

Let M be the non-projective summand of $\Lambda/p_n\Lambda$. By Lemma 3.4, we know $\operatorname{pd}_\Lambda M \geq n+1$. Since $\operatorname{pd}_\Lambda M$ is finite, we get

$$\operatorname{dell}\Lambda^{\mathrm{op}} = \operatorname{dell}_{\Lambda^{\mathrm{op}}} S_{\tilde{v}} \leq n+1 \leq \operatorname{pd}_\Lambda M \leq \operatorname{findim}\Lambda,$$

completing the proof. □

We reiterate that the result in Proposition 3.5 can be observed by using the property of the delooping level, but the proof above demonstrates intuitively why the equality $\operatorname{dell}\Lambda^{\mathrm{op}} = \operatorname{Findim}\Lambda$ holds through reversing maximal sequences of right completions. The key condition in the proof is that all sequences of right completions there have finite length since every module has finite projective dimension. We extend the idea of this proof to left and right serial algebras in the next section. Proposition 3.5 also recovers the result that $\operatorname{findim}\Lambda = \operatorname{Findim}\Lambda$ if Λ is a monomial algebra whose quiver is acyclic and that the finitistic dimension can be achieved by the quotient of a projective by a principal ideal generated by some path in the quiver.

4. Left and Right Serial Quiver Algebras

We begin the section by describing the quivers of left and right serial algebras.

LEMMA 4.1. *Suppose every vertex in a connected quiver has outdegree at most one. If the quiver has a cycle, then the cycle must be an oriented cycle. Moreover, the quiver has at most one cycle. The same is true if every vertex in a connected quiver has indegree at most one.*

PROOF. For both cases, it is clear that all cycles in the quiver must have straight orientation to keep all indegrees or outdegrees at most one. If the quiver has two cycles or more, then the vertices that connect the cycles will have indegree or outdegree larger than one. □

Therefore, the quiver of a left serial quiver algebra is either a tree or a cycle with straight orientation in which each vertex can have additional outgoing arrows that are part of a tree, and the trees are not connected to each other by any arrow. Similarly, the quiver of a right serial quiver algebra is the same except there can be additional incoming arrows out of each vertex in the cycle. If there is an oriented cycle in the quiver of a left or right serial quiver algebra, we call the vertices in the cycle the **cyclic part** of the quiver. The other vertices are called the **tree part** of the quiver.

We extend the validity of $\operatorname{dell}\Lambda^{\mathrm{op}} = \operatorname{ddell}\Lambda^{\mathrm{op}} = \operatorname{Findim}\Lambda$ to right serial quiver algebras Λ and show that the two upper bounds are not necessarily tight if Λ is a left serial algebra. We prove the result about right serial quiver algebras in Theorem 4.2 and recover the result that the right finitistic dimensions $\operatorname{findim}\Lambda = \operatorname{Findim}\Lambda$ are equal if Λ is right serial [**11**]. Note that the convention for path concatenation in [**11**] is the opposite to ours, so their results for left serial algebras are for right serial algebras in our context. The author in [**11**] also provides a method to calculate the finitistic dimensions through uniserial ideals and points out that the calculation only depends on the quiver and the relations. Our conclusion agrees with these statements and at the same time shows the delooping level and the derived delooping level serve as another tractable way to calculate the finitistic dimensions.

THEOREM 4.2. *Let $\Lambda = \mathbb{K}Q/I$ be a right serial quiver algebra. Then the right little and big finitistic dimensions of Λ are equal to the left delooping level and derived delooping level of Λ, i.e.,*

$$\operatorname{findim}\Lambda = \operatorname{Findim}\Lambda = \operatorname{dell}\Lambda^{\mathrm{op}} = \operatorname{ddell}\Lambda^{\mathrm{op}} < \infty.$$

PROOF. Since right serial quiver algebras are monomial, the four quantities are all finite. Also note that each vertex in Q (resp. Q^{op}) has at most one outgoing (resp. incoming) arrow. As in the proof of Proposition 3.5, it suffices to show $\operatorname{dell}\Lambda^{\mathrm{op}} \leq \operatorname{findim}\Lambda$.

Let $S = S_{\tilde{v}}$ be a simple module in $\operatorname{mod}\Lambda^{\mathrm{op}}$ such that $\operatorname{dell}_{\Lambda^{\mathrm{op}}} S = \operatorname{dell}\Lambda^{\mathrm{op}}$. We know $\operatorname{findim}\Lambda = \operatorname{Findim}\Lambda = 0$ if and only if $\operatorname{dell}\Lambda^{\mathrm{op}} = 0$, so the theorem is true when $\operatorname{dell}\Lambda^{\mathrm{op}}$ is 0. The theorem is also true when $\operatorname{dell}\Lambda^{\mathrm{op}} = 1$ since $\operatorname{Findim}\Lambda$ can only be 0 or 1 in that case, and $\operatorname{Findim}\Lambda = 0$ implies $\operatorname{dell}\Lambda^{\mathrm{op}} = 0$. We will prove the statement for $\operatorname{dell}\Lambda^{\mathrm{op}} \geq 3$ (corresponding to $n \geq 2$ in (4.1)), so the remaining $\operatorname{dell}\Lambda^{\mathrm{op}} = 2$ case will follow, which we also explain in Remark 4.3. Consider any arrow $\widetilde{y_0}$ starting at $\tilde{v}$ so that there exists a sequence of right completions

$$\widetilde{y_0}\widetilde{x_1}\widetilde{y_1}\widetilde{x_2}\widetilde{y_2}\cdots\widetilde{x_{n-1}}\widetilde{y_{n-1}}\widetilde{p_n}. \tag{4.1}$$

The sequence (4.1) is factored in the same way as (3.8) in the proof of Proposition 3.5, where $\widetilde{y_{n-1}p_n}$ and $\widetilde{y_i x_{i+1} y_{i+1}}$ for $i = 0, 1, \ldots, n-2$ are minimal relations in $B_{\tilde{I}}$. Since $\operatorname{dell}\Lambda^{\mathrm{op}} = \operatorname{dell}_{\Lambda^{\mathrm{op}}} S = n+1 \geq 3$, we may choose (4.1) to correspond to the information that there exists a summand of the form $\widetilde{x_{n-1}y_{n-1}}\Lambda^{\mathrm{op}}$ of $\Omega^n_{\Lambda^{\mathrm{op}}} S$ that is not $(n+1)$-deloopable and after we take another syzygy, the summand $\widetilde{p_n}\Lambda^{\mathrm{op}}$ of $\Omega^{n+1}_{\Lambda^{\mathrm{op}}} S$ is at least $(n+2)$-deloopable. We also know $\widetilde{y_0}\Lambda^{\mathrm{op}} \overset{\oplus}{\hookrightarrow} \Omega_{\Lambda^{\mathrm{op}}} S$ and $\widetilde{x_i}\widetilde{y_i}\Lambda^{\mathrm{op}} \overset{\oplus}{\hookrightarrow} \Omega^{i+1}_{\Lambda^{\mathrm{op}}} S$ for $1 \leq i \leq n-1$.

Consider the reverse of (4.1) in Q written as

$$p_n y_{n-1} x_{n-1} \cdots y_1 x_1 y_0. \tag{4.2}$$

We show that $x_1 y_0 \Lambda$ is always projective in $\operatorname{mod}\Lambda$. First, if $n = 2$, $\operatorname{dell}\Lambda^{\mathrm{op}} = 3$, but $x_1 y_0 \Lambda$ is not projective, then there exists a right completion p for $x_1 y_0$ in Q. We factor $x_1 = z_1 z_2$ such that $z_2 y_0 p$ is a minimal relation in B_I. Then we have a sequence of right completions $p_2 y_1 z_1 z_2 y_0 p$ in Q, and its reverse in Q^{op} is

$$\widetilde{p}\widetilde{y_0}\widetilde{z_2}\widetilde{z_1}\widetilde{y_1}\widetilde{p_2}, \tag{4.3}$$

where the minimal relations are $\widetilde{p}\widetilde{y_0}\widetilde{z_2}$, $\widetilde{y_0}\widetilde{z_2}\widetilde{z_1}\widetilde{y_1}$, and $\widetilde{y_1}\widetilde{p_2}$. So, we find that $\widetilde{z_1}\widetilde{y_1}\Lambda^{\mathrm{op}}$ is a summand of $\Omega^3_{\Lambda^{\mathrm{op}}}(\Lambda^{\mathrm{op}}/\widetilde{p}\Lambda^{op})$. On the other hand, $\widetilde{z_2}\widetilde{z_1}\widetilde{y_1}\Lambda^{\mathrm{op}}$ is a summand of $\Omega^2_{\Lambda^{\mathrm{op}}} S_{\tilde{v}}$. We showed in the proof of Lemma 3.4 that there is no minimal relation in (4.3) starting from any arrow in $\widetilde{z_2}\widetilde{z_1}$, so the summands of $\widetilde{z_2}\widetilde{z_1}\widetilde{y_1}\Lambda^{\mathrm{op}}$ and $\widetilde{z_1}\widetilde{y_1}\Lambda^{\mathrm{op}}$ that are supported in the vertices in $\widetilde{p_2}$ are equal. We can apply this argument to any such sequence in the form (4.3) where $\widetilde{y_0}$ is an arrow going out of $\tilde{v}$. This shows $\operatorname{dell}_{\Lambda^{\mathrm{op}}} S_{\tilde{v}} \leq 2$, contradicting the assumption.

Now we continue to the case $n \geq 3$ and $\operatorname{dell}\Lambda^{\mathrm{op}} \geq 4$. Suppose for a contradiction that $x_1 y_0 \Lambda$ is not projective. Then in the same way there exist a nonzero path p and a factorization $x_1 = z_1 z_2$ such that $z_2 y_0 p$ is a minimal relation in B_I. Reversing to Q^{op}, we get the sequence of right completions

$$\widetilde{p}\widetilde{y_0}\widetilde{z_2}\widetilde{z_1}\widetilde{y_1}\widetilde{x_2}\widetilde{y_2}\cdots\widetilde{x_{n-1}}\widetilde{y_{n-1}}\widetilde{p_n}. \tag{4.4}$$

Let M be the non-projective summand of $\Lambda^{\mathrm{op}}/\widetilde{p}\Lambda^{\mathrm{op}} \in \operatorname{mod}\Lambda^{\mathrm{op}}$. We get that the summand $\widetilde{x_2}\widetilde{y_2}\Lambda^{\mathrm{op}}$ of $\Omega^3_{\Lambda^{\mathrm{op}}} S$ is a summand of the fourth syzygy of M. This also contradicts the assumption that $\operatorname{dell}_{\Lambda^{\mathrm{op}}} S \geq 4$. Therefore, $x_1 y_0 \Lambda$ is always projective in $\operatorname{mod}\Lambda$.

Let $N \in \operatorname{mod}\Lambda$ be the non-projective summand of $\Lambda/p_n\Lambda$. Note that since Λ is right serial, there is no other sequence of right completions starting with any arrow in (4.2) except for the sequence (4.2) itself. From the relations in (4.2), we find that $\Omega_\Lambda^n N = x_2 y_1 \Lambda$ is not projective, but $\Omega_\Lambda^{n+1} N = x_1 y_0 \Lambda$ is projective. Thus, we complete the proof because

$$\operatorname{findim}\Lambda \geq \operatorname{pd}_\Lambda N = n+1 = \operatorname{dell}\Lambda^{\mathrm{op}} \geq \operatorname{ddell}\Lambda^{\mathrm{op}} \geq \operatorname{Findim}\Lambda \geq \operatorname{findim}\Lambda.$$

□

REMARK 4.3. Note that in the proof of Theorem 4.2, we did not need to explicitly prove the case $\operatorname{dell}\Lambda^{\mathrm{op}} = 2$. In that case, the sequence of right completions in Q^{op} is simply $\widetilde{y_0}\widetilde{p_1}$. Using the same argument as in the proof, we can show $y_0\Lambda$ must be projective in $\operatorname{mod}\Lambda$ to ensure $\operatorname{dell}\Lambda^{\mathrm{op}} > 1$. Since the argument is essentially verbatim, we did not include it in the proof, but the same connection between the delooping level and projective dimension still exists for $\operatorname{dell}\Lambda^{\mathrm{op}} = 2$.

The theorem immediately shows a type of modules whose projective dimension equals the finitistic dimension.

COROLLARY 4.4. *If $\Lambda = \mathbb{K}Q/I$ is a right serial quiver algebra, then the right finitistic dimensions* $\operatorname{findim}\Lambda$ *and* $\operatorname{Findim}\Lambda$ *are equal. In particular, this number can be achieved by the projective dimension of a finitely generated module of the form $\Lambda/p\Lambda$ for some path p.*

The theorem also recovers the result that $\operatorname{dell}\Lambda = \operatorname{Findim}\Lambda^{\mathrm{op}} = \operatorname{findim}\Lambda^{\mathrm{op}}$ and the finitistic dimension is left-right symmetric for Nakayama algebras Λ.

COROLLARY 4.5. *If Λ is a Nakayama algebra, then*

$$\begin{aligned}\operatorname{ddell}\Lambda = \operatorname{ddell}\Lambda^{\mathrm{op}} = \operatorname{dell}\Lambda = \operatorname{dell}\Lambda^{\mathrm{op}} &= \operatorname{findim}\Lambda \\ &= \operatorname{findim}\Lambda^{\mathrm{op}} = \operatorname{Findim}\Lambda = \operatorname{Findim}\Lambda^{\mathrm{op}}.\end{aligned}$$

PROOF. Since Λ is right serial, we have $\operatorname{findim}\Lambda = \operatorname{Findim}\Lambda = \operatorname{dell}\Lambda^{\mathrm{op}} = \operatorname{ddell}\Lambda^{\mathrm{op}}$. Suppose the sequences that achieves the delooping level and maximum projective dimension are

$$\widetilde{y_0}\widetilde{x_1}\widetilde{y_1}\cdots\widetilde{x_{n-1}}\widetilde{y_{n-1}}\widetilde{p_n}$$

in Q^{op} and

$$p_n y_{n-1} x_{n-1} y_{n-2} \cdots y_1 x_1 y_0$$

in Q. The latter sequence shows $\operatorname{pd}_\Lambda(\Lambda/p_n\Lambda) = n+1$ and $x_1y_0\Lambda$ is projective. Factoring the path p_n into $y_n x_n$ such that y_n is an arrow, we rewrite the sequence of right completions in Q as

$$y_n x_n y_{n-1} x_{n-1} y_{n-2} \cdots y_1 x_1 y_0. \tag{4.5}$$

Then the simple module $S = S_{s(y_0)}$ has finite projective dimension $n+1$ since

- $\Omega_\Lambda S = y_n\Lambda$
- $\Omega_\Lambda^2 S = x_n y_{n-1}\Lambda$
- $\Omega_\Lambda^n S = x_2 y_1 \Lambda$, which is not projective
- $\Omega_\Lambda^{n+1} S = x_1 y_0 \Lambda$, which is projective

This implies $\operatorname{findim}\Lambda \geq n+1 = \operatorname{dell}\Lambda^{\mathrm{op}} \geq \operatorname{dell}\Lambda$, but since the argument is symmetric, we also have $\operatorname{dell}\Lambda \geq \operatorname{dell}\Lambda^{\mathrm{op}}$, completing the proof. □

For left serial quiver algebras Λ, $\operatorname{dell}\Lambda = \operatorname{Findim}\Lambda^{\mathrm{op}}$ is not necessarily true, even for representation-finite algebras. We provide a sufficient condition for when the equality holds and an example (Example 4.7) for when the equality fails if the sufficient condition is not satisfied.

THEOREM 4.6. *Let $\Lambda = \mathbb{K}Q/I$ be a left serial quiver algebra. If every simple Λ^{op}-module S with $\operatorname{dell}_{\Lambda^{\mathrm{op}}} S = \operatorname{dell} \Lambda^{\mathrm{op}}$ has its corresponding sequence of right completions entirely supported on the cyclic part or the tree part of the quiver, then the right little and big finitistic dimensions of Λ are equal to the left delooping level and derived delooping level of Λ, i.e.,*

$$\operatorname{findim} \Lambda = \operatorname{Findim} \Lambda = \operatorname{dell} \Lambda^{\mathrm{op}} = \operatorname{ddell} \Lambda^{\mathrm{op}} < \infty.$$

PROOF. The proof is similar to that of Theorem 4.2. It suffices to prove $\operatorname{dell} \Lambda^{\mathrm{op}} \leq \operatorname{findim} \Lambda$. Let $S = S_{\tilde{v}}$ be a simple module in $\operatorname{mod} \Lambda^{\mathrm{op}}$ such that $\operatorname{dell}_{\Lambda^{\mathrm{op}}} S = \operatorname{dell} \Lambda^{\mathrm{op}} = n+1$, which corresponds to the sequence of right completions

$$\widetilde{y_0}\widetilde{x_1}\widetilde{y_1}\widetilde{x_2}\widetilde{y_2}\cdots\widetilde{x_{n-1}}\widetilde{y_{n-1}}\widetilde{p_n}, \tag{4.6}$$

where this sequence is unique since each vertex in Q^{op} has at most one outgoing arrow. The cases when $\operatorname{dell} \Lambda^{\mathrm{op}}$ is zero, one, or two are treated the same way as in the proof of Theorem 4.2.

For $n \geq 2$ and $\operatorname{dell} \Lambda^{\mathrm{op}} \geq 3$, the same argument shows $x_1 y_0 \Lambda$ is a projective Λ-module. We reverse (4.6) to get a sequence of right completions in Q

$$p_n y_{n-1} x_{n-1} \cdots y_1 x_1 y_0. \tag{4.7}$$

Let $M = \Lambda / p_n \Lambda \in \operatorname{mod} \Lambda$. We know $\operatorname{pd}_\Lambda M \geq n+1$. Since there is no restriction on the number of arrows going out of each vertex in Q, $\operatorname{pd}_\Lambda M$ could be infinite. However, we assumed that all vertices in (4.7) are all supported either in the cyclic part or in the tree part. In both cases, other sequences of right completions starting with p_n have finite length as they can only stay in the tree part. Therefore, $n+1 \leq \operatorname{pd}_\Lambda M < \infty$. □

The sufficient condition that is in the statement of Theorem 4.6 is clearly not a necessary condition, since the algebra can have finite global dimension without satisfying the condition. If the condition is not satisfied, $\operatorname{dell} \Lambda^{\mathrm{op}} = \operatorname{Findim} \Lambda$ does not hold even for representation-finite algebras. We demonstrate it in the next example and show that $\operatorname{ddell} \Lambda = \operatorname{Findim} \Lambda^{\mathrm{op}}$ in that example.

EXAMPLE 4.7. Let $Q =$

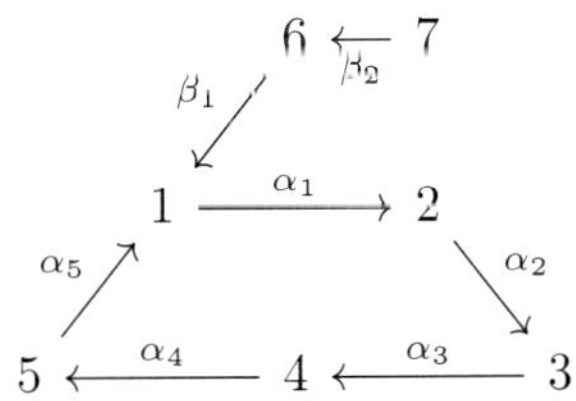

and $\Lambda = \mathbb{K}Q/I$ a monomial algebra such that the indecomposable projective modules of Λ and Λ^{op} are

$$\Lambda\text{-modules: } \begin{smallmatrix}1\\2\\3\end{smallmatrix}\ \ \begin{smallmatrix}2\\3\\4\end{smallmatrix}\ \ \begin{smallmatrix}3\\4\\5\\1\end{smallmatrix}\ \ \begin{smallmatrix}4\\5\\1\\2\end{smallmatrix}\ \ \begin{smallmatrix}5\\1\\2\\3\end{smallmatrix}\ \ \begin{smallmatrix}6\\1\end{smallmatrix}\ \ \begin{smallmatrix}7\\6\end{smallmatrix};\qquad \Lambda^{\mathrm{op}}\text{-modules: } \begin{smallmatrix}5\\4\\3\end{smallmatrix}\ \ 1\ \ 6\ \ \begin{smallmatrix}2\\1\\5\\4\end{smallmatrix}\ \ \begin{smallmatrix}3\\2\\1\\5\end{smallmatrix}\ \ \begin{smallmatrix}4\\3\\2\end{smallmatrix}\ \ \begin{smallmatrix}5\\4\\3\end{smallmatrix}\ \ \begin{smallmatrix}6\\7\end{smallmatrix}\ \ 7.$$

We can show that Λ is representation-finite, so it is straightforward to find $\operatorname{Findim} \Lambda = 1$ and $\operatorname{Findim} \Lambda^{\mathrm{op}} = 2$. Each vertex in Q has at most one outgoing arrow, so Λ is right serial and Λ^{op} is left serial.

Λ right serial: Starting with $\operatorname{dell} \Lambda^{\mathrm{op}}$, we find that the simple module supported on vertex 1 in $\operatorname{mod} \Lambda^{\mathrm{op}}$ is the only simple with nonzero delooping level. In the truncated projective resolution

$$\text{in } \operatorname{mod}\Lambda^{\mathrm{op}} : 0 \to \begin{smallmatrix}5\\4\\3\end{smallmatrix} \oplus \widetilde{S_6} \to \begin{smallmatrix} & 1 & \\ 5 & & 6\\ 4 & & \\ 3 & & \end{smallmatrix} \to \widetilde{S_1} \to 0,$$

the summand $\widetilde{S_6}$ of $\Omega_{\Lambda^{\mathrm{op}}}\widetilde{S_1}$ is a second syzygy of $\begin{smallmatrix}3\\2\end{smallmatrix}$, so $\operatorname{dell}\Lambda^{\mathrm{op}} = 1 = \operatorname{Findim}\Lambda$. This corresponds to a trivial case in Theorem 4.2.

$\underline{\Lambda^{\mathrm{op}} \text{ left serial:}}$ Now we show $3 = \operatorname{dell}\Lambda > \operatorname{Findim}\Lambda^{\mathrm{op}} = \operatorname{ddell}\Lambda = 2$. For $\operatorname{dell}\Lambda$, S_5 and S_7 are the two simple modules with nonzero delooping level. It is clear that $\operatorname{dell}_\Lambda S_5 = 1$. The truncated projective resolution of S_7 below shows $\operatorname{dell}_\Lambda S_7 = 3$ (4.8)

$$0 \longrightarrow \Omega^3_\Lambda S_7 = \begin{smallmatrix}2\\3\end{smallmatrix} \longrightarrow \begin{smallmatrix}1\\2\\3\end{smallmatrix} \longrightarrow \begin{smallmatrix}6\\1\end{smallmatrix} \longrightarrow \begin{smallmatrix}7\\6\end{smallmatrix} \longrightarrow S_7 \longrightarrow 0,$$

$$\Omega^2_\Lambda S_7 = S_1 \qquad\qquad \Omega_\Lambda S_7 = S_6$$

since $\Omega^3_\Lambda S_7$ is infinitely deloopable

- $\Omega_\Lambda S_7 = S_6$ is not a direct summand of $\Omega^2\mho^2 S_6 = 0$
- $\Omega^2_\Lambda S_7 = S_1$ is not a direct summand of $\Omega^3\mho^3 S_1 = \begin{smallmatrix}5\\1\end{smallmatrix}$
- $\Omega^3_\Lambda S_7 = \begin{smallmatrix}2\\3\end{smallmatrix} = \Omega^4_\Lambda \begin{pmatrix}4\\5\\1\end{pmatrix}.$

The sequence of right completions corresponding to (4.8) is $\beta_2\beta_1\alpha_1$ with $\beta_2\beta_1$ and $\beta_1\alpha_1$ being the minimal relations. Note that the vertices in the sequence are in both the cyclic and tree part of the quiver, so the condition in Theorem 4.6 is not satisfied. The candidate Λ^{op}-module considered in Theorem 4.6 is $\Lambda^{\mathrm{op}}/\widetilde{\alpha_1}\Lambda^{\mathrm{op}} = \widetilde{S_2}$. However, since the reversed sequence of right completions $\widetilde{\alpha_1}\widetilde{\beta_1}\widetilde{\beta_2}$ has into the cyclic part another branch $\widetilde{\alpha_1}\widetilde{\alpha_5}\widetilde{\alpha_4}\widetilde{\alpha_3}\widetilde{\alpha_2}\widetilde{\alpha_1}\widetilde{\alpha_5}\widetilde{\alpha_4}\widetilde{\alpha_3}\cdots$ that goes on infinitely, $\operatorname{pd}_{\Lambda^{\mathrm{op}}}\widetilde{S_2} = \infty$. The reverse of the subsequence $\beta_2\beta_1$ which is completely in the tree part leads to the module $\Lambda^{\mathrm{op}}/\widetilde{\beta_1}\Lambda^{\mathrm{op}}$, which has projective dimension 2 and the projective resolution

$$0 \to S_7 \to \begin{smallmatrix}6\\7\end{smallmatrix} \to \begin{smallmatrix} & 1 & \\ 5 & & 6\\ 4 & & \\ 3 & & \end{smallmatrix} \to \begin{smallmatrix}1\\5\\4\\3\end{smallmatrix} \to 0.$$

However, $\operatorname{ddell}\Lambda = \operatorname{Findim}\Lambda^{\mathrm{op}} = 2$ using the following exact sequence

$$0 \to \begin{smallmatrix}5\\1\end{smallmatrix} \to \begin{smallmatrix}5 & & 6\\ & 1 & \end{smallmatrix} \to \begin{smallmatrix}7\\6\end{smallmatrix} \to S_7 \to 0$$

shows $\operatorname{ddell} \Lambda = 2 = \operatorname{Findim} \Lambda^{\mathrm{op}}$, since

$$\begin{matrix} 5 \\ 1 \end{matrix} = \Omega^3 \begin{pmatrix} 4 \\ 5 \\ 1 \end{pmatrix} \Rightarrow 3\text{-}\operatorname{dell} \begin{pmatrix} 5 \\ 1 \end{pmatrix} \leq 0, \quad \Omega \begin{pmatrix} 5 & & 6 \\ & 1 & \end{pmatrix} = \begin{matrix} 1 \\ 2 \\ 3 \end{matrix} \Rightarrow 2\text{-}\operatorname{dell} \begin{pmatrix} 5 & & 6 \\ & 1 & \end{pmatrix} \leq 1,$$

$$1\text{-}\operatorname{dell} \begin{pmatrix} 7 \\ 6 \end{pmatrix} \leq 2.$$

We end the paper with some open questions. In the monomial example where $\operatorname{dell} \Lambda^{\mathrm{op}} - \operatorname{Findim} \Lambda$ can be made arbitrarily large [**2**], the corresponding quiver has five arrows going out of the "source" vertex. In our example where their difference is 1, each vertex has at most two outgoing arrows. We ask whether the relationship among $\operatorname{dell} \Lambda^{\mathrm{op}}$, $\operatorname{ddell} \Lambda^{\mathrm{op}}$, and $\operatorname{Findim} \Lambda$ can be quantified by the maximum incoming/outgoing arrows out of each vertex.

QUESTION 4.8. Given a quiver, there are many ways to manipulate the relations to create monomial algebras with desired finitistic dimensions [**10**]. Similarly, if Λ is a monomial algebra, can we quantify the differences $\operatorname{dell} \Lambda^{\mathrm{op}} - \operatorname{Findim} \Lambda$ and $\operatorname{ddell} \Lambda^{\mathrm{op}} - \operatorname{Findim} \Lambda$ in terms of the underlying quiver of Λ and its relations?

QUESTION 4.9. For left serial algebras Λ, is there a class of examples where $\operatorname{dell} \Lambda^{\mathrm{op}} - \operatorname{Findim} \Lambda$ and/or $\operatorname{ddell} \Lambda^{\mathrm{op}} - \operatorname{Findim} \Lambda$ becomes arbitrarily large? This would simplify the example in [**2**].

References

[1] Ibrahim Assem, Daniel Simson, and Andrzej Skowroński, *Elements of the representation theory of associative algebras. Vol. 1*, London Mathematical Society Student Texts, vol. 65, Cambridge University Press, Cambridge, 2006. Techniques of representation theory, DOI 10.1017/CBO9780511614309. MR2197389

[2] Marcos Barrios, Marcelo Lanzilotta, and Gustavo Mata, *Delooping levels*, Algebr. Represent. Theory **29** (2026), no. 1, 203–219, DOI 10.1007/s10468-026-10380-x. MR5030914

[3] Diego Bravo, Marcelo Lanzilotta, Octavio Mendoza, and José Vivero, *Generalised Igusa-Todorov functions and Lat-Igusa-Todorov algebras*, J. Algebra **580** (2021), 63–83, DOI 10.1016/j.jalgebra.2021.02.036. MR4243357

[4] Vincent Gélinas, *The finitistic dimension of an Artin algebra with radical square zero*, Proc. Amer. Math. Soc. **149** (2021), no. 12, 5001–5012, DOI 10.1090/proc/15441. MR4327410

[5] Vincent Gélinas, *The depth, the delooping level and the finitistic dimension*, Adv. Math. **394** (2022), Paper No. 108052, 34, DOI 10.1016/j.aim.2021.108052. MR4355734

[6] E. L. Green, D. Happel, and D. Zacharia, *Projective resolutions over Artin algebras with zero relations*, Illinois J. Math. **29** (1985), no. 1, 180–190. MR769766

[7] Edward L. Green, Ellen Kirkman, and James Kuzmanovich, *Finitistic dimensions of finite-dimensional monomial algebras*, J. Algebra **136** (1991), no. 1, 37–50, DOI 10.1016/0021-8693(91)90062-D. MR1085118

[8] Ruoyu Guo, *Symmetry of derived delooping level*, Algebr. Represent. Theory **28** (2025), no. 4, 1125–1137, DOI 10.1007/s10468-025-10355-4. MR4953583

[9] Ruoyu Guo and Kiyoshi Igusa, *Derived delooping levels and finitistic dimension*, Adv. Math. **464** (2025), Paper No. 110152, 28, DOI 10.1016/j.aim.2025.110152. MR4863732

[10] Dieter Happel and Dan Zacharia, *Algebras of finite global dimension*, Algebras, quivers and representations, Abel Symp., vol. 8, Springer, Heidelberg, 2013, pp. 95–113, DOI 10.1007/978-3-642-39485-0_5. MR3183882

[11] Birge Zimmermann-Huisgen, *Syzygies and homological dimensions over left serial rings*, Methods in module theory (Colorado Springs, CO, 1991), Lecture Notes in Pure and Appl. Math., vol. 140, Dekker, New York, 1993, pp. 161–174. MR1203806

[12] Birge Zimmermann-Huisgen, *Predicting syzygies over monomial relations algebras*, Manuscripta Math. **70** (1991), no. 2, 157–182, DOI 10.1007/BF02568368. MR1085630

[13] Birge Zimmermann-Huisgen, *Homological domino effects and the first finitistic dimension conjecture*, Invent. Math. **108** (1992), no. 2, 369–383, DOI 10.1007/BF02100610. MR1161097

[14] Kiyoshi Igusa and Gordana Todorov, *On the finitistic global dimension conjecture for Artin algebras*, Representations of algebras and related topics, Fields Inst. Commun., vol. 45, Amer. Math. Soc., Providence, RI, 2005, pp. 201–204. MR2146250

[15] Kiyoshi Igusa and Dan Zacharia, *Syzygy pairs in a monomial algebra*, Proc. Amer. Math. Soc. **108** (1990), no. 3, 601–604, DOI 10.2307/2047776. MR1031675

[16] Christian U. Jensen and Helmut Lenzing, *Homological dimension and representation type of algebras under base field extension*, Manuscripta Math. **39** (1982), no. 1, 1–13, DOI 10.1007/BF01312441. MR672397

[17] Luke Kershaw and Jeremy Rickard, *A finite dimensional algebra with infinite delooping level*, Ann. Represent. Theory **1** (2024), no. 1, 61–65. MR4877092

[18] Henning Krause, *On the symmetry of the finitistic dimension*, C. R. Math. Acad. Sci. Paris **361** (2023), 1449–1453, DOI 10.5802/crmath.481. MR4683322

[19] Claus Michael Ringel, *The finitistic dimension of a Nakayama algebra*, J. Algebra **576** (2021), 95–145, DOI 10.1016/j.jalgebra.2021.01.040. MR4227084

[20] Ralf Schiffler, *Quiver representations*, CMS Books in Mathematics/Ouvrages de Mathématiques de la SMC, Springer, Cham, 2014, DOI 10.1007/978-3-319-09204-1. MR3308668

[21] Emre Sen, *Delooping level of Nakayama algebras*, Arch. Math. (Basel) **117** (2021), no. 2, 141–146, DOI 10.1007/s00013-021-01622-z. MR4290320

[22] Hongbo Shi, *Finitistic dimension of monomial algebras*, J. Algebra **264** (2003), no. 2, 397–407, DOI 10.1016/S0021-8693(03)00074-7. MR1981412

[23] Jiaqun Wei, *Finitistic dimension and Igusa-Todorov algebras*, Adv. Math. **222** (2009), no. 6, 2215–2226, DOI 10.1016/j.aim.2009.07.008. MR2562782

415 South Street, Department of Mathematics, Brandeis University, Waltham, Massachusetts 02453

Current address: 1015 Floyd Avenue, Department of Mathematics and Applied Mathematics, Virginia Commonwealth University, Richmond, Virginia, 23284

Email address: `guor4@vcu.edu`

Contemporary Mathematics
Volume **845**, 2026
https://doi.org/10.1090/conm/845/16956

Continuous representations and applications

Fang Li, Jiongkai Pan, and Shijie Zhu

Abstract. This is an expository report on the recent development of continuous representations and their applications in data analysis. On one hand, we recall the technical foundations of homological persistence theory which originates from quiver representations. We discuss $\mathbb{R}$-representations and their applications, as well as $\mathbb{S}^1$-representations with an approach of covering theory. On the other hand, we focus on the recent development of the applications of representation theory in topological data analysis, including rank decompositions and homological invariants. We also show a generalized result about morphisms between thin modules over locally finite posets. Lastly, some further problems in the trends are summarized.

Contents

1. Introduction

1.1. Background. Poset representations are applied intensively in topological data analysis. Its technical foundation originates from the theory of quiver representations, developed by Gabriel [**Gab72**] in 1970s and many others.

Recall that a quiver $Q = (Q_0, Q_1, s, t)$ is a directed graph with Q_0 the set of vertices and Q_1 the set of arrows. For an arrow α, $s(\alpha)$ is the source and $t(\alpha)$ is the sink of α. When Q is finite, a quiver algebra of Q is the path algebra kQ modulo some admissible ideal I. Later the theory was extended for infinite quivers

2020 *Mathematics Subject Classification.* Primary 55N31, 16G20; Secondary 18G25.
Key words and phrases. continuous representations, persistence modules, barcode, homological invariants, covering theory.
FL is supported by the NSF of China (No. 12131015).
The corresponding author, JP, is supported by the NSF of China (No. 12131015).
SZ is supported by the NSF of China (No. 12201321).

(see [**RdB02, BLP13**]). Recall that a (possibly infinite) quiver Q is called **locally finite** if for each $x \in Q_0$, there are only finitely many arrows going to and coming from x; **interval finite** if for any vertices x, y, there are finitely many paths from x to y; and **strongly locally finite** if Q is both locally finite and interval finite.

The basic idea of persistence comes from topology: If there is a filtration of topological spaces

$$X_0 \subseteq X_1 \subseteq \cdots \subseteq X_n,$$

taking its homology over a field k yields a sequence of linear maps between k-vector spaces

$$H_*(X_0) \to H_*(X_1) \to \cdots \to H_*(X_n),$$

which is called a persistence module. In particular, given a continuous map $f : X \to \mathbb{R}$, a sublevel set is $X_t := f^{-1}(-\infty, t]$. All the sublevel sets form a filtration of the topological space X, where its corresponding persistence module is called the sublevel persistence [**Oud15**].

Persistent homology captures the persistence of topological features such as connected components, holes, voids, etc. in the filtration. The persistence of features is quantified by birth and death parameter values. This can be summarized compactly by the so-called persistence diagram, which is the multiset of birth-death pairs which is called a *barcode* drawn on the plane with multiplicity. The following is usually refered to as the barcode decomposition theorem of persistence modules.

THEOREM 1.1.1 ([**CB15, BCB18**]). *Pointwise finite-dimensional persistence modules over a totally ordered set decompose into interval modules (see §2.1).*

Algebraically, the persistence diagram can be explained as resulting from a structure theorem (the Krull-Remak-Schmidt-Azumaya theorem and Gabriel's Theorem) of persistence modules, which can also be regarded as representations of certain quivers. A persistence module is nothing but a representation on a totally ordered set, which is similar to a representation of type A quiver with linear orientation.

For a continuous map $f : X \to \mathbb{R}$, instead of sublevel sets, one can also consider the level sets $X_t = f^{-1}(t)$ and $X_{[t,t']} = f^{-1}([t, t'])$. The level persistence involves the collection of induced inclusion of linear maps:

$$H_*(X_t) \to H_*(X_{[t,t']}) \leftarrow H_*(X_{t'}).$$

A typical example of level persistence is the zigzag persistence introduced by Carlsson and de Silva [**CdS10**]. On representation theory side, the zigzag persistence corresponds to representations of type A quiver with zigzag orientation.

There are also some other variations of persistences. In [**BD13**], Burghelea and Dey considered level persistences for circle valued maps $f : X \to \mathbb{S}^1$. That is, setting $X_\theta = f^{-1}(\theta)$ for $\theta \in \mathbb{S}^1$, consider inclusion of linear maps:

$$H_*(X_\theta) \to H_*(X_{[\theta,\theta']}) \leftarrow H_*(X_{\theta'}).$$

Hence, we have a persistence module on $\mathbb{S}^1$. In representation theory, $\mathbb{S}^1$-persistences corresponds to the representations of $\tilde{A}$-type quivers, which fits in a larger class of gentle algebras. Its indecomposable representations has been fully classified as strings and bands. In [**HR24**], the authors gave the decomposition theorem for $\mathbb{S}^1$-persistence modules: each $\mathbb{S}^1$-persistence modules decomposes as a sum of barcodes (strings) and finitely many Jordan-cells (bands). We will revisit $\mathbb{S}^1$-persistence modules using covering theory in Section 3.

There are many instances in topological data analysis, where one wishes to build a filtration of topological spaces using two or more parameters, which are expected to provide richer information compared to the one-parameter setting. See [**BL23**, Section 1.1] for examples. This leads to the concept of multi-parameter persistent homology, originally formulated in [**CZ07**], which can be regarded as representations on $\mathbb{R}^n$. Unfortunately, some major difficulties arise when trying to generalize the concept of barcode to multi-parameter persistence modules: Except in small cases, this amounts to studying algebras of "wild representation type", i.e. algebras for which a complete description of all indecomposable modules is impossible. Since it is no longer computationally feasible to describe arbitrary persistence modules in terms of direct sums of indecomposables in this setting, one instead can turn to the study of *invariants*, which play a role analogous to barcodes in 1-parameter persistence and are descriptors that can capture as much of the modules' structure as possible. Another way to avoid this problem is to consider only a restricted class of persistence modules.

The first two simple invariants are the Hilbert function and the the rank invariant. Let $\mathbb{P}$ be a poset. The *Hilbert function* of a pointwise finite dimensional persistence module $M : \mathbb{P} \to k\text{-}\mathrm{Vec}$ is the function $\underline{dim} : \mathbb{P} \to \mathbb{N}$ sending z to $\dim M_z$. The *rank invariant* of M notated as Rk records the rank of the linear map $M_a \to M_b$ for every $a \leq b \in \mathbb{P}$. Clearly, this is a refinement of the Hilbert function.

The natural generalization of persistence diagram from standard persistence to multi-parameter persistence relies on the idea that persistence diagram in standard persistence can be extracted from the rank invariant by inclusion-exclusion principle, which is discovered as Möbius inversion in multiparameter persistence by Patel [**Pat18**]. However, this generalization needs the freedom of choosing the indexing poset $\mathbb{P}$ of $M : \mathbb{P} \to k\text{-}\mathrm{Vec}$ in defining Patel's generalized persistence diagram of M. Of particular importance is the fact that the generalized persistence diagram of M is defined regardless of whether M is interval decomposable or not. Based on these considerations, Kim and Mémoli [**KM21**]defined the generalized rank invariant, which is defined as the isomorphism class of the image of the canonical limit-to-colimit map $\varprojlim M \to \varinjlim M$.

In standard persistence, persistence diagram $\mathrm{Dgm}\, M$ provides a compact encoding of the rank invariant, more precisely:

$$\mathrm{Rk}\, M = \sum_{I \in \mathrm{Dgm}\, M} \mathrm{Rk}\, \mathbf{k}_I = \mathrm{Rk}\left(\bigoplus_{I \in \mathrm{Dgm}\, M} \mathbf{k}_I \right),$$

where each interval $I \in \mathrm{Dgm}\, M$ is considered with multiplicity in the sums, and where $\mathbf{k}_I$ denotes the interval module supported on I. Note that the coefficients in the $\mathbb{Z}$-linear combination are all positive. One naturally hopes that, in the context of multiparameter persistence modules, a similar decomposition formula could also be established for the rank invariant. Motivated by this idea, Botnan, Oppermann and Oudot [**BOO24**] explored the signed rank decomposition.

Definition 1.1.2 ([**BOO24**]). Given a collection $\mathcal{I}$ of intervals in $\mathbb{P}$ and a function $r : \mathcal{I} \to \mathbb{Z}$, a *(signed) rank decomposition of* r *over* $\mathcal{I}$ is given by the following kind of identity:

$$r = \mathbf{Rk}_{\mathcal{I}} k_{\mathcal{R}} - \mathbf{Rk}_{\mathcal{I}} k_{\mathcal{S}},$$

where $\mathcal{R}$ and $\mathcal{S}$ are multi-sets of elements taken from $\mathcal{I}$ and $k_{\mathcal{R}} = \bigoplus_{R\in\mathcal{R}} k_R$, $k_{\mathcal{S}} = \bigoplus_{S\in\mathcal{S}} k_S$ lie in $rep_{\mathcal{I}}\mathbb{P}$ (note that elements $R \in \mathcal{R}$ and $S \in \mathcal{S}$ are considered with multiplicity in the direct sums). By extension, we call the pair $(\mathcal{R}, \mathcal{S})$ itself a rank decomposition of r over $\mathcal{I}$. It is a *minimal rank decomposition* if $\mathcal{R}$ and $\mathcal{S}$ are disjoint as multi-sets.

The flexible choice of r, along with the expression of results in terms of exact categories and relative homological algebra in Botnan's work [**BOO24**], has inspired numerous subsequent research contributions.

The authors in [**BOO24**] proved that: For a finite poset $\mathbb{P}$, when $\mathcal{E}_{\mathrm{Rk}}$ has enough projectives and injectives, and every object has a finite projective and a finite injective resolution. The map $\mathrm{Rk} : K_0(\mathcal{E}_{\mathrm{Rk}}) \to \mathbb{Z}^{\{\leq \mathbb{P}\}}$ is an isomorphism. Moreover, the following three sets are bases for $K_0(\mathcal{E}_{\mathrm{Rk}})$.

$$\{[k_{\langle i,j\rangle}] \mid i \leq j \in \mathbb{P}\},\ \{[k_{\langle i,j\langle}] \mid i < j \in \mathbb{P}\cup\{\infty\}\},\ \text{and}\ \{[k_{\rangle i,j\rangle}] \mid i < j \in \mathbb{P}\cup\{-\infty\}\}.$$

For instance, it has become apparent that tools from relative homological algebra (RHA), and specifically the theory of exact categories, are useful in understanding old and defining new invariants. Blanchette, Brüstle and Hanson [**BBH24**] outline motivations for using RHA to define new class of invariants called homological invariants and explain its relationship to other constructions in the literature. In particular, they give a concrete description of of the Hom-space between spread modules [**BBH24**, Proposition 5.5]. As a generalization of this proposition, we give a concrete description of connected modules in locally finite setting.

THEOREM 1.1.3 (Cf. Theorem 5.3.1). *Let $\mathbb{P}$ be a locally finite poset. Let M and N be two thin modules supported on connected subsets A and B respectively. Denote by $Y_1, Y_2, \cdots, Y_n$ the connected components of $A \cap B$ satisfying*

$$M_1 \triangleq \{x | x \in A, x \notin B, \exists y \in Y_i, y \ \ cover \ \ x\} = \emptyset$$

$$M_2 \triangleq \{y | y \in B, y \notin A, \exists x \in Y_i, y \ \ cover \ \ x\} = \emptyset.$$

Then $\dim_k \mathrm{Hom}(M, N) = n$. Here thin module means that for $\forall v \in \mathbb{P}$, $\dim(M_v) \leq 1$ and the structure map is identity map or zero.

As a complement to [**BBH24**], [**BBH23**] discusses applications of exact structures and relative homological algebra to the study of invariants of multiparameter persistence modules, and propose many open questions through discussions in the framework of representation theory, such as the explicit descriptions of irreducible morphisms between relative projectives, and a way to "lift" the theory to certain infinite posets.

Now we turn to another way: consider only a restricted class of persistence modules. A fruitful idea in this direction has been to use homological approximations. In a nutshell: given a persistence module $M : \mathbb{R}^n \to k\text{-}\mathrm{Vec}$, take a resolution $P_\bullet$ of M by projectives, then build a descriptor from the supports of the direct summands of the P_i's. When M is finitely presentable (fp), the P_i's can be taken to be free modules, and Hilbert's Syzygy theorem ensures that M admits finite such resolutions, meaning that there are only finitely many non-zero P_i's in the exact sequence, and that each one of these terms decomposes as a finite direct sum of interval modules supported on principal upsets. The problem with free resolutions is that they offer little variety as to the shapes of the intervals involved in the signed barcodes, which have to be principal upsets.

The signed barcode introduced in [**BOO24**] as a generalization of the usual (unsigned) barcodes from 1-parameter persistence were shown to be amenable to interpretation in similar ways. Motivated by this idea, the use of larger collections of intervals has since been proposed, including: the whole collection of upsets [**Mil25**], the hooks or the rectangles [**BOO24**], the collection of single-source intervals [**BBH24**], or even the whole collection of intervals [**AENY23a**]. The main idea is as follows: given a collection $\mathcal{I}$ of intervals containing principal upsets, one restricts attention to those short exact sequences for which the Hom-functor, $\mathrm{Hom}(X, -)$ remains exact for every $X \in \mathcal{X}$, denoted as $\varepsilon_{\mathcal{X}}$. Under some mild conditions, by using standard tools from homological algebra a persistent module M admits finite resolutions by projectives relative to $\varepsilon_{\mathcal{X}}$ (also called $\mathcal{X}$-resolutions).

A natural follow-up open question is how to compute finite $\mathcal{X}$-resolutions, or more precisely the supports of the summands appearing in their various non-zero terms. The problem has some interesting recent contributions using Koszul complexes [**Asa24**] but still a long way to go towards scalable methods. In Section 7, we will discuss the latest developments about this.

1.2. Organisation of this paper. This paper is organized in the following way. In Section 2, we recall the definitions of persistence modules over a poset and exihibit two examples: the real-line persistence modules $\mathbb{R}$-pwf and zigzag persistence modules $(\mathbb{R}, \preceq)$-pwf. In Section 3, we review the basics of the classical covering theory and then introduce the definitions of $\mathbb{S}^1$-persistence modules using covering theory. In Section 4, we introduce generalized concepts in persistence theory in locally finite setting as applications of poset theory. In Section 5, we introduced homological invariants theory which uses relative homological theory and representation theory as a tool to study multiparameter persistence. In Section 6, we discuss some open questions and latest developments about representation theory and persistence theory. In Section 7, we illustrate some relevant concepts as an appendix.

2. Poset representations

In this section, we review the basics of poset representations, which are widely used in persistence theory. Poset representations have some similar properties as quiver representations, such as Krull-Schmidt decompositions. It fits into a more general context of representations of continuous thread quivers studied recently by Paquette, Rock and Yıldırım [**PRY24**].

2.1. Persistence modules. Let $\mathcal{C}$ be a small category. A persistence module over $\mathcal{C}$ is a covariant functor $M : \mathcal{C} \to k\text{-}\mathrm{Vec}$. A morphism of persistence modules is a natural transformation $f : M \to N$. That is, a morphism is a collection of k-linear maps $f(x)$ for each $x \in \mathcal{C}$ such that for each morphism $g_{xy} : x \to y$, the following diagram commutes:

$$\begin{array}{ccc} M(x) & \xrightarrow{M(g_{xy})} & M(y) \\ {\scriptstyle f(x)}\downarrow & & \downarrow{\scriptstyle f(y)} \\ N(x) & \xrightarrow[N(g_{xy})]{} & N(y). \end{array}$$

We say that M is pointwise finite-dimensional if each $M(x)$ is finite-dimensional for each $x \in \mathcal{C}$. Denote by $\mathcal{C}$-Mod and $\mathcal{C}$-pwf the category of persistence modules and pointwise finite-dimensional persistence modules over $\mathcal{C}$.

Recall that an abelian category is **Krull-Remak-Schmidt-Azumaya** if every object decomposes uniquely up to isomorphism into a (possibly infinite) direct sum of indecomposables and each indecomposable has a local endomorphism ring. It is well known that over a totally ordered set, an indecomposable persistence module with local endomorphism ring is an interval module. So, by Theorem 1.1.1, the following theorem which asserts that the category of persistence modules is Krull-Schmidt-Remak-Azumaya.

THEOREM 2.1.1 (Krull-Schmidt-Remak-Azumaya). [**BCB18**] *Any pointwise finite-dimensional persistence module is a direct sum of indecomposable modules with local endomorphism ring.*

Let $\mathbb{P}$ be a poset. Then $\mathbb{P}$ is a category with objects the elements of $\mathbb{P}$ and there is a unique morphism $g_{xy} : x \to y$ if and only if $x \leq y$. The identity morphisms are $1_x = g_{xx}$ and the compositions satisfy $g_{yz} \circ g_{xy} = g_{xz}$. It follows that

$$\mathrm{Hom}_{\mathbb{P}}(x, y) = \begin{cases} \{g_{xy}\} & x \leq y \\ \emptyset & x \nleq y \end{cases}.$$

Let us now restrict to the case when $\mathbb{P}$ is finite. Then the *incidence algebra* of $\mathbb{P}$ is the finite-dimensional k-algebra $I(\mathbb{P}) := kQ/J$, where Q is the Hasse quiver of $\mathbb{P}$ and J is the two-sided ideal generated by all commutative relations. Moreover, it is well known that the category mod$I(\mathbb{P})$ of finite-dimensional (right) $I(\mathbb{P})$-modules is isomorphic to that of pfd persistence modules over $\mathbb{P}$. Therefore, we will identify these categories throughout this paper.

DEFINITION 2.1.2. A subset $I \subseteq \mathbb{P}$ is convex if $p, r \in \mathbb{P}$ and $p \leq q \leq r$ implies that $q \in \mathbb{P}$. By **Con**($\mathbb{P}$), we denote the collection of all connected subposets of $\mathbb{P}$. Furthermore, I is connected if for every $p, q \in \mathbb{P}$ there exists a sequence $\{r_i\}_{i=0}^n \subseteq I$ such that $r_0 = p, r_n = q$ and $r_i \leq r_{i+1}$ or $r_i \geq r_{i+1}$ for all $0 \leq i < n$. An interval is a non-empty, connected, and convex set. By **Int**($\mathbb{P}$),we denote the collection of all intervals of $\mathbb{P}$.

Let $\mathbb{P}$ be a poset and let $p, q \in \mathbb{P}$. We say that p *covers* q if $q < p$ and there is no $r \in \mathbb{P}$ such that $q < r < p$. If either [p covers q] or [q covers p], then we write $p \simeq q$.

For an interval $I \subseteq \mathbb{P}$, the **interval module** $\mathbf{k}_I$ is a persistence module:

$$\mathbf{k}_I(t) = \begin{cases} k & t \in I \\ 0 & \text{otherwise} \end{cases} \qquad \mathbf{k}_I(g_{st}) = \begin{cases} 1 & s \leq t, s, t \in I \\ 0 & \text{otherwise.} \end{cases}$$

2.2. Some examples of poset representations.

EXAMPLE 2.2.1. Let $\mathbb{R}$ be the poset of real numbers with the usual partial order $\leq$. There is a nonzero homomorphism between interval modules $f : \mathbf{k}_{[1,3]} \to \mathbf{k}_{[0,2]}$ with

$$f(s) = \begin{cases} 0 = \mathbf{k}_{[1,3]}(s) \xrightarrow{0} \mathbf{k}_{[0,2]}(s) = k & 0 \leq s < 1 \\ k = \mathbf{k}_{[1,3]}(s) \xrightarrow{=} \mathbf{k}_{[0,2]}(s) = k & 1 \leq s \leq 2 \\ k = \mathbf{k}_{[1,3]}(s) \xrightarrow{0} \mathbf{k}_{[0,2]}(s) = 0 & 2 < s \leq 3 \end{cases}$$

We see that $\mathrm{Im}\, f = \mathbf{k}_{[1,2]}$, $\mathrm{Ker}\, f = \mathbf{k}_{(2,3]}$ and $\mathrm{Coker}\, f = \mathbf{k}_{[0,1)}$.

The notion of zigzag persistent modules was introduced by Carlsson and de Silva in [**CdS10**]. In [**BCB18**] and [**IRT23**], a Krull-Schmidt-Remak-Azumaya type theorem was proved for pointwise finite dimensional persistence modules of the continuous zigzag (or continuous type A).

In addition to the usual partial order, the set $\mathbb{R}$ can be equipped with zigzag partial orders by some discrete subsets of sink and source points. Here, we briefly recall the definition of continuous quivers of type A. For more details, the reader may refer to [**IRT23**, Definition 1.1.1].

DEFINITION 2.2.2. Let $S = \{\cdots, s_{-1}, s_0, s_1, \cdots\}$, where $s_k < s_{k+1}$, be a discrete subset of $\mathbb{R} \cup \{\pm\infty\}$, with no finite accumulation points. The partial order $\preceq_S$ (or simply $\preceq$) with respect to S is given by

$$\begin{cases} x \preceq_S y & s_{2n} \leq x \leq y \leq s_{2n+1} \\ x \succeq_S y & s_{2n-1} \leq x \leq y \leq s_{2n}. \end{cases}$$

The poset $(\mathbb{R}, \preceq_S)$ is also called a continuous quiver of type A.

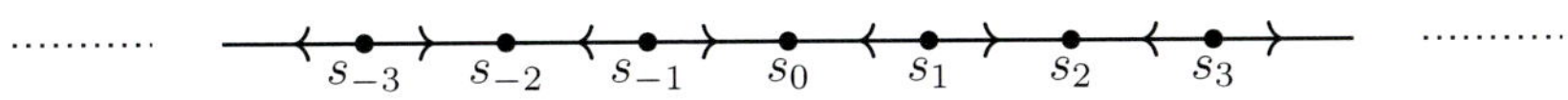

FIGURE 1. A continuous quiver of type A

REMARK 2.2.3. For each subset $S \subseteq \mathbb{R} \cup \{\pm\infty\}$, there are precisely two partial orders $\preceq_S$ can be defined, they are reverse to each other. For instance, when $S = \{s_{-1} = -\infty, s_0 = +\infty\}$, the partial order $\preceq_S$ is the same as the usual partial order $\leq$. When $S = \{s_0 = -\infty, s_1 = +\infty\}$, the partial order $\preceq_S$ is the reverse to $\leq$.

Let $(\mathbb{R}, \preceq)$ be a continuous quiver of type A. Then (pointwise finite dimensional) representations of $(\mathbb{R}, \preceq)$ are defined similarly. For any interval I in $(\mathbb{R}, \preceq)$, let $\mathbf{k}_I$ be the **interval module** of $(\mathbb{R}, \preceq)$-pwf given as follows.

$$\mathbf{k}_I(x) = \begin{cases} k & x \in I \\ 0 & \text{otherwise} \end{cases} \qquad \mathbf{k}_I(x, y) = \begin{cases} 1_k & y \preceq x \text{ and } x, y \in I \\ 0 & \text{otherwise} \end{cases}$$

Notice that I is an interval of the poset $(R, \preceq)$ in the sense of Definition 2.1.2. Meanwhile, since the poset $(R, \preceq)$ is not totally ordered, unless $\preceq$ is the same or reverse of the usual partial order $\leq$, Theorem 1.1.1 does not apply for the indecomposable decompositions of pointwise finite-dimensional persistent modules over $(R, \preceq)$. However, we have the following results:

THEOREM 2.2.4 ([**IRT23**, Theorem A,B]). *The interval modules $\mathbf{k}_I$ are indecomposable and any pointwise one-dimensional indecomposable representation of $(\mathbb{R}, \preceq)$ is isomorphic to some interval module. Any pointwise finite-dimensional $(\mathbb{R}, \preceq)$-representation decomposes uniquely up to isomorphism into a direct sum of interval modules. In particular,* $(\mathbb{R}, \preceq)$-pwf *is Krull-Remak-Schmidt-Azumaya.*

2.3. Applications. In statistics, clustering methods deal with the problems of classifying objects such that those with similar features are grouped together. It happens that the grouping usually depends on the choice of some smoothing parameter. For point cloud M, i.e. a finite set of points in a metric space, hierachical clustering method considers a filtration of topological spaces $X_t = \bigcup_{x\in M} B_t(x)$, $t \geq 0$, where $B_t(x)$ stands for the closed ball of radius t centering at x. Taking the homology with coefficients in k yields the persistence module $H_*(X_t; k)$ in $\mathbb{R}$-pwf. Each barcode $\mathbf{k}_{[a,b)}$ of the persistence module illustrates the homological feature which is born at time a and dies at time b. The hirachical clustering method enjoys some outstanding merits:

(1) There is no need to determine what is the "appropriate" value for the parameter t for a certain data set.
(2) In its barcode decomposition, the lengths of bars indicate the significance of the homological feature of the data.
(3) There exist fast algorithms for which the persistence modules can be obtained by computers.
(4) The barcode is a complete invariant for the persistence module.
(5) The barcode is stable with respect to the input data.

However, the hirachical clustering method or the sublevel persistence can only deal with the situations when there is a filtration of topological spaces. In some senarios, the properties of a point cloud are related with the probability density from which the data are sampled. To cope with such problems, zigzag persistences was introduced by Carlsson and de Silva [**CdS10**], typically in consideration of the persistences of topological space seqeunces as:

$$X^p_{r_1} \to X^p_{r_1} \cup X^p_{r_2} \leftarrow X^p_{r_2} \to X^p_{r_2} \cup X^p_{r_3} \leftarrow X^p_{r_3} \to \cdots \leftarrow X^p_{r_n}$$

where $X^p_{r_i}$ denotes $p\%$ of the point cloud when measured according to parameter value r_i, with parameters $r_1 < r_2 < \cdots < r_n$. Notice that in this case, $X^p_{r_i}$ is not always a subspace of $X^p_{r_{i+1}}$. So one cannot use the sublevel persistence to measure the evolution of the topology.

There is another situation when zigzag persistence comes in handy. If the point cloud is too large to compute one may first consider taking some samples $X_1, \cdots, X_n$ from the total data set X and calculate their homological features, and then analyze and combine these local information together to obtain the global features. For instance, if each sample detects certain feature, then we may conclude that the total point cloud has the same feature as well. The zigzag persistence can serve this purpose well. Consider the union sequence:

$$X_1 \to X_1 \cup X_2 \leftarrow X_2 \to X_2 \cup X_3 \leftarrow X_3 \to \cdots \leftarrow X_n$$

From the persistence module obtained by taking the homology of this sequence, a barcode $\mathbf{k}_{[i,j)}$ indicates samples $X_i, \cdots, X_j$ share the same feature.

3. Covering theory and representations of the circle

An important technique to go from $\mathbb{R}$-representations to $\mathbb{S}^1$-representations is covering theory, introduced by Garbriel [**Gab80**] and developed later by many others such as Roiter [**GR92**], Dowbor, Skrownski [**DS87**] and Asashiba [**Asa97**]. In this section, we first recall some essentials of the covering theory and then we will apply the covering theory to study $\mathbb{S}^1$-representations. This is somewhat a different approach than [**HR24**], where the authors use the one point compactification of $\mathbb{R}$.

3.1. Group actions and orbit categories.

DEFINITION 3.1.1. Let k be a field. A small category $\mathcal{C}$ is called a **k-category** if (1) Every set $\mathrm{Hom}_{\mathcal{C}}(x, y)$ is a k-vector space; (2) The composition $\mathrm{Hom}_{\mathcal{C}}(y, z) \times \mathrm{Hom}_{\mathcal{C}}(x, y) \to \mathrm{Hom}(x, z)$ is k-bilinear.

DEFINITION 3.1.2. A k-category $\mathcal{C}$ is **locally finite dimensional (or a spectroid)** if (1) different objects are non-isomorphic; (2) $\mathrm{End}_{\mathcal{C}}(x)$ is a local k-algebra for any object $x \in \mathcal{C}$; (3) $\dim_k \mathrm{Hom}_{\mathcal{C}}(x, y) < \infty$, $\forall x, y \in \mathcal{C}$.

A spectroid $\mathcal{C}$ is **locally bounded** if for each $x \in \mathcal{C}$, $\{y \in \mathcal{C} | \mathrm{Hom}_{\mathcal{C}}(x, y) \neq 0$ and $\mathrm{Hom}_{\mathcal{C}}(y, x) \neq 0\}$ is a finite set.

Let k be a field. For a locally bounded quiver Q (with some admissible ideal I), denote by $\mathcal{C}_Q$ the k-category whose objects are vertices of Q and $\mathrm{Hom}_{\mathcal{C}_Q}(x, y)$ paths from x to y (modulo I). If Q is strongly locally finite, $\mathcal{C}_Q$ is a spectroid.

Let G be a group. In general, a G-action on a k-category $\mathcal{C}$ is a group homomorphism $a : G \to \mathrm{Aut}(\mathcal{C})$. We say G acts **freely** on $\mathcal{C}$ if $g \cdot x \neq x$ for all $x \in \mathcal{C}$ whenever $g \neq 1$. A G-action is **locally bounded** if for any x, $y \in \mathcal{C}$, $\{g | \mathrm{Hom}_{\mathcal{C}}(g \cdot x, y) \neq 0\}$ is finite.

Let $\mathcal{C}$ be a k-category with a free and locally bounded G-action. The orbit category $\mathcal{C}/G$ is defined as the following: the objects of $\mathcal{C}/G$ are the G orbits $\bar{x} := Gx$ for $x \in \mathcal{C}$ and

$$\mathrm{Hom}_{\mathcal{C}/G}(\bar{x}, \bar{y}) = \left\{ f = (f_{xy}) \in \prod_{x \in Gx, y \in Gy} \mathrm{Hom}_{\mathcal{C}}(x, y) | g f_{xy} = f_{gxgy} \right\}.$$

EXAMPLE 3.1.3. Let Q be the quiver $\cdots 2 \xrightarrow{\alpha_2} 1 \xrightarrow{\alpha_1} 0 \xrightarrow{\alpha_0} 1 \cdots$ with admissible ideal $I = \langle \alpha_i \alpha_{i+1} \rangle$. The free group $\mathbb{Z}$ acts on Q by $n \cdot i = i + n$.

On the other hand, R is the quiver algebra generated by $\overset{\alpha}{\circlearrowleft} x$ with admissible ideal $J = \langle \alpha^2 \rangle$.

Then one can check that the orbit category $\mathcal{C}_Q/G \cong \mathcal{C}_R$.

In general, let $\mathcal{C}$ and $\mathcal{D}$ be spectroids. A k-linear functor $F : \mathcal{C} \to \mathcal{D}$ between two k-categories is said to be a covering functor if for every two objects a, b of $\mathcal{D}$, F induces isomorphisms $\bigoplus_{F(z)=a} \mathrm{Hom}_{\mathcal{C}}(z, y) \to \mathrm{Hom}_{\mathcal{D}}(a, b)$ and $\bigoplus_{F(z)=b} \mathrm{Hom}_{\mathcal{C}}(x, z) \to \mathrm{Hom}_{\mathcal{D}}(a, b)$ where x, y are objects of $\mathcal{C}$ with $a = F(x), b = F(y)$. A covering functor $F : \mathcal{C} \to \mathcal{D}$ is called a **Galois covering** if F is surjective on objects and there exists a group G of k-linear automorphisms of $\mathcal{C}$ which acts freely on (objects of) $\mathcal{C}$, such that $F \circ g = F$ for any $g \in G$ and G acts transitively on $F^{-1}(a)$ for each object $a \in \mathcal{D}$. From the definition, we know that if $\mathcal{C}$ is a spectroid and G acts freely and locally bounded on $\mathcal{C}$, then the canonical projection $\pi : \mathcal{C} \to \mathcal{C}/G$ is a Galois covering.

3.2. Pushdown and pullup. Let $\mathcal{C}$ be a k-category. By a left (right) $\mathcal{C}$-module, we mean a k-linear covariant (contravariant) functor $M : \mathcal{C} \to k$-Vec. Denote by $\mathcal{C}$-Mod the category of right $\mathcal{C}$-modules. A right $\mathcal{C}$ module M is finitely generated if there is an epimorphism $\bigoplus_{i=1}^{n}(-,x_i) \to M$, where $x_i \in \mathcal{C}$. Denote by $\mathcal{C}$-mod the subcategory of finitely generated right $\mathcal{C}$ modules. A $\mathcal{C}$ module M is projective if $M \cong (-,x)$. For a $\mathcal{C}$ module M, denote by ${}^gM$ the shifted module $M \circ g$.

For a Galois covering $\pi : \mathcal{C} \to \mathcal{C}/G$, to describe the relation between $\mathcal{C}$-Mod and $\mathcal{C}/G$-Mod, we need to introduce the important pushdown and pullup functors.

The pullup functor $F^\bullet : \mathcal{C}/G\text{-Mod} \to \mathcal{C}\text{-Mod}$ is defined by $F^\bullet(M) = M \circ \pi$.

The pushdown functor $F_\bullet : \mathcal{C}\text{-Mod} \to \mathcal{C}/G\text{-Mod}$ is defined as the following: (1) $F_\bullet M$ is a contravariant functor such that $F_\bullet M(\bar{x}) = \bigoplus_{x\in\bar{x}} Mx$; for a morphism $f = (f_{xy}) : \bar{x} \to \bar{y}$, $F_\bullet M(f) : \bigoplus_{y\in\bar{y}} My \to \bigoplus_{x\in\bar{x}} Mx$ is given by $F_\bullet M(f) = (Mf_{xy})$. For moprhisms $\alpha : M \to N$ in $\mathcal{C}$, the morphism $F_\bullet(\alpha) : F_\bullet(M) \to F_\bullet(N)$ is a natural transformation of functors in $\mathcal{C}/G$-Mod, which is defined by:

$$\begin{array}{ccc} P_\bullet M(\bar{x}) & \xrightarrow{(P_\bullet\alpha)_{\bar{x}}} & P_\bullet N(\bar{x}) \\ \| & & \| \\ \bigoplus_{x\in\bar{x}} Mx & \xrightarrow{\bigoplus_{x\in\bar{x}}\alpha_x} & \bigoplus_{x\in\bar{x}} Nx \end{array}$$

Although we use a functorial approach here, it is usually helpful to view the $\mathcal{C}$ modules as representations rather than functors. The definition of pullup and pushdow functors may look abstract. But roughly speaking, pushdown $F_\bullet$ means folding a module and pullup $F^\bullet$ means taking preimages. We are going to illustrate it in a concrete example below.

EXAMPLE 3.2.1. Let $\mathcal{C}$ and $\mathcal{C}/G$ be the same as in Example 3.1.3. Let $M = \cdots \to 0 \to k \xrightarrow{id} k \to 0 \to \cdots$ be a $\mathcal{C}$ module, then $F_\bullet M$ is the $\mathcal{C}/G$ module $k^2 \circlearrowleft^{J}$, where $J = \begin{bmatrix} 0 & 1 \\ 0 & 0 \end{bmatrix}$.

$F^\bullet F_\bullet M = \cdots \to k^2 \xrightarrow{J} k^2 \xrightarrow{J} k^2 \xrightarrow{J} k^2 \to \cdots$.

If S is the (unique) simple $\mathcal{C}/G$ module, then $F^\bullet S = \cdots \to k \xrightarrow{0} k \xrightarrow{0} k \xrightarrow{0} k \to \cdots$.

PROPOSITION 3.2.2 ([**Asa22**]). *Let $\mathcal{C}$ be a spectroid with a free locally bounded group action G. Then the pushdown functor $F_\bullet : \mathcal{C}\text{-Mod} \to \mathcal{G}/G\text{-Mod}$ and the pullup functor $F^\bullet : \mathcal{C}/G\text{-Mod} \to \mathcal{C}$ satisfy the following properties:*

(1) *$(F_\bullet, F^\bullet)$ is an adjoint pair.*
(2) *$F^\bullet F_\bullet M = \bigoplus_{g\in G} {}^gM$.*
(3) *$F_\bullet$ preserves projectivity, and hence finitely generated modules.*
(4) *$F_\bullet$ preserves indecomposability.*

Since the pushdown $F_\bullet$ preserves finitely generated modules, it is a functor $F_\bullet : \mathcal{C}\text{-mod} \to \mathcal{C}/G\text{-mod}$. It is not necessarily dense. Denote by $\mathcal{C}/G\text{-mod}_1 := \operatorname{Im} F_\bullet$ which is called $\mathcal{C}/G$ modules of first kind. If a $\mathcal{C}/G$ module M has no direct summands in $\mathcal{C}/G\text{-mod}_1$, it is of second kind, and denote by $\mathcal{C}/G\text{-mod}_2$ the subcategory containing $\mathcal{C}/G$ modules of second kind.

Dowbor and Skrownski [**DS87**] describes the indecomposable $\mathcal{C}/G$ modules of first and second kind. An indecomposable $\mathcal{C}$ module M is called **weakly-G-periodic** if $suppM$ is infinite but $suppM/G_M$ is finite, where $G_M = \{g | gM = M\}$.

PROPOSITION 3.2.3. *Let $M \in \mathcal{C}/G$-mod. Then*

(1) *$M \in \mathcal{C}/G$-mod$_1$ if and only if $F^{\bullet}(X) = \oplus_i N_i$, where N_i are finite dimensional.*
(2) *$M \in \mathcal{C}/G$-mod$_2$ if and only if $F^{\bullet}(X) = \oplus_i N_i$, where N_i are weakly-G-periodic.*

3.3. k-linear enrichment. Let k be a field and $\mathcal{C}$ a small category. Denote by $k\mathcal{C}$ the k-linear enrichment of $\mathcal{C}$, which has the same objects as $\mathcal{C}$, but the Hom-sets $\mathrm{Hom}_{k\mathcal{C}}(x, y)$ are k-vector space generated by $f \in \mathrm{Hom}_{\mathcal{C}}(x, y)$ for any objects $x, y \in \mathcal{C}$. The compositions of morphisms are bilinear. It follows that $k\mathcal{C}$ is a k-category.

PROPOSITION 3.3.1. *There is a categorical isomorphism* $\mathcal{C}$-Mod $\cong k\mathcal{C}$-Mod.

PROOF. Define $\Phi : \mathcal{C}\text{-Mod} \to k\mathcal{C}\text{-Mod}$ by $\Phi(M)(x) = M(x)$ and $\Phi(M)(\sum s_i f_i) = \sum s_i M(f_i)$, where $f_i : x \to y$ are morphisms in $\mathcal{C}$. And for $\alpha : M \to N$, define $\Phi(M)(\alpha)_x = \alpha_x$. It is clear that $\Phi(M)$ is a k-linear functor.

Conversely, define $\Psi : k\mathcal{C}\text{-Mod} \to \mathcal{C}\text{-Mod}$ as the restriction functor (on the hom-sets). It is easy to see that Φ, Ψ are mutually inverse. □

3.4. The category $\mathbb{S}^1$. Now we proceed to discuss the $\mathbb{S}^1$-representations. Throughout, we denote the standard covering map $p : \mathbb{R} \to \mathbb{S}^1$ by $p(x) := [x] = e^{2\pi i x}$.

A subset of $S \subseteq \mathbb{R}$ is called 1-equivariant (or simply equivariant), if $x \in S \cap \mathbb{R}$ implies $x \pm 1 \in S$. In particular $\emptyset$ is equivariant by convention. Furthermore, if S is a subset of $\mathbb{R}$ with no accumulation point, then by nested interval principal, $S \cap [0, 1)$ is a finite set.

An orientation set S is a discrete equivariant subset of $\mathbb{R}$ with no finite accumulation points and $|S \cap [0, 1)|$ is even. Now we introduce the definition of category $(\mathbb{S}^1, S)$ following [**HR24**].

DEFINITION 3.4.1. Let S an oriatation set. The category $(\mathbb{S}^1, S)$ contains points on $\mathbb{S}^1$ as objects, there is a generating morphism $g_{xy} : x \to y$ if $x = [a]$, $y = [b]$ for some $a \preceq_S b$ and $|a - b| < 1$ in the poset $(\mathbb{R}, \preceq_S)$ (identify $(\mathbb{R}, \emptyset) = (\mathbb{R}, \leq)$). The generating morphisms g_{xx} are identity.

We make some remarks to clarify the definition.

REMARK 3.4.2.

(1) When $S \neq \emptyset$, it follows that $|S \cap [0, 1)| \geq 2$. Hence intervals $(a, b) \cap S \neq \emptyset$ when $|a - b| \geq 1$, which means a and b are imcomparable in $(\mathbb{R}, \preceq_S)$. So the condition $|a - b| < 1$ is redundant for the case $S \neq \emptyset$.
(2) When $S = \emptyset$, we usually refer to $(\mathbb{S}^1, \emptyset)$ simply as the category $\mathbb{S}^1$, where generating morphisms g_{xy} are defined for $x = [a], y = [b]$ for $0 \leq b - a < 1$. In particular, for any $y \neq x$ in $\mathbb{S}^1$, we have a morphism $\omega_x := g_{yx} \circ g_{xy}$. It follows that
$$\mathrm{Hom}_{\mathbb{S}^1}(x, y) = \left\{ g_{xy} \circ \omega_x^n = \omega_y^n \circ g_{xy} : n \in \mathbb{N} \right\}.$$
(3) Assume $S \cap [0, 1) = \{s_0, s_1\}$. There are two generating morphisms $g^{\downarrow}, g^{\uparrow}$: $[s_1] \to [s_0]$ which can be distinguished as below: for $s_0 < t < s_1$, the composition $g_{[t][s_0]} g_{[s_1][t]} = g^{\downarrow}$ and for $s_1 < t < s_0 + 1$, the composition

$g_{[t][s_0]}g_{[s_1][t]} = g^{\uparrow}$. Notice that $|\operatorname{Hom}_{(\mathbb{S}^1,S)}(x,y)| \leq 2$, and the equality holds if and only if $x = [s_1]$ and $y = [s_0]$.

(4) When $|S \cap [0,1)| > 2$, $|\operatorname{Hom}_{(\mathbb{S}^1,S)}(x,y)| \leq 1$ for any $x, y \in (\mathbb{S}^1, S)$.

The category $(\mathbb{S}^1, S)$, rather its k-linear enrichment, can be illustrated by covering theory. Let S be an orientation set. Consider the free group $\mathbb{Z}$ acts on the category $k(\mathbb{R}, \preceq_S)$ by $n \cdot i = i + n$. It is easy to check that the action is locally bounded. Hence we have:

PROPOSITION 3.4.3. *The orbit category* $k(\mathbb{R}, \preceq_S)/\mathbb{Z} \cong k(\mathbb{S}^1, S)$.

3.5. $\mathbb{S}^1$-persistence modules. In light of Proposition 3.3.1 and Proposition 3.4.3, we can characterize $(\mathbb{S}^1, S)$-modules using covering theory.

Let S be an orientation set. Denote by $(\mathbb{R}, \preceq_S)$-Mod and $(\mathbb{S}^1, S)$-Mod the category of all modules on $(\mathbb{R}, \preceq_S)$ and $(\mathbb{S}^1, S)$ respectively. Hence the covering functor $p : (\mathbb{R}, \preceq_S) \to (\mathbb{S}^1, S)$, $p(x) = [x]$ gives rise to two functors between module categories:

(1) the pull-up functor $p^\bullet : (\mathbb{S}^1, S)\text{-Mod} \to (\mathbb{R}, \preceq_S)\text{-Mod}$ sending $M \mapsto M \circ p$,
(2) the push-down functor $p_\bullet : (\mathbb{R}, \preceq_S)\text{-Mod} \to (\mathbb{S}^1, S)\text{-Mod}$ which satisfies $p_\bullet M([x]) = \oplus_{i \in \mathbb{Z}} M(x + i)$.

Let $\mathbf{k}_I$ be an interval module in $(R, \preceq_S)$-pwf. By Proposition 3.2.2, its image under the pushdown functor, denoted by $\overline{\mathbf{k}}_I = p_\bullet \mathbf{k}_I$, is still indecomposable pointwise finite-dimensional in $(\mathbb{S}^1, S)$-Mod, which is called a **bar**. A **barcode** is a direct sum of bars. Barcodes are just first kind modules in the orbit category $(\mathbb{R}, \preceq_S)/\mathbb{Z}$.

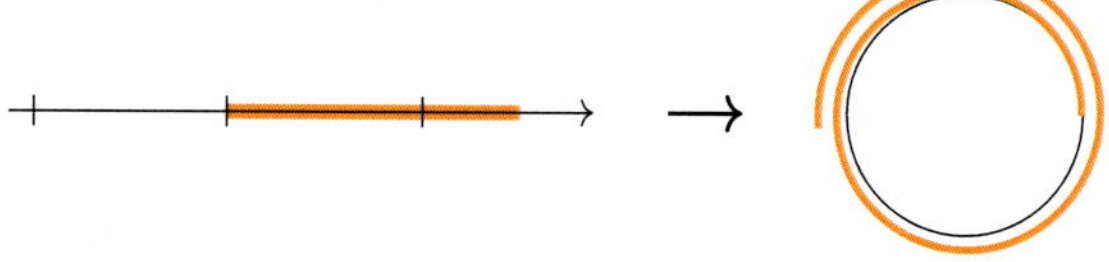

FIGURE 2. Push-down of the interval $\mathbb{R}$-module $M_{[0,1.5]}$ to the $\mathbb{S}^1$-module $\overline{M}_{[0,1.5]}$.

In addition, the second kind modules are called **Jordan-cells**. These are indecomposable modules J satisfying (1) $\dim J_{[x]} = d < \infty$ is a constant for all x, (2) $J(g_{[x][y]})$ is an isomorphism for all generating morphism $g_{[x][y]}$.

EXAMPLE 3.5.1. Let $S = \frac{1}{4}\mathbb{Z}$ be the orientation set and $\lambda \in k^\times$. Define an $(\mathbb{S}^1, S)$ persistence module J as the following:

$$J_{[x]} = k^2 \qquad J(g_{[x][y]}) = \begin{cases} \begin{pmatrix} \lambda & 1 \\ 0 & \lambda \end{pmatrix} & \frac{1}{4} = x < y \leq \frac{1}{2} \\ 1_{k^2} & \text{otherwise.} \end{cases}$$

It follows that J is a 2-dimensional Jordan cell (see Figure 3).

It turns out that an indecomposable pointwise finite-dimensional $(\mathbb{S}^1, S)$-persistence module is either a bar or a Jordan cell. Denoted by $(\mathbb{S}^1, S)$-pwf the category of pointwise finite dimensional persistence modules over $(\mathbb{S}^1, S)$.

THEOREM 3.5.2 ([**HR24**]). *Any object in* $(\mathbb{S}^1, S)$-pwf *decomposes into a direct sum of barcodes and finitely-many Jordan-cells. In particular,* $(\mathbb{S}^1, S)$-pwf *is Krull–Remak–Schmidt–Azumaya.*

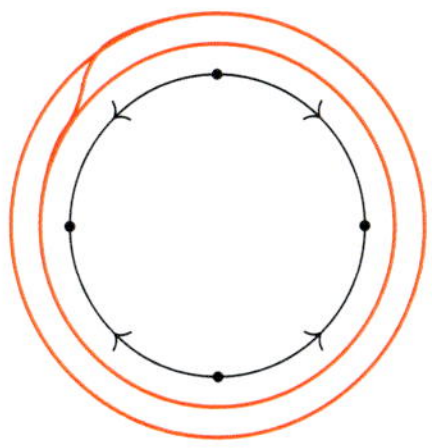

FIGURE 3. Jordan cell J for Example 3.5.1

4. Applications of poset representations

4.1. Generalized bar code. This section presents the definition of generalized barcodes and the Möbius inversion. It further explains how the generalized barcode and Möbius inversion can be utilized to generalize barcodes within the context of a locally finite poset.

DEFINITION 4.1.1. Consider $\mathbb{P}$ as a partially ordered set and $M \in \mathbb{P}$-mod. For a given interval $I \subseteq \mathbb{P}$, the generalized rank of M over I, denoted as $\mathrm{Rk}_I\, M$, is expressed as:

$$\mathrm{Rk}_I\, M = \mathrm{rank}\left[\varinjlim M|_I \to \varprojlim M|_I\right].$$

If $\mathcal{I}$ represents a collection of intervals, the generalized rank invariant of M for $\mathcal{I}$ is a mapping $\mathrm{Rk}_{\mathcal{I}}\, M : \mathcal{I} \to \mathbb{N} \cup \{\infty\}$, where $\mathrm{Rk}_{\mathcal{I}}\, M(I) = \mathrm{Rk}_I\, M$.

The example below illustrates that the generalized rank invariant is either equivalent to or a finer refinement of the rank invariant.

EXAMPLE 4.1.2.

(i) (*1-parameter persistence*) Consider the poset formed by the integers $\mathbb{Z}$. Here, $\mathrm{Con}(\mathbb{Z}) = \mathrm{Int}(\mathbb{Z})$, and the generalized rank invariant for a module $F : \mathbb{Z} \to k\text{-}\mathrm{Vec}$ aligns with the standard rank invariant.This result arises from the following fact: for any $a, b \in \mathbb{Z}$ such that $a \leq b$, the value of $\mathrm{Rk}_{[a,b]}\, F$ equals the rank of the restriction $F|_{[a,b]}$.This statement holds true even if the target category of F is not k-Vec.

(ii) (*Zigzag persistence*) Consider the zigzag poset $\mathbb{ZZ}$. It is noted that $\mathrm{Con}(\mathbb{ZZ}) = \mathrm{Int}(\mathbb{ZZ})$.The generalized rank invariant of a diagram $F : \mathbb{ZZ} \to k\text{-}\mathrm{Vec}$ provides more information compared to the rank invariant. Interestingly, even though $\mathrm{Rk}(F)$ is defined without explicit reference to whether F is interval decomposable, it turns out that the barcode of F can be derived from $\mathrm{Rk}(F)$.This implies that the generalized rank invariant serves as a complete invariant for zigzag modules; in other words, any two zigzag modules sharing the same rank invariant are isomorphic.This result generalizes the completeness of the rank invariant for $\mathbb{Z}$-indexed persistence.

(iii) (*Multiparameter persistence*) Consider the two-dimensional grid poset $\mathbb{Z}^2$ and a functor $F : \mathbb{Z}^2 \to k\text{-}\mathrm{Vec}$. For any $(a,b),(c,d) \in \mathbb{Z}^2$ such that $(a,b) \leq (c,d)$, the rank of $\varphi_F((a,b),(c,d))$ is equal to the rank of the restricted diagram $F|_{[(a,b),(c,d)]}$. However, the standard rank invariant does not capture the rank of $F|_I$ when I is not in the form $[(a,b),(c,d)]$.This indicates that the generalized rank invariant refines the standard rank invariant for 2-parameter persistence modules $\mathbb{Z}^2 \to k\text{-}\mathrm{Vec}$. The same reasoning extends to diagrams $\mathbb{Z}^d \to k\text{-}\mathrm{Vec}$ for $d > 2$.

DEFINITION 4.1.3. A poset $\mathbb{P}$ is said to be locally finite if for all $p, q \in \mathbb{P}$ with $p \leq q$, the set $[p, q] := \{r \in \mathbb{P} : p \leq r \leq q\}$ is finite.

Let $\mathbb{P}$ be a locally finite poset. For $I \in \mathrm{Con}(\mathbb{P})$, the neighborhood of I is defined as:

$$\mathrm{nbd}(I) := \{p \in \mathbb{P} \setminus I : \text{there exists } q \in I \text{ such that } q \simeq p\}.$$

The *perimeter* o_I of I is the number of elements in $\mathrm{nbd}(I)$.

A locally finite poset $\mathbb{P}$ is said to be essentially finite if every $I \in \mathrm{Con}(\mathbb{P})$ has a finite perimeter.

Now, suppose $\mathbb{P}$ is a connected and essentially finite poset, and let $I \in \mathrm{Con}(\mathbb{P})$. For a fixed $n \in \mathbb{N}$, the set I^n is defined to consist of all subsets $J \subset \mathbf{P}$ such that $I \subset J \subset I \cup \mathrm{nbd}(I)$ and $|J \cap \mathrm{nbd}(I)| = n$. In other words, every $J \in I^n$ is obtained by adding exactly n elements from $\mathrm{nbd}(I)$ to I.

Given a module $F : \mathbb{Z} \to k\text{-}\mathrm{Vec}$ and an interval $[a, b] \in \mathrm{Int}(\mathbb{Z}) = \mathrm{Con}(\mathbb{Z})$ with $a < b$, recall that in the standard persistence diagram of F, the multiplicity of $(a, b) \in \mathbb{Z}^2$ is defined as:

$$\mathrm{rk}(F)([a,b]) - \mathrm{rk}(F)([a-1,b]) - \mathrm{rk}(F)([a,b+1]) + \mathrm{rk}(F)([a-1,b+1]).$$

This formula serves as a motivation for:

DEFINITION 4.1.4 (Generalized persistence diagram of $\mathbb{P} \to k\text{-}\mathrm{Vec}$). Let $\mathbb{P}$ be a connected, essentially finite poset. Given a persistence module $F : \mathbb{P} \to k\text{-}\mathrm{Vec}$, the persistence diagram $\mathrm{dgm}^{\mathbb{P}}(F) : \mathrm{Con}(\mathbb{P}) \to \mathbb{Z}$ of F is defined by mapping each $I \in \mathrm{Con}(\mathbb{P})$ to

$$\text{(2)} \quad \mathrm{dgm}^{\mathbb{P}}(F)(I) := \mathrm{rk}(F)(I) - \sum_{J \in I^1} \mathrm{rk}(F)(J) + \sum_{K \in I^2} \mathrm{rk}(F)(K) - \cdots + (-1)^{o_I} \sum_{L \in I^{o_I}} \mathrm{rk}(F)(L).$$

The persistence diagram $\mathrm{dgm}^{\mathbb{P}}(F)$ is the Möbius inversion of $\mathrm{rk}(F)$ over the poset $\mathrm{Con}^{\mathrm{op}}(\mathbb{P})$. By applying the principle of inclusion-exclusion, the barcode of interval-decomposable persistence modules can be derived using formula (2):

THEOREM 4.1.5 (Generalized persistence diagram recovers the barcode). **[KM21]** *Let $\mathbb{P}$ be any connected, essentially finite poset, and let $F : \mathbb{P} \to k\text{-}\mathrm{Vec}$ be interval decomposable. For $I \in \mathrm{Con}(\mathbb{P})$, the value obtained from equation (2) corresponds to the multiplicity of I in* $\mathrm{barc}^{\mathbb{P}}(F)$.*(This implies that if $I \in \mathrm{Con}(\mathbb{P}) \setminus \mathrm{Int}(\mathbb{P})$, the value is zero.)*

REMARK 4.1.6. Since the barcode of an interval decomposable module $F : \mathbb{P} \to k\text{-}\mathrm{Vec}$ serves as a complete invariant of F, Theorem 4.1.5 establishes that the generalized rank invariant of F (as defined in Definition 4.1.1) is also a complete invariant for F.

REMARK 4.1.7. Theorem 4.1.5 also leads to the following conclusions: 1. If there exists $I \in \mathrm{Con}(\mathbb{P}) \setminus \mathrm{Int}(\mathbb{P})$ such that $\mathrm{dgm}^{\mathbb{P}}(G)(I) \neq 0$, then G is not interval decomposable. 2. If there exists $J \in \mathrm{Int}(\mathbb{P})$ such that $\mathrm{dgm}^{\mathbb{P}}(G)(J) < 0$, then G is not interval decomposable. 3. The converse of the statement in item (2) does not hold.

Möbius inversion. Let $\mathbf{Q}$ be a locally finite poset. The Möbius function $\mu_{\mathbf{Q}} : \mathbf{Q} \times \mathbf{Q} \to \mathbb{Z}$ of $\mathbf{Q}$ is defined recursively as:

$$\mu_{\mathbf{Q}}(p,q) = \begin{cases} 1, & p = q, \\ -\sum_{p \leq r < q} \mu_{\mathbf{Q}}(p,r), & p < q, \\ 0, & \text{otherwise.} \end{cases}$$

Assuming $\mathrm{Con}(\mathbf{P})$ is a locally finite poset, formula (2) is the Möbius inversion of the rank invariant $\mathrm{rk}(F)$ over the poset $\mathrm{Con}(\mathbf{P})^{\mathrm{op}}$:

PROPOSITION 4.1.8 ([**KM21**]). *Let $\mathbb{P}$ be a connected, essentially finite poset such that $\mathrm{Con}(\mathbb{P})$ is locally finite. Let $F : \mathbb{P} \to \mathcal{C}$ and let $\mu : \mathrm{Con}(\mathbb{P})^{\mathrm{op}} \times \mathrm{Con}(\mathbb{P})^{\mathrm{op}} \to \mathbb{Z}$ be the Möbius function of the poset $\mathrm{Con}(\mathbb{P})^{\mathrm{op}}$. Then, for $I \in \mathrm{Con}(\mathbb{P})^{\mathrm{op}}$,*

$$\mathrm{dgm}_{\mathbb{P}}(F)(I) = \sum_{J \supseteq I} \mathrm{rk}(F)(J) \cdot \mu(J, I).$$

4.2. Signed rank decomposition. In this section, we define signed rank decomposition and demonstrate the existence and uniqueness of such a decomposition. Additionally, we present an explicit formula for computing signed rank decomposition using Möbius inversion. We also highlight the categorical enrichment of signed rank decomposition, which motivates the use of relative homological theory and representation theory in the study of multiparameter persistence modules.

DEFINITION 4.2.1 ([**BOO24**]). Consider a collection $\mathcal{I}$ of intervals in $\mathbb{P}$ and a function $r : \mathcal{I} \to \mathbb{Z}$. A (signed) rank decomposition of r over $\mathcal{I}$ is expressed as:

$$r = \mathrm{Rk}_{\mathcal{I}}\, \mathbf{k}_{\mathcal{R}} - \mathrm{Rk}_{\mathcal{I}}\, \mathbf{k}_{\mathcal{S}},$$

where $\mathcal{R}$ and $\mathcal{S}$ are multi-sets of elements selected from $\mathcal{I}$, and $\mathbf{k}_{\mathcal{R}}$ and $\mathbf{k}_{\mathcal{S}}$ are elements in $\mathrm{rep}_{\mathcal{I}}\, \mathbb{P}$. Specifically, $\mathbf{k}_{\mathcal{R}} = \bigoplus_{R \in \mathcal{R}} \mathbf{k}_R$ and $\mathbf{k}_{\mathcal{S}} = \bigoplus_{S \in \mathcal{S}} \mathbf{k}_S$ where elements $R \in \mathcal{R}$ and $S \in \mathcal{S}$ are included with their multiplicities in the direct sums. The pair $(\mathcal{R}, \mathcal{S})$ is referred to as a rank decomposition of r over $\mathcal{I}$. If $\mathcal{R}$ and $\mathcal{S}$ are disjoint as multisets, the decomposition is considered minimal.

THEOREM 4.2.2 ([**BOO24**] [Existence]). *Let $\mathcal{I}$ represent a locally finite collection of intervals in $\mathbb{P}$. Any function $r : \mathcal{I} \to \mathbb{Z}$ with locally finite support can be uniquely expressed as a (possibly infinite, but pointwise finite) $\mathbb{Z}$-linear combination of the functions $\mathrm{Rk}_I\, \mathbf{k}_I$, where $I \in \mathcal{I}$.*

A significant implication of the property $\mathrm{Rk}_{\mathcal{I}}\, \mathbf{k}_R(I) = \mathbb{1}_{R \supseteq I}$ is that adding an equal number of copies of an interval $I \in \mathcal{I}$ to both $\mathcal{R}$ and $\mathcal{S}$ does not affect the difference $\mathrm{Rk}_{\mathcal{I}}\, \mathbf{k}_{\mathcal{R}} - \mathrm{Rk}_{\mathcal{I}}\, \mathbf{k}_{\mathcal{S}}$. Consequently, rank decompositions cannot be unique. This leads to the concept of minimal rank decompositions, which require the multisets $\mathcal{R}, \mathcal{S}$ to be disjoint. Under this condition, the signed rank decomposition can be shown to be unique.

COROLLARY 4.2.3 (Uniqueness). *For any finite poset $\mathbb{P}$ and any map*

$$r : \{\leq_{\mathbb{P}}\} \to \mathbb{Z} \text{ where } \{\leq_{\mathbb{P}}\} = \{(a,b) \in \mathbb{P} \times \mathbb{P} \mid a \leq b\}$$

there exists a unique pair $\mathcal{R}, \mathcal{S}$ of disjoint finite multi-sets of closed segments such that

$$r = \mathbf{Rk}\, \mathbf{k}_{\mathcal{R}} - \mathbf{Rk}\, \mathbf{k}_{\mathcal{S}}.$$

An application to rank decompositions. Here, $\operatorname{mult}_I \mathcal{R}$ denotes the multiplicity of the element I in a given multi-set $\mathcal{R}$. In the context of locally finite settings, the multiplicities of elements in the minimal rank decomposition of a map $r : \mathcal{I} \to \mathbb{Z}$ can be determined using an explicit inclusion-exclusion formula, which essentially corresponds to computing the Möbius inverse of r.

PROPOSITION 4.2.4 ([**BOO24**]). *Suppose $\mathcal{I}$ is a collection of locally finite intervals in $\mathbb{P}$, and $r : \mathcal{I} \to \mathbb{Z}$ is a function with locally finite support. A pair of pointwise finite multi-sets $(\mathcal{R}, \mathcal{S})$ of elements in $\mathcal{I}$ forms a rank decomposition of r if and only if the condition*

$$(r * \mu)(I) \doteq \sum_{J \supseteq I} r(J)\mu(J, I) = \operatorname{mult}_I \mathcal{R} - \operatorname{mult}_I \mathcal{S} \quad \forall I \in \mathcal{I}.$$

In particular, the minimal rank decomposition of r is constructed as follows: I^n represents a multi-set containing n copies of the interval I,and

$$\mathcal{R} = \bigsqcup_{I \in \mathcal{I}, (r*\mu)(I) > 0} I^{(r*\mu)(I)} \quad \text{and} \quad \mathcal{S} = \bigsqcup_{I \in \mathcal{I}, (r*\mu)(I) < 0} I^{(r*\mu)(I)}.$$

4.3. The finite grid case. In this situation, the poset is a finite grid $G = \prod_{i=1}^d [[1, n_i]]$, where $[[1, n_i]]$ denotes the set $\{i \in \mathbb{Z} | 1 \le i \le n_i\}$. We provide explicit formulas for computing signed rank decompositions. Theorem 4.2.2 is reformulated as follows:.

COROLLARY 4.3.1. *Let $\mathcal{I}$ represent any arbitrary collection of intervals in a finite grid $G = \prod_{i=1}^d [[1, n_i]] \subset \mathbb{R}^d$. The generalized rank invariant $\operatorname{Rk}_{\mathcal{I}} M$ of any pfd persistence module M defined over G has a unique minimal rank decomposition $(\mathcal{R}, \mathcal{S})$ over $\mathcal{I}$.*

By defining $\mathcal{I}$ as the collection of all closed rectangles in the grid G, we obtain the following reformulation of Corollary 4.2.3.

COROLLARY 4.3.2. *The usual rank invariant of any pfd persistence module M indexed over a finite grid $G = \prod_{i=1}^d [[1, n_i]] \subset \mathbb{R}^d$ has a unique minimal rank decomposition $(\mathcal{R}, \mathcal{S})$, where $\mathcal{R}$ and $\mathcal{S}$ are finite multi-sets of (closed) rectangles in G.*

To compute the minimal rank decompositions, we can use the formula from Proposition 4.2.4, which is reformulated as follows for the case of the usual rank invariant of a pfd persistence module M indexed over a finite grid $G = \prod_{i=1}^d [[1, n_i]] \subset \mathbb{R}^d$:

$$\forall s \le t \in G, \quad \alpha_{(s,t)} = \sum_{\substack{s' \le s \\ \|s'-s\|_\infty \le 1}} \sum_{\substack{t' \ge t \\ \|t'-t\|_\infty \le 1}} (-1)^{\|s'-s\|_1 + \|t'-t\|_1} \operatorname{Rk} M(s', t'). \tag{4.1}$$

In the case $d = 1$, the well-known inclusion-exclusion formula relates the persistence diagram of a one-parameter persistence module to its rank invariant [**CSEH05**]. For $d = 2$, the inclusion-exclusion formula is used to compute the multiplicities of summands in rectangle-decomposable two-parameter persistence modules [**BLO22**].

In this section, we concentrate on the usual rank invariant, where the associated intervals in the ground poset $\mathbb{P}$ are segments. We introduce an exact structure on the category of pointwise finite-dimensional representations of $\mathbb{P}$ by examining sequences that preserve ranks. This provides a categorical enhancement of Theorem 4.2.2, demonstrating that, for any given representation, the same rank invariant can be expressed as the difference of two multi-sets of segments.

DEFINITION 4.3.3. A short exact sequence $0 \to A \to B \to C \to 0$ in $\operatorname{rep}\mathbb{P}$ is called *rank-exact* if $\operatorname{Rk} B = \operatorname{Rk} A + \operatorname{Rk} C$. The collection of rank-exact sequences is denoted by $\mathcal{E}_{\mathrm{Rk}}$.

DEFINITION 4.3.4. Let $\mathbb{P}$ be a poset. A subset $A \subseteq \mathbb{P}$ is called an *antichain* if no two distinct elements of A are related. If $A = \{a\}$ is a singleton, we often denote $A = a$. For two antichains A and B, we write $A \leq B$ if, for every $a \in A$, there exists $b_a \in B$ such that $a \leq b_a$, and for every $b \in B$, there exists $a_b \in A$ such that $a_b \leq b$. Denote:

(1) $\langle A, \infty\rangle = \{c \in \mathbb{P} \mid A \leq c\}$. These subsets are referred to as closed upsets
(2) $\langle A, B\rangle = \{c \in \mathbb{P} \mid A \leq c \leq B\}$. If $|A| = 1 = |B|$, $\langle A, B\rangle$ is called a segment.
(3) $\langle A, B\langle = \{c \in \mathbb{P} \mid A \leq c \not\leq B\}$. If $|A| = 1 = |B|$, $\langle A, B\langle$ is called a hook.
(4) $\rangle A, B\rangle = \{c \in \mathbb{P} \mid A \not\leq c \leq B\}$. If $|A| = 1 = |B|$, $\rangle A, B\rangle$ is called a cohook.

THEOREM 4.3.5 ([**BOO24**]). *The collection $\mathcal{E}_{\mathrm{Rk}}$ of rank-exact sequences establishes an exact category structure on* $\operatorname{rep}\mathbb{P}$. *In this exact category, the projective objects are direct sums of objects of the form $\mathbf{k}_{\langle i,j\langle}$, where $i < j \in \mathbb{P} \cup \{\infty\}$. Dually, the injective objects are direct sums of objects of the form $\mathbf{k}_{\rangle i,j\rangle}$, where $i < j \in \mathbb{P} \cup \{-\infty\}$.*

THEOREM 4.3.6 ([**BOO24**]). *Let $\mathbb{P}$ be a finite poset. Then $\mathcal{E}_{\mathrm{Rk}}$ contains enough projective and injective objects, and every object in $\mathcal{E}_{\mathrm{Rk}}$ admits both a finite projective resolution and a finite injective resolution. The map*

$$\operatorname{Rk} : K_0(\mathcal{E}_{\mathrm{Rk}}) \to \mathbb{Z}^{\{\leq \mathbb{P}\}}$$

is an isomorphism.

Moreover, the following three sets form bases for $K_0(\mathcal{E}_{\mathrm{Rk}})$.

$\{[k_{\langle i,j\rangle}] \mid i \leq j \in \mathbb{P}\}$, $\{[k_{\langle i,j\langle}] \mid i < j \in \mathbb{P} \cup \{\infty\}\}$, *and* $\{[k_{\rangle i,j\rangle}] \mid i < j \in \mathbb{P} \cup \{-\infty\}\}$.

Theorem 4.3.6 can be regarded as a categorical refinement of Theorem 4.2.2. It states that any vector in $\mathbb{Z}^{\{\leq \mathbb{P}\}}$ can be uniquely expressed as a $\mathbb{Z}$-linear combination of the ranks of segment representations. Additionally, it reveals that this uniqueness arises from an isomorphism to the Grothendieck group of the exact category.

The existence of signed rank decompositions implies that every object in the category has both a finite projective resolution and a finite injective resolution. Furthermore, the uniqueness of signed rank decompositions corresponds to the existence of a minimal relative-projective resolution.

This result also demonstrates that two representations cannot "coincidentally" share the same rank invariant. If two representations A and B do share the same rank invariant, then $[A] = [B]$ in $K_0(\mathcal{E}_{\mathrm{Rk}})$.

5. Homological invariants

5.1. Some background on relative homological algebras. This section revisits the concepts of covers, resolutions, and Grothendieck groups within the context of an exact structure. The relevant symbols and concepts can be found in Appendix 7.2. Our discussion is primarily based on the framework presented in

[**BBH24**, Section 4.1], but we restrict our focus to the category $\mathbb{P}$-Mod, where $\mathbb{P}$ is an arbitrary poset. Additionally, we fix a class $\mathcal{X} \subseteq \mathbb{P}$-Mod such that $\text{proj}(\mathbb{P}\text{-Mod}) \subseteq \text{add}(\mathcal{X})$.

DEFINITION 5.1.1. Let $M \in \mathbb{P}$-Mod. An *add*($\mathcal{X}$)*-precover* (also referred to as a *right add*($\mathcal{X}$)*-approximation*) of M is an object $U \in \text{add}(\mathcal{X})$ along with an $\mathcal{F}_{\mathcal{X}}$-admissible epimorphism $f : U \to M$.

The map f is called an *add*($\mathcal{X}$)*-cover* (or a *minimal right add*($\mathcal{X}$)*-approximation*) of M if, in addition, every endomorphism $g : U \to U$ such that $f = f \circ g$ is an automorphism.

We say that add($\mathcal{X}$) is *precovering* if every object in $\text{rep}\,\mathbb{P}$ admits an *add*($\mathcal{X}$)*-precover.* Furthermore, if $\text{add}(\mathcal{X}) = \text{proj}(\mathcal{F}_{\mathcal{X}})$, we say that $\mathcal{F}_{\mathcal{X}}$ has enough projectives.

REMARK 5.1.2. The category $\mathbb{P}$-Mod has *add*($\mathcal{X}$)-precovers for any finite set $\mathcal{X}$.

Since $\mathbb{P}$-Mod is Krull-Remak-Schmidt-Azumaya, the following is well-known.

LEMMA 5.1.3. *Let $M \in \mathbb{P}$-Mod. Then M has an* add*($\mathcal{X}$)-cover if and only if M admits an* add*($\mathcal{X}$)-precover. Furthermore, if $f : U \to M$ and $g : U \to M$ are both* add*($\mathcal{X}$)-covers of M, then there exists an isomorphism $h : U \to V$ such that $f = g \circ h$.*

By definition, the standard exact structure $\mathcal{E}_{\max} = \mathcal{F}_{\text{proj}(\mathbb{P}\text{-Mod})}$ on $\mathbb{P}$-Mod has enough projectives. In this setting, any epimorphism from an object in $\text{proj}(\mathbb{P}\text{-Mod})$ to M is a $\text{proj}(\mathbb{P}\text{-Mod})$-precover. However, for other choices of $\mathcal{X}$, there may exist epimorphisms from *add*($\mathcal{X}$) to M that are not $\mathcal{F}_{\mathcal{X}}$-admissible, see e.g. [**BBH24**, Example 4.3(2)] .

We now present the following definitions, each of which corresponds to the classical notions from homological algebra when $\text{add}(\mathcal{X}) = \text{proj}(\mathbb{P}\text{-Mod})$.

DEFINITION 5.1.4. Assume that add($\mathcal{X}$) is precovering and let $M \in \mathbb{P}$-Mod.

(1) An *add*($\mathcal{X}$)*-resolution* of M is an $\mathcal{F}_{\mathcal{X}}$-exact sequence of the form $U_\bullet = (\cdots \to U_2 \overset{q_2}{\to} U_1 \overset{q_1}{\to} U_0)$ such that $M = \text{coker}(q_1)$. We call $U_\bullet$ a *minimal add*($\mathcal{X}$)*-resolution* if all of the q_i and the quotient map $U_0 \twoheadrightarrow M$ are *add*($\mathcal{X}$)-covers of their respective images. Further, $U_\bullet$ is *finite* if there exists some j such that $U_i = 0$ for all $i > j$.
(2) Let $U_\bullet$ be a minimal *add*($\mathcal{X}$)-resolution of M. The *add*($\mathcal{X}$)*-dimension* of M is
$$\dim_{\mathcal{X}}(M) = \min\{j \mid U_{j+1} = 0\} \cup \{\infty\}.$$
(3) The *add*($\mathcal{X}$)*-global dimension* of $I(\mathbb{P})$ is given by
$$\dim_{\mathcal{X}}(k\mathbb{P}) = \sup_{M \in \mathbb{P}\text{-Mod}} \dim_{\mathcal{X}}(M).$$
Moreover, the *add*($\mathcal{X}$)-global dimension of $k\mathbb{P}$ is *realisably-infinite* if there exists $M \in \mathbb{P}$-Mod such that $\dim_{\mathcal{X}}(M) = \infty$.

REMARK 5.1.5. Note that if $\mathbb{P}$ is finite, then the *add*($\mathcal{X}$)-global dimension of $I(\mathbb{P})$ is infinite if and only if it is realisably infinite.

DEFINITION 5.1.6. Let F be the free abelian group generated by the symbols $[M]$ for each isomorphism class of $M \in \mathbb{P}$-Mod. The subgroup $H_{\mathcal{X}}$ is defined as being generated by:

$$\{[M] - [N] + [L] \mid M \to N \to L \text{ is } \mathcal{F}_{\mathcal{X}}\text{-exact}\}.$$

The quotient $K_0(\mathbb{P}, \mathcal{X}) = F/H_{\mathcal{X}}$ is referred to as the *Grothendieck group* of $\mathbb{P}$ relative to $\mathcal{X}$. For $M \in \mathbb{P}$-Mod, we denote by $[M]_{\mathcal{X}}$ (or just $[M]$ when $\mathcal{X}$ is clear from context) the representative of M in the Grothendieck group $K_0(\mathbb{P}, \mathcal{X})$. The representative of M in the *Grothendieck group* $K_0(\mathbb{P}, \mathcal{X})$ is denoted as $[M]_{\mathcal{X}}$, or simply $[M]$ if $\mathcal{X}$ is clear from the context.

The proof of [**BBH24**, Proposition 4.9] can be readily generalized to establish the following result.

PROPOSITION 5.1.7. *Assume that* add$(\mathcal{X})$ *is precovering and that the* add$(\mathcal{X})$*-global dimension of* $I(\mathbb{P})$ *is not realisably infinite. Then:*

(1) *The group* $K_0(\mathbb{P}, \mathcal{X})$ *is a free abelian group with a basis given by* $\{[U] \mid U \in \text{ind}(\text{add}(\mathcal{X}))\}$
(2) *For any* $M \in \mathbb{P}$-Mod *and any finite* add$(\mathcal{X})$*-resolution* $U_\bullet$ *of* M, *the following holds:*

$$[M]_{\mathcal{X}} = \sum_{i=0}^{\infty} (-1)^i [U_i]_{\mathcal{X}}.$$

DEFINITION 5.1.8. Let $\mathcal{C} \subseteq \mathbb{P}$-Mod be an additive subcategory, and let $K_0^{\text{split}}(\mathcal{C})$ denote the free abelian group with a basis ind$(\mathcal{C})$.

(1) An *additive invariant*(referred to simply as an *invariant*) on $\mathcal{C}$ is a group homomorphism $p : K_0^{\text{split}}(\mathcal{C}) \to G$, where G is an abelian group. Alternatively, it can be described as a function $p : \mathcal{C} \to G$ that is constant on isomorphism classes and satisfies $p(M \oplus N) = p(M) + p(N)$.
(2) Let $p : K_0^{\text{split}}(\mathcal{C}) \to G$ and $p' : K_0^{\text{split}}(\mathcal{C}) \to G'$ be two additive invariants. We say that p is *equivalent* to p' if $\ker(p) = \ker(p')$ and that p is *finer* than p' if $\ker(p) \subsetneq \ker(p')$.
(3) An additive invariant $p : K_0^{\text{split}}(\mathcal{C}) \to G$ is said to be *complete* on $\mathcal{C}$ if $\ker(p) = 0$. Alternatively, p is equivalent to the identity on $K_0^{\text{split}}(\mathcal{C})$, and therefore $p(M) = p(N)$ if and only if $M \cong N$.

DEFINITION 5.1.9. Let $p : K_0^{\text{split}}(\mathcal{C}) \to \mathbb{Z}^n$ be an invariant, and let $\mathcal{X}$ be a finite set of indecomposable modules.

(1) p is called a *dim-hom invariant relative to* $\mathcal{X}$ if it is equivalent to the invariant defined by the map:

$$K_0^{\text{split}}(\mathcal{C}) \to \mathbb{Z}^{|\mathcal{X}|}, \quad M \mapsto (\dim \text{Hom}_\Lambda(R, M))_{R \in \mathcal{X}}.$$

(2) p is said to be *homological relative to* $\mathcal{X}$ if all the following conditions are satisfied:
   (a) All indecomposable projective modules are in $\mathcal{X}$.
   (b) The add$(\mathcal{X})$-dimension of $\mathcal{C}$ is not realisably infinite.
   (c) p is equivalent to the invariant given by the quotient map

$$K_0^{\text{split}}(\mathcal{C}) \to K_0(\mathcal{C}, \mathcal{X}) \cong \mathbb{Z}^{|\mathcal{X}|}.$$

(3) p is referred to as a *dim-hom invariant* (resp. *homological invariant*) if it is a dim-hom (resp. homological) invariant relative to some $\mathcal{X}$.

REMARK 5.1.10. The association $M \mapsto [M]_{\mathcal{X}}$ can be interpreted as a method of "collapsing" a minimal $add(\mathcal{X})$-resolution of M into an invariant. There are several alternative ways to achieve this, with one notable method being the use of relative Betti numbers. This approach captures the direct sum decomposition of each term in a minimal $add(\mathcal{X})$-resolution and keeps them separate, instead of combining them into an alternating sum. For further discussion on this approach, see [**CGR**$^+$**25**].

EXAMPLE 5.1.11.

(1) If $\mathbb{P}$-Mod has finite global dimension, the dimension vector is a homological invariant relative to $\mathcal{X} = \text{ind}(\text{proj}(\mathbb{P}\text{-Mod}))$.
(2) It is shown in [**BOO24**, Section 4] that the rank invariant is a homological invariant relative to $\mathcal{X}$, which is the union of $\text{ind}(\text{proj}(\mathbb{P}\text{-Mod}))$ and the set of hook modules. Additionally, it is demonstrated that the exact structure $\mathcal{F}_{\mathcal{X}}$ is composed precisely of those exact sequences over which the association $M \mapsto \text{rk}(M)$ is additive.
(3) If $\mathbb{P}$ is finite and:
    (i) $\mathcal{X}$ consists only of single-source spread modules, or
    (ii) $\mathcal{X}$ consists only of upset modules,
    then the association $M \mapsto [M]_{\mathcal{X}}$ is a homological invariant relative to $\mathcal{X}$. See also [**Mil25**, Theorem 6.12], which explores $\mathcal{X}$ as the set of all upset modules over arbitrary posets and a broader category than $\mathbb{P}$-Mod.
(4) Let $\mathbb{P} = \{1, \dots, n\} \times \{1, \dots, m\}$ be the product of two finite totally ordered sets, and let $\mathcal{X}$ be the set of interval modules. Then the association $M \mapsto [M]_{\mathcal{X}}$ is a homological invariant relative to $\mathcal{X}$, as shown in [**AENY23a**, Proposition 4.5].

While we have examples of homological invariants, the natural question is to identify precisely which sets of interval modules yield homological invariants.

THEOREM 5.1.12 ([**BBH24**]). *Let $\mathbb{P}$ be a finite poset.*

(1) *Let $\mathcal{X}$ be a set of single-source interval modules over $I(\mathbb{P})$ that includes the indecomposable projectives. Then, the quotient map*

$$K_0^{split}(I(\mathbb{P})) \to K_0(I(\mathbb{P}), \mathcal{X})$$

is a homological invariant.
(2) *Let $\mathcal{X}$ be a set of connected upset modules over $I(\mathbb{P})$ that includes the indecomposable projectives. Then, the quotient map*

$$K_0^{split}(I(\mathbb{P})) \to K_0(I(\mathbb{P}), \mathcal{X})$$

is a homological invariant.

Unfortunately, for arbitrary posets, it is not possible to generalize Theorem 5.1.12 to arbitrary choices of $\mathcal{X}$, as the following example demonstrates.

EXAMPLE 5.1.13. Let $\mathbb{P}$ be the poset represented by the given Hasse diagram.

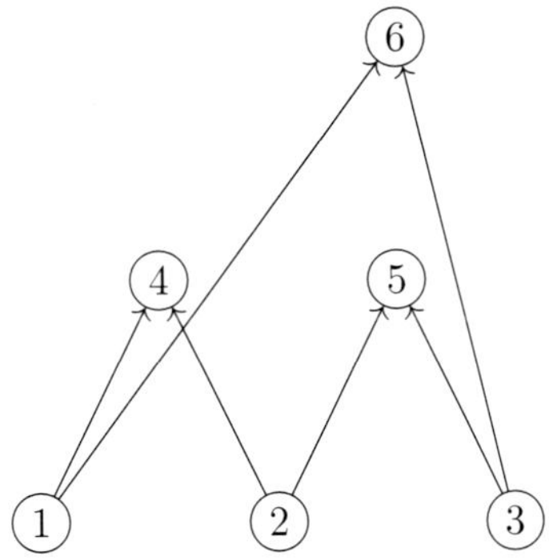

Note that the incidence algebra $I(\mathbb{P})$ in this case is a path algebra of type $\tilde{A}_5$. Now define:

$$\mathcal{X} = \{\text{indecomposable projectives}\} \cup \{M_{[\{1,3\},\{5,6\}]}, M_{[\{1,2\},\{4,6\}]}, M_{[\{2,3\},\{4,5\}]}\}.$$

The Gabriel quiver of $\Gamma \triangleq \mathrm{End}_{I(\mathcal{P})}(T)^{\mathrm{op}}$ is then as follows:

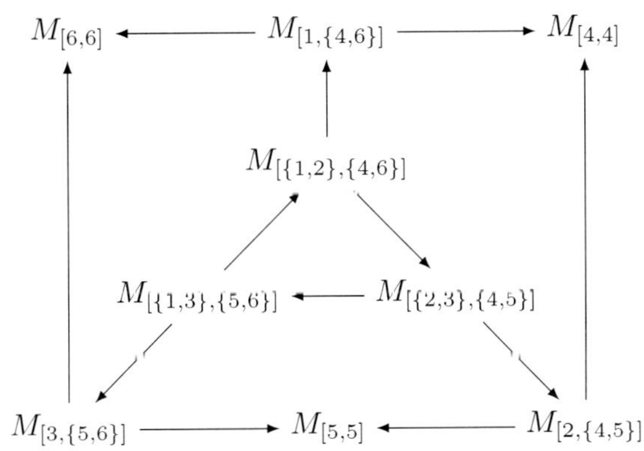

The oriented 3-cycle at the center of the quiver satisfies the condition that the composition of any two arrows is zero. Moreover, the three non-oriented pentagons (e.g., from $M_{[\{1,3\},\{5,6\}]}$ to $M_{[6,6]}$) are subject to commutative relations. We claim that Γ has infinite global dimension. To verify this, consider a interval module $M_{[A,B]} \in \mathcal{X}$. Denote the projective Γ-module corresponding to the vertex labeled by $M_{[A,B]}$ as $P_{[A,B]}$. Now, let N represent the indecomposable Γ-module supported at the vertices associated with $M_{[\{1,3\},\{5,6\}]}$, $M_{[3,\{5,6\}]}$, and $M_{[5,5]}$. In this setup, N has a minimal projective resolution:

$$\cdots \to P_{[\{1,2\},\{4,6\}]} \to P_{[\{1,3\},\{5,6\}]} \to P_{[\{2,3\},\{4,5\}]} \to P_{[\{1,2\},\{4,6\}]} \to N \to 0.$$

The kernel of the first map is the Γ-module supported at the vertices labeled by $M_{[\{1,2\},\{4,6\}]}$, $M_{[1,\{4,6\}]}$, and $M_{[6,6]}$. Given the symmetry of the quiver, it is evident that this constitutes the projective resolution of N.

Furthermore, it is clear that the $\mathcal{X}$-dimension of $\mathrm{mod} I(\mathcal{P})$ with respect to $\mathcal{X}$ is infinite. Indeed, the following can be readily identified as a minimal $\mathcal{X}$-resolution of the interval representation $M_{[1,6]}$:

$$\cdots \to M_{[\{1,2\},\{4,6\}]} \to M_{[\{1,3\},\{5,6\}]} \to M_{[\{2,3\},\{4,5\}]} \to M_{[\{1,2\},\{4,6\}]} \to M_{[1,6]} \to 0.$$

Although Example 5.1.13 demonstrates that Theorem 5.1.12 cannot be fully generalized, it naturally leads to the following question.

QUESTION 5.1.14. Let $\mathbb{P}$ be any finite poset, and let $\mathcal{X}$ be the set of all connected interval modules. Is the quotient map

$$K_0^{\text{split}}(I(\mathbb{P})) \to K_0(I(\mathbb{P}), \mathcal{X})$$

a homological invariant?

REMARK 5.1.15. This question has been positively answered by Asashiba, Escolar, Nakashima, and Yoshiwaki in [**AENY23a**]. Their proof relies on the well-known result that for any finite-dimensional algebra A and any generator-cogenerator $G \in \operatorname{mod} A$, there exists G' such that

(i) $\operatorname{End}_A(G \oplus G')^{op}$ has finite global dimension and
(ii) every indecomposable direct summand of G' is a submodule of an indecomposable direct summand of G.

Using this, they show that if $\mathcal{X}$ is the set of all connected intervals, then every submodule of an object in $\mathcal{X}$ lies in $\operatorname{add}(\mathcal{X})$.

Finally, we conclude with a brief discussion of signed decompositions (see Definition 4.2.1) of homological invariants. The following result immediately follows from Proposition 5.1.7

COROLLARY 5.1.16. *Suppose $p : K_0^{split}(\mathbb{P}\text{-Mod}) \to G$ is a homological invariant relative to $\mathcal{X}$. Then p is complete on* $\operatorname{add}(\mathcal{X})$. *Furthermore, for $M \in \mathbb{P}\text{-Mod}$, write*

$$[M]_{\mathcal{X}} = \sum_{X \in \mathcal{X}} m_X \cdot [X]_{\mathcal{X}}$$

and define:

$$M_{\mathcal{X}}^{+} = \bigoplus_{X \in \mathcal{X}} X^{\max\{0, m_X\}}, \quad M_{\mathcal{X}}^{-} = \bigoplus_{X \in \mathcal{X}} X^{\max\{0, -m_X\}}.$$

Then $(M_{\mathcal{X}}^{+}, M_{\mathcal{X}}^{-})$ is the unique minimal signed ($\operatorname{add}(\mathcal{X})$*)-decomposition of p.*

REMARK 5.1.17. While the above corollary guarantees the uniqueness and provides a method for computing minimal signed decompositions of homological invariants, it may still be useful to explore other signed decompositions. For example, in [**OS24**, Section 8.3] and [**BOOS24**, Sections 6 and 7], the use of non-minimal signed decompositions is necessary. Such decompositions can still be derived from minimal *add*$(\mathcal{X})$-resolutions (and consequently from relative Betti numbers).

5.2. Homoligical invariants in persistence theory. We discuss examples of homological invariants for persistence modules over certain incidence algebras, providing many concrete examples.

PROPOSITION 5.2.1. *Let $M \in \operatorname{mod} I(\mathbb{P})$. Then the following are equivalent.*

(1) *M is a single-source interval module.*
(2) *There exists an indecomposable projective $P \in \operatorname{mod} I(\mathbb{P})$ and a (possibly empty) set $\mathcal{Q}$ of proper projective submodules of P such that $M \cong P/(\sum_{Q \in \mathcal{Q}} Q)$.*
(3) *The projective cover of M is indecomposable.*

We now examine the relationship between homological invariants and other invariants in persistence theory. It is well known that the dimension vector is a homological invariant relative to the set of projective modules. We focus on the barcode, the rank invariant, the generalized rank invariant, the signed barcode, and the generalized persistence diagrams.

The barcode. We assume that the Hasse quiver of $\mathbb{P} = (\mathbb{P}, \preceq)$ is Dynkin type A.

PROPOSITION 5.2.2. *Suppose that the Hasse quiver of $\mathbb{P}$ is Dynkin type A. Then the barcode*

$$\text{bar} : K_0^{split} \to \mathbb{Z}^{\binom{|\mathbb{P}|}{2}}$$

is a homological invariant.

The rank invariant. We now focus on $\mathbb{P}$, which denotes an arbitrary finite poset.

An extra element ∞ is appended to $\mathbb{P}$, with $a < \infty$ for all $a \in \mathbb{P}$. For any $a < b \in \mathbb{P} \cup \{\infty\}$, the persistence module $N_{\langle a,b\rangle}$, defined by

$$N_{\langle a,b\rangle}(c) = \begin{cases} K & a \leq c \text{ and } b \nleq c \\ 0 & \text{otherwise} \end{cases}$$

and

$$N_{\langle a,b\rangle}(c,d) = \begin{cases} 1_K & N_{\langle a,b\rangle}(c) = K = N_{\langle a,b\rangle}(d) \\ 0 & \text{otherwise,} \end{cases}$$

is called the hook module from a to b. Note that, if $b \neq \infty$, then $N_{\langle a,b\rangle}$ corresponds to the quotient of the projective at a by the projective at b. Likewise, if $b = \infty$, then $N_{\langle a,\infty\rangle}$ is simply the projective at a. In particular, hook modules are single-source interval modules

One of the main results of [**BOO24**], specialized to finite posets and restated in the context of this paper, is stated below.

THEOREM 5.2.3 ([**BOO24**], Theorems 4.4, 4.8 and 4.10). *Let $\mathbb{P}$ be a finite poset and let $\mathcal{H}$ be the set of hook modules in* mod $I(\mathbb{P})$. *Then the rank invariant is homological with respect to $\mathcal{H}$.*

REMARK 5.2.4. One outcome of this result, explicitly stated in [**BOO24**, Theorem 4.4] is that the exact sequences $0 \to L \to M \to N \to 0$ in $\mathcal{E}_{\mathcal{H}}$ are characterized by the property that $\text{rk}(M) = \text{rk}(L) + \text{rk}(N)$. In other words, the set of exact sequences on which $\text{rk}(-)$ is additive form an exact structure, and this exact structure is precisely $\mathcal{E}_{\mathcal{H}}$.

The single-source spread invariant is finer than the rank invariant.

THEOREM 5.2.5. *Let $\mathbb{P}$ be a finite poset. Then the single-source homological spread invariant is finer than the rank invariant on* mod $I(\mathbb{P})$. *Furthermore, these invariants are equivalent if and only if for all $a \in \mathbb{P}$ the upset $[a,\infty]$ is totally ordered.*

A detailed explanation is provided to illustrate why the signed barcode introduced in [**BOO24**] and the homological invariant relative to $\mathcal{I}$ are not equivalent.

EXAMPLE 5.2.6 ([**BOO24**, Example 7.6]). Let $\mathbb{P}$ be a finite poset which contains a unique maximal element. Let $\mathcal{H}$ represent the set of hook modules, and let $\mathcal{I}$ denote the set of interval modules. Since $\mathbb{P}$ has a unique maximal element, we note that $\mathcal{I}$ contains all indecomposable projectives. As every interval is a single-source spread, the quotient map

$$K_0^{\text{split}}(\mathcal{I}(\mathbb{P})) \to K_0(\mathcal{I}(\mathbb{P}))$$

is therefore a homological invariant by Theorem 5.1.12(1). Additionally, the quotient map

$$K_0^{\text{split}}(\mathcal{I}(\mathbb{P})) \to K_0(\mathcal{I}(\mathbb{P}), \mathcal{H})$$

corresponds to the rank invariant and is also homological by Theorem 5.2.3.

While these two invariants share the same rank (determined by the number of pairs $a \leq b \in \mathcal{P}$), they are not equivalent. For instance, consider $\mathbb{P} = \{(0,0), (0,1), (1,0), (1,1)\}$, which forms a 2×2 grid. We proceed by analyzing the modules.

$$\begin{array}{ccc} K & \longrightarrow & 0 \\ {\scriptstyle (1\ 0)}\uparrow & & \uparrow \\ K^2 & \xrightarrow{(0\ 1)} & K \end{array} \qquad\qquad \begin{array}{ccc} K & \longrightarrow & 0 \\ {\scriptstyle (1\ 0)}\uparrow & & \uparrow \\ K^2 & \xrightarrow{(1\ 0)} & K \end{array}$$

$$M \qquad\qquad\qquad\qquad M'$$

We observe that $\underline{dim}(M) = \underline{dim}(M')$ and $\underline{\mathrm{rk}}(M) = \underline{\mathrm{rk}}(M')$. However, M is a direct sum of interval modules, while M' has an $\mathcal{I}$-resolution given by:

$$0 \to M_{[(1,1),(1,1)]} \to M_{[(0,0),(1,1)]} \oplus M_{[(0,0),(0,0)]} \to M' \to 0.$$

Since this resolution has a non-zero term in degree 1, this means $[M]_{\mathcal{I}} \neq [M']_{\mathcal{I}}$.

REMARK 5.2.7 ([**BOO24**, Remark 7.7]). Let $\mathcal{H}$ and $\mathcal{I}$ be as defined in Example 5.2.6. Denote the two quotient maps by

$$p : K_0^{\text{split}}(\mathcal{I}(\mathbb{P})) \to K_0(\mathcal{I}(\mathbb{P}), \mathcal{H}) \quad \text{and} \quad q : K_0^{\text{split}}(\mathcal{I}(\mathbb{P})) \to K_0(\mathcal{I}(\mathbb{P}), \mathcal{I})$$

Recall that the signed barcode of a persistence module M is defined in [**BOO24**] by applying Möbius inversion to the classical rank invariant. The resulting invariant is contained in $K_0(\mathcal{I}(\mathbb{P}), \mathcal{I})$ and is equivalent to p. More specifically, there exists an isomorphism

$$\sigma : K_0(\mathcal{I}(\mathbb{P}), \mathcal{H}) \to K_0(\mathcal{I}(\mathbb{P}), \mathcal{I})$$

such that $\sigma \circ p(M)$ represents the signed barcode of M. Determining whether the signed barcode and the homological invariant relative to $\mathcal{I}$ coincide is equivalent to asking whether $\sigma \circ p = q$.

However, if this equality were true, p and q would be equivalent, contradicting Example 5.2.6. In other words, there exists a persistence module M such that $[M]_{\mathcal{I}}$ and the signed barcode of M do not coincide.

Generalized rank invariant. Again $\mathbb{P}$ denotes an arbitrary finite poset in this section. Given a set $\mathcal{I}$ of connected interval modules, we recall the "$\mathcal{I}$-rank invariant" $\mathrm{Rk}_{\mathcal{I}}$ in Definition 4.1.1.

EXAMPLE 5.2.8 ([**BOO24**, Proposition 7.8, Remark 7.9]). Let $\mathbb{P} = \{1,2\} \times \{1,2,3\}$ represent the 2×3 grid. Let $\mathcal{R}$ denote the set of connected intervals in $\mathbb{P}$, and let $\mathcal{X}$ represent the set of connected interval modules over $I(\mathbb{P})$. We assert that the generalized rank invariant $\underline{\mathrm{rk}}(-,\mathcal{R})$ is not a dim-hom invariant relative to $\mathcal{X}$.

Proof: Let M be the module defined as follows:

$$\begin{array}{ccccc} K & \xrightarrow{1} & K & \longrightarrow & 0 \\ {\scriptstyle 1}\uparrow & & \uparrow{\scriptstyle (1\ 0)} & & \uparrow \\ K & \xrightarrow{\binom{1}{1}} & K^2 & \xrightarrow{(0\ 1)} & K \end{array}$$

Over this poset, indecomposable modules are uniquely determined by their dimension vectors. For instance, the module M can be represented by the vector: $\begin{bmatrix} 1 & 1 & 0 \\ 1 & 2 & 1 \end{bmatrix}$

To derive a contradiction, we compute the "generalized persistence diagram" obtained by applying Möbius inversion to the invariant $\mathrm{rk}(M,\mathcal{R})$. According to [**KM21**, Proposition 3.19] or [**AENY23b**, Definition 5.9], this computation yields the following.

$$\delta(M,\mathcal{R}) = \begin{bmatrix} 1 & 0 & 0 \\ 1 & 1 & 1 \end{bmatrix} + \begin{bmatrix} 1 & 1 & 0 \\ 1 & 1 & 0 \end{bmatrix} + \begin{bmatrix} 0 & 0 & 0 \\ 0 & 1 & 0 \end{bmatrix} - \begin{bmatrix} 1 & 0 & 0 \\ 1 & 1 & 0 \end{bmatrix}.$$

Now denote

$$N := M \oplus \begin{bmatrix} 1 & 0 & 0 \\ 1 & 1 & 0 \end{bmatrix}, \quad L := \begin{bmatrix} 1 & 0 & 0 \\ 1 & 1 & 1 \end{bmatrix} \oplus \begin{bmatrix} 1 & 1 & 0 \\ 1 & 1 & 0 \end{bmatrix} \oplus \begin{bmatrix} 0 & 0 & 0 \\ 0 & 1 & 0 \end{bmatrix}.$$

Recalling that $\delta(X,\mathcal{R}) = X$ whenever X is a direct sum of spread modules (see [**KM21**, Theorem 3.14]), we conclude that $\delta(N,\mathcal{R}) = \delta(L,\mathcal{R})$. On the other hand, denote $X = \begin{bmatrix} 1 & 0 & 0 \\ 1 & 1 & 0 \end{bmatrix}$ and observe that

$$\dim \mathrm{Hom}_{I(\mathbb{P})}(X,N) = 1, \quad \dim \mathrm{Hom}_{I(\mathbb{P})}(X,L) = 0.$$

Thus the dim-hom invariant induced by $\mathcal{X}$ can distinguish N and L, while the invariant $\delta(-,\mathcal{R})$ cannot. Since the latter is equivalent to the generalized rank invariant $\mathrm{rk}(-,\mathcal{R})$, this proves the result. $\square$

REMARK 5.2.9. Let $\mathbb{P}$ be a poset in which the 2×3 grid is embedded. Let $\mathcal{R}$ denote the set of connected intervals in $\mathbb{P}$, and let $\mathcal{X}$ represent the set of connected interval modules over $I(\mathbb{P})$. Then, the generalized rank invariant $\mathrm{rk}(-,\mathcal{R})$ is not a dim-hom invariant relative to $\mathcal{X}$.

To summarize, let $\mathbb{P}$, $\mathcal{R}$, and $\mathcal{X}$ be as in Remark 5.2.9. We have demonstrated that the generalized rank invariant $\mathrm{rk}(-,\mathcal{R})$, consequently the signed barcode $\delta(-,\mathcal{R})$, is not a dim-hom invariant relative to $\mathcal{X}$. This particular choice of $\mathcal{X}$ is especially natural (see Question 5.1.14). Indeed, if the quotient map $p : K_0^{\mathrm{split}}(I(\mathbb{P})) \to K_0(I(\mathbb{P}),\mathcal{X})$ is a homological invariant, then both p and $\mathrm{rk}(-,\mathcal{R})$ assign a persistence module M to a "signed approximation" of M by the set of spread modules. Corollary 5.2.9 implies that these approximations do not coincide, even when a change of basis is allowed.

While limiting $\mathcal{X}$ to only interval modules is a natural choice, it remains an open question whether the generalized rank invariant can be classified as either homological or a dim-hom invariant relative to some other set $\mathcal{Y}$. Using a brute-force calculation, we demonstrated that this is not the case when $\mathbb{P}$ is the 2×3 grid, leading to the following conjecture.

CONJECTURE 5.2.10. Let $\mathbb{P}$ be a locally finite poset in which the 2×3 grid is embedded, and let $\mathcal{R}$ denote the set of connected intervals in $\mathbb{P}$. Then the generalized rank invariant $\mathrm{rk}(-, \mathcal{R})$ is neither a homological invariant nor a dim-hom invariant (relative to any $\mathcal{Y}$).

5.3. A generalized result about morphisms between thin modules. The main purpose of this subsection is to show the following result, which is a generalization of [**BBH24**, Proposition 5.5].

THEOREM 5.3.1. *Let $\mathbb{P}$ be a locally finite poset. Let M and N be two thin modules supported on connected subsets A and B respectively. Denote by $Y_1, Y_2, \cdots, Y_n$ the connected components of $A \cap B$ satisfying $M_1 \triangleq \{x \mid x \in A, x \notin B, \exists y \in Y_i, y$ cover $x\} = \emptyset$ and $M_2 \triangleq \{y \mid y \in B, y \notin A, \exists x \in Y_i, y$ cover $x\} = \emptyset$. Then $\dim_k \mathrm{Hom}(M, N) = n$. Here thin module means that for $\forall v \in \mathbb{P}$, $\dim(M_v) \leq 1$ and the structure map is identity map or zero.*

Given two thin modules M and N, notice that any morphism f from M to N within the Hom-set is exclusively constrained by the following 16 commutative diagrams and their finite combinations.

We then have the two lemmas telling us information about morphism sets:

$$\begin{array}{ccc} 0(M_\mu) & \xrightarrow{0} & 0(M_v) \\ \downarrow{\scriptstyle 0} & & \downarrow{\scriptstyle 0} \\ 0(N_\mu) & \xrightarrow{0} & 0(N_v) \end{array}$$
(A)

$$\begin{array}{ccc} 0(M_\mu) & \xrightarrow{0} & K(M_v) \\ \downarrow{\scriptstyle 0} & & \downarrow{\scriptstyle 0} \\ 0(N_\mu) & \xrightarrow{0} & 0(N_v) \end{array}$$
(B)

$$\begin{array}{ccc} K(M_\mu) & \xrightarrow{0} & 0(M_v) \\ \downarrow{\scriptstyle 0} & & \downarrow{\scriptstyle 0} \\ 0(N_\mu) & \xrightarrow{0} & 0(N_v) \end{array}$$
(C)

$$\begin{array}{ccc} 0(M_\mu) & \xrightarrow{0} & 0(M_v) \\ \downarrow{\scriptstyle 0} & & \downarrow{\scriptstyle 0} \\ K(N_\mu) & \xrightarrow{0} & 0(N_v) \end{array}$$
(D)

$$\begin{array}{ccc} 0(M_\mu) & \xrightarrow{0} & K(M_v) \\ \downarrow{\scriptstyle 0} & & \downarrow{\scriptstyle \lambda_1} \\ 0(N_\mu) & \xrightarrow{0} & K(N_v) \end{array}$$
(E)

$$\begin{array}{ccc} K(M_\mu) & \xrightarrow{1} & K(M_v) \\ \downarrow{\scriptstyle 0} & & \downarrow{\scriptstyle 0} \\ 0(N_\mu) & \xrightarrow{0} & 0(N_v) \end{array}$$
(F)

$$\begin{array}{ccc} K(M_\mu) & \xrightarrow{0} & 0(M_v) \\ \downarrow{\scriptstyle \lambda_2} & & \downarrow{\scriptstyle 0} \\ K(N_\mu) & \xrightarrow{0} & 0(N_v) \end{array}$$
(G)

$$\begin{array}{ccc} 0(M_\mu) & \xrightarrow{0} & 0(M_v) \\ \downarrow{\scriptstyle 0} & & \downarrow{\scriptstyle 0} \\ K(N_\mu) & \xrightarrow{1} & K(N_v) \end{array}$$
(H)

$$\begin{array}{ccc} 0(M_\mu) & \xrightarrow{0} & K(M_v) \\ \downarrow{\scriptstyle 0} & & \downarrow{\scriptstyle 0} \\ K(N_\mu) & \xrightarrow{0} & 0(N_v) \end{array}$$
(I)

$$\begin{array}{ccc} K(M_\mu) & \xrightarrow{0} & 0(M_v) \\ \downarrow{\scriptstyle 0} & & \downarrow{\scriptstyle 0} \\ 0(N_\mu) & \xrightarrow{0} & K(N_v) \end{array}$$
(J)

$$\begin{array}{ccc} K(M_\mu) & \xrightarrow{1} & K(M_v) \\ \downarrow{\scriptstyle 0} & & \downarrow{\scriptstyle 0} \\ 0(N_\mu) & \xrightarrow{0} & K(N_v) \end{array}$$
(K)

$$\begin{array}{ccc} K(M_\mu) & \xrightarrow{0} & 0(M_v) \\ \downarrow{\scriptstyle 0} & & \downarrow{\scriptstyle 0} \\ K(N_\mu) & \xrightarrow{1} & K(N_v) \end{array}$$
(L)

$$\begin{array}{ccc} K(M_\mu) & \xrightarrow{1} & K(M_v) \\ \downarrow{\scriptstyle \lambda_3} & & \downarrow{\scriptstyle 0} \\ K(N_\mu) & \xrightarrow{0} & 0(N_v) \end{array}$$
(M)

$$\begin{array}{ccc} 0(M_\mu) & \xrightarrow{0} & K(M_v) \\ \downarrow{\scriptstyle 0} & & \downarrow{\scriptstyle \lambda_4} \\ K(N_\mu) & \xrightarrow{1} & K(N_v) \end{array}$$
(N)

$$\begin{array}{ccc} 0(M_\mu) & \xrightarrow{0} & 0(M_v) \\ \downarrow{\scriptstyle 0} & & \downarrow{\scriptstyle 0} \\ 0(N_\mu) & \xrightarrow{0} & 0(N_v) \end{array}$$
(O)

$$\begin{array}{ccc} K(M_\mu) & \xrightarrow{1} & K(M_v) \\ \downarrow{\scriptstyle \lambda_5} & & \downarrow{\scriptstyle \lambda_5} \\ K(N_\mu) & \xrightarrow{1} & K(N_v) \end{array}$$
(P)

LEMMA 5.3.2. *Let M and N be two thin modules, any $f: M \to N$ must satisfy $f_v = 0$ for $v \notin supM \cap supN$.*

PROOF. When $v \notin supM \cap supN$, the possible commutative diagrams that can occur are commutative diagrams (A), (B), (C), (D), (F), (H), (I), (J) and (O). All the commutative diagrams above imply that $f_x = 0$. □

LEMMA 5.3.3. *Consider a connected component of* $\sup M \cap \sup N$ *denoted by Y_i. For any elements x and y within Y_i, if f_x and f_y are non-zero, then f_x must equal f_y. (f remains constant on Y_i.)*

PROOF. Let $f: M \to N$, and let $x, y \in P$ lie in the same connected component Y_i of $\sup M \cap \sup N$. Then x and y are connected by a path in the Hasse quiver of Y_i. Assume, without loss of generality, that the path connecting x and y is represented as $x - u_1 - u_2 - \ldots - u_n - y$. It follows that the connection between x and y is governed by a series of constraints similar to those depicted in commutative diagram (P), thereby deducing that f_x must be equal to f_y. □

In light of Lemma 5.3.2 and Lemma 5.3.3, it is established that if f_x is non-zero, then x necessarily resides within the connected component of the intersection $\sup M \cap \sup N$.

In accordance with commutative diagram (P) and the construction of a spread module, it holds that f_x equals λ times the identity, denoted by $\lambda \mathbf{1}$, for some λ in the field K, when f_x is non-zero.

Subsequently, we delve into the necessary and sufficient condition for f_x potentially to be non-zero when x resides within the connected component Y_i. By the previous paragraph, our analysis must extend to the examination of the remaining commutative diagrams labeled (E), (G), (K), (L), (M) and (N), subsequent to the exclusion of diagrams (A) through (D), (F), (H), (I), (J), (O) and (P).

LEMMA 5.3.4. *The map f_x is non-zero if and only if the two sets M_1, M_2 are empty.*

PROOF. Firstly, it is observed that within the commutative diagrams (E), (G), (K), (L), (M) and (N), only (K) and (L) necessitate that the mapping f_ω must be the zero map. Therefore, f_x may be non-zero only if the commutative diagrams (K) and (L) are absent.

Next, we turn our attention to the examination of commutative diagrams (K) and (L).

(1) Upon considering commutative diagram (K), one may observe that $\mu \in M, \mu \notin N$ with $v \in Y_i$, therefore the absence of diagram (K) is equivalent to stating that $M_1 = \emptyset$.
(2) Upon considering commutative diagram (L), one may observe that $v \in M, v \notin N$ with $\mu \in Y_i$, therefore the absence of diagram (L) is equivalent to stating that $M_2 = \emptyset$.

In conclusion, from the analysis above, we deduce that f_x can be non-zero if and only if both $M_1 = \emptyset$ and $M_2 = \emptyset$. □

Now Theorem 5.3.1 follows immediately from Lemma 5.3.3 and Lemma 5.3.4.

Lastly, we give an example showing that our theorem is a generalization of [**BBH24**, Proposition 5.5].

EXAMPLE 5.3.5. Let $\mathbb{P}$ be the lattice $\mathbb{Z}^2$. Let M, N be two thin modules such that

$\sup M = \langle\{(0,2),(3,0)\},+\infty\langle$, $\sup N = \langle -\infty, \{(1,2),(3,0)\}\langle$ are in the blue and orange region respectively.

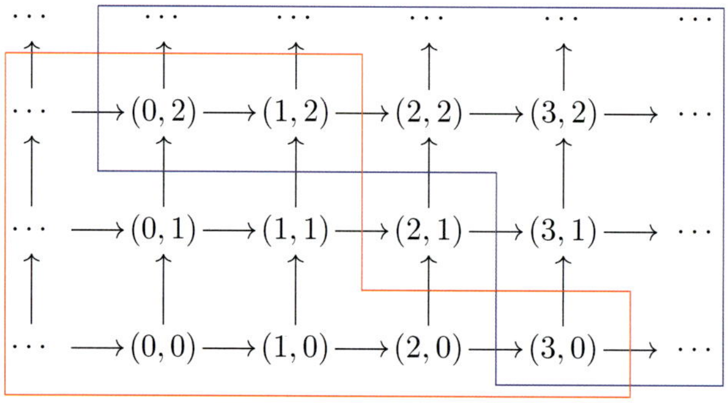

Note that in [**BBH24**, Proposition 5.5], the proposition is only applied for intervals where minimal sets and maximal sets are finite sets and only applied when $\mathbb{P}$ is finite. However, we can use the theorem 5.3.1 in this example: first, note that there are only two connected components of $\sup M \cap \sup N$, so $\dim Hom(M,N) \leq 2$, which is $\{(0,2),(1,2)\}$ and $\{(3,0)\}$; Next check whether the two connected components satisfy the condition in Theorem 5.3.1, which is true. So $\dim Hom(M,N) = 2$.

REMARK 5.3.6. Note that in Theorem 5.3.1, we only need modules M and N being connected. We point out that because M and N can be seen as $I(\mathbb{P}) := kQ/J$-module where the ideal is generated by all commutative relations, the condition "connected" is equivalent to "connected and convex". However, if we take M and N as kQ-module, then connected subsets strictly contain connected and convex subsets, in which sense Theorem 5.3.1 is indeed a generalization of [**BBH24**, Proposition 5.5].

6. Further questions and discussion

6.1. Dicussion 1: What is the relationship between homological invariants and dim-hom invariants? In this section, we explore the relationship between these two invariants, as introduced in Definition 5.1.9. Before delving into the discussion, we introduce the concept of projectivization, which simplifies the computation of $\mathcal{X}$-resolutions for a finite set $\mathcal{X}$ of indecomposable objects by reducing it to computing projective resolutions over an endomorphism algebra. This framework allows us to demonstrate that, for many choices of $\mathcal{X}$, the homological and dim-hom invariants relative to $\mathcal{X}$ coincide.

Now, consider the module $T = T_{\mathcal{X}} := \bigoplus_{R\in\mathcal{X}} R$. Define $\Gamma = \Gamma_{\mathcal{X}} := \mathrm{End}_{I(\mathcal{P})}(T)^{\mathrm{op}}$, a finite-dimensional k-algebra under the composition of morphisms, equipped with the standard vector space structure. Let Q denote the Gabriel quiver of Γ, whose vertices correspond to $\mathcal{X}$. For any $R, R' \in \mathcal{X}$, the number of arrows $R \to R'$ in Q matches the number of linearly independent $\mathcal{X}$-irreducible morphisms $R' \to R$ in $\mathrm{Hom}_{I(\mathcal{P})}(X,Y)$.

The association $M \mapsto \mathrm{Hom}_{I(\mathcal{P})}(T,M)$ defines a functor from $\mathrm{mod}\,\Lambda$ to $\mathrm{mod}\,\Gamma$. The action of Γ on $\mathrm{Hom}_{I(\mathcal{P})}(T,M)$ is given by precomposition. This functor establishes an additive equivalence between $\mathrm{add}(\mathcal{X})$ and the category of projective

Γ-modules. Specifically, for an (indecomposable) module $R \in \mathcal{X}$, the Γ-module $\operatorname{Hom}_{I(\mathcal{P})}(T, R)$ is isomorphic to the projective module (over Γ) at the vertex in the quiver Q corresponding to R.

Regarding the relationship between homological invariants and dim-hom invariants, we arrive at the following conclusion.

THEOREM 6.1.1 ([**BBH24**]). *Let $\mathcal{X}$ be a finite set of indecomposable $I(\mathcal{P})$-modules that includes all indecomposable projectives, and assume that the global dimension of $\Gamma = \Gamma_{\mathcal{X}}$ is finite. Let $p : K_0^{split}(I(\mathcal{P})) \to K_0(I(\mathcal{P}), \mathcal{X})$ be the canonical quotient map and let $q : K_0^{split}(I(\mathcal{P})) \to \mathbb{Z}^{|\mathcal{X}|}$ be the dim-hom invariant given by $q(M) = (\dim \operatorname{Hom}_{I(\mathcal{P})}(R, M))_{R \in \mathcal{X}}$. Then p and q are equivalent invariants. In particular, any invariant that is either homological or a dim-hom invariant relative to $\mathcal{X}$ is both homological and a dim-hom invariant relative to $\mathcal{X}$.*

REMARK 6.1.2. This result suggests the possibility that all homological invariants are dim-hom invariants. Specifically, if we replace the finiteness condition (2)(b) in Definition 5.1.9 with the condition that the global dimension of $\Gamma_{\mathcal{X}}$ is finite, this conclusion would follow from Theorem 6.1.1.

So far, there is no known example where the $\mathcal{X}$-dimension of $I(\mathcal{P})$ is finite while the global dimension of Γ is infinite. It would not be surprising if such an example does not exist, given that tilting preserves finite global dimension (see [**KK20**]). Blanchette and Brüstle hypothesize that these two finiteness conditions are equivalent, implying that all homological invariants are indeed dim-hom invariants.

CONJECTURE 6.1.3 ([**BBH24**]). *All homological invariants are in fact dim-hom invariants.*

6.2. Discussion 2: How to compute finite $\mathcal{X}$-resolutions? If a module M admits finite resolutions by projectives relative to $\mathcal{X}$ (referred to as $\mathcal{X}$-resolutions), then the minimal $\mathcal{X}$-resolutions yield a unique signed barcode. In this barcode, each bar corresponds to the support of some element in $\mathcal{X}$ and thus represents an interval from the collection $\mathcal{I}$.

A key question that arises is how to compute finite $\mathcal{X}$-resolutions, or more specifically, how to determine the supports of the summands that appear in the various non-zero terms of these resolutions. In the following, we discuss and address issues related to the computation of finite dimensions.

QUESTION 6.2.1 ([**LLO25**]). Given a fixed collection $\mathcal{I}$ of intervals, its corresponding class $\mathcal{X}$ of indecomposable projectives, and a fixed natural number k, can we efficiently determine whether a given finitely presented persistence module M admits an $\mathcal{X}$-resolution of length at most k?

The functor $M \mapsto \operatorname{Hom}_{I(\mathcal{P})}(T, M)$ can be utilized in an algorithm to compute the value of the invariant $[M]_{\mathcal{X}}$ when Γ has finite global dimension. Specifically, one can first compute a minimal projective resolution of the Γ-module $\operatorname{Hom}_{I(\mathbb{P})}(T, M)$. The class of M in the Grothendieck group can then be determined by taking the alternating sum of the terms in this resolution. The advantage of this approach is that it avoids the direct computation of an $\mathcal{X}$-resolution, replacing it with the simpler process of computing a projective resolution.

Another approach to improving the efficiency of approximation computations is to reduce the size of $\mathcal{X}$ based on the module being approximated. The following proposition demonstrates how this can be achieved when all quotients of the modules in $\mathcal{X}$ belong to $\operatorname{add}(\mathcal{X})$. It is important to note that this does not imply that $\operatorname{add}(\mathcal{X})$ is closed under all quotients. For instance, if $\mathcal{X}$ is the set of all single-source spread modules, then by Proposition 5.2.1, all quotients of modules in $\mathcal{X}$ will be in $\operatorname{add}(\mathcal{X})$. However, $\operatorname{add}(\mathcal{X})$ is not closed under arbitrary quotients, as this would imply $\operatorname{add}(\mathcal{X}) = \operatorname{mod} I(\mathbb{P})$. In fact, every module is a quotient of its projective cover, which will always lie in $\operatorname{add}(\mathcal{X})$.

PROPOSITION 6.2.2 ([**BBH24**]). *Let M be an arbitrary $I(\mathbb{P})$-module. Let $\mathcal{X}$ be a set of interval modules and assume that every quotient of a module in $\mathcal{X}$ lies in $\operatorname{add}(\mathcal{X})$. Denote by $\mathcal{Y} = \{R \in \mathcal{X} \mid \operatorname{supp}(R) \subseteq \operatorname{supp}(M)\}$. Then every $\mathcal{X}$-approximation of M is a also $\mathcal{Y}$-approximation. In particular, the minimal $\mathcal{X}$-approximation of M coincides with the minimal $\mathcal{Y}$-approximation of M.*

REMARK 6.2.3. The preceding proposition suggests that much of the theory developed in this paper may extend to finite-dimensional representations of locally finite posets, even when these posets are not finite. Here, "finite-dimensional" refers to the total dimension being finite, rather than pointwise finite-dimensionality. Notably, such representations exhibit finite support. Given an infinite set of finite-dimensional interval modules $\mathcal{X}$, Proposition 6.2.2 can be applied to construct $\mathcal{X}$ approximations for arbitrary finite-dimensional representations. However, indecomposable projective representations are generally not finite-dimensional. If $\mathcal{X}$ is infinite, then the module $T_{\mathcal{X}} = \bigoplus_{R \in \mathcal{X}} R$ will also not be finite-dimensional. Consequently, the extent to which the theory introduced here applies to locally finite posets remains an open question and warrants further exploration.

From this remark, we are led to the following question.

QUESTION 6.2.4. To what extent can the theory presented in this paper be applied to finite-dimensional representations of locally finite posets?

Next, we address the question of whether M is projective relative to $\mathcal{X}$, or equivalently, whether M is $\mathcal{X}$-decomposable, meaning that it can be expressed as a direct sum of elements from $\mathcal{X}$.

Botnan, Lebovici, and Oudot explored this question for the poset $\mathbb{R}^2$, focusing on the cases where $\mathcal{I}$ is either the collection of rectangles (i.e., Cartesian products of two intervals in $\mathbb{R}$) or the collection of blocks (i.e., rectangles representing infinite horizontal or vertical bands, co-directed upsets, or co-directed downsets):

THEOREM 6.2.5 ([**BLO23**]). *A finitely presented (fp) persistence module $M : \mathbb{R}^2 \to k\text{-}\mathrm{Vec}$ is block-decomposable (resp. rectangle-decomposable) if and only if its restriction to any commutative square $\{r, r'\} \times \{s, s'\}$ is block-decomposable (resp. rectangle-decomposable) for every $r < r' \in \mathbb{R}$ and $s < s' \in \mathbb{R}$. This result remains valid if M is only pointwise finite-dimensional (pfd) or if $\mathbb{R}^2$ is replaced by the Cartesian product of any two totally ordered sets.*

In [**LLO25**], Theorem 6.2.5 is extended as follows:

THEOREM 6.2.6 ([**LLO25**]). *A pointwise finite-dimensional (pfd) persistence module $M : T_1 \times \cdots \times T_n \to k\text{-}\mathrm{Vec}$, defined over the Cartesian product of n totally ordered sets, is block-decomposable if and only if its restrictions to all commutative cubes $\{t_1, t_1'\} \times \cdots \times \{t_n, t_n'\}$, where $t_i < t_i' \in T_i$, are block-decomposable.*

Question 6.2.1 remains a largely open problem, with some notable recent advancements using Koszul complexes [**CGR$^+$25**], though scalable methods are still far off. We now present some results on the question, focusing on the relative Koszul complex.

In [**CGR$^+$25**], for an indecomposable module $R \in \mathcal{X}$, the author introduces the minimal $\mathcal{X}$-relative Koszul coresolution $\mathcal{K}^{\cdot}(R)$ of R, which is represented as a sequence of modules in $\mathbb{P}$-Mod

$$0 \to R \xrightarrow{d^0} X^1 \xrightarrow{d^1} X^2 \xrightarrow{d^2} \dots \tag{6.1}$$

The author demonstrates that, for a finitely generated $\mathbb{P}$-module M, the $\mathcal{X}$-relative i-th Betti number of M at R is equal to the k-dimension of the i-th homology of the $\mathcal{X}$-relative Koszul complex $\mathcal{K}^{\cdot}(R) := Hom_{I(\mathbb{P})}(\mathcal{K}^{\cdot}(R), M)$ at R for all $i \geq 0$.

This result can also be applied to explore the minimal interval resolution or coresolution of a persistence module M. For example, it can be used to verify the interval decomposability of M and compute its interval approximation. For further details, refer to [**CGR$^+$25**].

7. Appendix

7.1. Limits and colimits. This section revisits the notions of limits and colimits [**ML71**]. Throughout, I will denote a small category.

DEFINITION 7.1.1. Let $F : I \to \mathcal{C}$ be a functor. A cone over F is a pair $(L, (\pi_x)_{x \in \text{ob}(I)})$, where L is an object in $\mathcal{C}$, and $(\pi_x)_{x \in \text{ob}(I)}$ is a collection of morphisms $\pi_x : L \to F(x)$ that commute with the arrows in the diagram of F. Specifically, if $g : x \to y$ is a morphism in I, then $\pi_y = F(g) \circ \pi_x$ in $\mathcal{C}$. This condition ensures that the diagram below is commutative.

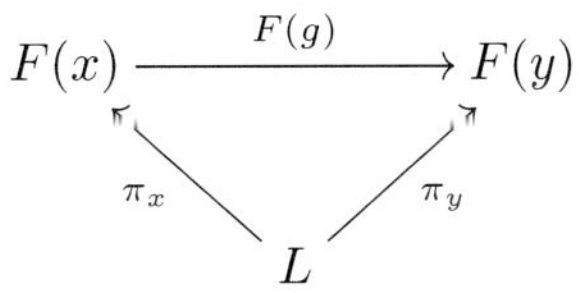

In Definition 7.1.1, the cone $(L, (\pi_x)_{x \in \text{ob}(I)})$ over F will sometimes be denoted simply by L, omitting the collection $(\pi_x)_{x \in \text{ob}(I)}$ of morphisms when there is no ambiguity. A limit of a diagram $F : I \to \mathcal{C}$ is a terminal object in the collection of all cones:

DEFINITION 7.1.2. Let $F : I \to \mathcal{C}$ be a functor. A *limit* of F is a cone over F, denoted by $\left(\varprojlim F, (\pi_x)_{x \in \text{ob}(I)}\right)$ or simply $\varprojlim F$, with the following *terminal property*: If there is another cone $\left(L', (\pi'_x)_{x \in \text{ob}(I)}\right)$ of F, then there is a unique morphism $u : L' \to \varprojlim F$ such that $\pi'_x = \pi_x \circ u$ for all $x \in \text{ob}(I)$.

REMARK 7.1.3. A diagram may not necessarily have a limit. However, if a diagram does have a limit, the terminal property ensures its uniqueness up to isomorphism. For this reason, a limit is often referred to as the limit of a diagram.

Cocones and colimits are defined in a dual manner.

REMARK 7.1.4. A diagram may not always have a colimit. However, if a diagram does possess a colimit, its initial property ensures uniqueness up to isomorphism. Therefore, the colimit is often referred to as the colimit of a diagram.

REMARK 7.1.5 (Restriction of an indexing poset). Let $\mathbb{P}$ be any poset and let $\mathbb{Q}$ be a subposet of $\mathbb{P}$. In categorical language, $\mathbb{Q}$ is a full subcategory of $\mathbb{P}$. Let $F : \mathbb{P} \to \mathcal{C}$ be a functor.

(1) Suppose that the limit of the restriction $F|_{\mathbb{Q}}$ exists. For any cone $\left(L', (\pi'_p)_{p\in \mathrm{ob}(\mathbb{P})}\right)$ over F, its restriction $\left(L', (\pi'_p)_{p\in \mathrm{ob}(\mathbb{Q})}\right)$ forms a cone over the restriction $F|_{\mathbb{Q}} : \mathbb{Q} \to \mathcal{C}$. Therefore, by the terminal property of the limit $\left(\varprojlim F|_{\mathbb{Q}}, (\pi_q)_{q\in \mathrm{ob}(\mathbb{Q})}\right)$, there exists the unique morphism $u : L' \to \varprojlim F|_{\mathbb{Q}}$ such that $\pi'_q = \pi_q \circ u$ for all $q \in \mathrm{ob}(\mathbb{Q})$.

(2) Suppose that the colimit of the restriction $F|_{\mathbb{Q}}$ exists. For any cocone $\left(C', (i'_p)_{p\in \mathrm{ob}(\mathbb{P})}\right)$ over F, its restriction $\left(C', (i'_p)_{p\in \mathrm{ob}(\mathbb{Q})}\right)$ forms a cocone over the restriction $F|_{\mathbb{Q}} : \mathbb{Q} \to \mathcal{C}$. Therefore, by the initial property of the colimit $\left(\varinjlim F|_{\mathbb{Q}}, (i_q)_{q\in \mathrm{ob}(\mathbb{Q})}\right)$, there exists the unique morphism $u : \varinjlim F|_{\mathbb{Q}} \to C'$ such that $i'_q = u \circ i_q$ for all $q \in \mathrm{ob}(\mathbb{Q})$.

7.2. Exact structure. The concept of an exact category was initially introduced by Quillen in [**Qui72**], and later, [**Kel90**] demonstrated the redundancy of certain axioms. This section reviews various definitions and results related to exact structures, particularly in the context of persistence modules and, more broadly, abelian categories.

Although the definition of an exact structure is generally formulated for additive categories, this discussion is confined to the special case of abelian categories for clarity. For additional background and historical context, we refer to [**BBGH22**, **Büh10**, **DRS+99**, **EJ00**] and the references therein.

In an abelian category $\mathcal{A}$, we denote monomorphisms as $X \rightarrowtail Y$ and epimorphisms as $Y \twoheadrightarrow Z$. The following standard terminology is also adopted.

DEFINITION 7.2.1. Let $\mathcal{E}$ be a class of short exact sequences in $\mathcal{A}$ that is closed under isomorphisms.

(1) A short exact sequence η in $\mathcal{A}$ is called $\mathcal{E}$-exact if $\eta \in \mathcal{E}$.

(2) Given a long exact sequence $\eta = (\dots X_0 \xrightarrow{f_0} X_1 \xrightarrow{f_1} X_2 \xrightarrow{f_2} \dots)$ in $\mathcal{A}$, we say that η is $\mathcal{E}$-exact if, for all i, the short exact sequences

$$\ker f_i \to X_i \to \operatorname{im} f_i$$

are $\mathcal{E}$-exact.

(3) For a morphism $A \xrightarrow{h} B$ in $\mathcal{A}$, we say that h is $\mathcal{E}$-admissible if the sequences $\ker h \to A \to \operatorname{im} h$ and $\operatorname{im} h \to B \to \operatorname{coker} h$ are both $\mathcal{E}$-exact.

We now revisit the definition of an exact structure. The axioms we follow are those presented in [**Büh10**, section2], which were proven in [**Kel90**] to be equivalent to the original axioms introduced by Quillen in his seminal work [**Qui72**].

REMARK 7.2.2. There are two common conventions in terminology found in the literature. One approach is to refer to $\mathcal{E}$ as an exact structure on $\mathcal{A}$, as adopted in this paper. The other is to refer to the pair $(\mathcal{A}, \mathcal{E})$ as an exact category. In this work, we use the term exact structure since multiple exact structures are often considered within the same abelian category.

REMARK 7.2.3. A direct consequence of the definition is that an exact structure $\mathcal{E}$ can be uniquely determined by any of the following classes:

(1) The class of short $\mathcal{E}$-exact sequences.
(2) The class of $\mathcal{E}$-admissible monomorphisms.
(3) The class of $\mathcal{E}$-admissible epimorphisms.

REMARK 7.2.4. An exact structure can alternatively be viewed as a "closed" additive sub-bifunctor of $\mathrm{Ext}_{\mathcal{A}}(-,-)$. For further details on this viewpoint, refer to [**BH61**, Theorem 1.1] and [**DRS$^+$99**, Corollary 1.6].

We now introduce a collection of exact structures, as established in [**AS93**, Section 1] for finite-dimensional algebras and in [**DRS$^+$99**, Section1.3] for arbitrary abelian categories.

DEFINITION 7.2.5. Let $\mathcal{X}$ be a class of objects in $\mathcal{A}$.

(1) The class of short exact sequences $\eta = (L \xrightarrow{f} M \xrightarrow{g} N)$ in $\mathcal{A}$, where the induced sequences
$$0 \to \mathrm{Hom}_{\mathcal{A}}(X, L) \xrightarrow{f_*} \mathrm{Hom}_{\mathcal{A}}(X, M) \xrightarrow{g_*} \mathrm{Hom}_{\mathcal{A}}(X, N) \to 0$$
are exact for all $X \in \mathcal{X}$, is denoted by $\mathcal{F}_{\mathcal{X}}$. Equivalently, $\eta \in \mathcal{F}_{\mathcal{X}}$ if and only if $g_* = \mathrm{Hom}_{\mathcal{A}}(X, g)$ is an epimorphism for all $X \in \mathcal{X}$.
(2) The class of short exact sequences $\eta = (L \xrightarrow{f} M \xrightarrow{g} N)$ in $\mathcal{A}$, where the induced sequences
$$0 \to \mathrm{Hom}_{\mathcal{A}}(N, X) \xrightarrow{g^*} \mathrm{Hom}_{\mathcal{A}}(M, X) \xrightarrow{f^*} \mathrm{Hom}_{\mathcal{A}}(L, X) \to 0$$
are exact for all $X \in \mathcal{X}$, is denoted by $\mathcal{F}^{\mathcal{X}}$. Equivalently, $\eta \in \mathcal{F}^{\mathcal{X}}$ if and only if $f^* = \mathrm{Hom}_{\mathcal{A}}(f, X)$ is a monomorphism for all $X \in \mathcal{X}$.

REMARK 7.2.6. We note that if $\mathcal{X}$ is additive then $\mathcal{F}_{\mathcal{X}} = \mathcal{F}_{\mathrm{ind}\,\mathcal{X}}$. It is therefore common practice to replace $\mathcal{X}$ with $\mathrm{ind}(\mathcal{X})$ when applying Definition 7.2.5.

PROPOSITION 7.2.7 ([**DRS$^+$99**, Proposition 1.7]). , [**AS93**, Proposition 1.7] *Let $\mathcal{X}$ be a class of objects in $\mathcal{A}$. Then $\mathcal{F}_{\mathcal{X}}$ is an exact structure.*

We typically characterize exact structures by referring to their corresponding relative projective objects, which are defined as follows:

DEFINITION 7.2.8. Let $\mathcal{E}$ be an exact structure on $\mathcal{A}$. We say an object $U \in \mathcal{A}$ is *$\mathcal{E}$-projective*, denoted by $U \in \mathrm{proj}(\mathcal{E})$, if for every $\mathcal{E}$-admissible epimorphisms $Y \overset{g}{\twoheadrightarrow} Z$, the induced map $g_* = \mathrm{Hom}_{\mathcal{A}}(U, g)$ is an epimorphism. Equivalently, $U \in \mathrm{proj}(\mathcal{E})$ if and only if $\mathrm{Hom}_{\mathcal{A}}(U, -)$ transforms $\mathcal{E}$-exact sequences to exact sequences.

From this definition, it follows immediately that for any exact structures $\mathcal{E}$ and $\mathcal{F}$, if $\mathcal{E} \supseteq \mathcal{F}$, then $\mathrm{proj}(\mathcal{A})) \subseteq \mathrm{proj}(\mathcal{E}) \subseteq \mathrm{proj}(\mathcal{F})$. We denote by $\mathcal{E} := \mathcal{E}_{\max}$ the largest exact structure on $\mathcal{A}$, provided that $\mathrm{proj}(\mathcal{A}) = \mathrm{proj}(\mathcal{E}_{\max})$. Additionally, $\mathcal{X} \subseteq \mathrm{proj}(\mathcal{F}_{\mathcal{X}})$ and $\mathrm{add}(\mathcal{X} \cup \mathrm{proj}(\mathcal{A}) \subseteq \mathrm{proj}(\mathcal{F}_{\mathcal{X}})$. As a partial converse, we have the following:

THEOREM 7.2.9 ([**AS93**, Proposition 1.10]). *Let Λ be an artin algebra and $\mathcal{A} = \mathrm{mod}\,\Lambda$, and let $\mathcal{X}$ be a class of objects in $\mathcal{A}$. Then* $\mathrm{add}(\mathcal{X} \cup \mathrm{proj}(\mathcal{A})) = \mathrm{proj}(\mathcal{F}_{\mathcal{X}})$.

REMARK 7.2.10. The proof of [**AS93**, Proposition 1.10] relies on the concept of almost split sequences, which are also referred to as Auslander-Reiten sequences. It is demonstrated in [**AS93**, Proposition 1.9] that an indecomposable nonprojective module $U \in \operatorname{mod}\Lambda$ (in the traditional sense) is $\mathcal{F}_{\mathcal{X}}$-projective if and only if the almost split sequence that terminates in U fails to be $\mathcal{F}_{\mathcal{X}}$-exact. As detailed in [**DRS+99**, Section 2], this finding extends to any Krull-Schmidt category that possesses almost split pairs. Nonetheless, there are numerous intriguing instances of abelian categories lacking almost split pairs, such as the category $\operatorname{rep}(\mathbb{R}^n)$.

COROLLARY 7.2.11 ([**AS93**, Proposition 1.10]). *Let $\mathbb{P}$ be a finite poset and let $\mathcal{X}$ be a class of objects in* $\operatorname{rep}\mathbb{P}$. *Then* $\operatorname{add}(\mathcal{X} \cup \operatorname{proj}(\operatorname{rep}\mathbb{P})) = \operatorname{proj}(\mathcal{F}_{\mathcal{X}})$.

We conclude with the following open question.

QUESTION 7.2.12. It is uncertain whether Corollary 7.2.11 applies to arbitrary infinite posets, as they typically lack almost split pairs. However, the authors in [**Asa24**] indicate that the outcome is valid for specific selections of $\mathcal{X}$ when $\mathbb{P}$ is a finite product of totally ordered sets.

References

[AENY23a] Hideto Asashiba, Emerson G. Escolar, Ken Nakashima, and Michio Yoshiwaki, *Approximation by interval-decomposables and interval resolutions of persistence modules*, J. Pure Appl. Algebra **227** (2023), no. 10, Paper No. 107397, 20, DOI 10.1016/j.jpaa.2023.107397. MR4576885

[AENY23b] Hideto Asashiba, Emerson G. Escolar, Ken Nakashima, and Michio Yoshiwaki, *On approximation of 2D persistence modules by interval-decomposables*, J. Comput. Algebra **6/7** (2023), Paper No. 100007, 21, DOI 10.1016/j.jaca.2023.100007. MR4886185

[AS93] M. Auslander and Ø. Solberg, *Relative homology and representation theory. I. Relative homology and homologically finite subcategories*, Comm. Algebra **21** (1993), no. 9, 2995–3031, DOI 10.1080/00927879308824717. MR1228751

[Asa97] Hideto Asashiba, *A covering technique for derived equivalence*, J. Algebra **191** (1997), no. 1, 382–415, DOI 10.1006/jabr.1997.6906. MR1444505

[Asa22] Hideto Asashiba, *Categories and representation theory—with a focus on 2-categorical covering theory*, Mathematical Surveys and Monographs, vol. 271, American Mathematical Society, Providence, RI, [2022] ©2022. Translated from the 2019 Japanese version, DOI 10.1090/surv/271. MR4486377

[Asa24] Hideto Asashiba, *Relative Koszul coresolutions and relative Betti numbers*, J. Pure Appl. Algebra **229** (2025), no. 3, Paper No. 107905, 33, DOI 10.1016/j.jpaa.2025.107905. MR4866517

[BBGH22] Rose-Line Baillargeon, Thomas Brüstle, Mikhail Gorsky, and Souheila Hassoun, *On the lattices of exact and weakly exact structures*, J. Algebra **612** (2022), 77–109, DOI 10.1016/j.jalgebra.2022.07.043. MR4484820

[BBH23] Benjamin Blanchette, Thomas Brüstle, and Eric J. Hanson, *Exact structures for persistence modules*, Representations of algebras and related topics, EMS Ser. Congr. Rep., Eur. Math. Soc., Zürich, [2025] ©2025, pp. 121–160. MR4993173

[BBH24] Benjamin Blanchette, Thomas Brüstle, and Eric J. Hanson, *Homological approximations in persistence theory*, Canad. J. Math. **76** (2024), no. 1, 66–103, DOI 10.4153/s0008414x22000657. MR4687766

[BCB18] Magnus Bakke Botnan and William Crawley-Boevey, *Decomposition of persistence modules*, Proc. Amer. Math. Soc. **148** (2020), no. 11, 4581–4596, DOI 10.1090/proc/14790. MR4143378

[BD13] Dan Burghelea and Tamal K. Dey, *Topological persistence for circle-valued maps*, Discrete Comput. Geom. **50** (2013), no. 1, 69–98, DOI 10.1007/s00454-013-9497-x. MR3070542

[BH61] M. C. R. Butler and G. Horrocks, *Classes of extensions and resolutions*, Philos. Trans. Roy. Soc. London Ser. A **254** (1961/62), 155–222, DOI 10.1098/rsta.1961.0014. MR188267

[BL23] Magnus Bakke Botnan and Michael Lesnick, *An introduction to multiparameter persistence*, Representations of algebras and related structures, EMS Ser. Congr. Rep., EMS Press, Berlin, [2023] ©2023, pp. 77–150. MR4693638

[BLO22] Magnus Bakke Botnan, Vadim Lebovici, and Steve Oudot, *On rectangle-decomposable 2-parameter persistence modules*, Discrete Comput. Geom. **68** (2022), no. 4, 1078–1101, DOI 10.1007/s00454-022-00383-y. MR4517095

[BLO23] Magnus B. Botnan, Vadim Lebovici, and Steve Oudot, *Local characterizations for decomposability of 2-parameter persistence modules*, Algebr. Represent. Theory **26** (2023), no. 6, 3003–3046, DOI 10.1007/s10468-022-10189-4. MR4681340

[BLP13] Raymundo Bautista, Shiping Liu, and Charles Paquette, *Representation theory of strongly locally finite quivers*, Proc. Lond. Math. Soc. (3) **106** (2013), no. 1, 97–162, DOI 10.1112/plms/pds039. MR3020740

[BOO24] Magnus Bakke Botnan, Steffen Oppermann, and Steve Oudot, *Signed barcodes for multi-parameter persistence via rank decompositions and rank-exact resolutions*, Found. Comput. Math. **25** (2025), no. 5, 1815–1874, DOI 10.1007/s10208-024-09672-9. MR4983902

[BOOS24] Magnus Bakke Botnan, Steffen Oppermann, Steve Oudot, and Luis Scoccola, *On the bottleneck stability of rank decompositions of multi-parameter persistence modules*, Adv. Math. **451** (2024), Paper No. 109780, 53, DOI 10.1016/j.aim.2024.109780. MR4761951

[Büh10] Theo Bühler, *Exact categories*, Expo. Math. **28** (2010), no. 1, 1–69, DOI 10.1016/j.exmath.2009.04.004. MR2606234

[CB15] William Crawley-Boevey, *Decomposition of pointwise finite-dimensional persistence modules*, J. Algebra Appl. **14** (2015), no. 5, 1550066, 8, DOI 10.1142/S0219498815500668. MR3323327

[CdS10] Gunnar Carlsson and Vin de Silva, *Zigzag persistence*, Found. Comput. Math. **10** (2010), no. 4, 367–405, DOI 10.1007/s10208-010-9066-0. MR2657946

[CGR⁺25] Wojciech Chachólski, Andrea Guidolin, Isaac Ren, Martina Scolamiero, and Francesca Tombari, *Koszul complexes and relative homological algebra of functors over posets*, Found. Comput. Math. **25** (2025), no. 4, 1121–1165, DOI 10.1007/s10208-024-09660-z. MR4951524

[CSEH05] David Cohen-Steiner, Herbert Edelsbrunner, and John Harer, *Stability of persistence diagrams*, Computational geometry (SCG'05), ACM, New York, 2005, pp. 263–271, DOI 10.1145/1064092.1064133. MR2460372

[CZ07] Gunnar Carlsson and Afra Zomorodian, *The theory of multidimensional persistence*, Computational geometry (SCG'07), ACM, New York, 2007, pp. 184–193, DOI 10.1145/1247009.1247105. MR2409104

[DRS⁺99] Peter Dräxler, Idun Reiten, Sverre O. Smalø, and Øyvind Solberg, *Exact categories and vector space categories*, Trans. Amer. Math. Soc. **351** (1999), no. 2, 647–682, DOI 10.1090/S0002-9947-99-02322-3. With an appendix by B. Keller. MR1608305

[DS87] Piotr Dowbor and Andrzej Skowroński, *Galois coverings of representation-infinite algebras*, Comment. Math. Helv. **62** (1987), no. 2, 311–337, DOI 10.1007/BF02564450. MR896100

[EJ00] Edgar E. Enochs and Overtoun M. G. Jenda, *Relative homological algebra*, De Gruyter Expositions in Mathematics, vol. 30, Walter de Gruyter & Co., Berlin, 2000, DOI 10.1515/9783110803662. MR1753146

[Gab72] Peter Gabriel, *Unzerlegbare Darstellungen. I* (German, with English summary), Manuscripta Math. **6** (1972), 71–103; correction, ibid. 6 (1972), 309, DOI 10.1007/BF01298413. MR332887

[Gab80] P. Gabriel, *The universal cover of a representation-finite algebra*, Springer Lect. Notes 903, Springer, 1980, 68–105.

[GR92] P. Gabriel and A. V. Roĭter, *Representations of finite-dimensional algebras*, Algebra, VIII, Encyclopaedia Math. Sci., vol. 73, Springer, Berlin, 1992, pp. 1–177. With a chapter by B. Keller. MR1239447

[HR24] Eric J. Hanson and Job Daisie Rock, *Decomposition of pointwise finite-dimensional $\mathbb{S}^1$ persistence modules*, J. Algebra Appl. **23** (2024), no. 3, Paper No. 2450054, 24, DOI 10.1142/S0219498824500543. MR4688822

[IRT23] Kiyoshi Igusa, Job D. Rock, and Gordana Todorov, *Continuous quivers of type A (I) foundations*, Rend. Circ. Mat. Palermo (2) **72** (2023), no. 2, 833–868, DOI 10.1007/s12215-021-00691-x. MR4559075

[Kel90] Bernhard Keller, *Chain complexes and stable categories*, Manuscripta Math. **67** (1990), no. 4, 379–417, DOI 10.1007/BF02568439. MR1052551

[KK20] Bernhard Keller and Henning Krause, *Tilting preserves finite global dimension*, C. R. Math. Acad. Sci. Paris **358** (2020), no. 5, 563–570, DOI 10.5802/crmath.72. MR4149855

[KM21] Woojin Kim and Facundo Mémoli, *Generalized persistence diagrams for persistence modules over posets*, J. Appl. Comput. Topol. **5** (2021), no. 4, 533–581, DOI 10.1007/s41468-021-00075-1. MR4334502

[LLO25] V. Lebovici, J. Lerch, and S. Oudot, *Local characterization of block-decomposability for multiparameter persistence modules*, Hal Open Science, hal-05026629f (2025).

[Mil25] Ezra Miller, *Homological algebra of modules over posets*, SIAM J. Appl. Algebra Geom. **9** (2025), no. 3, 483–524, DOI 10.1137/22M1516361. MR4941899

[ML71] Saunders MacLane, *Categories for the working mathematician*, Graduate Texts in Mathematics, Vol. 5, Springer-Verlag, New York-Berlin, 1971. MR354798

[OS24] Steve Oudot and Luis Scoccola, *On the stability of multigraded Betti numbers and Hilbert functions*, SIAM J. Appl. Algebra Geom. **8** (2024), no. 1, 54–88, DOI 10.1137/22M1489150. MR4695718

[Oud15] Steve Y. Oudot, *Persistence theory: from quiver representations to data analysis*, Mathematical Surveys and Monographs, vol. 209, American Mathematical Society, Providence, RI, 2015, DOI 10.1090/surv/209. MR3408277

[Pat18] Amit Patel, *Generalized persistence diagrams*, J. Appl. Comput. Topol. **1** (2018), no. 3-4, 397–419, DOI 10.1007/s41468-018-0012-6. MR3975559

[PRY24] C. Paquette, J. D. Rock, and E. Yıldırım, *Categories of generalized thread quivers*, `arXiv:2410.14656` (2024).

[Qui72] Daniel Quillen, *Higher algebraic K-theory. I*, Algebraic K-theory, I: Higher K-theories (Proc. Conf., Battelle Memorial Inst., Seattle, Wash., 1972), Lecture Notes in Math., Vol. 341, Springer, Berlin-New York, 1973, pp. 85–147. MR338129

[RdB02] I. Reiten and M. Van den Bergh, *Noetherian hereditary abelian categories satisfying Serre duality*, J. Amer. Math. Soc. **15** (2002), no. 2, 295–366, DOI 10.1090/S0894-0347-02-00387-9. MR1887637

Fang Li, School of Mathematical Sciences, Zhejiang University, Hangzhou, Zhejiang, People's Republic of China 310058
Email address: `fangli@zju.edu.cn`

Jiongkai Pan, School of Mathematical Sciences, Zhejiang University, Hangzhou, Zhejiang, People's Republic of China 310058
Email address: `panjiongkai@zju.edu.cn`

Shijie Zhu, School of Mathematics and Statistics, Nantong University, Nantong, Jiangsu, People's Republic of China, 226019
Email address: `shijiezhu@ntu.edu.cn`

Contemporary Mathematics
Volume **845**, 2026
https://doi.org/10.1090/conm/845/16957

Representations of queer supergroup Q(2)

František Marko and Alexandr N. Zubkov

ABSTRACT. The purpose of this paper is to explicitly describe the induced supermodules $H^0_-(\lambda)$ for the queer supergroup $\mathbf{Q}(2)$, characters of simple supermodules $L_-(\lambda)$, simple composition series of $H^0_-(\lambda)$ and blocks of $\mathbf{Q}(2)$ over a field of positive characteristic $p > 2$.

Introduction

The representation theory of the queer supergroup $\mathbf{Q}(n)$ over the field of characteristic zero has been extensively studied by Penkov and Serganova. The modular representation theory of queer supergroup $\mathbf{Q}(n)$ was advanced by the works of Brundan and Kleshchev ([**3**] and [**2**]).

The purpose of this paper is to study the representation theory of the supergroup $\mathbf{Q}(2)$ explicitly. We compute the basis of the induced and simple supermodules, establish the strong linkage principle for $\mathbf{Q}(2)$, and compute the blocks of $\mathbf{Q}(2)$.

The paper is organized as follows.

The first three sections give the background information for the representation theory of $\mathbf{Q}(2)$. In Section 1 we introduce the queer supergroups $\mathbf{Q}(2)$, its coordinate superalgebra $k[\mathbf{Q}(2)]$, maximal supertorus $\mathbf{H}(2)$, Borel subsupergroups $\mathbf{B}^+$, $\mathbf{B}^-$, their unipotent radicals $\mathbf{U}^+$, $\mathbf{U}^-$, the even part $\mathbf{U}^+_{ev}$ of $\mathbf{U}^+$, and their basic properties. In Section 2, we discuss representations of $\mathbf{H}(2)$. Section 3 introduces some results in the representation theory of $\mathbf{Q}(2)$.

The results in Sections 4 through 7 lead to a description of a basis of the induced supermodule $H^0(\lambda)$ for antidominant weights λ. The reason for considering antidominant weights instead of dominant ones is that we start with some classical result of Pommerening [**8**] to compute certain invariants with respect to the action of $\mathbf{U}^+$.

In Section 4, we describe a basis of $\mathbf{U}^+_{ev}$-invariant elements of $k[\mathbf{Q}(2)]$ using the results from [**8**] expressed using standard bitableaux T. The tableau T' assigned to a standard bitableau T is quasi-standard, and the $\mathbf{U}^+$-invariants of the coordinate algebra $\Bbbk[\mathbf{Q}(2)]$ are spanned by such tableau T' that satisfy some additional conditions. We list the generators of $\Bbbk[\mathbf{Q}(2)]^{\mathbf{U}^+_{ev}}$ in Proposition 4.3 and prove their independence in Proposition 4.7.

2020 *Mathematics Subject Classification.* Primary 17Bxx, 20G05.

In Section 5 we determine a basis $\mathfrak{B}$ of $\mathbf{U}^+$-invariant elements of $\Bbbk[\mathbf{Q}(2)]$. First, in Proposition 5.1, we determine the action of $e'_{\epsilon_1-\epsilon_2}$ on generators of $\Bbbk[\mathbf{Q}(2)]^{\mathbf{U}^+_{ev}}$ computed in Section 4. The basis of $\mathbf{U}^+$-invariants is computed in Proposition 5.2 by a lengthy process of comparison of coefficients.

In Section 6, we describe the action of $\mathbf{H}(2)$ on $\mathfrak{B}$. The action of h'_1 and h'_2 on generators of $\mathbf{U}^+_{ev}[\mathbf{Q}(2)]$ is established first, and then the action on $\mathfrak{B}$ is computed in Proposition 6.3 and 6.4.

In Section 7, we utilize results from Sections 4, 5, and 6 to describe a basis of induced supermodules $H^0(\lambda)$ for antidominant weights λ. The result is recorded in Theorem 7.1. We consider those elements from $\Bbbk[\mathbf{Q}(2)]^{\mathbf{U}^+}$ that are annihilated by the action of h'_1 and h'_2. We organize it by weight spaces and compute the invariants as the nullspaces of the corresponding matrices.

Section 8 records the action of elements $e_{\epsilon_2-\epsilon_1}$ and $e'_{\epsilon_2-\epsilon_1}$ from $\mathbf{U}^-$ on the basis $\mathfrak{B}$ of $\mathbf{U}^+$-invariant elements of $\Bbbk[\mathbf{Q}(2)]$.

In Section 9, we use the results from Section 8 to compute the action $e_{\epsilon_2-\epsilon_1}$ and $e'_{\epsilon_2-\epsilon_1}$ on the weight spaces of $H^0(\lambda)$ and determine the characters of simple supermodules corresponding to p-restricted dominant weights.

In Section 10, we switch from anti-dominant weights to dominant weights. We apply the Steinberg tensor product theorem for $\mathbf{Q}(n)$ from [**3**] to move beyond p-restricted dominant weights to dominant weights. An essential tool in this section is the description of irreducible $\mathbf{SL(2)}$-composition factors of induced $\mathbf{SL}(2)$-modules used in [**6**]. We determine simple composition factors of induced supermodules $H^0_-(\lambda)$ for dominant weights λ and establish the strong linkage principle for $\mathbf{Q}(2)$.

In Section 11, combinatorial and number-theoretical arguments are used to determine the blocks of $\mathbf{Q}(2)$ in Proposition 11.6.

Several examples illustrating arguments from Sections 10 and 11 are given in Section 12.

Finally, the paper concludes with an Appendix (Section 13), where we have collected some tedious computations and proofs of Propositions 5.1, 5.2, 6.3, 6.4, Theorem 7.1, and Propositions 8.1 and 8.2.

1. Queer supergroup $\mathbf{Q}(2)$

For the background content of this section, the reader is referred to [**3, 7, 9**]. Let $\Bbbk$ be a ground field of characteristic zero or $p > 2$.

A *queer supergroup* $\mathbf{Q}(2)$ is a group functor from the category of superalgebras $\mathsf{SAlg}_\Bbbk$ over $\Bbbk$ to the category of groups such that

$$\mathbf{Q}(2)(A) = \{\begin{pmatrix} S & S' \\ -S' & S \end{pmatrix} \mid S \in \mathrm{GL}(2)(A_0), S' \in \mathrm{Mat}_2(A_1)\},$$

where A_0 and A_1 are even and odd parts of $A \in \mathsf{SAlg}_\Bbbk$.

The supergroup $\mathbf{Q}(2)$, considered as a super-subgroup of $\mathbf{GL}(2|2)$, is closed, and therefore it is an algebraic supergroup.

The coordinate superalgebra $\Bbbk[\mathbf{Q}(2)]$ is isomorphic to

$$A(n) = \Bbbk[s_{ij}, s'_{ij} \mid 1 \le i,j \le 2]_{\det((s_{ij})_{1\le i,j\le 2})},$$

and its Hopf superalgebra structure is defined by

$$\Delta_{A(n)}(s_{ij}) = \sum_{1\le k\le 2} (s_{ik}\otimes s_{kj} - s'_{ik}\otimes s'_{kj}),\ \Delta_{A(2)}(s'_{ij}) = \sum_{1\le k\le 2} (s_{ik}\otimes s'_{kj} + s'_{ik}\otimes s_{kj}),$$

$$\epsilon_{A(2)}(s_{ij}) = \delta_{ij},\ \epsilon_{A(2)}(s'_{ij}) = 0, 1 \leq i, j \leq 2.$$

Let V be the standard $\mathbf{GL}(2|2)$-supermodule with the homogeneous basis

$$v_1, v_2, v'_1, v'_2, |v_i| = 0, |v'_i| = 1, 1 \leq i \leq 2.$$

As super subgroup of $\mathbf{GL}(2|2)$, $\mathbf{Q}(2)$ acts on V as

$$g(v_j) = \sum_{1\leq i\leq 2} v_i \otimes g_{ij} - \sum_{1\leq i\leq 2} v'_i \otimes g'_{ij},$$

$$g(v'_j) = \sum_{1\leq i\leq 2} v_i \otimes g'_{ij} + \sum_{1\leq i\leq 2} v'_i \otimes g_{ij}, 1 \leq j \leq 2.$$

The category of $\mathbf{Q}(2)$-supermodules admits the *parity change functor* Π. If W is a $\mathbf{Q}(2)$-supermodule, hence a $\Bbbk[\mathbf{Q}(2)]$-supercomodule, then ΠW has the same structure as a $\Bbbk[\mathbf{Q}(2)]$-comodule as W, but $(\Pi W)_i = W_{i+1 \pmod 2}$ for $i = 0, 1$.

There are two *standard* Borel supersubgroups $\mathbf{B}^+$ and $\mathbf{B}^-$ such that $g \in \mathbf{Q}(2)(A)$ belongs to $\mathbf{B}^+(A)$ (g belongs to $\mathbf{B}^-(A)$, respectively) if and only if the blocks S and S' of g are upper (lower, respectively) triangular matrices. It is easy to see that $g \in \mathbf{B}^+(A)$ if and only if g stabilizes the maximal flag

$$\mathcal{F}^+ : V_1 = (\Bbbk v_1 + \Bbbk v'_1) \subseteq V_2 = V.$$

Symmetrically, $g \in \mathbf{B}^-(A)$ if and only if g stabilizes the opposite (maximal) flag

$$\mathcal{F}^- : W_1 = (\Bbbk v_2 + \Bbbk v'_2) \subseteq W_2 = V.$$

The supergroup $\mathbf{Q}(2)$ contains the *maximal supertorus* $\mathbf{H}(2)$ such that $h \in \mathbf{Q}(2)(A)$ belongs to $\mathbf{H}(n)(A)$ if and only if

$$h(\Bbbk v_i + \Bbbk v'_i) \subseteq (\Bbbk v_i + \Bbbk v'_i) \otimes A, 1 \leq i \leq 2.$$

Then $\mathbf{H}(2)_{ev} = \mathbf{T}(2)$ is isomorphic to the standard maximal torus of $\mathbf{Q}(2)_{ev} \simeq \mathbf{GL}(2)$. The supergroup $\mathbf{H}(2)$ is isomorphic to the direct product of two copies of $\mathbf{Q}(1)$.

The character group $X(\mathbf{T}(2)) \simeq \mathbb{Z}^2$ has the standard basis ϵ_1, ϵ_2, where $\epsilon_i : \mathbf{T}(2) \to \mathbb{G}_m$ picks out the i-th diagonal entry.

Let $\mathfrak{m}$ denote $\ker \epsilon_{A(2)}$. Then the elements $s_{ij} - \delta_{ij}, s'_{ij}, 1 \leq i, j \leq 2$ form a homogeneous basis of the superspace $\mathfrak{m}/\mathfrak{m}^2$. The Lie superalgebra $\mathrm{Lie}(\mathbf{Q}(2))$ is naturally identified with $(\mathfrak{m}/\mathfrak{m}^2)^*$ as a superspace. It has a basis consisting of elements e_{ij}, e'_{ij} dual to the above basis of $\mathfrak{m}/\mathfrak{m}^2$. Set $\Delta = \{\pm(\epsilon_1 - \epsilon_2)\}$. The following lemma is obvious.

LEMMA 1.1. *The Lie superalgebra* $\mathbf{q}(2) = \mathrm{Lie}(\mathbf{Q}(2))$ *has the root decomposition* $\oplus_{\alpha\in\Delta\sqcup 0}\mathbf{q}(2)^\alpha$ *such that*

$$\mathbf{q}(2)^\alpha = \Bbbk e_{ij} + \Bbbk e'_{ij} \textit{ for } \alpha = \epsilon_i - \epsilon_j, i \neq j,$$

and

$$\mathbf{q}(2)^0 = \sum_{1\leq i\leq 2} (\Bbbk e_{ii} + \Bbbk e'_{ii}) = \mathrm{Lie}(\mathbf{H}(2)).$$

Moreover, we have

$$\mathrm{Lie}(\mathbf{B}^+) = \oplus_{\alpha\in\Delta^+\sqcup 0}\mathbf{q}(2)^\alpha,\ \mathrm{Lie}(\mathbf{B}^+) = \oplus_{\alpha\in\Delta^-\sqcup 0}\mathbf{q}(2)^\alpha,$$

where $\Delta^+ = \{\epsilon_1 - \epsilon_2\}$ *and* $\Delta^- = \{\epsilon_2 - \epsilon_1\}$.

As it was noted in [**3**], e_{ij} and e'_{ij} can be identified with matrices $E_{i,j}+E_{2+i,2+j}$ and $E_{i,2+j}+E_{2+i,j}$, respectively, accordingly to their induced action on V. We also denote them by e_α and e'_α, respectively, if $\alpha=\epsilon_i-\epsilon_j, 1\leq i\neq j\leq 2$. The elements e_{ii} and e'_{ii} are denoted by h_i and h'_i, respectively. The root $\epsilon_1-\epsilon_2$ is *positive*, and $\epsilon_2-\epsilon_1$ is *negative*.

The distribution superalgebra $\mathrm{Dist}(\mathbf{Q}(2))$ has a PBW basis consisting of certain ordered products

$$\prod_{1\leq i\neq j\leq 2} e_{ij}^{(m_{ij})}\prod_{1\leq i\neq j\leq 2}(e'_{ij})^{d_{ij}}\prod_{1\leq i\leq 2}\binom{h_i}{m_i}\prod_{1\leq i\leq 2}(h'_i)^{d_i},$$

where $m_{ij}, m_i\geq 0, 0\leq d_i, d_{ij}\leq 1$ for $1\leq i\neq j\leq 2$.

PROPOSITION 1.2. *There is*

(1) $\mathbf{B}^\pm\simeq \mathbf{H}(2)\ltimes \mathbf{U}^\pm$, *where* $\mathbf{U}^\pm$ *is the centralizer of the flag* $\mathcal{F}^\pm$ *in the supergroup* $\mathbf{Q}(2)$, *and also in the unipotent radical of* $\mathbf{B}^\pm$*;*

(2) $\mathrm{Dist}(\mathbf{H}(2))$ *and* $\mathrm{Dist}(\mathbf{B}^\pm)$ *are generated by the elements*

$$\binom{h_i}{m_i}, h'_i;\ m_i\geq 0, 1\leq i\leq 2,$$

and

$$e_\alpha^{(m_\alpha)}, e'_\alpha, \binom{h_i}{m_i}, h'_i;\ m_\alpha, m_i\geq 0, 1\leq i\leq 2, \alpha\in\Delta^\pm$$

respectively.

PROOF. The induced action of $\mathbf{B}^\pm$ on the superspace

$$\oplus_{1\leq k\leq 2}V_k(\mathcal{F}^\pm)/V_{k-1}(\mathcal{F}^\pm)$$

naturally defines a supergroup morphism $\mathbf{B}^\pm\to\mathbf{Q}(1)\times\mathbf{Q}(1)\simeq\mathbf{H}(2)$ such that its restriction on $\mathbf{H}(2)\leq\mathbf{B}^\pm$ is just identity map. Thus (i) follows. The proof of (ii) can be copied from [**9**, Section 3.4]. □

Throughout the article, we use that the Hopf superalgebra of the direct product of two (affine algebraic) supergroups $G\times H$ is isomorphic to $\Bbbk[G]\otimes\Bbbk[H]$, where

$$\Delta_{\Bbbk[G\times H]}(a\otimes b)=\sum(-1)^{|a_2||b_1|}a_1b_1\otimes a_2b_2,$$

$$\epsilon_{\Bbbk[G\times H]}(a\otimes b)=\epsilon_{\Bbbk[G]}(a)\epsilon_{\Bbbk[H]}(b), s_{\Bbbk[G\times H]}(a\otimes b)=s_{\Bbbk[G]}(a)\otimes s_{\Bbbk[H]}(b),$$

and

$$\Delta_{\Bbbk[G]}(a)=\sum a_1\otimes a_2,$$

$$\Delta_{\Bbbk[H]}(b)=\sum b_1\otimes b_2, a\in\Bbbk[G], b\in\Bbbk[H].$$

For any (affine algebraic) supergroup G the natural pairing $\mathrm{Dist}(G)\times\Bbbk[G]\to\Bbbk$ is denoted by

$$(\phi, f)\mapsto<\phi, f>, \phi\in\mathrm{Dist}(G), f\in\Bbbk[G].$$

2. Representations of the supergroup $\mathbf{H}(2)$

The representation theory of $\mathbf{H}(2)$ is discussed in [**3**, Lemma 6.4] and [**4**, Lemma 6.8] (see also [**7**, Proposition 5.1]). We recall results that will be needed later.

Following [**4**], we call an irreducible $\mathbf{H}(n)$-supermodule M *absolutely irreducible*, if M is irreducible as $\mathbf{H}(n)$-module. If M is reducible as $\mathbf{H}(n)$-module, it is called *self-associate*.

2.1. The case $n = 1$. The Hopf superalgebra $B = \Bbbk[\mathbf{Q}(1)]$ is isomorphic to $\Bbbk[s, s']_s$, where $|s| = 0, |s'| = 1$, and

$$\Delta_B(s) = s \otimes s - s' \otimes s', \ \Delta_B(s') = s \otimes s' + s' \otimes s,$$

$$\epsilon_B(s) = 1, \ \epsilon_B(s') = 0, \ s_B(s) = s^{-1}, \ s_B(s') = -s^{-2}s'.$$

It is well known that $\Bbbk[\mathbf{Q}(1)]$ is a direct sum of left indecomposable (injective) super submodules

$$B_m = \Bbbk s^m + \Bbbk s^{m-1}s'$$

for $m \in \mathbb{Z}$. We have

$$\Delta_B(s^m) = s^m \otimes s^m - m s^{m-1}s' \otimes s^{m-1}s'$$

and

$$\Delta_B(s^{m-1}s') = s^{m-1}s' \otimes s^m + s^m \otimes s^{m-1}s'.$$

If $p \nmid m$, then $B_m = \mathfrak{u}(m)$ is irreducible $H(1)$-supermodule of dimension two and formal character $2x_1^m$. The $H(1)$-supermodule $\mathfrak{u}(m)$ is self-associate. Indeed, if $w_+ = s_1^{\lambda_1} + \sqrt{\lambda_1}s_1^{\lambda_1-1}s_1'$ and $w_- = s_1^{\lambda_1} - \sqrt{\lambda_1}s_1^{\lambda_1-1}s_1'$, then $\mathfrak{u}(m) = \Bbbk w_+ \oplus \Bbbk w_-$ is a direct sum of one-dimensional $H(1)$ -modules.

If $p|m$, then B_m has the composition series with two quotients. The socle of B_m is $\Bbbk s^m = \mathfrak{u}(m)$ and the top of B_m is $\Bbbk s^{m-1}s' = \Pi\mathfrak{u}(m)$. Since $\mathfrak{u}(m)$ has dimension one and formal character x_1^m, it is absolutely irreducible.

2.2. The case $n = 2$. Note that $\Bbbk[\mathbf{H}(2)] \simeq \Bbbk[\mathbf{Q}(1)]^{\otimes 2}$ is isomorphic to the Hopf superalgebra $B(2) = \Bbbk[s_1, s_2, s_1', s_2']_{s_1 s_2}$, where

$$\Delta_{B(n)}(s_i) = s_i \otimes s_i - s_i' \otimes s_i', \ \Delta_{B(n)}(s_i') = s_i \otimes s_i' + s_i' \otimes s_i,$$

$$\epsilon_{B(n)}(s_i) = 1, \ \epsilon_{B(n)}(s_i') = 0, \ s_{B(n)}(s_i) = s_i^{-1}, \ s_{B(n)}(s_i') = -s_i^{-2}s_i', 1 \le i \le 2.$$

For every i, both elements s_i and s_i' have the weight ϵ_i, hence a product

$$t_\delta^\lambda = \prod_{1 \le i \le 2} s_i^{m_i} \prod_{1 \le i \le 2} (s_i')^{d_i} = s^{\lambda-\delta}(s')^\delta$$

has the weight $\lambda = (m_1, m_2) + \delta$, where $\delta = (d_1, d_2)$. The superspace $B(2)_\lambda$ has a basis $t_\delta^\lambda = s^{\lambda-\delta}(s')^\delta$, indexed by the 2-tuples $\delta \in \{0,1\}^2$. In what follows we identify $\{0,1\}^n$ with the additive group $\mathbb{Z}_2^n$.

If it does not lead to confusion, we denote t_δ^λ just by t_δ. Note that $|t_\delta| = |\delta| = \delta_1 + \delta_2 \pmod 2$.

For any weight $\mu \in X(\mathbf{T}(2))_{\ge 0}$ let $\binom{h}{\mu}$ denote the element

$$\prod_{1 \le i \le 2} \binom{h_i}{\mu_i}.$$

By the general remark from [**3**, page 72], the element $\binom{h}{\mu}$ acts on the left $\mathrm{Dist}(\mathbf{H}(2))$-supermodule $B(2)_\lambda$ by the rule

$$\binom{h}{\mu}(t_\delta) = \binom{\lambda}{\mu} t_\delta.$$

Besides, each h_i' acts on $B(n)$ as an odd superderivation such that

$$h_i'(s_j) = \delta_{ij}s_j', \ h_i'(s_j') = \delta_{ij}s_j, 1 \le j \le n.$$

Using these formulas we obtain that each h'_i acts on $B(n)_\lambda$ as

$$h'_i(t^\lambda_\delta) = (-1)^{\sum_{1\leq k<i}\delta_k}\lambda_i t^\lambda_{\delta+\epsilon_i} + (-1)^{\sum_{1\leq k<i}\delta_k}\delta_i t^\lambda_{\delta+\epsilon_i} \tag{2.1}$$

for every $\delta\in\mathbb{Z}_2^2$.

As a left $\mathrm{Dist}(\mathbf{Q}(2))$-supermodule $\Bbbk[\mathbf{Q}(2)]$ is a direct sum of injective supermodules $B(2)_\lambda\simeq B_{\lambda_1}\otimes B_{\lambda_2}$. We compute

$$\begin{aligned} &h'_1(t_0)=\lambda_1 t_{(10)}, && h'_1(t_{(10)})=t_0, && h'_1(t_{(01)})=\lambda_1 t_{(11)}, && h'_1(t_{(11)})=t_{(01)};\\ &h'_2(t_0)=\lambda_2 t_{(01)}, && h'_2(t_{(10)})=-\lambda_2 t_{(11)}, && h'_2(t_{(01)})=t_0, && h'_2(t_{(11)})=-t_{(10)}.\end{aligned}$$

1) If $p|\lambda_1, p|\lambda_2$, then $B(2)_\lambda$ has the composition series of $\mathbf{H}(2)$-subsupermodules

$$\begin{array}{ccccc} & & \langle t_{(11)}\rangle & & \\ & / & & \backslash & \\ \langle t_{(10)}\rangle & & & & \langle t_{(01)}\rangle \\ & \backslash & & / & \\ & & \langle t_{(00)}\rangle & & \end{array}$$

with factor supermodules

$$\begin{array}{ccccc} & & \mathfrak{u}(\lambda_1)\otimes\mathfrak{u}(\lambda_2) & & \\ & / & & \backslash & \\ \Pi(\mathfrak{u}(\lambda_1)\otimes\mathfrak{u}(\lambda_2)) & & & & \Pi(\mathfrak{u}(\lambda_1)\otimes\mathfrak{u}(\lambda_2)) \\ & \backslash & & / & \\ & & \mathfrak{u}(\lambda_1)\otimes\mathfrak{u}(\lambda_2) & & \end{array},$$

where $\mathfrak{u}(\lambda_1)\otimes\mathfrak{u}(\lambda_2)$ is the one-dimensional, hence absolutely irreducible $\mathbf{H}(2)$-supersubmodule $\mathfrak{u}(\lambda)$ of $B(2)_\lambda$, generated by the element t_0. The formal character of $\mathfrak{u}(\lambda)$ is $x^\lambda=x_1^{\lambda_1}x_2^{\lambda_2}$.

2) If $p\nmid\lambda_1, p|\lambda_2$, then $B(2)_\lambda$ has the composition series of $\mathbf{H}(2)$-subsupermodules

$$\begin{array}{c} B(2)_\lambda \\ | \\ \Bbbk t_0+\Bbbk t_{(10)} \end{array}$$

with factor supermodules

$$\begin{array}{c} \Pi(B_{\lambda_1}\otimes\mathfrak{u}(\lambda_2)) \\ | \\ B_{\lambda_1}\otimes\mathfrak{u}(\lambda_2) \end{array},$$

where $B_{\lambda_1}\otimes\mathfrak{u}(\lambda_2)$ is the irreducible $\mathbf{H}(2)$-supermodule $\mathfrak{u}(\lambda)$, generated by the elements t_0 and $t_{(10)}$. The formal character of $\mathfrak{u}(\lambda)$ is $2x^\lambda$. The supermodule $\mathfrak{u}(\lambda)$ is self-associate because $\Bbbk t_0+\Bbbk t_{(10)}=\Bbbk w_+\oplus\Bbbk w_-$ as $\mathbf{H}(2)$-modules, where $w_+=t_0+\sqrt{\lambda_1}t_{(10)}$ and $w_-=t_0-\sqrt{\lambda_1}t_{(10)}$.

3) The case $p|\lambda_1, p\nmid\lambda_2$ is symmetric to 2).

4) If $p\nmid\lambda_1, p\nmid\lambda_2$, then set $a=\sqrt{\lambda_1}$, $b=\sqrt{\lambda_2}$, $\iota=\sqrt{-1}$,

$$w_0=\frac{t_0}{b}+a\iota t_{(11)}, w_1=\frac{at_{(10)}}{b}+\iota t_{(01)}, u_0=\frac{t_0}{b}-a\iota t_{(11)} \text{ and } u_1=\frac{at_{(10)}}{b}-\iota t_{(01)}.$$

Then

$$h'_1(w_0)=aw_1, h'_1(w_1)=aw_0, h'_2(w_0)=-b\iota w_1, h'_2(w_1)=b\iota w_0,$$

$$h'_1(u_0)=au_1, h'_1(u_1)=au_0, h'_2(u_0)=b\iota u_1 \text{ and } h'_2(u_1)=-b\iota u_0,$$

which shows that $M_1 = \Bbbk u_0 + \Bbbk u_1$ and $M_2 = \Bbbk w_0 + \Bbbk w_1$ are obviously irreducible $\mathbf{H}(2)$-supermodules. The formal characters are equal to $2x^{\lambda}$. Since both t_0 and $t_{(11)}$ are even, $\mathfrak{u}(\lambda) \simeq M_1$ is absolutely irreducible.

The morphism given by $u_0 \mapsto iw_1$, $u_1 \mapsto iw_0$ induces an isomophism $M_1 \simeq \Pi M_2$. Thus $M_1 \simeq \mathfrak{u}(\lambda)$, and $B(2)_\lambda \simeq \mathfrak{u}(\lambda) \oplus \Pi\mathfrak{u}(\lambda)$ is semisimple.

3. Representation theory of $\mathbf{Q}(2)$

Since every $\mathbf{H}(2)$-supermodule $\mathfrak{u}(\lambda)$ can be regarded as a $\mathbf{B}^{\pm}$-supermodule, we define the induced supermodules

$$H^0_{\pm}(\lambda) = \mathrm{ind}_{\mathbf{B}^{\pm}}^{\mathbf{Q}(2)} \mathfrak{u}(\lambda).$$

Recall that for any (affine) supergroup G and its (closed) supersubgroup H, the induced supermodule $\mathrm{ind}_H^G W$, where W is an H-supermodule, can be identified wih the G-supersubmodule $(W \otimes \Bbbk[G])^H$ of $W \otimes \Bbbk[G]$ (cf. [**10**, Proposition 3.3]). More precisely, $W \otimes \Bbbk[G]$ has a structure of H-supermodule via the diagonal action of H, where the action on the second tensor factor is induced by the *left regular* action of G on $\Bbbk[G]$.

In terms of coaction, $W \otimes K[G]$ is regarded as $\Bbbk[H]$-supercomodule via the $\Bbbk[H]$-supercomodule map

$$\tau(w \otimes f) = \sum (-1)^{|f_2|(|f_1|+|g_2|)} w_1 \otimes f_2 \otimes g_2 s_{\Bbbk[H]}(\overline{f_1}), \text{ where}$$

$$w \in W, f \in \Bbbk[G], \tau_W(w) = \sum w_1 \otimes g_2, \Delta_{\Bbbk[G]}(f) = \sum f_1 \otimes f_2$$

and $\overline{f_1}$ is the image of f_1 with respect to the Hopf supealgebra morphism $\Bbbk[G] \to \Bbbk[H]$. Finally, the G-supermodule structure of $W \otimes \Bbbk[G]$ is defined by the supercomodule map $\mathrm{id}_W \otimes \Delta_{\Bbbk[G]}$, or, equivalently, by the trivial action on W and by the *right regular* action on $\Bbbk[G]$.

LEMMA 3.1. *The induced action of* $\mathrm{Dist}(H)$ *on* $W \otimes \Bbbk[G]$ *is defined as*

$$\phi(w \otimes f) = \sum (-1)^{|w||\phi_2|} \phi_1(w) \otimes \phi_2(f), \ \Delta_{\mathrm{Dist}(H)}(\phi) = \sum \phi_1 \otimes \phi_2,$$

where the action of $\mathrm{Dist}(H)$ *on* $\Bbbk[G]$ *is induced by the left regular action*

$$f \mapsto \sum (-1)^{|f_1||f_2|} f_2 \otimes s_{\Bbbk[G]}(f_1),$$

such that $\phi_2(f) = \sum f_2 < \phi_2, s_{\Bbbk[H]}(\overline{f_1}) >$.

PROOF. We have

$$\phi(w \otimes f) = \sum (-1)^{|f_2|(|f_1|+|g_2|)+|\phi|(|w_1|+|f_2|)} w_1 \otimes f_2 \otimes < \phi, g_2 s_{\Bbbk[H]}(\overline{f_1}) > .$$

Note that $< \phi, g_2 s_{\Bbbk[H]}(\overline{f_1}) > \neq 0$ implies $|\phi| = |f_1| + |g_2|$, hence

$$\phi(w \otimes f) = \sum (-1)^{|w_1|(|f_1|+|g_2|)} w_1 \otimes f_2 \otimes < \phi, g_2 s_{\Bbbk[H]}(\overline{f_1}) > .$$

The definition of comultiplication in $\mathrm{Dist}(H)$ infers

$$< \phi, g_2 s_{\Bbbk[H]}(\overline{f_1}) > = \sum (-1)^{|\phi_2||g_2|} < \phi_1, g_2 >< \phi_2, s_{\Bbbk[H]}(\overline{f_1}) >,$$

where $\sum \phi_1 \otimes \phi_2 = \Delta_{\mathrm{Dist}(H)}(\phi)$. Using the same trick with parities, we obtain

$$\phi(w \otimes f) = \sum (-1)^{|w||\phi_2|} ((-1)^{|w_1||\phi_1|} w_1 < \phi_1, g_2 > \otimes f_2 < \phi_2, s_{\Bbbk[H]}(\overline{f_1}) >) =$$

$$\sum (-1)^{|w||\phi_2|} \phi_1(w) \otimes \phi_2(f).$$

□

We apply the previous general considerations to the case $G = \mathbf{Q}(2)$, and use this description of the $\mathbf{Q}(2)$-supersubmodule $(\mathfrak{u}(\lambda)\otimes \Bbbk[\mathbf{Q}(2)])^{\mathbf{B}^{\pm}}$ in Section 7.

Let α denote $\epsilon_1-\epsilon_2$. Recall that $\mathrm{Dist}(\mathbf{Q}(2))$ is generated by the elements

$$e_{\pm\alpha}^{(m_{\pm\alpha})}, e'_{\pm\alpha}, \binom{h_s}{m_s}, h'_s;\ m_\alpha, m_s \geq 0, 1\leq s\leq 2,$$

and $\mathrm{Dist}(\mathbf{B}^+)$ and $\mathrm{Dist}(\mathbf{B}^-)$ are generated by the elements

$$e_{\alpha}^{(m_{\alpha})}, e'_{\alpha}, \binom{h_s}{m_s}, h'_s;\ m_\alpha, m_s \geq 0, 1\leq s\leq 2,$$

and by the elements

$$e_{-\alpha}^{(m_{-\alpha})}, e'_{-\alpha}, \binom{h_s}{m_s}, h'_s;\ m_{-\alpha}, m_s \geq 0, 1\leq s\leq 2,$$

respectively.

Observe that all elements $e_{\pm\alpha}, h_s$ and $e'_{\pm\alpha}, h'_s$ are even and odd primitive elements of the Hopf superalgebra $\mathrm{Dist}(\mathbf{Q}(2))$ respectively. In particular, its comultiplication map acts as

$$e_{\pm\alpha}^{(m)} \mapsto \sum_{0\leq k\leq m} e_{\pm\alpha}^{(k)}\otimes e_{\pm\alpha}^{(m-k)}$$

and

$$\binom{h_s}{m} \mapsto \sum_{0\leq k\leq m}\binom{h_s}{k}\otimes\binom{h_s}{m-k}, 1\leq s\leq 2.$$

The action of $\mathrm{Dist}(\mathbf{B}^{\pm})$ on $\mathfrak{u}(\lambda)$ is defined via superalgebra epimorphisms

$$\mathrm{Dist}(\mathbf{B}^{\pm})\to \mathrm{Dist}(\mathbf{H}(2))$$

induced by the maps

$$e_{\pm\alpha}^{(m_{\pm\alpha})}\mapsto 0, e'_{\pm\alpha}\mapsto 0, \binom{h_s}{m_s}\mapsto\binom{h_s}{m_s}, h'_s\mapsto h'_s,$$

provided $m_{\pm\alpha}>0$.

Recall that the antipode of a distribution superalgebra $\mathrm{Dist}(G)$ is defined as $s_{\mathrm{Dist}(G)}(\phi)=\phi s_{\Bbbk[G]}$. If ϕ is a primitive element of $\mathrm{Dist}(G)$, then $s_{\mathrm{Dist}(G)}(\phi)=-\phi$. Note also that for the left regular action, we have

$$\phi(f_1f_2)=\sum(-1)^{|\phi_2||f_1|}\phi_1(f_1)\phi_2(f_2),$$

where $\phi\in\mathrm{Dist}(G)$ and $f_1,f_2\in\Bbbk[G]$. Note also that $\epsilon_{\Bbbk[G]}$, the unit of $\mathrm{Dist}(G)$, acts as the identity map. In particular, if ϕ is a primitive element, then

$$\phi(f_1f_2)=\phi(f_1)f_2+(-1)^{|\phi||f_1|}f_1\phi(f_2),$$

which means ϕ acts as a superderivation. If $G=\mathbf{Q}(2)$, then the left regular action of the elements e_α and $e'_{-\alpha}$ is defined by

$$e_\alpha(s_{uv})=-\delta_{u1}s_{2v}, \text{ and } e_\alpha(s'_{uv})=-\delta_{u1}s'_{2v}, \tag{3.1}$$

$$e'_\alpha(s_{uv})=\delta_{u1}s'_{2v}, \text{ and } e'_\alpha(s'_{uv})=-\delta_{u1}s_{2v}, \tag{3.2}$$

and symmetrically

$$e_{-\alpha}(s_{uv})=-\delta_{u2}s_{1v}, \text{ and } e_{-\alpha}(s'_{uv})=-\delta_{u2}s'_{1v}, \tag{3.3}$$

$$e'_{-\alpha}(s_{uv})=\delta_{u2}s'_{1v}, \text{ and } e'_{-\alpha}(s'_{uv})=-\delta_{u2}s_{1v}, 1\leq u,v\leq 2. \tag{3.4}$$

Since all $e^{(k)}_{\pm\alpha}$ and $e'_{\pm\alpha}$ vanish on $\mathfrak{u}(\lambda)$ if $k>0$, Lemma 3.1 implies

$$e^{(m_{\pm\alpha})}_{\pm\alpha}(u_\delta\otimes f)=u_\delta\otimes e^{(m_{\pm\alpha})}_{\pm\alpha}(f) \text{ and } e'_{\pm\alpha}(u_\delta\otimes f)=u_\delta\otimes e'_{\pm\alpha}(f), f\in\Bbbk[\mathbf{Q}(2)].$$

4. A basis of $\mathbf{U}^+_{ev}$-invariant elements of $\Bbbk[\mathbf{Q}(2)]$

We consider the left regular action of $\mathbf{U}^+$. Recall that $\alpha=\epsilon_1-\epsilon_2$. The formulas (3.1) and (3.2) imply

$$e_\alpha(s_{12})=-s_{22},\ e_\alpha(s_{11})=-s_{21},\ e_\alpha(s'_{12})=-s'_{22},\ e_\alpha(s'_{11})=-s'_{21},$$
$$e'_\alpha(s_{12})=s'_{22},\ e'_\alpha(s_{11})=s'_{21},\ e'_\alpha(s'_{12})=-s_{22},\ e'_\alpha(s'_{11})=-s_{21}.$$

LEMMA 4.1. *The supergroup $\mathbf{U}^+$ is isomorphic to the direct product $G_a\times G_a^-$ of two unipotent supergroups of super dimensions $(1|0)$ and $(0|1)$, respectively. In particular, $\mathbf{U}^+_{ev}\simeq G_a$.*

PROOF. $\Bbbk[\mathbf{U}^+]$ is freely generated by the two primitive elements s_{12} and s'_{12}. □

Let $\mathbf{U}^+_{odd}$ denote the second factor G_a^- in the above decomposition. Lemma 4.1 implies that $\mathrm{Dist}(\mathbf{U}^+)\simeq\mathrm{Dist}(\mathbf{U}^+_{ev})\otimes\mathrm{Dist}(\mathbf{U}^+_{odd})$. In particular, it is a commutative (and cocommutative) Hopf superalgebra. Moreover, $\mathrm{Dist}(\mathbf{U}^+_{ev})$ is isomorphic to the (super)subalgebra of $\mathrm{Dist}(\mathbf{U}^+)$, generated by the elements $e^{(k)}_\alpha$ for $k\geq 0$. Similarly, $\mathrm{Dist}(\mathbf{U}^+_{odd})$ is isomorphic to the (super)subalgebra of $\mathrm{Dist}(\mathbf{U}^+)$, generated by the element e'_α.

Let d denote the determinant $\det((s_{uv})_{u,v\in\{1,2\}})$. It is obviously an $\mathbf{U}^+_{ev}$-invariant. Moreover, we have

$$\Bbbk[\mathbf{Q}(2)]=\varinjlim_{k\geq 0}\frac{\Bbbk[s_{kl},s'_{kl}\mid k,l\in\{1,2\}]}{d^k},$$

hence

$$\Bbbk[\mathbf{Q}(2)]^{\mathbf{U}^+_{ev}}=\varinjlim_{k\geq 0}\frac{\Bbbk[s_{kl},s'_{kl}\mid k,l\in\{1,2\}]^{\mathbf{U}^+_{ev}}}{d^k}.$$

The group $\mathbf{U}^+_{ev}$ can be identified with the the upper triangular subgroup $\mathbf{U}^+(2)$ of $\mathbf{GL}(2)$, so that for any $l\in[1,2]$ the *column spaces* $\sum_{k\in\{1,2\}}\Bbbk s_{kl}$ and $\sum_{k\in\{1,2\}}\Bbbk s'_{kl}$ are isomorphic to the standard two dimensional $\mathbf{U}^+(2)$-module, say M. Thus

$$\Bbbk[s_{kl},s'_{kl}\mid k,l\in\{1,2\}]\simeq\mathrm{Sym}(M\oplus M)\otimes\Lambda(M\oplus M)\simeq$$
$$\mathrm{Sym}(M\oplus M)\otimes(\oplus_{0\leq a,b\leq 2}\Lambda^a(M)\otimes\Lambda^b(M)).$$

Since $char\,\Bbbk\neq 2$, the canonical surjective morphism of $\mathbf{U}^+(2)$-modules $M\otimes M\to\Lambda^2(M)$ is split, and its right inverse is defined as

$$m\wedge n\mapsto\frac{1}{2}(m\otimes n-n\otimes m) \text{ for } m,n\in M.$$

Thus any component $(\mathrm{Sym}(M\oplus M)\otimes\Lambda^a(M)\otimes\Lambda^b(M))^{\mathbf{U}^+(2)}$ is the epimorphic image of $(\mathrm{Sym}(M\oplus M)\otimes M^{\otimes(a+b)})^{\mathbf{U}^+(2)}$. The latter, in turn, is isomorphic to the homogeneous component of $\mathrm{Sym}(M^{\oplus(2+a+b)})^{\mathbf{U}^+(2)}$ of degree 1 in the last $a+b$ *vector variables*.

The basis of $(\mathrm{Sym}(M^{\oplus N})^{\mathbf{U}^+(2)}$ has been calculated in [**8**] for arbitrary $N\geq 1$. Let X denote the generic matrix $(x_{kl})_{k=1,2;1\leq l\leq N}$. The left translation $X\mapsto g^{-1}X, g\in\mathbf{U}^+(2)$, induces the structure of $\mathbf{U}^+(2)$-module on $\sum_{k=1,2;1\leq l\leq 6}\Bbbk x_{kl}$, in such a way that it is naturally isomorphic to $M^{\oplus N}$. In particular, $\mathrm{Sym}(M^{\oplus N}) \simeq \Bbbk[x_{kl} \mid k = 1,2; 1 \le l \le N]$ as an $\mathbf{U}^+(2)$-module.

Let $(12|l_1l_2)$ and $(2|l)$ denote the minors $x_{1l_1}x_{2l_2} - x_{2l_1}x_{1l_2}$ and x_{2l} respectively, $1 \le l_1 < l_2 \le N, 1 \le l \le N$. A product

$$\prod_{1\le a\le s} (12|l_{1,a}l_{2,a}) \prod_{1\le b\le t} (2|l_b)$$

is identified with the *bitableau*

$$T = \begin{pmatrix} 12 & | & l_{1,1}l_{2,1} \\ \dots & | & \dots \\ 12 & | & l_{1,s}l_{2,s} \\ 2 & | & l_1 \\ \dots & | & \dots \\ 2 & | & l_t \end{pmatrix}.$$

We need to stress that there are no components of type $(1|l_a)$.

T is called *standard*, provided the numbers along rows (on both sides, individually) are strictly increasing and the numbers along columns are not decreasing. Then [**8**, Theorem 4.2] states that $\Bbbk[x_{kl} \mid k = 1,2; 1 \le l \le N]^{\mathbf{U}^+(2)}$ is spanned by the standard bitableaux.

Let us extend the above notations for the superalgebra $\Bbbk[s_{kl}, s'_{kl} \mid k,l \in \{1,2\}]$ as $(k|l) = s_{kl}$ and $(k|l') = s'_{kl}$, $k,l \in \{1,2\}$. Moreover, we define a *superminor* $(12|uv) = (1|u)(2|v) - (-1)^{|u||v|}(1|v)(2|u)$, where $u,v \in \{1,2,1',2'\}$ and $|1| = |2| = 0, |1'| = |2'| = 1$.

LEMMA 4.2. *The superalgebra $\Bbbk[s_{kl}, s'_{kl} \mid k,l \in \{1,2\}]^{\mathbf{U}^+_{ev}}$ is generated by the elements*

$$d = (12|12),\ (12|11'),\ (12|12'),\ (12|21'),\ (12|22'),\ (12|1'1'),\ (12|1'2'),\ (12|2'2')$$

and $(2|l)$ for $l \in \{1,2,1',2'\}$.

PROOF. Let us consider the component $(\mathrm{Sym}(M \oplus M) \otimes \Lambda^2(M) \otimes \Lambda^2(M))^{\mathbf{U}^+(2)}$. By the above, one has to consider the homogeneous component of $\Bbbk[x_{kl} \mid k = 1,2; 1 \le l \le N]^{\mathbf{U}^+(2)}$ of degree 1 in the entries of the last four columns of X. The latter is spanned by the standard bitableaux T whose right half-tableau contains each number $3 \le l \le 6$ at most once. Then, up to a sign, the image of T in $\Bbbk[s_{kl}, s'_{kl} \mid k,l \in \{1,2\}]^{\mathbf{U}^+_{ev}}$ can be obtained by sending the entries of the first two columns of X to the corresponding entries of the columns of $S = (s_{kl})_{k,l\in\{1,2\}}$, and by sending the entries of third and fourth (respectively, fifth and sixth) columns of X to the corresponding entries of the first (respectively, the second) column of $S' = (s'_{kl})_{k,l\in\{1,2\}}$.

We illustrate this in the following examples. Consider

$$T = \begin{pmatrix} 12 & | & 12 \\ \dots & | & \dots \\ 12 & | & 12 \\ 12 & | & 35 \\ 12 & | & 46 \end{pmatrix},$$

where the row $(12 \mid 12)$ appears with the multiplicity k. The subtableau

$$T' = \begin{pmatrix} 12 & \mid & 12 \\ \dots & \mid & \dots \\ 12 & \mid & 12 \end{pmatrix}$$

is mapped to d^k, and we can ignore it in further computations. The subtableau

$$T = \begin{pmatrix} 12 & \mid & 35 \\ 12 & \mid & 46 \end{pmatrix} =$$

$$x_{13}x_{25}x_{14}x_{26} - x_{13}x_{25}x_{16}x_{24} - x_{15}x_{23}x_{14}x_{26} + x_{15}x_{23}x_{16}x_{24}$$

is mapped to

$$x_{13} \otimes x_{14} \otimes x_{25} \otimes x_{26} - x_{13} \otimes x_{24} \otimes x_{25} \otimes x_{16} - x_{23} \otimes x_{14} \otimes x_{15} \otimes x_{26} + x_{23} \otimes x_{24} \otimes x_{15} \otimes x_{16}$$

in $M^{\otimes 4}$, and finally, to

$$s'_{11}s'_{11}s'_{22}s'_{22} - s'_{11}s'_{21}s'_{22}s'_{12} - s'_{21}s'_{11}s'_{12}s'_{22} + s'_{21}s'_{21}s'_{12}s'_{12} =$$
$$-(s'_{11}s'_{22}s'_{11}s'_{22} - s'_{11}s'_{22}s'_{21}s'_{12} - s'_{21}s'_{12}s'_{11}s'_{22} + s'_{21}s'_{12}s'_{21}s'_{12}) =$$
$$-(s'_{11}s'_{22} - s'_{21}s'_{12})(s'_{11}s'_{22} - s'_{21}s'_{12}) =$$
$$-\begin{pmatrix} 12 & \mid & 1'2' \\ 12 & \mid & 1'2' \end{pmatrix}.$$

The factor (-1) appears because of the disorder $5, 4$ on the right side. For example, the tableau

$$T = \begin{pmatrix} 12 & \mid & 34 \\ 12 & \mid & 56 \end{pmatrix}$$

is mapped to

$$\begin{pmatrix} 12 & \mid & 1'1' \\ 12 & \mid & 2'2' \end{pmatrix}.$$

In other words, the image of any tableau

$$T = \begin{pmatrix} 12 & \mid & l_{1,1}l_{2,1} \\ \dots & \mid & \dots \\ 12 & \mid & l_{1,s}l_{2,s} \\ 2 & \mid & l_1 \\ \dots & \mid & \dots \\ 2 & \mid & l_t \end{pmatrix}$$

is a bitableau

$$T' = \pm \begin{pmatrix} 12 & \mid & k_{1,1}k_{2,1} \\ \dots & \mid & \dots \\ 12 & \mid & k_{1,s}k_{2,s} \\ 2 & \mid & k_1 \\ \dots & \mid & \dots \\ 2 & \mid & k_t \end{pmatrix},$$

where each k_{uv} or k_s is the image of l_{uv} or l_s respectively, under the map $1 \mapsto 1, 2 \mapsto 2; 3, 4 \mapsto 1'; 5, 6 \mapsto 2'$. The same arguments work for other components. $\square$

Note that if we totally order the set $\{1,2,1',2'\}$ as $1<2<1'<2'$, then the above map $\{1,2,3,4,5,6\}\to\{1,2,1',1'\}$ is monotonic. The images of standard bitableaux T are bitableaux T' that are quasi-standard in the sense that the indices along rows and columns are not decreasing, and there are no rows 11 or 22. If the right-hand side of T' contains 11 or 22, then the corresponding bideterminant vanishes. Since the specialization of indices is monotonic, $\Bbbk[\mathbf{Q}(2)]^{\mathbf{U}^+_{ev}}$ is generated by the bideterminants corresponding to quasi-standard tableaux T'.

If the symbol $1'$ appears at least three times (or the symbol $2'$ appears at least three times) in a quasi-standard bitableau T', then the corresponding bideterminant also vanishes. Indeed, if we compute the corresponding polynomial in the variables s and s', then every monomial in this polynomial contains a product of of variables $s'_{11'}$ and $s'_{21'}$ of length three (of the variables $s'_{12'}$ and $s'_{22'}$, respectively,). It is clear that these monomials will always vanish.

Thus the invariants $\Bbbk[s_{kl},s'_{kl}\mid k,l\in\{1,2\}]^{\mathbf{U}^+_{ev}}$ are spanned by bideterminants corresponding to quasi-standard bitableaux T', where the multiplicities of $1'$ and $2'$ are at most 2.

Denote

$$det=C(1,2)=s_{11}s_{22}-s_{12}s_{21}, C(1,1')=s_{11}s'_{21}-s'_{11}s_{21}, C(1,2')=s_{11}s'_{22}-s'_{12}s_{21},$$

$$C(2,1')=s_{12}s'_{21}-s'_{11}s_{22}, C(2,2')=s_{12}s'_{22}-s'_{12}s_{22}, C(1',2')=s'_{11}s'_{22}+s'_{12}s'_{21}.$$

Note that $C(1',1')=2s'_{11}s'_{21}$ and $C(2',2')=2s'_{12}s'_{22}$.

For nonnegative integers a,b,c, denote the bideterminants

$$B(a,b,c)=(s_{11}s_{22}-s_{12}s_{21})^a s_{21}^b s_{22}^c.$$

PROPOSITION 4.3. *The following elements are generators of the invariants of $\Bbbk[\mathbf{Q}(2)]$ under the action of the group $\mathbf{U}^+_{ev}$: Here a,b,c are integers, and $b,c\geq 0$.*

$$x_1=B(a,b,c),$$

$$\begin{aligned}
&w_1=B(a,b,c)s'_{21}, w_2=B(a,b,c)s'_{22}, w_3=B(a,b,c)(s_{11}s'_{21}-s_{21}s'_{11}),\\
&w_4=B(a,b,c)(s_{11}s'_{22}-s_{21}s'_{12}), w_5=B(a,0,c)(s_{12}s'_{21}-s_{22}s'_{11}),\\
&w_6=B(a,0,c)(s_{12}s'_{22}-s_{22}s'_{12}),
\end{aligned}$$

$$\begin{aligned}
v_1 =&B(a,b,c)s'_{21}s'_{22}, v_2=B(a,b,c)s'_{12}s'_{22}, v_3=B(a,b,c)s'_{11}s'_{21},\\
v_4 =&B(a,b,c)C(1,1')C(1,2')=B(a,b,c)(s_{11}^2s'_{21}s'_{22}-s_{11}s_{21}s'_{11}s'_{22}+s_{11}s_{21}s'_{12}s'_{21}\\
&+s_{21}^2s'_{11}s'_{12}),\\
v_5 =&B(a,b,c)C(1,1')s'_{22}=B(a,b,c)(s_{11}s'_{21}s'_{22}-s_{21}s'_{11}s'_{22}),\\
v_6 =&B(a,b,c)C(1,2')s'_{21}=B(a,b,c)(-s_{11}s'_{21}s'_{22}-s_{21}s'_{12}s'_{21}),\\
v_7 =&B(a,0,c)C(1,1')C(2,2')=B(a,0,c)(s_{11}s_{12}s'_{21}s'_{22}-s_{12}s_{21}s'_{11}s'_{22}+s_{11}s_{22}s'_{12}s'_{21}\\
&+s_{21}s_{22}s'_{11}s'_{12}),\\
v_8 =&B(a,0,c)C(2,1')C(2,2')=B(a,0,c)(s_{12}^2s'_{21}s'_{22}-s_{12}s_{22}s'_{11}s'_{22}+s_{12}s_{22}s'_{12}s'_{21}\\
&+s_{22}^2s'_{11}s'_{12}),\\
v_9 =&B(a,0,c)C(2,1')s'_{22}=B(a,0,c)(s_{12}s'_{21}s'_{22}-s_{22}s'_{11}s'_{22}),\\
v_{10} =&B(a,0,c)C(2,2')s'_{21}=B(a,0,c)(-s_{12}s'_{21}s'_{22}-s_{22}s'_{12}s'_{21}),\\
v_{11} =&B(a,0,0)C(1',2')=B(a,0,0)(s'_{11}s'_{22}+s'_{12}s'_{21}),
\end{aligned}$$

$$
\begin{aligned}
z_1 =&B(a,b,c)s'_{11}s'_{21}s'_{22},\ z_2 = B(a,b,c)s'_{12}s'_{21}s'_{22},\\
z_3 =&B(a,b,c)C(1,1')C(1,2')s'_{21} = B(a,b,c)(s_{11}s_{21}s'_{11}s'_{21}s'_{22} + s_{21}^2 s'_{11}s'_{12}s'_{21}),\\
z_4 =&B(a,b,c)C(1,1')C(1,2')s'_{22} = B(a,b,c)(s_{11}s_{21}s'_{12}s'_{21}s'_{22} + s_{21}^2 s'_{11}s'_{12}s'_{22}),\\
z_5 =&B(a,0,c)C(1,1')C(2,2')s'_{21} = B(a,0,c)(s_{12}s_{21}s'_{11}s'_{21}s'_{22} + s_{21}s_{22}s'_{11}s'_{12}s'_{21}),\\
z_6 =&B(a,0,c)C(1,1')C(2,2')s'_{22} = B(a,0,c)(s_{11}s_{22}s'_{12}s'_{21}s'_{22} + s_{21}s_{22}s'_{11}s'_{12}s'_{22}),\\
z_7 =&B(a,0,c)C(2,1')C(2,2')s'_{21} = B(a,0,c)(s_{12}s_{22}s'_{11}s'_{21}s'_{22} + s_{22}^2 s'_{11}s'_{12}s'_{21}),\\
z_8 =&B(a,0,c)C(2,1')C(2,2')s'_{22} = B(a,0,c)(s_{12}s_{22}s'_{12}s'_{21}s'_{22} + s_{22}^2 s'_{11}s'_{12}s'_{22}),\\
z_9 =&B(a,0,0)C(1,2')C(1',2') = B(a,0,0)(s_{21}s'_{11}s'_{12}s'_{22} + s_{11}s'_{12}s'_{21}s'_{22}),\\
z_{10} =&B(a,0,0)C(2,2')C(1',2') = B(a,0,0)(s_{22}s'_{11}s'_{12}s'_{22} + s_{12}s'_{12}s'_{21}s'_{22}),\\
z_{11} =&B(a,0,0)C(1,1')C(1',2') = B(a,0,0)(-s_{11}s'_{11}s'_{21}s'_{22} - s_{21}s'_{11}s'_{12}s'_{21}),\\
z_{12} =&B(a,0,0)C(2,1')C(1',2') = B(a,0,0)(-s_{12}s'_{11}s'_{21}s'_{22} - s_{22}s'_{11}s'_{12}s'_{21}),\\
y_1 =&B(a,b,c)s'_{11}s'_{12}s'_{21}s'_{22}.
\end{aligned}
$$

The left and right weights of the above elements are given as

$$
\begin{aligned}
lw(x_1) =&(-a,-a-b-c),\\
lw(w_1) =&lw(w_2) = (-a,-a-b-c-1),\\
lw(w_3) =&lw(w_4) = lw(v_2) = lw(v_3) = (-a-1,-a-b-c-1),\\
lw(w_5) =&lw(w_6) = (-a-1,-a-c-1),\\
lw(v_1) =&(-a,-a-b-c-2),\\
lw(v_4) =&lw(v_7) = lw(v_8) = lw(y_1) = (-a-2,-a-b-c-2),\\
lw(v_5) =&lw(v_6) = lw(z_1) = lw(z_2) = (-a-1,-a-b-c-2),\\
lw(v_9) =&lw(v_{10}) = (-a-1,-a-c-2),\\
lw(v_{11}) =&(-a-1,-a-1),\\
lw(z_3) =&lw(z_4) = (-a-2,-a-b-c-3),\\
lw(z_5) =&lw(z_6) = lw(z_7) = lw(z_8) = (-a-2,-a-c-3),\\
lw(z_9) =&lw(z_{10}) = lw(z_{11}) = lw(z_{12}) = (-a-2,-a-2),
\end{aligned}
$$

and

$$
\begin{aligned}
rw(x_1) =&(-a-b,-a-c),\\
rw(w_1) =&(-a-b-1,-a-c),\\
rw(w_2) =&(-a-b,-a-c-1),\\
rw(w_3) =&(-a-b-2,-a-c),\\
rw(w_4) =&rw(v_1) = (-a-b-1,-a-c-1),\\
rw(w_5) =&(-a-1,-a-c-1),\\
rw(w_6) =&(-a,-a-c-2),\\
rw(v_2) =&(-a-b,-a-c-2),\\
rw(w_3) =&(-a-b-2,-a-c),\\
rw(v_4) =&(-a-b-3,-a-c-1),\\
rw(v_5) =&rw(v_6) = rw(z_1) = (-a-b-2,-a-c-1),
\end{aligned}
$$

$$
\begin{aligned}
rw(v_7) =& rw(v_8) = (-a-2, -a-c-2),\\
rw(v_9) =& rw(v_{10}) = (-a-1, -a-c-2),\\
rw(v_{11}) =& (-a-1, -a-1),\\
rw(z_2) =& (-a-b-1, -a-c-2),\\
rw(z_3) =& (-a-b-4, -a-c-1),\\
rw(z_4) =& (-a-b-3, -a-c-2),\\
rw(z_5) =& (-a-3, -a-c-2),\\
rw(z_6) =& rw(z_7) = (-a-2, -a-c-3),\\
rw(z_8) =& (-a-1, -a-c-4),\\
rw(z_9) =& rw(z_{12}) = rw(y_1) = (-a-2, -a-2),\\
rw(z_{10}) =& (-a-1, -a-3),\\
rw(z_{11}) =& (-a-3, -a-1).
\end{aligned}
$$

PROOF. First, we prove that the elements listed above generate all $\mathbf{U}^+_{ev}$-invariants. We work with tableaux, but at the end, we can multiply the resulting expressions by any power of det; therefore, a can be any integer.

Since the shape of T' is the same as the shape of T, and it consists of a rows of entries 12, followed by rows of entries 2, we can code all bitableux just by the right standard tableau T with respect to the ordering $1 < 2 < 1' < 2'$, we can write each row of this tableau in parentheses.

Since each tableau T is a linear combination of quasi-standard tableaux, we can assume that a generator corresponds to a quasi-standard tableau T.

If T does not contain any of the symbols $1', 2'$, then $t = 0$ and T is of the form $T = (12)^a(1)^b(2)^c$ while the corresponding bideterminant is $x_1 = B(a, b, c)$.

If T contains exactly one of the symbols $1', 2'$, then there are the following possibilities for T and the corresponding bideterminant.

$$
\begin{aligned}
&T = (12)^a(1)^b(2)^c(1') && w_1 = B(a,b,c)s'_{21};\\
&T = (12)^a(1)^b(2)^c(2') && w_2 = B(a,b,c)s'_{22};\\
&T = (12)^a(11')(1)^b(2)^c && w_3 = B(a,b,c)C(1,1');\\
&T = (12)^a(12')(1)^b(2)^c && w_4 = B(a,b,c)C(1,2');\\
&T = (12)^a(2,1')(2)^c && w_5 = B(a,0,c)C(2,1');\\
&T = (12)^a(2,2')(2)^c && w_6 = B(a,0,c)C(2,2').
\end{aligned}
$$

If T contains exactly two of the symbols $1', 2'$, then there are the following possibilities for T and the corresponding bideterminant.

$$
\begin{aligned}
&T = (12)^a(1)^b(2)^c(1')(2') && v_1 = B(a,b,c)s'_{21}s'_{22};\\
&T = (12)^a(2'2') && 2v_2 = 2B(a,0,0)s'_{12}s'_{22};\\
&T = (12)^a(12')(22')(2)^c && v_2 = B(a+1,0,c)s'_{12}s'_{22};\\
&T = (12)^a(12')(1)^b(2)^c(2') && -v_2 = -B(a,b+1,c)s'_{12}s'_{22};\\
&T = (12)^a(22')(2)^c(2') && -v_2 = -B(a,0,c+1)s'_{12}s'_{22};
\end{aligned}
$$

$$
\begin{aligned}
&T = (12)^a(1'1') && 2v_3 = 2B(a,0,0)s'_{11}s'_{21};\\
&T = (12)^a(11')(21')(2)^c && v_3 = B(a+1,0,c)s'_{11}s'_{21};\\
&T = (12)^a(11')(1)^b(2)^c(1') && -v_3 = -B(a,b+1,c)s'_{11}s'_{21};\\
&T = (12)^a(21')(2)^c(1') && -v_3 = -B(a,0,c+1)s'_{11}s'_{21};\\
&T = (12)^a(11')(12')(1)^b(2)^c && v_4 = B(a,b,c)C(1,1')C(1,2');\\
&T = (12)^a(11')(22')(2)^c && v_7 = B(a,0,c)C(1,1')C(2,2');\\
&T = (12)^a(21')(22')(2)^c && v_8 = B(a,0,c)C(2,1')C(2,2');\\
&T = (12)^a(11')(1)^b(2)^c(2') && v_5 = B(a,b,c)C(1,1')s'_{22};\\
&T = (12)^a(12')(1)^b(2)^c(1') && v_6 = B(a,b,c)C(1,2')s'_{21};\\
&T = (12)^a(21')(2)^c(2') && v_9 = B(a,0,c)C(2,1')s'_{22};\\
&T = (12)^a(22')(2)^c(1') && v_{10} = B(a,0,c)C(2,2')s'_{21};\\
&T = (12)^a(1'2') && v_{11} = B(a,0,0)C(1',2')s'_{21}.
\end{aligned}
$$

If T contains exactly three of the symbols $1', 2'$, then there are the following possibilities for T and the corresponding bideterminant.

$$
\begin{aligned}
&T = (12)^a(1'2')(1') && -z_1 = -B(a,0,0)s'_{11}s'_{21}s'_{22};\\
&T = (12)^a(11')(1)^b(2)^c(1')(2') && -z_1 = -B(a,b+1,c)s'_{11}s'_{21}s'_{22};\\
&T = (12)^a(21')(2)^b(1')(2') && -z_1 = -B(a,0,c+1)s'_{11}s'_{21}s'_{22};\\
&T = (12)^a(1'2')(2') && z_2 = B(a,0,0)s'_{12}s'_{21}s'_{22};\\
&T = (12)^a(12')(1)^a(2)^b(1')(2') && -z_2 = -B(a,b+1,c)s'_{12}s'_{21}s'_{22};\\
&T = (12)^a(22')(2)^b(1')(2') && -z_2 = -B(a,0,c+1)s'_{12}s'_{21}s'_{22};\\
&T = (12)^a(11')(12')(1)^b(2)^c(1') && z_3 = B(a,b,c)C(1,1')C(1,2')s'_{21};\\
&T = (12)^a(11')(12')(1)^b(2)^c(2') && z_4 = B(a,b,c)C(1,1')C(1,2')s'_{22};\\
&T = (12)^a(11')(22')(2)^c(1') && z_5 = B(a,0,c)C(1,1')C(2,2')s'_{21};\\
&T = (12)^a(11')(22')(2)^c(2') && z_6 = B(a,0,c)C(1,1')C(2,2')s'_{22};\\
&T = (12)^a(21')(22')(2)^b(1') && z_7 = B(a,0,c)C(2,1')C(2,2')s'_{21};\\
&T = (12)^a(21')(22')(2)^b(2') && z_8 = B(a,0,c)C(2,1')C(2,2')s'_{22};\\
&T = (12)^a(12')(1'2') && z_9 = B(a,0,0)C(1,2')C(1',1');\\
&T = (12)^a(11')(12')(22') && z_9 = B(a+1,0,0)(s_{11}s'_{21} - s'_{11}s_{21})s'_{12}s'_{22};\\
&T = (12)^a(22')(1'2') && z_{10} = B(a,0,0)C(2,2')C(1',2');\\
&T = (12)^a(11')(1'2') && z_{11} = B(a,0,0)C(1,1')C(1',2');\\
&T = (12)^a(21')(1'2') && z_{12} = B(a,0,0)C(2,1')C(1',2');\\
&T = (12)^a(11')(21')(22') && z_{12} = B(a+1,0,0)(s_{12}s'_{22} - s'_{12}s_{22})s'_{11}s'_{21}.
\end{aligned}
$$

If T contains exactly four of the symbols $1', 2'$, then there are the following possibilities for T and the corresponding bideterminant.

$$T = (12)^a(11')(12')(1)^b(2)^c(1')(2') \qquad y_1 = B(a, b+2, c)s'_{11}s'_{12}s'_{21}s'_{22};$$
$$T = (12)^a(11')(22')(2)^c(1')(2') \qquad y_1 = B(a, 1, c+1)s'_{11}s'_{12}s'_{21}s'_{22};$$
$$T = (12)^a(21')(22')(2)^c(1')(2') \qquad y_1 = B(a, 0, c+2)s'_{11}s'_{12}s'_{21}s'_{22};$$
$$T = (12)^a(11')(1'2')(2') \qquad y_1 = B(a, 1, 0)s'_{11}s'_{12}s'_{21}s'_{22};$$
$$T = (12)^a(21)(1'2')(2') \qquad y_1 = B(a, 0, 1)s'_{11}s'_{12}s'_{21}s'_{22};$$
$$T = (12)^a(1'2')(2'2') \qquad -2y_1 = -2B(a, 0, 0)s'_{11}s'_{12}s'_{21}s'_{22}.$$

Another reasoning is that since the element $s'_{11}s'_{12}s'_{21}s'_{22}$ is invariant, the generators in the case of exactly four of the symbols $1', 2'$ are $y_1 = B(a, b, c)s'_{11}s'_{12}s'_{21}s'_{22}$.

It is clear that if the number of symbols $1', 2'$ is more than four, then the corresponding expression vanishes. □

Now we show that the above generators are linearly independent. We can group the arguments by generators with the same number of odd factors.

If there are no odd factors, it is a classical result that the bideterminants $B(a, b, c)$ are linearly independent.

The generators with exactly four odd factors are of type $B(a, b, c)s'_{11}s'_{12}s'_{21}s'_{22}$, and they are linearly independent because the even bideterminants $B(a, b, c)$ are.

LEMMA 4.4. *The elements w_1 through w_6 are linearly independent.*

PROOF. For exactly one odd factor, we group the generators into four summands, the first one is a multiple of s'_{11}, the second one is a multiple of s'_{12}, the third is a multiple of s'_{21}, and the fourth one is a multiple of s'_{22}.

There are three generators w_1, w_3 and w_5, that have nonzero coefficients at s'_{21}, given by the even bideterminants $B(a_1, b_1, c_1)$, $B(a_3, b_3, c_3)s_{11}$ and $B(a_5, 0, c_5)s_{12}$. These bideterminants correspond to standard even bitableaux that are distinct because the first one does not contain any row of type $(1|*)$, and the second and third ones contain only one row of this type. The second bideterminant contains row $(1|1)$, and the third one contains row $(1|2)$. Therefore, these even bideterminants are linearly independent, and we infer that w_1, w_3, w_5 are linearly independent.

Analogously, there are three generators w_2, w_4 and w_6 nonzero coefficients at s'_{22}, given by the even bideterminants $B(a_2, b_2, c_2)$, $B(a_4, b_4, c_4)s_{11}$ and $B(a_6, 0, c_6)s_{12}$. These bideterminants correspond to standard even bitableaux that are distinct because the first one does not contain any row $(1|1)$ or $(1|2)$, and the second and third ones contain only one row of this type. The second bideterminant contains row $(1|1)$, and the third one contains row $(1|2)$. Therefore, these even bideterminants are linearly independent, and we infer that w_2, w_4, w_6 are linearly independent.

Putting it together, we conclude that the above generators, with exactly one odd factor, are linearly independent. □

Consider the generators from our list that are multiples of exactly two odd elements.

LEMMA 4.5. *The elements v_1 through v_{11} are linearly independent.*

PROOF. Since the factors $s'_{12}s'_{22}, s'_{11}s'_{21}$ appear only in the elements v_2 and v_3, these elements are not involved in any dependency relation. The remaining factors are

$$s'_{21}s'_{22},\ s'_{11}s'_{22},\ s'_{12}s'_{21},\ s'_{11}s'_{12}.$$

Consider the coefficient at the factor $s'_{11}s'_{22}$ in any possible dependency relation on the remaining elements. This factor appears in the elements v_4, v_5, v_7, v_8, v_9 and v_{11}. The corresponding coefficients up to the sign $\pm$ are

$$\begin{aligned}
&v_4 : B(a,b,c)s_{11}s_{21} = B(a,b+1,c)s_{11};\\
&v_5 : B(a,b,c)s_{21} = B(a,b+1,c);\\
&v_7 : B(a,0,c)s_{12}s_{21} = B(a+1,0,c) - B(a,0,c+1)s_{11};\\
&v_8 : B(a,0,c)s_{12}s_{22} = B(a,0,c+1)s_{12};\\
&v_9 : B(a,0,c)s_{22} = -B(a,0,c+1);\\
&v_{11} : B(a,0,0).
\end{aligned}$$

The expressions on the right-hand side are such (linear) combinations of (standard bitableaux) basic elements of $\Bbbk[s_{kl} \mid 1 \leq k,l \leq 2]$, which are linearly independent. Thus the elements v_4, v_5, v_7, v_8, v_9 and v_{11} are not involved in any dependence relation. It remains to consider the elements v_1, v_6 and v_{10}.

As above, let us consider the coefficient of a dependency relation at $s'_{21}s'_{22}$. We have coefficients up to the sign $\pm$:

$$\begin{aligned}
&v_1 : B(a,b,c);\\
&v_6 : B(a,b,c)s_{11};\\
&v_{10} : B(a,0,c)s_{12}.
\end{aligned}$$

Analogous arguments conclude the proof. □

Finally, consider the case of generators with exactly three odd factors.

LEMMA 4.6. *The generators z_1 through z_{12} are linearly independent.*

PROOF. The only generators that have nonzero coefficient at $s'_{11}s'_{12}s'_{21}$ are z_3, z_5, z_7, z_{11} and z_{12}. The corresponding coefficients, up to the sign $\pm$, are listed below and correspond to standard tableaux.

$$\begin{aligned}
&z_3 : B(a,b+2,c),\\
&z_5 : B(a,1,c+1),\\
&z_7 : B(a,0,c+2),\\
&z_{11} : -B(a,1,0),\\
&z_{12} : -B(a,0,1).
\end{aligned}$$

Therefore, in any linear dependency relation, the coefficients at z_3, z_5, z_7, z_{11} and z_{12} vanish.

The only generators that have nonzero coefficient at $s'_{11}s'_{12}s'_{22}$ are z_4, z_6, z_8, z_9 and z_{10}. The corresponding coefficients are listed below, and they correspond to standard tableaux.

$$z_4 : B(a, b+2, c),$$
$$z_6 : B(a, 1, c+1),$$
$$z_8 : B(a, 0, c+2),$$
$$z_9 : B(a, 1, 0),$$
$$z_{10} : B(a, 0, 1).$$

Therefore, the coefficients at z_4, z_6, z_8, z_9, and z_{10} vanish in any linear dependency relation.

Since z_1 and z_2 are also linearly independent, the result follows. □

We have proved the following proposition.

PROPOSITION 4.7. *The elements x_1, $w_1, \ldots w_6$, $v_1, \ldots, v_{11}$, $z_1, \ldots, z_{12}$ and y_1 are linearly independent.*

REMARK 4.8. There are relations between standard bideterminants. One of the simple relations is $C(1,2')C(2,2') = C(1,2)s'_{12}s'_{22}$ showing that standard bideterminants of different shapes can be the same. However, we found no relations between standard determinants representing the invariant elements.

5. A basis $\mathfrak{B}$ of $\mathbf{U}^+$-invariant elements of $\Bbbk[\mathbf{Q}(2)]$)

The action of e'_α on the basis elements of $\mathbf{U}^+_{ev}$-invariants of $\Bbbk[\mathbf{Q}(2)]$ from Proposition 4.3 is given as follows:

PROPOSITION 5.1.

$$\begin{aligned}
e'_\alpha(x_1) =& a w_1(a-1,b,c+1) - a w_2(a-1,b+1,c);\\
e'_\alpha(w_1) =& a v_1(a-1,b+1,c);\\
e'_\alpha(w_2) =& a v_1(a-1,b,c+1);\\
e'_\alpha(w_3) =& a v_5(a-1,b+1,c) + a v_3(a-1,b+1,c+1) + x_1(a,b+2,c);\\
e'_\alpha(w_4) =& - a v_6(a-1,b,c+1) - a v_2(a-1,b+2,c) + v_1(a,b,c) + x_1(a,b+1,c+1);\\
e'_\alpha(w_5) =& - (a+1) v_1(a,0,c) + a v_5(a-1,0,c+1) + a v_3(a-1,0,c+2)\\
&+ x_1(a,1,c+1);\\
e'_\alpha(w_6) =& a v_{10}(a-1,0,c+1) - a v_2(a-1,1,c+1) + x_1(a,0,c+2);
\end{aligned}$$

$$
\begin{aligned}
e'_\alpha(v_1) =&0;\\
e'_\alpha(v_2) =&-az_2(a-1,b,c+1)-w_2(a,b,c+1);\\
e'_\alpha(v_3) =&-az_1(a-1,b+1,c)-w_1(a,b+1,c);\\
e'_\alpha(v_4) =&az_3(a-1,b,c+1)-az_4(a-1,b+1,c)+z_1(a,b+1,c)+w_4(a,b+2,c)\\
&-w_3(a,b+1,c+1);\\
e'_\alpha(v_5) =&az_1(a-1,b+1,c+1)+w_2(a,b+2,c);\\
e'_\alpha(v_6) =&az_2(a-1,b+2,c)+w_1(a,b+1,c+1);\\
e'_\alpha(v_7) =&az_5(a-1,0,c+1)-az_4(a-1,0,c+1)-w_2(a+1,1,c)+w_4(a,1,c+1)\\
&-w_3(a,0,c+2);\\
e'_\alpha(v_8) =&az_7(a-1,0,c+1)-az_6(a-1,0,c+1)+az_2(a,0,c+1)+z_2(a,0,c+1)\\
&+w_4(a,0,c+2)-w_2(a+1,0,c+1)-w_5(a,0,c+2);\\
e'_\alpha(v_9) =&az_1(a-1,0,c+2)+w_2(a,1,c+1);\\
e'_\alpha(v_{10}) =&az_2(a-1,1,c+1)+w_1(a,0,c+2);\\
e'_\alpha(v_{11}) =&az_1(a-1,0,1)-az_2(a-1,1,0)-w_2(a,1,0)-w_1(a,0,1);
\end{aligned}
$$

$$
\begin{aligned}
e'_\alpha(z_1) =&-v_1(a,b+1,c);\\
e'_\alpha(z_2) =&-v_1(a,b,c+1);\\
e'_\alpha(z_3) =&ay_1(a-1,b+3,c)+v_6(a,b+2,c)+v_3(a,b+2,c+1);\\
e'_\alpha(z_4) =&ay_1(a-1,b+2,c+1)-v_5(a,b+1,c+1)-v_2(a,b+3,c);\\
e'_\alpha(z_5) =&ay_1(a-1,2,c+1)+v_1(a+1,1,c)+v_6(a,1,c+1)+v_3(a,1,c+2);\\
e'_\alpha(z_6) =&ay_1(a-1,1,c+2)-v_5(a,0,c+2)-v_2(a,2,c+1);\\
e'_\alpha(z_7) =&ay_1(a-1,1,c+2)+v_1(a+1,0,c+1)+v_6(a,0,c+2)+v_3(a,0,c+3);\\
e'_\alpha(z_8) =&ay_1(a-1,0,c+3)-v_9(a,0,c+2)-v_2(a,1,c+2);\\
e'_\alpha(z_9) =&ay_1(a-1,1,1)-v_2(a,2,0)+v_5(a,0,1);\\
e'_\alpha(z_{10}) =&ay_1(a-1,0,2)-v_9(a,0,1)-v_2(a,1,1);\\
e'_\alpha(z_{11}) =&-ay_1(a-1,2,0)-v_6(a,1,0)-v_3(a,1,1);\\
e'_\alpha(z_{12}) =&-ay_1(a-1,1,1)-v_{10}(a,0,1)-v_3(a,0,2);\\
e'_\alpha(y_1) =&-z_2(a,b+1,c)+z_1(a,b,c+1).
\end{aligned}
$$

For the reader's convenience, we moved the computational proof of this proposition to the Appendix.

To find the $\mathbf{U}^+$-invariants of $\Bbbk[\mathbf{Q}(2)]$, we determine those linear combinations of the basis elements from Proposition 4.3 on which e'_α acts trivially.

Since every element $v_1(a,b,c)$ is $\mathbf{U}^+$-invariant, we will consider an $\mathbf{U}^+_{ev}$-invariant element

$$\eta = \sum_{\text{admissible } a,b,c} (c(x_1)x_1 + \sum_{i=1}^{6} c(w_i)w_i + \sum_{i=2}^{11} c(v_i)v_i + \sum_{i=1}^{12} c(z_i)z_i + c(y_1)y_1),$$

where $c(x_1) = c(x_1(a,b,c))$, $c(w_i) = c(w_i(a,b,c))$ for $i=1,2,3,4$, $c(w_i) = c(w_i(a,0,c))$ for $i=5,6$ and so on.

Then η is $\mathbf{U}^+$-invariant if and only if $e'_\alpha(\eta) = 0$. We will expand $e'_\alpha(\eta)$ using the formulas from Proposition 5.1 to get defining relations between coefficients of η.

PROPOSITION 5.2. *The basis of* $\mathbf{U}^+$*-invariants consists of the following elements:*

- $a_1 = x_1(pt, 0, 0)$;
- $a_2 = x_1(a+1, 0, c+1) - (a+1)v_{10}(a, 0, c) + (a+1)v_9(a, 0, c)$
  $- a(a+1)y_1(a-1, 0, c+1))$;
- $a_3 = x_1(a+1, b+1, 0) + (a+1)v_5(a, b, 0) - (a+1)v_6(a, b, 0)$
  $- a(a+1)y_1(a-1, b+1, 0)$;
- $a_4 = v_3(a, b, 1) + v_6(a, b, 0) + ay_1(a-1, b+1, 0)$;
- $a_5 = x_1(a+1, b+1, c+1) + (a+1)v_5(a, b, c+1) - (a+1)v_6(a, b, c+1)$
  $- a(a+1)y_1(a-1, b+1, c+1)$;
- $a_6 = v_2(a, b+2, c) + v_5(a, b, c+1) - ay_1(a-1, b+1, c+1)$;
- $a_7 = v_3(a, b, c+2) + v_6(a, b, c+1) + ay_1(a-1, b+1, c+1)$;
- $a_8 = v_1(a, b, c)$;
- $a_9 = v_2(a, 1, c) + v_9(a, 0, c) - ay_1(a-1, 0, c+1)$;
- $b_1 = w_1(pt, b, c)$;
- $b_2 = w_2(pt, b, c)$;
- $b_3 = w_2(a+2, 0, c) + (a+2)z_2(a+1, 0, c)$;
- $b_4 = w_4(a+1, 0, c+1) - w_5(a+1, 0, c+1) + (2a+4)z_2(a+1, 0, c)$
  $- (a+1)z_6(a, 0, c) + (a+1)z_7(a, 0, c)$;
- $b_5 = w_1(a+2, 0, 0) + (a+2)z_1(a+1, 0, 0)$;
- $b_6 = - w_3(a+1, 0, 1) + w_4(a+1, 1, 0) + (a+2)z_1(a+1, 0, 0)$
  $- (a+1)z_4(a, 0, 0) + (a+1)z_5(a, 0, 0)$;
- $b_7 = w_1(a+2, b+1, 0) + (a+2)z_1(a+1, b+1, 0)$;
- $b_8 = - w_3(a+1, b+1, 1) + w_4(a+1, b+2, 0) + z_1(a+1, b+1, 0)$
  $+ (a+1)z_3(a, b, 1) - (a+1)z_4(a, b+1, 0)$;
- $b_9 = w_1(a+2, 0, c+1) + (a+2)z_2(a+1, 1, c)$;
- $b_{10} = w_2(a+2, 1, c) + (a+2)z_2(a+1, 1, c)$;
- $b_{11} = - w_3(a+1, 0, c+2) + w_4(a+1, 1, c+1) + (a+2)z_2(a+1, 1, c)$
  $- (a+1)z_4(a, 0, c+1) + (a+1)z_5(a, 0, c+1)$;
- $b_{12} = z_1(a+1, 0, c+1) - z_2(a+1, 1, c)$;
- $b_{13} = w_1(a+2, b+1, c+1) + (a+2)z_2(a+1, b+2, c)$;
- $b_{14} = w_2(a+2, b+2, c) + (a+2)z_2(a+1, b+2, c)$;
- $b_{15} = - w_3(a+1, b+1, c+2) + w_4(a+1, b+2, c+1) + z_2(a+1, b+2, c)$
  $+ (a+1)z_3(a, b, c+2) - (a+1)z_4(a, b+1, c+1)$;
- $b_{16} = z_1(a+1, b+1, c+1) - z_2(a+1, b+2, c)$.

The proof of Proposition 5.2 is provided in the Appendix.

We consolidate the above elements as follows.

PROPOSITION 5.3. *The basis* $\mathfrak{B}$ *of* $\mathbf{U}^+$*-invariant elements of* $\Bbbk[\mathbf{Q}(2)]$ *consists of the following elements*

- $f_1(pt,0,0) = x_1(pt,0,0)$;
- $f_2(a,0,c) = x_1(a+1,0,c+1) - (a+1)v_{10}(a,0,c) + (a+1)v_9(a,0,c)$
  $- a(a+1)y_1(a-1,0,c+1))$;
- $f_3(a,b,c) = x_1(a+1,b+1,c) + (a+1)v_5(a,b,c) - (a+1)v_6(a,b,c)$
  $- a(a+1)y_1(a-1,b+1,c)$;
- $f_4(a,b,c) = v_3(a,b,c+1) + v_6(a,b,c) + ay_1(a-1,b+1,c)$;
- $f_5(a,b,c) = v_2(a,b+2,c) + v_5(a,b,c+1) - ay_1(a-1,b+1,c+1)$;
- $f_6(a,b,c) = v_1(a,b,c)$;
- $f_7(a,0,c) = v_2(a,1,c) + v_9(a,0,c) - ay_1(a-1,0,c+1)$;
- $g_1(a,0,c) = w_2(a+1,0,c) + (a+1)z_2(a,0,c)$;
- $g_2(a,0,c) = w_4(a,0,c+1) - w_5(a,0,c+1) + (2a+2)z_2(a,0,c)$
  $- az_6(a-1,0,c) + az_7(a-1,0,c)$;
- $g_3(a,b,0) = w_1(a+1,b,0) + (a+1)z_1(a,b,0)$;
- $g_4(a,0,0) = - w_3(a+1,0,1) + w_4(a+1,1,0) + (a+2)z_1(a+1,0,0)$
  $(a+1)z_4(a,0,0) + (a+1)z_5(a,0,0)$;
- $g_5(a,b,0) = - w_3(a+1,b+1,1) + w_4(a+1,b+2,0) + z_1(a+1,b+1,0)$
  $+ (a+1)z_3(a,b,1) - (a+1)z_4(a,b+1,0)$;
- $g_6(a,b,c) = w_1(a+1,b,c+1) + (a+1)z_2(a,b+1,c)$;
- $g_7(a,b,c) = w_2(a+1,b+1,c) + (a+1)z_2(a,b+1,c)$;
- $g_8(a,b,c) = z_1(a,b,c+1) - z_2(a,b+1,c)$;
- $g_9(a,0,c) = - w_3(a+1,0,c+2) + w_4(a+1,1,c+1) + (a+2)z_2(a+1,1,c)$
  $(a+1)z_4(a,0,c+1) + (a+1)z_5(a,0,c+1)$;
- $g_{10}(a,b,c) = - w_3(a+1,b+1,c+2) + w_4(a+1,b+2,c+1)$
  $+ z_2(a+1,b+2,c) + (a+1)z_3(a,b,c+2) - (a+1)z_4(a,b+1,c+1)$;

PROOF. We combine a_3 and a_5 to f_3; a_4 and a_7 to f_4; b_5 and b_7 to g_3; b_9 and b_{13} to g_6; b_{10} and b_{14} to g_7; and b_{12} and b_{16} to g_8.

We have also used the equalities $g_3(pt-1,b,0) = w_1(pt,b,0)$, $g_6(pt-1,b,c) = w_1(pt,b,c+1)$; and $g_1(pt-1,0,c) = w_2(pt,0,c)$, $g_7(pt-1,b,c) = w_2(pt,b+1,c)$. $\square$

6. The action of $\mathbf{H}(2)$ on $\mathfrak{B}$

Observe that $\mathbf{B}^{\pm} = \mathbf{H}(2) \ltimes \mathbf{U}^{\pm}$. Moreover, the torus $\mathbf{T}(2)$ is a central subgroup of $\mathbf{H}(2)$, hence

$$(\mathfrak{u}_\lambda \otimes \Bbbk[\mathbf{Q}(2)])^{\mathbf{B}^+} = (\mathfrak{u}_\lambda \otimes \Bbbk[\mathbf{Q}(2)]^{\mathbf{U}^+})^{\mathbf{H}(2)} =$$
$$((\mathfrak{u}_\lambda \otimes \Bbbk[\mathbf{Q}(2)]^{\mathbf{U}^+})^{\mathbf{T}(2)})^{\mathbf{H}(2)/\mathbf{T}(2)} =$$
$$(\mathfrak{u}_\lambda \otimes_\lambda \Bbbk[\mathbf{Q}(2)]^{\mathbf{U}^+})^{\mathbf{H}(2)/\mathbf{T}(2)}.$$

So, it remains to calculate the invariants of the action of the elements h'_t on the subspace of $\mathfrak{u}_\lambda \otimes_\lambda (\Bbbk[\mathbf{Q}(2)]^{\mathbf{U}^+})$, where ${}_\lambda(\Bbbk[\mathbf{Q}(2)]^{\mathbf{U}^+})$ is the subspace of $\Bbbk[\mathbf{Q}(2)]^{\mathbf{U}^+}$ consisting of vectors of the (left) weight λ.

LEMMA 6.1. *The left regular action of h'_t on $\Bbbk[\mathbf{Q}(2)]$ is given by*

$$h'_t(s_{kl}) = \delta_{kt}s'_{tl},\ h'_t(s'_{kl}) = -\delta_{kt}s_{tl}, 1 \leq k, l \leq 2.$$

The action of any h'_t on an element $v \otimes f \in \mathfrak{u}(\lambda) \otimes \Bbbk[\mathbf{O}(2)]$ is given by

$$h'_t(v \otimes f) = h'_t(v) \otimes f + (-1)^{|v|} v \otimes h'_t(f),$$

where each $h'_t(v)$ is calculated in Subsection 2.2 as well as each $h'_t(f)$ can be calculated concerning the following two rules:

1) h'_t acts as a left odd superderivation;

2) on generators s_{kl} and s'_{kl} it acts as in Lemma 6.1.

Let us explain the notations for the weightspaces ${}_\lambda(\Bbbk[\mathbf{Q}(2)]$ and $\Bbbk[\mathbf{Q}(2)]_\lambda$. The weight space ${}_\lambda(\Bbbk[\mathbf{Q}(2)]$ is calculated with respect to the left regular action of the torus, that means that the generators s_{ij} and s'_{ij} have the *left weight* $-\epsilon_i$. Symmetrically, the weight space $\Bbbk[\mathbf{Q}(2)]_\lambda$ is calculated with respect to the right regular action, that means the *right weight* of s_{ij} and s'_{ij} is ϵ_j. The structure of the induced supermodule $(\mathfrak{u}_\lambda \otimes \Bbbk[\mathbf{Q}(2)]^{\mathbf{B}^+}$ is given by the right regular action on $\Bbbk[\mathbf{Q}(2)]$. Hence the weight spaces of the induced supermodules are exactly $((\mathfrak{u}_\lambda \otimes \Bbbk[\mathbf{Q}(2)]^{\mathbf{B}^+})_\mu$ for various weights μ. Consequently, $((\mathfrak{u}_\lambda \otimes \Bbbk[\mathbf{Q}(2)]^{\mathbf{B}^+})_\lambda$ has the dimension $\dim \mathfrak{u}_\lambda$.

The action of h'_1 and h'_2 on various generators is given as follows.

PROPOSITION 6.2.

$$\begin{aligned}
h'_1(x_1(a,b,c)) =& a det^{a-1} s_{21}^b s_{22}^{c+1} s'_{11} - a det^{a-1} s_{21}^{b+1} s_{22}^c s'_{12};\\
h'_1(w_1(a,b,c)) =& a det^{a-1} s_{21}^b s_{22}^{c+1} s'_{11} s'_{21} - a det^{a-1} s_{21}^{b+1} s_{22}^c s'_{12} s'_{21};\\
h'_1(w_2(a,b,c)) =& a det^{a-1} s_{21}^b s_{22}^{c+1} s'_{11} s'_{22} - a det^{a-1} s_{21}^{b+1} s_{22}^c s'_{12} s'_{22};\\
h'_1(w_3(a,b,c)) =& a det^{a-1} s_{11} s_{21}^b s_{22}^{c+1} s'_{11} s'_{21} - a det^{a-1} s_{11} s_{21}^{b+1} s_{22}^c s'_{12} s'_{21}\\
&- a det^{a-1} s_{21}^{b+2} s_{22}^c s'_{11} s'_{12} + det^a s_{11} s_{21}^{b+1} s_{22}^c + det^a s_{21}^b s_{22}^c s'_{11} s'_{21};\\
h'_1(w_4(a,b,c)) =& a det^{a-1} s_{11} s_{21}^b s_{22}^{c+1} s'_{11} s'_{22} - a det^{a-1} s_{11} s_{21}^{b+1} s_{22}^c s'_{12} s'_{22}\\
&- a det^{a-1} s_{21}^{b+1} s_{22}^{c+1} s'_{11} s'_{12} + det^a s_{12} s_{21}^{b+1} s_{22}^c + det^a s_{21}^b s_{22}^c s'_{11} s'_{22};\\
h'_1(w_5(a,0,c)) =& a det^{a-1} s_{12} s_{22}^{c+1} s'_{11} s'_{21} - a det^{a-1} s_{12} s_{21}^1 s_{22}^c s'_{12} s'_{21}\\
&- a det^{a-1} s_{21}^1 s_{22}^{c+1} s'_{11} s'_{12} + det^a s_{11} s_{22}^{c+1} + det^a s_{22}^c s'_{12} s'_{21};\\
h'_1(v_1(a,b,c)) =& a det^{a-1} s_{21}^b s_{22}^{c+1} s'_{11} s'_{21} s'_{22} - a det^{a-1} s_{21}^{b+1} s_{22}^c s'_{12} s'_{21} s'_{22};\\
h'_1(v_2(a,b,c)) =& a det^{a-1} s_{21}^b s_{22}^{c+1} s'_{11} s'_{12} s'_{22} - det^a s_{12} s_{21}^b s_{22}^c s'_{22};
\end{aligned}$$

$$\begin{aligned}
h_1'(v_3(a,b,c)) =& adet^{a-1}s_{21}^{b+1}s_{22}^{c}s_{11}'s_{12}'s_{21}' - det^a s_{11}s_{21}^{b}s_{22}^{c}s_{21}';\\
h_1'(v_5(a,b,c)) =& adet^{a-1}s_{11}s_{21}^{b}s_{22}^{c+1}s_{11}'s_{21}'s_{22}' - adet^{a-1}s_{11}s_{21}^{b+1}s_{22}^{c}s_{12}'s_{21}'s_{22}'\\
&- adet^{a-1}s_{21}^{b+2}s_{22}^{c}s_{11}'s_{12}'s_{22}' + det^a s_{11}s_{21}^{b+1}s_{22}^{c}s_{22}'\\
&+ det^a s_{21}^{b}s_{22}^{c}s_{11}'s_{21}'s_{22}';\\
h_1'(v_6(a,b,c)) =& - adet^{a-1}s_{11}s_{21}^{b}s_{22}^{c+1}s_{11}'s_{21}'s_{22}' + adet^{a-1}s_{11}s_{21}^{b+1}s_{22}^{c}s_{12}'s_{21}'s_{22}'\\
&- adet^{a-1}s_{21}^{b+1}s_{22}^{c+1}s_{11}'s_{12}'s_{21}' + det^a s_{12}s_{21}^{b+1}s_{22}^{c}s_{21}'\\
&- det^a s_{21}^{b}s_{22}^{c}s_{11}'s_{21}'s_{22}';\\
h_1'(v_9(a,0,c)) =& adet^{a-1}s_{12}s_{22}^{c+1}s_{11}'s_{21}'s_{22}' - adet^{a-1}s_{12}s_{21}^{1}s_{22}^{c}s_{12}'s_{21}'s_{22}'\\
&- adet^{a-1}s_{21}^{1}s_{22}^{c+1}s_{11}'s_{12}'s_{22}' + det^a s_{11}s_{22}^{c+1}s_{22}' + det^a s_{22}^{c}s_{12}'s_{21}'s_{22}';\\
h_1'(v_{10}(a,0,c)) =& - adet^{a-1}s_{12}s_{22}^{c+1}s_{11}'s_{21}'s_{22}' + adet^{a-1}s_{12}s_{21}^{1}s_{22}^{c}s_{12}'s_{21}'s_{22}'\\
&- adet^{a-1}s_{22}^{c+2}s_{11}'s_{12}'s_{21}' + det^a s_{12}s_{22}^{c+1}s_{21}' - det^a s_{22}^{c}s_{12}'s_{21}'s_{22}';\\
h_1'(z_1(a,b,c)) =& adet^{a-1}s_{21}^{b+1}s_{22}^{c}s_{11}'s_{12}'s_{21}'s_{22}' - det^a s_{11}s_{21}^{b}s_{22}^{c}s_{21}'s_{22}';\\
h_1'(z_2(a,b,c)) =& adet^{a-1}s_{21}^{b}s_{22}^{c+1}s_{11}'s_{12}'s_{21}'s_{22}' - det^a s_{12}s_{21}^{b}s_{22}^{c}s_{21}'s_{22}';\\
h_1'(z_3(a,b,c)) =& adet^{a-1}s_{11}s_{21}^{b+2}s_{22}^{c}s_{11}'s_{12}'s_{21}'s_{22}' + det^a s_{12}s_{21}^{b+2}s_{22}^{c}s_{11}'s_{21}'\\
&- det^a s_{11}s_{21}^{b+2}s_{22}^{c}s_{12}'s_{21}' - det^a s_{11}^{2}s_{21}^{b+1}s_{22}^{c}s_{21}'s_{22}';\\
h_1'(z_4(a,b,c)) =& adet^{a-1}s_{11}s_{21}^{b+1}s_{22}^{c+1}s_{11}'s_{12}'s_{21}'s_{22}' + det^a s_{12}s_{21}^{b+2}s_{22}^{c}s_{11}'s_{22}'\\
&- det^a s_{11}s_{21}^{b+2}s_{22}^{c}s_{12}'s_{22}' - det^a s_{11}s_{12}s_{21}^{b+1}s_{22}^{c}s_{21}'s_{22}'\\
&+ det^a s_{21}^{b+1}s_{22}^{c}s_{11}'s_{12}'s_{21}'s_{22}';\\
h_1'(z_5(a,0,c)) =& adet^{a-1}s_{12}s_{21}^{2}s_{22}^{c}s_{11}'s_{12}'s_{21}'s_{22}' + det^a s_{12}s_{21}^{1}s_{22}^{c+1}s_{11}'s_{21}'\\
&- det^a s_{11}s_{21}^{1}s_{22}^{c+1}s_{12}'s_{21}' - det^a s_{11}s_{12}s_{21}^{1}s_{22}^{c}s_{21}'s_{22}'\\
&- det^a s_{21}^{1}s_{22}^{c}s_{11}'s_{12}'s_{21}'s_{22}';\\
h_1'(z_6(a,0,c)) =& adet^{a-1}s_{11}s_{22}^{c+2}s_{11}'s_{12}'s_{21}'s_{22}' + det^a s_{12}s_{21}^{1}s_{22}^{c+1}s_{11}'s_{22}'\\
&- det^a s_{11}s_{21}^{1}s_{22}^{c+1}s_{12}'s_{22}' - det^a s_{11}s_{12}s_{22}^{c+1}s_{21}'s_{22}'\\
&+ det^a s_{22}^{c+1}s_{11}'s_{12}'s_{21}'s_{22}';\\
h_1'(z_7(a,0,c)) =& adet^{a-1}s_{12}s_{21}^{1}s_{22}^{c+1}s_{11}'s_{12}'s_{21}'s_{22}' + det^a s_{12}s_{22}^{c+2}s_{11}'s_{21}'\\
&- det^a s_{11}s_{22}^{c+2}s_{12}'s_{21}' - det^a s_{11}s_{12}s_{22}^{c+1}s_{21}'s_{22}'\\
&- det^a s_{22}^{c+1}s_{11}'s_{12}'s_{21}'s_{22}';\\
h_1'(y_1(a,b,c)) =& - det^a s_{11}s_{21}^{b}s_{22}^{c}s_{12}'s_{21}'s_{22}' + det^a s_{12}s_{21}^{b}s_{22}^{c}s_{11}'s_{21}'s_{22}'
\end{aligned}$$

and

$$\begin{aligned}
h_2'(x_1(a,b,c)) =& adet^{a-1}s_{11}s_{21}^{b}s_{22}^{c}s_{22}' - adet^{a-1}s_{12}s_{21}^{b}s_{22}^{c}s_{21}' + bdet^a s_{21}^{b-1}s_{22}^{c}s_{21}'\\
&+ cdet^a s_{21}^{b}s_{22}^{c-1}s_{22}';\\
h_2'(w_1(a,b,c)) =& - adet^{a-1}s_{11}s_{21}^{b}s_{22}^{c}s_{21}'s_{22}' - cdet^a s_{21}^{b}s_{22}^{c-1}s_{21}'s_{22}' - det^a s_{21}^{b+1}s_{22}^{c};\\
h_2'(w_2(a,b,c)) =& - adet^{a-1}s_{12}s_{21}^{b}s_{22}^{c}s_{21}'s_{22}' + bdet^a s_{21}^{b-1}s_{22}^{c}s_{21}'s_{22}' - det^a s_{21}^{b}s_{22}^{c+1};
\end{aligned}$$

$$\begin{aligned}
h_2'(w_3(a,b,c)) = &- a\mathit{det}^{a-1} s_{11}^2 s_{21}^b s_{22}^c s_{21}' s_{22}' - c\mathit{det}^a s_{11} s_{21}^b s_{22}^{c-1} s_{21}' s_{22}' \\
&+ a\mathit{det}^{a-1} s_{11} s_{21}^{b+1} s_{22}^c s_{11}' s_{22}' - a\mathit{det}^{a-1} s_{12} s_{21}^{b+1} s_{22}^c s_{11}' s_{21}' \\
&+ b\mathit{det}^a s_{21}^b s_{22}^c s_{11}' s_{21}' + c\mathit{det}^a s_{21}^{b+1} s_{22}^{c-1} s_{11}' s_{22}' - \mathit{det}^a s_{11} s_{21}^{b+1} s_{22}^c \\
&+ \mathit{det}^a s_{21}^b s_{22}^c s_{11}' s_{21}';
\end{aligned}$$

$$\begin{aligned}
h_2'(w_4(a,b,c)) = &- a\mathit{det}^{a-1} s_{11} s_{12} s_{21}^b s_{22}^c s_{21}' s_{22}' + b\mathit{det}^a s_{11} s_{21}^{b-1} s_{22}^c s_{21}' s_{22}' \\
&+ a\mathit{det}^{a-1} s_{11} s_{21}^{b+1} s_{22}^c s_{12}' s_{22}' - a\mathit{det}^{a-1} s_{12} s_{21}^{b+1} s_{22}^c s_{12}' s_{21}' \\
&+ b\mathit{det}^a s_{21}^b s_{22}^c s_{12}' s_{21}' + c\mathit{det}^a s_{21}^{b+1} s_{22}^{c-1} s_{12}' s_{22}' - \mathit{det}^a s_{11} s_{21}^b s_{22}^{c+1} \\
&+ \mathit{det}^a s_{21}^b s_{22}^c s_{12}' s_{21}';
\end{aligned}$$

$$\begin{aligned}
h_2'(w_5(a,0,c)) = &- a\mathit{det}^{a-1} s_{11} s_{12} s_{22}^c s_{21}' s_{22}' - c\mathit{det}^a s_{12} s_{22}^{c-1} s_{21}' s_{22}' \\
&+ a\mathit{det}^{a-1} s_{11} s_{22}^{c+1} s_{11}' s_{22}' - a\mathit{det}^{a-1} s_{12} s_{22}^{c+1} s_{11}' s_{21}' + c\mathit{det}^a s_{22}^c s_{11}' s_{22}' \\
&- \mathit{det}^a s_{12} s_{21}^1 s_{22}^c + \mathit{det}^a s_{22}^c s_{11}' s_{22}';
\end{aligned}$$

$$h_2'(v_1(a,b,c)) = \mathit{det}^a s_{21}^b s_{22}^{c+1} s_{21}' - \mathit{det}^a s_{21}^{b+1} s_{22}^c s_{22}';$$

$$\begin{aligned}
h_2'(v_2(a,b,c)) = &a\mathit{det}^{a-1} s_{12} s_{21}^b s_{22}^c s_{12}' s_{21}' s_{22}' - b\mathit{det}^a s_{21}^{b-1} s_{22}^c s_{12}' s_{21}' s_{22}' \\
&+ \mathit{det}^a s_{21}^b s_{22}^{c+1} s_{12}';
\end{aligned}$$

$$\begin{aligned}
h_2'(v_3(a,b,c)) = &a\mathit{det}^{a-1} s_{11} s_{21}^b s_{22}^c s_{11}' s_{21}' s_{22}' + c\mathit{det}^a s_{21}^b s_{22}^{c-1} s_{11}' s_{21}' s_{22}' \\
&+ \mathit{det}^a s_{21}^{b+1} s_{22}^c s_{11}';
\end{aligned}$$

$$\begin{aligned}
h_2'(v_5(a,b,c)) = &- a\mathit{det}^{a-1} s_{12} s_{21}^{b+1} s_{22}^c s_{11}' s_{21}' s_{22}' \\
&+ b\mathit{det}^a s_{21}^b s_{22}^c s_{11}' s_{21}' s_{22}' - \mathit{det}^a s_{21}^{b+1} s_{22}^{c+1} s_{11}' \\
&+ \mathit{det}^a s_{11} s_{21}^b s_{22}^{c+1} s_{21}' - \mathit{det}^a s_{11} s_{21}^{b+1} s_{22}^c s_{22}' + \mathit{det}^a s_{21}^b s_{22}^c s_{11}' s_{21}' s_{22}';
\end{aligned}$$

$$\begin{aligned}
h_2'(v_6(a,b,c)) = &- a\mathit{det}^{a-1} s_{11} s_{21}^{b+1} s_{22}^c s_{12}' s_{21}' s_{22}' - c\mathit{det}^a s_{21}^{b+1} s_{22}^{c-1} s_{12}' s_{21}' s_{22}' \\
&- \mathit{det}^a s_{21}^{b+2} s_{22}^c s_{12}' - \mathit{det}^a s_{11} s_{21}^b s_{22}^{c+1} s_{21}' + \mathit{det}^a s_{11} s_{21}^{b+1} s_{22}^c s_{22}';
\end{aligned}$$

$$\begin{aligned}
h_2'(v_9(a,0,c)) = &- a\mathit{det}^{a-1} s_{12} s_{22}^{c+1} s_{11}' s_{21}' s_{22}' - \mathit{det}^a s_{22}^{c+2} s_{11}' + \mathit{det}^a s_{12} s_{22}^{c+1} s_{21}' \\
&- \mathit{det}^a s_{12} s_{21}^1 s_{22}^c s_{22}';
\end{aligned}$$

$$\begin{aligned}
h_2'(v_{10}(a,0,c)) = &- a\mathit{det}^{a-1} s_{11} s_{22}^{c+1} s_{12}' s_{21}' s_{22}' - c\mathit{det}^a s_{22}^c s_{12}' s_{21}' s_{22}' - \mathit{det}^a s_{21}^1 s_{22}^{c+1} s_{12}' \\
&- \mathit{det}^a s_{12} s_{22}^{c+1} s_{21}' + \mathit{det}^a s_{12} s_{21}^1 s_{22}^c s_{22}' - \mathit{det}^a s_{22}^c s_{12}' s_{21}' s_{22}'
\end{aligned}$$

$$h_2'(z_1(a,b,c)) = - \mathit{det}^a s_{21}^b s_{22}^{c+1} s_{11}' s_{21}' + \mathit{det}^a s_{21}^{b+1} s_{22}^c s_{11}' s_{22}';$$

$$h_2'(z_2(a,b,c)) = - \mathit{det}^a s_{21}^b s_{22}^{c+1} s_{12}' s_{21}' + \mathit{det}^a s_{21}^{b+1} s_{22}^c s_{12}' s_{22}';$$

$$\begin{aligned}
h_2'(z_3(a,b,c)) = &- a\mathit{det}^{a-1} s_{11} s_{21}^{b+2} s_{22}^c s_{11}' s_{12}' s_{21}' s_{22}' - c\mathit{det}^a s_{21}^{b+2} s_{22}^{c-1} s_{11}' s_{12}' s_{21}' s_{22}' \\
&- \mathit{det}^a s_{21}^{b+3} s_{22}^c s_{11}' s_{12}' - \mathit{det}^a s_{11} s_{21}^{b+1} s_{22}^{c+1} s_{11}' s_{21}' \\
&+ \mathit{det}^a s_{11} s_{21}^{b+2} s_{22}^c s_{11}' s_{22}';
\end{aligned}$$

$$\begin{aligned}
h_2'(z_4(a,b,c)) = &- a\mathit{det}^{a-1} s_{12} s_{21}^{b+2} s_{22}^c s_{11}' s_{12}' s_{21}' s_{22}' + b\mathit{det}^a s_{21}^{b+1} s_{22}^c s_{11}' s_{12}' s_{21}' s_{22}' \\
&- \mathit{det}^a s_{21}^{b+2} s_{22}^{c+1} s_{11}' s_{12}' - \mathit{det}^a s_{11} s_{21}^{b+1} s_{22}^{c+1} s_{12}' s_{21}' \\
&+ \mathit{det}^a s_{11} s_{21}^{b+2} s_{22}^c s_{12}' s_{22}' \\
&+ 2\mathit{det}^a s_{21}^{b+1} s_{22}^c s_{11}' s_{12}' s_{21}' s_{22}';
\end{aligned}$$

$$\begin{aligned}
h_2'(z_5(a,0,c)) =& - adet^{a-1}s_{11}s_{21}^1s_{22}^{c+1}s_{11}'s_{12}'s_{21}'s_{22}' - cdet^a s_{21}^1 s_{22}^c s_{11}'s_{12}'s_{21}'s_{22}'\\
&- det^a s_{21}^2 s_{22}^{c+1}s_{11}'s_{12}' - det^a s_{12}s_{21}^1 s_{22}^{c+1}s_{11}'s_{21}'\\
&+ det^a s_{12}s_{21}^2 s_{22}^c s_{11}'s_{22}'\\
&- det^a s_{21}^1 s_{22}^c s_{11}'s_{12}'s_{21}'s_{22}';\\
h_2'(z_6(a,0,c)) =& - adet^{a-1}s_{12}s_{21}s_{22}^{c+1}s_{11}'s_{12}'s_{21}'s_{22}' - det^a s_{21}^1 s_{22}^{c+2}s_{11}'s_{12}'\\
&- det^a s_{11}s_{22}^{c+2}s_{12}'s_{21}' + det^a s_{11}s_{21}^1 s_{22}^{c+1}s_{12}'s_{22}'\\
&+ det^a s_{22}^{c+1}s_{11}'s_{12}'s_{21}'s_{22}';\\
h_2'(z_7(a,0,c)) =& - adet^{a-1}s_{11}s_{22}^{c+2}s_{11}'s_{12}'s_{21}'s_{22}' - cdet^a s_{22}^{c+1}s_{11}'s_{12}'s_{21}'s_{22}'\\
&- det^a s_{21}^1 s_{22}^{c+2}s_{11}'s_{12}' - det^a s_{12}s_{22}^{c+2}s_{11}'s_{21}' + det^a s_{12}s_{21}^1 s_{22}^{c+1}s_{11}'s_{22}'\\
&- 2det^a s_{22}^{c+1}s_{11}'s_{12}'s_{21}'s_{22}';\\
h_2'(y_1(a,b,c)) =& - det^a s_{21}^{b+1}s_{22}^c s_{11}'s_{12}'s_{22}' + det^a s_{21}^b s_{22}^{c+1}s_{11}'s_{12}'s_{21}'.
\end{aligned}$$

The action of h_1' on the $\mathbf{U}^+$-invariants is recorded as follows.

PROPOSITION 6.3.

$$\begin{aligned}
h_1'(f_1(pt,0,0)) =&0;\\
h_1'(f_2(a,0,c)) =&(a+1)g_2(a,0,c);
\end{aligned}$$

If $b,c>0$, then

$$h_1'(f_3(a,b,c)) = (a+1)g_{10}(a-1,b-1,c-1) + (a+1)g_6(a,b,c-1) + (a+1)(a+2)g_8(a,b,c-1).$$

If $c>0$ and $b=0$, then

$$h_1'(f_3(a,0,c) = (a+1)g_9(a-1,0,c-1) + (a+1)g_6(a,0,c-1) + (a+1)(2a+2)g_8(a,0,c-1).$$

If $b>0$ and $c=0$, then

$$h_1'(f_3(a,b,0)) = (a+1)g_5(a-1,b-1,0) + (a+1)g_3(a,b,0).$$

If $b=c=0$, then

$$h_1'(f_3(a,0,0)) = (a+1)g_4(a-1,0,0) + (a+1)g_3(a,0,0).$$

If $c=0$, then

$$h_1'(f_4(a,b,0)) = -g_3(a,b,0).$$

If $c>0$, then

$$h_1'(f_4(a,b,c)) = -g_6(a,b,c-1) - (a+1)g_8(a,b,c-1).$$

$$
\begin{aligned}
h_1'(f_5(a,b,c)) =& g_7(a,b,c) + (a+1)g_8(a,b,c);\\
h_1'(f_6(a,b,c)) =& ag_8(a-1,b,c);\\
h_1'(f_7(a,0,c)) =& g_1(a,0,c);\\
h_1'(g_1(a,0,c)) =& -(a+1)f_7(a,0,c)\\
h_1'(g_2(a,0,c)) =& -f_2(a,0,c)\\
h_1'(g_3(a,b,0)) =& (a+1)f_4(a,b,0);\\
h_1'(g_4(a,0,0)) =& -f_3(a+1,0,0) - (a+2)f_4(a+1,0,0);\\
h_1'(g_5(a,b,0)) =& -f_3(a+1,b+1,0) - (a+2)f_4(a+1,b+1,0);\\
h_1'(g_6(a,b,c)) =& (a+1)f_4(a,b,c+1) + (a+1)f_6(a+1,b,c);\\
h_1'(g_7(a,b,c)) =& -(a+1)f_5(a,b,c) + (a+1)f_6(a+1,b,c)\\
h_1'(g_8(a,b,c)) =& -f_6(a+1,b,c);\\
h_1'(g_9(a,0,c)) =& -f_3(a+1,0,c+1) - (a+2)f_4(a+1,0,c+1)\\
&+ (a+2)f_6(a+2,0,c);\\
h_1'(g_{10}(a,b,c)) =& -f_3(a+1,b+1,c+1) - (a+2)f_4(a+1,b+1,c+1)\\
&+ f_6(a+2,b+1,c).
\end{aligned}
$$

For the reader's convenience, we moved the computational proof of this proposition to the Appendix.

Regarding the action of h_2', we have the following.

PROPOSITION 6.4.

$$
\begin{aligned}
h_2'(f_1(pt,0,0) =& 0;\\
h_2'(f_2(a,0,c)) =& (2a+c+3)g_1(a,0,c) - (a+1)g_2(a,0,c);
\end{aligned}
$$

If $b,c>0$ *then*

$$
\begin{aligned}
h_2'(f_3(a,b,c)) =& -(a+1)g_{10}(a-1,b-1,c-1) + (a+b+2)g_6(a,b,c-1)\\
&+ cg_7(a,b,c-1) + (a+1)(a+b+1)g_8(a,b,c-1).
\end{aligned}
$$

If $c>0$, *then*

$$
\begin{aligned}
h_2'(f_3(a,0,c)) =& -(a+1)g_9(a-1,0,c-1) + (a+2)g_6(a,0,c-1)\\
&+ cg_7(a,0,c-1) + (a+1)g_8(a,0,c-1).
\end{aligned}
$$

If $b>0$, *then*

$$h_2'(f_3(a,b,0)) = -(a+1)g_5(a-1,b-1,0) + (a+b+2)g_3(a,b,0).$$

Also,

$$h_2'(f_3(a,0,0)) = -(a+1)g_4(a-1,0,0) + (a+2)g_3(a,0,0).$$

If $b,c>0$, *then*

$$h_2'(f_4(a,b,c)) = g_{10}(a-1,b-1,c-1) + (c+1)g_8(a,b,c-1).$$

If $b>0$, *then*

$$h_2'(f_4(a,b,0)) = g_5(a-1,b-1,0).$$

If $c>0$, *then*

$$h_2'(f_4(a,0,c)) = g_9(a-1,0,c-1) + (c+a+1)g_8(a,0,c-1).$$

Also,

$$h_2'(f_4(a,0,0) = g_4(a-1,0,0).$$

If $b>0$*, then*

$$h_2'(f_5(a,b,c)) = -g_{10}(a-1,b-1,c) + (b+a+1)g_8(a,b,c).$$

Also,

$$h_2'(f_5((a,0,c)) = -g_9(a-1,0,c) + g_8(a,0,c).$$

$$\begin{aligned}
h_2'(f_6(a,b,c)) =& g_6(a-1,b,c) - g_7(a-1,b,c);\\
h_2'(f_7(a,0,c)) =& g_1(a,0,c) - g_2(a,0,c);\\
h_2'(g_1(a,0,c)) =& - f_2(a,0,c) + (a+1)f_7(a,0,c);\\
h_2'(g_2(a,0,c)) =& (2a+c+3)f_7(a,0,c) - f_2(a,0,c);\\
h_2'(g_3(a,b,0)) =& - f_3(a,b,0) - (a+1)f_4(a,b,0);\\
h_2'(g_4(a,0,0)) =& - (a+3)f_4(a+1,0,0);\\
h_2'(g_5(a,b,0)) =& - (a+b+4)f_4(a+1,b+1,0);\\
h_2'(g_6(a,b,c)) =& [-f_3(a,b,c+1) + (a+1)f_5(a,b,c)] - (c+1)f_6(a+1,b,c);\\
h_2'(g_7(a,b,c)) =& [-f_3(a,b,c+1) + (a+1)f_5(a,b,c)] + (a+b+2)f_6(a+1,b,c);\\
h_2'(g_8(a,b,c)) =& - f_4(a,b,c+1) - f_5(a,b,c);\\
h_2'(g_9(a,0,c)) =& (a+c+3)f_5(a+1,0,c) - f_4(a+1,0,c+1);\\
h_2'(g_{10}(a,b,c)) =& (c+2)f_5(a+1,b+1,c) - (a+b+3)f_4(a+1,b+1,c+1).
\end{aligned}$$

For the reader's convenience, we moved the computational proof of this proposition to the Appendix.

7. Basis of induced supermodules $H^0(\lambda)$ for antidominant λ

In this section, we initially assume that the characteristic of $\Bbbk$ is zero. We compute a basis of the induced supermodule $H^0(\lambda)$ for antidominant weights λ. Then, considering the explicit elements of this basis, we observe that after the process of modular reduction, they remain linearly independent over any ground field $\Bbbk$ of characteristic $p>2$.

For the description of $H^0(\lambda) = (\mathfrak{u}(\lambda)\otimes\Bbbk[\mathbf{Q}(2)])^{\mathbf{B}^+}$, we apply the results derived in Section 3.

We list the weights of elements from $\mathfrak{B}$. The left weights are:

$$\begin{aligned}
lwt(f_1(pt,0,0)) =& lwt(x_1(pt,0,0)) &=& - (pt,pt)\\
lwt(f_2(a,0,c)) =& lwt(x_1(a+1,0,c+1)) &=& - (a+1,a+c+2)\\
lwt(f_3(a,b,c)) =& lwt(x_1(a+1,b+1,c)) &=& - (a+1,a+b+c+2)\\
lwt(f_4(a,b,c)) =& lwt(v_6(a,b,c)) &=& - (a+1,a+b+c+2)\\
lwt(f_5(a,b,c)) =& lwt(v_5(a,b,c+1)) &=& - (a+1,a+b+c+3)\\
lwt(f_6(a,b,c)) =& lwt(v_1(a,b,c)) &=& - (a,a+b+c+2)\\
lwt(f_7(a,0,c)) =& lwt(v_2(a,1,c)) &=& - (a+1,a+c+2)
\end{aligned}$$

$$\begin{aligned}
lwt(g_1(a,0,c)) =& lwt(w_2(a+1,0,c)) &&=-(a+1,a+c+2)\\
lwt(g_2(a,0,c)) =& lwt(z_2(a,0,c)) &&=-(a+1,a+c+2)\\
lwt(g_3(a,b,0)) =& lwt(w_1(a+1,b,0)) &&=-(a+1,a+b+2)\\
lwt(g_4(a,0,0)) =& lwt(w_3(a+1,0,1)) &&=-(a+2,a+3)\\
lwt(g_5(a,b,0)) =& lwt(z_1(a+1,b+1,0)) &&=-(a+2,a+b+4)\\
lwt(g_6(a,b,c)) =& lwt(w_1(a+1,b,c+1)) &&=-(a+1,a+b+c+3)\\
lwt(g_7(a,b,c)) =& lwt(w_2(a+1,b+1,c)) &&=-(a+1,a+b+c+3)\\
lwt(g_8(a,b,c)) =& lwt(z_1(a,b,c+1)) &&=-(a+1,a+b+c+3)\\
lwt(g_9(a,0,c)) =& lwt(z_2(a+1,1,c)) &&=-(a+2,a+c+4)\\
lwt(g_{10}(a,b,c)) =& lwt(z_2(a+1,b+2,c)) &&=-(a+2,a+b+c+5).
\end{aligned}$$

The right weights are

$$\begin{aligned}
rwt(f_1(pt,0,0)) =& rwt(x_1(pt,0,0)) &&=(pt,pt)\\
rwt(f_2(a,0,c)) =& rwt(x_1(a+1,0,c+1)) &&=(a+1,a+c+2)\\
rwt(f_3(a,b,c)) =& rwt(x_1(a+1,b+1,c)) &&=(a+b+2,a+c+1)\\
rwt(f_4(a,b,c)) =& rwt(v_6(a,b,c)) &&=(a+b+2,a+c+1)\\
rwt(f_5(a,b,c)) =& rwt(v_5(a,b,c+1)) &&=(a+b+2,a+c+2)\\
rwt(f_6(a,b,c)) =& rwt(v_1(a,b,c)) &&=(a+b+1,a+c+1)\\
rwt(f_7(a,0,c)) =& rwt(v_2(a,1,c)) &&=(a+1,a+c+2)\\
rwt(g_1(a,0,c)) =& rwt(z_2(a,0,c)) &&=(a+1,a+c+2)\\
rwt(g_2(a,0,c)) =& rwt(z_2(a,0,c)) &&=(a+1,a+c+2)\\
rwt(g_3(a,b,0)) =& rwt(w_1(a+1,b,0)) &&=(a+b+2,a+1)\\
rwt(g_4(a,0,0)) =& rwt(w_3(a+1,0,1)) &&=(a+3,a+2)\\
rwt(g_5(a,b,0)) =& rwt(z_1(a+1,b+1,0)) &&=(a+b+4,a+2)\\
rwt(g_6(a,b,c)) =& rwt(w_1(a+1,b,c+1)) &&=(a+b+2,a+c+2)\\
rwt(g_7(a,b,c)) =& rwt(w_2(a+1,b+1,c)) &&=(a+b+2,a+c+2)\\
rwt(g_8(a,b,c)) =& rwt(z_1(a,b,c+1)) &&=(a+b+2,a+c+2)\\
rwt(g_9(a,0,c)) =& rwt(z_2(a+1,1,c)) &&=(a+3,a+c+3)\\
rwt(g_{10}(a,b,c)) =& rwt(z_2(a+1,b+2,c)) &&=(a+b+4,a+c+3).
\end{aligned}$$

Now we list all elements of $\mathfrak{B}$ of the of the left weight $-\lambda$, where $\lambda=(\lambda_1,\lambda_2)$ is antidominant.

$\underline{\lambda_1=k, \lambda_2=k}$

The $\mathbf{U}^+$-invariants of the left weight $(-k,-k)$ consists of multiples of $f_1(k,0,0)$ if p divides k, and the space is trivial if p does not divide k.

$\underline{\lambda_1=k, \lambda_2=k+1}$

Basis of $\mathbf{U}^+$-invariants with the left weight $(-k,-k-1)$ consists of

$$\begin{aligned}
B_{(k,k+1)} =& \{f_2(k-1,0,0), f_3(k-1,0,0), f_4(k-1,0,0), f_7(k-1,0,0),\\
& g_1(k-1,0,0), g_2(k-1,0,0), g_3(k-1,0,0), g_4(k-2,0,0)\}.
\end{aligned}$$

This set breaks into two subsets

$$\{f_2(k-1,0,0), f_7(k-1,0,0), g_1(k-1,0,0), g_2(k-2,0,0)\}$$

of elements of right weight $(k, k+1)$, and

$$\{f_3(k-1,0,0), f_4(k-1,0,0), g_3(k-1,0,0), g_4(k-2,0,0)\}$$

of elements of right weight $(k+1, k)$. Only the weight $(k, k+1)$ is antidominant.

$\underline{\lambda_1 = k, \lambda_2 = k+2}$

Basis of $\mathbf{U}^+$-invariants with the left weight $(-k, -k-2)$ consists of

$$\begin{aligned} B_{(k,k+2)} =\{& f_2(k-1,0,1), f_3(k-1,1,0), f_3(k-1,0,1),\\ & f_4(k-1,1,0), f_4(k-1,0,1), f_7(k-1,0,1),\\ & g_1(k-1,0,1), g_2(k-1,0,1), g_3(k-1,1,0),\\ & f_5(k-1,0,0), f_6(k,0,0), g_5(k-2,0,0), g_6(k-1,0,0),\\ & g_7(k-1,0,0), g_8(k-1,0,0), g_9(k-2,0,0)\}. \end{aligned}$$

This set breaks into subsets

$$\{f_2(k-1,0,1), f_7(k-1,0,1), g_1(k-1,0,1), g_2(k-1,0,1)\}$$

of elements of right weight $(k, k+2)$,

$$\{f_3(k-1,1,0), f_4(k-1,1,0), g_3(k-1,1,0), g_5(k-2,0,0)\}$$

of elements of right weight $(k+2, k)$,

$$\begin{aligned} \{& f_3(k-1,0,1), f_4(k-1,0,1), f_5(k-1,0,0), f_6(k,0,0),\\ & g_6(k-1,0,0), g_7(k-1,0,0), g_8(k-1,0,0), g_9(k-2,0,0)\} \end{aligned}$$

of elements of right weight $(k+1, k+1)$.

Only the weights $(k, k+2)$ and $(k+1, k+1)$ are antidominant.

$\underline{\lambda_1 = k,\ \lambda_2 = k+q,\ q \geq 3}$

Basis of $\mathbf{U}^+$-invariants with the left weight $(-k, -k-q)$ consists of

$$\begin{aligned} B_{(k,k+2)} =\{& f_2(k-1,0,q-1), f_3(k-1,u,v), f_4(k-1,u,v) \text{ where } u+v=q-1,\\ & f_7(k-1,0,q-1), g_1(k-1,0,q-1), g_2(k-1,0,q-1), g_3(k-1,q-1,0),\\ & f_5(k-1,u,v) \text{ where } u+v=q-2, f_6(k,u,v) \text{ where } u+v=q-2,\\ & g_5(k-2,q-2,0), g_6(k-1,u,v), g_7(k-1,u,v),\\ & g_8(k-1,u,v) \text{ where } u+v=q-2, g_9(k-2,0,q-2),\\ & g_{10}(k-2,u,v) \text{ where } u+v=q-3.\} \end{aligned}$$

This set breaks into subsets

$$S_{(k,k+q)} = \{f_2(k-1,0,q-1), f_7(k-1,0,q-1), g_1(k-1,0,q-1), g_2(k-1,0,q-1)\}$$

of elements of right weight $(k, k+q)$,

$$S_{(k+q,k)} = \{f_3(k-1,q-1,0), f_4(k-1,q-1,0), g_3(k-1,q-1,0), g_5(k-2,q-2,0)\}$$

of elements of right weight $(k+q, k)$,

$$\begin{aligned} S_{(k+1,k+q-1)} = \{& f_3(k-1,0,q-1), f_4(k-1,0,q-1), f_5(k-1,0,q-2),\\ & f_6(k,0,q-2), g_6(k-1,0,q-2), g_7(k-1,0,q-2),\\ & g_8(k-1,0,q-2), g_9(k-2,0,q-2)\} \end{aligned}$$

of weight $(k+1, k+q-1)$,

$$\begin{aligned}S_{(k+q-v,k+v)} = \{&f_3(k-1,q-1-v,v), f_4(k-1,q-1-v,v),\\ &f_5(k-1,q-1-v,v-1), f_6(k,q-v-1,v-1),\\ &g_6(k-1,q-1-v,v-1), g_7(k-1,q-1-v,v-1),\\ &g_8(k-1,q-1-v,v-1), g_{10}(k-2,q-v-2,v-1)\}\end{aligned}$$

of elements of right weight $(k+q-v, k+v)$ for $0 < v < q-1$.

To determine the elements of the induced supermodule $H^0(k, k+q)$, we need to determine which $\mathbf{U}^+$-invariants of $\Bbbk[\mathbf{Q}(2)]$ are annihilated by the action of h'_1 and h'_2. We separate the elements of $\mathfrak{B}$ by their weights.

THEOREM 7.1. *Let $\lambda = (k, k+q)$ be an antidominant weight and denote $e = \sqrt{k}, l = \sqrt{k+q}i$.*

The K-basis of the induced supermodule $H^0(\lambda)$ is described as follows.

- $q = 0$

  If $p|k$, then $H^0(k,k)$ is one-dimensional simple supermodule spanned by $f_1(k,0,0)$, $ch(H^0(k,k)) = x_1^k x_2^k$.

  If $p \nmid k$, then $H^0(k,k) = 0$.

- $q = 1$

  If $p|k$ and $p \nmid k+1$, then $H^0(k,k+1)$ is four-dimensional simple supermodule such that the weight space corresponding to

  $\mu = (k, k+1)$ has a basis

$$\{-u_0 \otimes f_2(k-1,0,0) + u_1 \otimes g_1(k-1,0,0), u_0 \otimes g_1(k-1,0,0) + u_1 \otimes f_2(k-1,0,0)\};$$

  $\mu = (k+1, k)$ has a basis

$$\{-u_0 \otimes f_3(k-1,0,0) + u_1 \otimes g_3(k-1,0,0), u_0 \otimes g_3(k-1,0,0) + u_1 \otimes f_3(k-1,0,0)\}.$$

  .

  We have $ch(H^0(k,k+1)) = 2x_1^k x_2^{k+1} + 2x_1^{k+1} x_2^k$.

  If $p|k+1$ and $p \nmid k$, then $H^0(k,k+1)$ is four-dimensional simple supermodule such that the weight space corresponding to

  $\mu = (k, k+1)$ has a basis

$$\begin{aligned}\{&u_0 \otimes (f_2(k-1,0,0) + f_7(k-1,0,0)) - u_1 \otimes (g_1(k-1,0,0) - g_2(k-1,0,0)),\\ &u_0 \otimes (g_1(k-1,0,0) - g_2(k-1,0,0) - u_1 \otimes (f_2(k-1,0,0) + f_7(k-1,0,0))\};\end{aligned}$$

  $\mu = (k+1, k)$ has a basis

$$\begin{aligned}\{&u_0 \otimes (f_3(k-1,0,0) - f_4(k-1,0,0)) + u_1 \otimes g_4(k-2,0,0),\\ &u_0 \otimes g_4(k-2,0,0) + u_1 \otimes (f_3(k-1,0,0) - f_4(k-1,0,0))\}.\end{aligned}$$

  .

  We have $ch(H^0(k,k+1)) = 2x_1^k x_2^{k+1} + 2x_1^{k+1} x_2^k$.

  If $p \nmid k$ and $p \nmid k+1$, then $H^0(k,k+1)$ is four-dimensional simple supermodule such that the weight space corresponding to $\mu = (k, k+1)$ has a basis

$$\begin{aligned}\{&[eu_0 \otimes g_2(k-1,0,0) + u_1 \otimes f_2(k-1,0,0)]\\ &+ (e+l)[eu_0 \otimes f_4(k-1,0,0)) + u_1 \otimes g_3(k-1,0,0)],\\ &[-u_0 \otimes f_2(k-1,0,0) + eu_1 \otimes g_2(k-1,0,0)]\\ &+ (l-e)[-eu_0 \otimes f_7(k-1,0,0) + u_1 \otimes g_1(k-1,0,0)]\};\end{aligned}$$

$\mu = (k+1, k)$ *has a basis*

$$\begin{aligned}&\{(e+l)[-u_0 \otimes g_3(k-1,0,0) + eu_1 \otimes f_4(k-1,0,0)]\\&\quad + [eu_0 \otimes (g_3(k-1,0,0) + g_4(k-2,0,0))\\&\quad + u_1 \otimes f_3(k-1,0,0)],\\&\quad (l-e)[eu_0 \otimes f_4(k-1,0,0)) + u_1 \otimes g_3(k-1,0,0)]\\&\quad + [-u_0 \otimes f_3(k-1,0,0) + eu_1 \otimes (g_3(k-1,0,0)\\&\quad + g_4(k-2,0,0))]\}.\end{aligned}$$

We have $ch(H^0(k,k+1)) = 2x_1^k x_2^{k+1} + 2x_1^{k+1}x_2^k$.

- $q \geq 2$

  $p|k,\ p|k+q$

  the weight space corresponding to
  $\mu = (k, k+q)$ *has a basis*

$$\{f_2(k-1,0,q-1)\},$$

$\mu = (k+1, k+q-1)$ *has a basis*

$$\begin{aligned}&\{u_0 \otimes [f_3(k-1,0,q-1) - f_6(k,0,q-2)],\\&u_0 \otimes [g_6(k-1,0,q-2) - g_7(k-1,0,q-2)],\\&u_1 \otimes [f_3(k-1,0,q-1) - f_6(k,0,q-2)],\\&u_1 \otimes [g_6(k-1,0,q-2) - g_7(k-1,0,q-2)]\},\end{aligned}$$

$\mu = (k+q-v, k+v)$ *for* $1 < v < q-1$ *has a basis*

$$\begin{aligned}&\{u_0 \otimes [f_3(k-1,0,q-1) + vf_6(k,0,q-2)],\\&u_0 \otimes [g_6(k-1,0,q-2) - g_7(k-1,0,q-2)],\\&u_1 \otimes [f_3(k-1,0,q-1) + vf_6(k,0,q-2)],\\&u_1 \otimes [g_6(k-1,0,q-2) - g_7(k-1,0,q-2)]\},\end{aligned}$$

$\mu = (k+q, k)$ *has a basis*

$$\{f_3(k-1,q-1,0)\}.$$

We have $ch(H^0(k,k+q)) = x_1^k x_2^{k+q} + 4\sum_{v-1}^{q-1} x_1^{k+v}x_2^{k+q-v} + x_1^{k+q}x_2^k$.

$p|k,\ p\nmid k+q$

the weight space corresponding to
$\mu = (k, k+q)$ *has a basis*

$$\begin{aligned}&\{u_0 \otimes g_1(k-1,0,q-1) + u_1 \otimes f_2(k-1,0,q-1);\\&r_2 = u_0 \otimes f_2(k-1,0,q-1) - qu_1 \otimes g-1(k-1,0,q-1)\},\end{aligned}$$

$\mu = (k+1, k+q-1)$ *has a basis*

$$\begin{aligned}&\{-u_0 \otimes g_6(k-1,0,q-2) + (1-q)u_0 \otimes g_7(k-1,0,q-2)\\&\quad - qu_1 \otimes f_3(k-1,0,q-1),\\&\quad - u_0 \otimes g_6(k-1,0,q-2) + u_0 \otimes g_7(k-2,0,q-2) - qu_1 \otimes f_6(k,0,q-2),\\&\quad u_0 \otimes f_3(k-1,0,q-1) + (-1+q)u_0 \otimes f_6(k,0,q-2) - qu_1 \otimes g_6(k-1,0,q-2),\\&\quad u_0 \otimes f_3(k-1,0,q-1) - u_0 \otimes f_6(k,0,q-2) - qu_1 \otimes g_7(k-1,0,q-2)\},\end{aligned}$$

$\mu = (k+q-v, k+v)$ for $1 < v < q-1$ has a basis

$$\begin{aligned}
\{ &- u_0 \otimes f_3(k-1, q-1-v, v) + (q-v)u_0 \otimes f_6(k, q-v-1, v-1) \\
&+ qu_1 \otimes g_7(k-1, q-1-v, v-1), \\
&- u_0 \otimes f_3(k-1, q-1-v, v) - vu_0 \otimes f_6(k, q-v-1, v-1) \\
&+ qu_1 \otimes g_6(k-1, q-1-v, v-1), \\
&u_0 \otimes g_6(k-1, q-1-v, v-1) - u_0 \otimes g_7(k-1, q-1-v, v-1) \\
&+ qu_1 \otimes f_6(k, q-v-1, v-1), \\
&(q-v)u_0 \otimes g_6(k-1, q-1-v, v-1) + vu_0 \otimes g_7(k-1, q-1-v, v-1) \\
&+ qu_1 \otimes f_3(k-1, q-1-v, v)\},
\end{aligned}$$

$\mu = (k+q, k)$ has a basis

$$\begin{aligned}
\{&u_0 \otimes f_3(k-1, 0, q-1) - qu_1 \otimes g_3(k-1, 0, q-1), \\
&u_0 \otimes g_3(k-1, 0, q-1) + qu_1 \otimes f_3(k-1, 0, q-1)\}.
\end{aligned}$$

We have $ch(H^0(k, k+q)) = 2x_1^k x_2^{k+q} + 4\sum_{v=1}^{q-1} x_1^{k+v} x_2^{k+q-v} + 2x_1^{k+q} x_2^k$.
<u>*$p \nmid k$, $p|k+q$*</u>
the weight space corresponding to
$\mu = (k, k+q)$ has a basis

$$\begin{aligned}
\{&u_0 \otimes (f_2(k-1, 0, q-1) - kf_7(k-1, 0, q-1)) \\
&+ ku_1 \otimes (g_1(k-1, 0, q-1) - g_2(k-1, 0, q-1)), \\
&u_0 \otimes (g_1(k-1, 0, q-1) - g_2(k-1, 0, q-1)) \\
&- u_1 \otimes (f_2(k-1, 0, q-1) - kf_7(k-1, 0, q-1)),
\end{aligned}$$

$\mu = (k+1, k+q-1)$ has a basis

$$\begin{aligned}
\{&[u_0 \otimes g_6(k-1, 0, q-2) + ku_0 \otimes g_8(k-1, 0, q-2) - ku_1 \otimes f_4(k-1, 0, q-1)] \\
&+ [-u_0 \otimes g_7(k-1, 0, q-2) - ku_0 \otimes g_8(k-1, 0, q-2) \\
&- ku_1 \otimes f_5(k-1, 0, q-2)], \\
&[-u_0 \otimes g_6(k-1, 0, q-2) - 2ku_0 \otimes g_8(k-1, 0, q-2) - u_0 \otimes g_9(k-2, 0, q-2) \\
&- u_1 \otimes f_3(k-1, 0, q-1)] \\
&+ [u_0 \otimes g_6(k-1, 0, q-2) + ku_0 \otimes g_8(k-1, 0, q-2) - ku_1 \otimes f_4(k-1, 0, q-1)] \\
&- (k+1)[-u_0 \otimes g_8(k-1, 0, q-2) - u_1 \otimes f_6(k, 0, q-2)], \\
&[-u_0 \otimes f_4(k-1, 0, q-1) - u_0 \otimes f_6(k, 0, q-2) - u_1 \otimes g_6(k-1, 0, q-2)] \\
&- [u_0 \otimes f_5(k-1, 0, q-2) - u_0 \otimes f_6(k, 0, q-2) - u_1 \otimes g_7(k-1, 0, q-2)], \\
&[u_0 \otimes f_6(k, 0, q-2) - ku_1 \otimes g_8(k-1, 0, q-2)] \\
&- [u_0 \otimes f_3(k-1, 0, q-1) + ku_0 \otimes f_4(k-1, 0, q-1) - ku_0 \otimes f_6(k, 0, q-2) \\
&- ku_1 \otimes g_9(k-2, 0, q-2)\},
\end{aligned}$$

$\mu = (k+q-v, k+v)$ for $1 < v < q-1$ has a basis

$$
\begin{aligned}
\{&[-u_0 \otimes g_6(k-1, q-1-v, v-1) - (1+k)u_0 \otimes g_8(k-1, q-1-v, v-1)] \\
&+ [u_0 \otimes g_6(k-1, q-1-v, v-1) + ku_0 \otimes g_8(k-1, q-1-v, v-1) \\
&- ku_1 \otimes f_4(k-1, q-1-v, v)] \\
&+ v[-u_0 \otimes g_8(k-1, q-1-v, v-1) - u_1 \otimes f_6(k, q-1-v, v-1)], \\
&[u_0 \otimes g_6(k-1, q-1-v, v-1) + ku_0 \otimes g_8(k-1, q-1-v, v-1) \\
&- ku_1 \otimes f_4(k-1, q-1-v, v)] \\
&+ [-u_0 \otimes g_7(k-1, q-1-v, v-1) - ku_0 \otimes g_8(k-1, q-1-v, v-1) \\
&- ku_1 \otimes f_5(k-1, q-1-v, v-1)], \\
&[-u_0 \otimes f_4(k-1, q-1-v, v) - u_0 \otimes f_6(k, q-v-1, v-1) \\
&- u_1 \otimes g_6(k-1, q-1-v, v-1)] \\
&- [u_0 \otimes f_5(k-1, q-1-v, v-1) - u_0 \otimes f_6(k, q-v-1, v-1) \\
&- u_1 \otimes g_7(k-1, q-1-v, v-1)], \\
&(1+v)[u_0 \otimes f_6(k, q-v-1, v-1) - ku_1 \otimes g_8(k-1, q-1-v, v-1)] \\
&+ [u_0 \otimes f_3(k-1, q-1-v, v-1) + ku_0 \otimes f_4(k-1, q-1-v, v) \\
&- u_0 \otimes f_6(k, q-v-1, v-1)]\},
\end{aligned}
$$

$\mu = (k+q, k)$ has a basis

$$
\begin{aligned}
\{&u_0 \otimes (f_2(k-1, 0, q-1) - kf_7(k-1, 0, q-1)) \\
&- ku_1 \otimes (g_1(k-1, 0, q-1) - g_2(k-1, 0, q-1)), \\
&u_0 \otimes (g_1(k-1, 0, q-1) - g_2(k-1, 0, q-1)) \\
&- u_1 \otimes (f_2(k-1, 0, q-1) - kf_7(k-1, 0, q-1))]\}.
\end{aligned}
$$

We have $ch(H^0(k, k+q)) = 2x_1^k x_2^{k+q} + 4\sum_{v=1}^{q-1} x_1^{k+v} x_2^{k+q-v} + 2x_1^{k+q} x_2^k$.

<u>$p \nmid k$, $p \nmid k+q$</u>

the weight space corresponding to

$\mu = (k, k+q)$ has a basis

$$
\begin{aligned}
\{&[u_0 \otimes f_2(k-1, 0, q-1) - eu_1 \otimes g_2(k-1, 0, q-1)] \\
&+ (l-e)[eu_0 \otimes f_7(k-1, 0, q-1) - u_1 \otimes g_1(k-1, 0, q-1)], \\
&(e+l)[u_0 \otimes g_1(k-1, 0, q-1) + eu_1 \otimes f_7(k-1, 0, q-1)] \\
&- [eu_0 \otimes g_2(k-1, 0, q-1) + u_1 \otimes f_2(k-1, 0, q-1)]\},
\end{aligned}
$$

$\mu = (k+1, k+q-1)$ *has a basis*

$$\begin{aligned}
\{ &- l[eu_0 \otimes g_6(k-1,0,q-2) + 2e^3u_0 \otimes g_8(k-1,0,q-2) \\
&+ eu_0 \otimes g_9(k-2,0,q-2) + u_1 \otimes f_3(k-1,0,q-1)] \\
&+ (1+e^2-el)[-u_0 \otimes g_6(k-1,0,q-2) - e^2u_0 \otimes g_8(k-1,0,q-2) \\
&+ eu_1 \otimes f_4(k-1,0,q-1)] \\
&+ (1+e^2+l^2)[u_0 \otimes g_7(k-1,0,q-2) + e^2u_0 \otimes g_8(k-1,0,q-2) \\
&+ eu_1 \otimes f_5(k-1,0,q-2)], \\
&- [eu_0 \otimes g_6(k-1,0,q-2) + 2e^3u_0 \otimes g_8(k-1,0,q-2) \\
&+ eu_0 \otimes g_9(k-2,0,q-2) \\
&+ u_1 \otimes f_3(k-1,0,q-1)] \\
&- (e+l)[-u_0 \otimes g_6(k-1,0,q-2) - e^2u_0 \otimes g_8(k-1,0,q-2) \\
&+ eu_1 \otimes f_4(k-1,0,q-1)] \\
&+ (1+e^2+l^2)[eu_0 \otimes g_8(k-1,0,q-2) + u_1 \otimes f_6(k,0,q-2)], \\
&- [eu_0 \otimes f_4(k-1,q-1-v,v) + eu_0 \otimes f_6(k,q-v-1,v-1) \\
&+ u_1 \otimes g_6(k-1,q-1-v,v-1)] \\
&+ [-eu_0 \otimes f_5(k-1,0,q-2) + eu_0 \otimes f_6(k,0,q-2) + u_1 \otimes g_7(k-1,0,q-2)] \\
&+ l[-u_0 \otimes f_6(k,0,q-2) + eu_1 \otimes g_8(k-1,0,q-2)], \\
&l[eu_0 \otimes f_4(k-1,0,q-1) + eu_0 \otimes f_6(k,0,q-2) + u_1 \otimes g_6(k-1,0,q-2)] \\
&+ (-1+el-l^2)[-u_0 \otimes f_6(k,0,q-2) + eu_1 \otimes g_8(k-1,0,q-2)] \\
&+ [-u_0 \otimes f_3(k-1,0,q-1) - e^2u_0 \otimes f_4(k-1,0,q-1) + e^2u_0 \otimes f_6(k,0,q-2) \\
&+ eu_1 \otimes g_9(k-2,0,q-2)]\},
\end{aligned}$$

$\mu = (k+q-v, k+v)$ *for* $1 < v < q-1$ *has a basis*

$$\begin{aligned}
\{&l[eu_0 \otimes g_6(k-1,q-1-v,v-1) + (e+e^3)u_0 \otimes g_8(k-1,q-1-v,v-1)] \\
&+ (el+l^2+v)[-u_0 \otimes g_6(k-1,q-1-v,v-1) \\
&- e^2u_0 \otimes g_8(k-1,q-1-v,v-1) + eu_1 \otimes f_4(k-1,q-1-v,v)] \\
&v[u_0 \otimes g_7(k-1,q-1-v,v-1) + e^2u_0 \otimes g_8(k-1,q-1-v,v-1) \\
&+ eu_1 \otimes f_5(k-1,q-1-v,v-1)], \\
&[-u_0 \otimes g_6(k-1,q-1-v,v-1) - e^2u_0 \otimes g_8(k-1,q-1-v,v-1) \\
&+ eu_1 \otimes f_4(k-1,q-1-v,v)] \\
&+ [u_0 \otimes g_7(k-1,q-1-v,v-1) + e^2u_0 \otimes g_8(k-1,q-1-v,v-1) \\
&+ eu_1 \otimes f_5(k-1,q-1-v,v-1)] \\
&- l[eu_0 \otimes g_8(k-1,q-1-v,v-1) + u_1 \otimes f_6(k,q-v-1,v-1)],
\end{aligned}$$

$$
\begin{aligned}
&l[eu_0 \otimes f_4(k-1,q-1-v,v) + eu_0 \otimes f_6(k,q-v-1,v-1) \\
&+ u_1 \otimes g_6(k-1,q-1-v,v-1)] \\
&+ (1+el+v)[-u_0 \otimes f_6(k,q-v-1,v-1) + eu_1 \otimes g_8(k-1,q-1-v,v-1)] \\
&+ [-u_0 \otimes f_3(k-1,q-1-v,v) - e^2u_0 \otimes f_4(k-1,q-1-v,v) \\
&+ u_0 \otimes f_6(k,q-v-1,v-1) \\
&+ eu_1 \otimes g_{10}(k-2,q-2-v,v-1)], \\
&- [eu_0 \otimes f_4(k-1,q-1-v,v) + eu_0 \otimes f_6(k,q-v-1,v-1) \\
&+ u_1 \otimes g_6(k-1,q-1-v,v-1)] \\
&+ [-eu_0 \otimes f_5(k-1,q-1-v,v-1) + eu_0 \otimes f_6(k,q-1-v,v-1) \\
&+ u_1 \otimes g_7(k-1,q-1-v,v-1)] \\
&+ l[-u_0 \otimes f_6(k,q-v-1,v-1) + eu_1 \otimes g_8(k-1,q-1-v,v-1)]\},
\end{aligned}
$$

$\mu = (k+q,k)$ has a basis

$$
\begin{aligned}
&\{[u_0 \otimes f_3(k-1,q-1,0) + e^2u_0 \otimes f_4(k-1,q-1,0) - eu_1 \otimes g_5(k-2,q-1,0)] \\
&- l[eu_0 \otimes f_4(k-1,q-1,0) + u_1 \otimes g_3(k-1,q-1,0)], \\
&(e+l)[u_0 \otimes g_3(k-1,q-1,0) - eu_1 \otimes f_4(k-1,q-1,0)] \\
&- [eu_0 \otimes g_3(k-1,q-1,0) + eu_0 \otimes g_5(k-2,q-1,0) + u_1 \otimes f_3(k-1,q-1,0)]\}.
\end{aligned}
$$

We have $ch(H^0(k,k+q)) = 2x_1^kx_2^{k+q} + 4\sum_{v=1}^{q-1} x_1^{k+v}x_2^{k+q-v} + 2x_1^{k+q}x_2^k$.

The lengthy proof of Theorem 7.1 is given in the Appendix.

If M is a $GL(2)$-module, we denote by $[M]$ the image of the module M in the Grothendieck group of $GL(2)$. If M is a $\mathbf{Q}(2)$-supermodule, and $M_{GL(2)}$ is M considered as a $GL(2)$-module, then we will write $[M] = [M_{GL(2)}]$.

COROLLARY 7.2. *The induced $\mathbf{Q}(2)$-supermodule $H^0(q,q)$ is irreducible and one-dimensional if p divides q, and vanishes otherwise.*

The induced $\mathbf{Q}(2)$-supermodule $H^0(q,q+1)$ is irreducible and has the character $2x_1^qx_2^{q+1} + 2x_1^{q+1}x_2^q$. Also, $[H^0(q,q+1)] = 2[H^0_{GL(2)}(q,q+1)]$.

Assume $k>1$. If $p \nmid q$ or $p \nmid q+k$, then the induced $\mathbf{Q}(2)$-supermodule $H^0(q,q+k)$ has the character $2x_1^qx_2^{k+q} + 4\sum_{v=1}^{q-1} x_1^{k+v}x_2^{k+q-v} + 2x_1^{k+q}x_2^k$. Also,

$$[H^0(q,q+k)] = 2[H^0_{GL(2)}(q,q+k)] + 2[H^0_{GL(2)}(q+1,q+k-1)].$$

If $p|q$ and $p|k$, then the induced $\mathbf{Q}(2)$-supermodule $H^0(q,q+k)$ has the character $x_1^qx_2^{k+q} + 4\sum_{v=1}^{q-1} x_1^{k+v}x_2^{k+q-v} + x_1^{k+q}x_2^k$. Also,

$$[H^0(q,q+k)] = [H^0_{GL(2)}(q,q+k)] + 3[H^0_{GL(2)}(q+1,q+k-1)].$$

Note that the statement that $p \nmid k$ implies $H^0(k,k) = 0$ also follows from Theorem 6.11 of [**3**].

8. Action of e_β and e'_β on the induced supermodule

Set $\beta = \epsilon_2 - \epsilon_1$.

The right action of e_β on generators of the induced supermodule is given in the following statement.

PROPOSITION 8.1.

$$
\begin{aligned}
f_1(pt,0,0)e_\beta &= 0;\\
f_2(a,0,c)e_\beta &= -(c+1)f_3(a,0,c) + 2(a+1)cf_6(a+1,0,c-1);\\
f_3(a,b,c)e_\beta &= -cf_3(a,b+1,c-1);\\
f_4(a,b,c)e_\beta &= -cf_4(a,b+1,c-1);\\
f_5(a,b,c)e_\beta &= -cf_5(a,b+1,c-1) + f_4(a,b+1,c);\\
f_6(a,b,c)e_\beta &= -cf_6(a,b+1,c-1);\\
f_7(a,b,c)e_\beta &= -cf_5(a,0,c-1) + cf_6(a+1,0,c-1) + f_4(a,0,c);\\
g_1(a,0,c)e_\beta &= -cg_7(a,0,c-1) - g_6(a,0,c-1) - (a+1)g_8(a,0,c-1) \text{ if } c>0;\\
&= -g_3(a,0,0) \text{ if } c=0;\\
g_2(a,0,c)e_\beta &= -(c+1)g_9(a-1,0,c-1) - (c+1)g_6(a,0,c-1)\\
&\quad - (2a+2)g_8(a,0,c-1) \text{ if } c>0;\\
&= -g_4(a-1,0,0) - g_3(a,0,0) \text{ if } c=0;\\
g_3(a,b,0)e_\beta &= 0;\\
g_4(a,0,0)e_\beta &= 0;\\
g_5(a,b,0)e_\beta &= 0;\\
g_6(a,b,c)e_\beta &= -cg_6(a,b+1,c-1) - g_6(a,b+1,c-1)\\
&\quad + (a+1)g_8(a,b+1,c-1) \text{ if } c>0;\\
&= -g_3(a,b+1,0) \text{ if } c=0;\\
g_7(a,b,c)e_\beta &= -cg_7(a,b+1,c-1) - g_6(a,b+1,c-1)\\
&\quad + (a+1)g_8(a,b+1,c-1) \text{ if } c>0;\\
&= -g_3(a,b+1,0) \text{ if } c=0;\\
g_8(a,b,c)e_\beta &= -cg_8(a,b+1,c-1);\\
g_9(a,0,c)e_\beta &= -cg_{10}(a,0,c-1) + (a+1)cg_8(a+1,1,c-1) - g_{10}(a,0,c-1)\\
&\quad - g_8(a+1,1,c-1) \text{ if } c>0;\\
&= -g_5(a,0,0) \text{ if } c=0;\\
g_{10}(a,b,c)e_\beta &= -cg_{10}(a,b+1,c-1) - g_{10}(a,b+1,c-1)\\
&\quad + g_8(a+1,b+1,c-1) \text{ if } c>0;\\
&= -g_5(a,b+1,0) \text{ if } c=0;
\end{aligned}
$$

The proof of this statement is given in the Appendix.

The right action of e'_β on generators of the induced supermodule is given in the following statement.

PROPOSITION 8.2.

$$
\begin{aligned}
f_1(pt,0,0)e'_\beta &= 0;\\
f_2(a,0,c)e'_\beta &= (2a+c+3)g_6(a,0,c-1)\\
&\qquad + (a+1)(2a+c+3)g_8(a,0,c-1)] \textit{ if } c>0;\\
&= (2a+3)g_3(a,0,0) \textit{ if } c=0;\\
f_3(a,b,c)e'_\beta &= cg_6(a,b+1,c-2)+(a+1)g_8(a,b+1,c-2) \textit{ if } c>1;\\
&= g_3(a,b+1,0) \textit{ if } c=1;\\
&= 0 \textit{ if } c=0;\\
f_4(a,b,c)e'_\beta &= 0;
\end{aligned}
$$

$$
\begin{aligned}
f_5(a,b,c)e'_\beta &= g_{10}(a-1,b,c-1)+(c+1)g_8(a,b+1,c-1) \textit{ if } c>0;\\
&= g_5(a-1,b,0) \textit{ if } c=0;\\
f_6(a,b,c)e'_\beta &= g_6(a-1,b+1,c-1)-ag_8(a-1,b+1,c-1) \textit{ if } c>0;\\
&= -g_3(a-1,b+1,0) \textit{ if } c=0;\\
f_7(a,0,c)e'_\beta &= g_9(a-1,0,c-1)+(2a+c+2)g_8(a,0,c-1)\\
&\qquad + g_6(a,0,c-1) \textit{ if } c>0;\\
&= g_4(a-1,0,0)+g_3(a,0,0) \textit{ if } c=0;\\
g_1(a,0,c)e'_\beta &= -f_3(a,0,c)-cf_6(a+1,0,c-1);\\
g_2(a,0,c)e'_\beta &= (2a+c+3)f_4(a,0,c);\\
g_3(a,b,0)e'_\beta &= 0;\\
g_4(a,0,0)e'_\beta &= 0;\\
g_5(a,b,0)e'_\beta &= 0;\\
g_6(a,b,c)e'_\beta &= (a+1)f_4(a,b+1,c);\\
g_7(a,b,c)e'_\beta &= -f_3(a,b+1,c)-cf_6(a+1,b+1,c-1);\\
g_8(a,b,c)e'_\beta &= -f_4(a,b+1,c);\\
g_9(a,0,c)e'_\beta &= (a+c+3)f_4(a+1,1,c);\\
g_{10}(a,b,c)e'_\beta &= (c+2)f_4(a+1,b+2,c);
\end{aligned}
$$

The proof of this statement is given in the Appendix.

9. Characters of simple supermodules for p-restricted anti-dominant weights $(k, k+q)$

In this section, we assume that $\lambda = (k, k+q)$ is an anti-dominant p-restricted weight. This means that $0 \leq q \leq p$ if p does not divide k, and $0 \leq q < p$ if p divides k.

9.1. **p divides $k, k+q$.** In this case, $q = 0$. Theorem 7.1 implies

PROPOSITION 9.1. *If p divides k, then the induced supermodule $H^0(k,k)$ is irreducible of character $x_1^k x_2^k$.*

9.2. $2 \leq q < p$ and $p \nmid k$ but $p|k+q$. The weight space for $(k, k+q)$ has a basis

$$\begin{aligned}\{s_1 &= u_0 \otimes (f_2(k-1,0,q-1) - kf_7(k-1,0,q-1))\\ &+ ku_1 \otimes (g_1(k-1,0,q-1) - g_2(k-1,0,q-1));\\ s_2 &= u_0 \otimes (g_1(k-1,0,q-1) - g_2(k-1,0,q-1))\\ &- u_1 \otimes (f_2(k-1,0,q-1) - kf_7(k-1,0,q-1))\}.\end{aligned}$$

The weight space for $(k+1, k+q-1)$ has a basis $\{r_2+r_3, r_1+r_2-(k+1)r_4, r_5-r_6, r_7-r_8\}$, where

$$\begin{aligned}r_1 =& - u_0 \otimes g_6(k-1,0,q-2) - 2ku_0 \otimes g_8(k-1,0,q-2)\\ &- u_0 \otimes g_9(k-2,0,q-2) - u_1 \otimes f_3(k-1,0,q-1),\\ r_2 =& u_0 \otimes g_6(k-1,0,q-2) + ku_0 \otimes g_8(k-1,0,q-2) - ku_1 \otimes f_4(k-1,0,q-1),\\ r_3 =& - u_0 \otimes g_7(k-1,0,q-2) - ku_0 \otimes g_8(k-1,0,q-2)\\ &- ku_1 \otimes f_5(k-1,0,q-2),\\ r_4 =& - u_0 \otimes g_8(k-1,0,q-2) - u_1 \otimes f_6(k,0,q-2),\\ r_5 =& - u_0 \otimes f_4(k-1,0,q-1) - u_0 \otimes f_6(k,0,q-2) - u_1 \otimes g_6(k-1,0,q-2),\\ r_6 =& u_0 \otimes f_5(k-1,0,q-2) - u_0 \otimes f_6(k,0,q-2) - u_1 \otimes g_7(k-1,0,q-2),\\ r_7 =& u_0 \otimes f_6(k,0,q-2) - ku_1 \otimes g_8(k-1,0,q-2),\\ r_8 =& u_0 \otimes f_3(k-1,0,q-1) + ku_0 \otimes f_4(k-1,0,q-1) - ku_0 \otimes f_6(k,0,q-2)\\ &- ku_1 \otimes g_9(k-2,0,q-2).\end{aligned}$$

Compute

$$\begin{aligned}&s_1 e_\beta =\\ &u_0 \otimes [-qf_3(k-1,0,q-1) + 2k(q-1)f_6(k,0,q-2) + k(q-1)f_5(k-1,0,q-2)\\ &- k(q-1)f_6(k,0,q-1) - kf_4(k-1,0,q-1)] - ku_1 \otimes [(q-1)g_7(k-1,0,q-2)\\ &+ g_6(k-1,0,q-2) + kg_8(k-1,0,q-2) - qg_9(k-2,0,q-2)\\ &- qg_6(k-1,0,q-2) - 2kg_8(k-1,0,q-2)]\\ &= u_0 \otimes [-qf_3(k-1,0,q-1) - kf_4(k-1,0,q-1) + k(q-1)f_5(k-1,0,q-2)\\ &+ k(q-1)f_6(k,0,q-1)] + u_1 \otimes [(q-1)kg_6(k-1,0,q-2)\\ &- (q-1)kg_7(k-1,0,q-2) + k^2 g_8(k-1,0,q-1) + kqg_9(k-2,0,q-2)]\end{aligned}$$

and

$$\begin{aligned}&(1-q)k(r_5-r_6) - k(r_7-r_8) =\\ &(1-q)k\{u_0 \otimes [-f_4(k-1,0,q-1) - f_5(k-1,0,q-2)]\\ &+ u_1 \otimes [-g_6(k-1,0,q-2) + g_7(k-1,0,q-2)]\}\\ &- k\{u_0 \otimes [-f_3(k-1,0,q-1) - kf_4(k-1,0,q-1) + (k+1)f_6(k,0,q-2)]\\ &+ u_1 \otimes [-kg_8(k-1,0,q-2) + kg_9(k-2,0,q-2)\}.\end{aligned}$$

This implies

$$s_1 e_\beta = (1-q)k(r_5-r_6) - k(r_7-r_8).$$

Additionally,

$$\begin{aligned}
&s_2e_\beta = \\
&u_0 \otimes [(q-1)g_6(k-1,0,q-2) - (q-1)g_7(k-1,0,q-2) \\
&+ kg_8(k-1,0,q-1) + qg_9(k-2,0,q-2)] \\
&+ u_1 \otimes [qf_3(k-1,0,q-1) + kf_4(k-1,0,q-1) - k(q-1)f_5(k-1,0,q-2) \\
&- k(q-1)f_6(k,0,q-1)]
\end{aligned}$$

and

$$\begin{aligned}
&k(r_1 + r_2 - (k+1)r_4) + (q-1)(r_2 + r_3) = \\
&kr_1 + (k+q-1)r_2 + (q-1)r_3 - k(k+1)r_4 = u_0 \otimes [(q-1)g_6(k-1,0,q-2) \\
&- (q-1)g_7(k-1,0,q-2) + kg_8(k-1,0,q-1) - kg_9(k-2,0,q-2)] \\
&+ u_1 \otimes [-kf_3(k-1,0,q-1) + kf_4(k-1,0,q-1) - (q-1)kf_5(k-1,0,q-2) \\
&+ k(k+1)f_6(k,0,q-1)].
\end{aligned}$$

Therefore,

$$s_2e_\beta = k(r_1 + r_2 - (k+1)r_4) + (q-1)(r_2 + r_3).$$

Furthermore, since

$$\begin{aligned}
&s_1e'_\beta = \\
&u_0 \otimes [kg_8(k-1,0,q-2) - kg_9(k-2,0,q-2)] \\
&+ ku_1 \otimes [-f_3(k-1,0,q-1) - kf_4(k-1,0,q-1) + (k-1)f_6(k,0,q-2)]
\end{aligned}$$

and

$$\begin{aligned}
&k(r_1 + r_2 - (k-1)r_4) = \\
&u_0 \otimes [kg_8(k-1,0,q-2) - kg_9(k-2,0,q-2)] \\
&+ u_1 \otimes [-kf_3(k-1,0,q-1) - k^2 f_4(k-1,0,q-1) + k(k-1)f_6(k,0,q-2)],
\end{aligned}$$

we obtain

$$s_1e'_\beta = k[r_1 + r_2 - (k-1)r_4].$$

Finally,

$$\begin{aligned}
&s_2e'_\beta = \\
&u_0 \otimes [-f_3(k-1,0,q-1) - kf_4(k-1,0,q-1) + (k-1)f_6(k,0,q-2)] \\
&- u_1 \otimes [kg_8(k-1,0,q-2) - kg_9(k-2,0,q-2)]
\end{aligned}$$

and

$$\begin{aligned}
&(r_7 - r_8) = \\
&u_0 \otimes [-f_3(k-1,0,q-1) - kf_4(k-1,0,q-1) + (k-1)f_6(k,0,q-2)] \\
&+ u_1 \otimes [-kg_8(k-1,0,q-2) + kg_9(k-2,0,q-2)]
\end{aligned}$$

imply that

$$s_2e'_\beta = (r_7 - r_8).$$

9.3. $2 \leq q < p$ **and** $p|k$, $p \nmid k+q$. The weight space for $(k, k+q)$ has a basis

$$\begin{aligned}\{s_1 &= u_0 \otimes g_1(k-1,0,q-1) + u_1 \otimes f_2(k-1,0,q-1),\\ s_2 &= u_0 \otimes f_2(k-1,0,q-1) - qu_1 \otimes g_1(k-1,0,q-1)\}.\end{aligned}$$

The weight space for $(k+1, k+q-1)$ has a basis

$$\{r_1, r_4, r_5, r_6\},$$

where

$$\begin{aligned}r_1 =& - u_0 \otimes g_6(k-1,0,q-2) + (1-q)u_0 \otimes g_7(k-1,0,q-2)\\ &- qu_1 \otimes f_3(k-1,0,q-1),\\ r_4 =& - u_0 \otimes g_6(k-1,0,q-2) + u_0 \otimes g_7(k-2,0,q-2) - qu_1 \otimes f_6(k,0,q-2),\\ r_5 =& u_0 \otimes f_3(k-1,0,q-1) + (-1+q)u_0 \otimes f_6(k,0,q-2)\\ &- qu_1 \otimes g_6(k-1,0,q-2),\\ r_6 =& u_0 \otimes f_3(k-1,0,q-1) - u_0 \otimes f_6(k,0,q-2) - qu_1 \otimes g_7(k-1,0,q-2).\end{aligned}$$

Compute

$$\begin{aligned}s_1 e_\beta =& u_0 \otimes [-(q-1)g_7(k-1,0,q-2) - g_6(k-1,0,q-2)]\\ &+ u_1 \otimes [-qf_3(k-1,0,q-1)] = r_1,\\ s_2 e_\beta =& u_0 \otimes [-qf_3(k-1,0,1)] + qu_1 \otimes [(q-1)g_7(k-1,0,q-2)\\ &+ g_6(k-1,0,q-2)] = -r_5 + (1-q)r_6,\\ s_1 e'_\beta =& - u_0 \otimes [f_3(k-1,0,q-1) + (q-1)f_6(k,0,q-2)]\\ &+ u_1 \otimes [qg_6(k-1,0,q-2)] = -r_5,\\ s_2 e'_\beta =& u_0 \otimes [qg_6(k-1,0,0)] + qu_1 \otimes [f_3(k-1,0,q-1)\\ &+ (q-1)f_6(k,0,q-2)] = -r_1 - (q-1)r_4.\end{aligned}$$

The above computations imply the following result.

PROPOSITION 9.2. *Assume p divides exactly one of k and $k+q$. Then the induced supermodule $H^0(k, k+q)$ is irreducible of character*

$$2x_1^k x_2^{k+q} + 4\sum_{v=1}^{q-1} x_1^{k+v} x_2^{k+q-v} + 2x_1^{k+q} x_2^k.$$

PROOF. When $q = 1$, then the statement follows from

$$[H^0(k, k+1)] = 2[L_{GL(2)}(k, k+1)].$$

If $q \geq 2$, then the earlier computations in 9.2 and 9.3 show that the weight space of $L(k, k+q)$ corresponding to the weight $(k+1, k+q-1)$ is four-dimensional. Since

$$[H^0(k, k+q)] = 2[L_{GL(2)}(k, k+q)] + 2[L_{GL(2)}(k+1, k+q-1)],$$

the statement follows. □

9.4. p **does not divide** $k, k+q$. Assume $q > 1$. We compute the action of e_β and e'_β on the generators of the weight space $(k, k+q)$, and describe the image inside the weight space of weight $(k+1, k+q-1)$.

The weight space corresponding to $\mu = (k, k+q)$ has a basis

$$\begin{aligned}\{s_1 &= [u_0 \otimes f_2(k-1,0,q-1) - eu_1 \otimes g_2(k-1,0,q-1)] \\ &+ (l-e)[eu_0 \otimes f_7(k-1,0,q-1) - u_1 \otimes g_1(k-1,0,q-1)], \\ s_2 &= (e+l)[u_0 \otimes g_1(k-1,0,q-1) + eu_1 \otimes f_7(k-1,0,q-1)] \\ &- [eu_0 \otimes g_2(k-1,0,q-1) + u_1 \otimes f_2(k-1,0,q-1)]\}.\end{aligned}$$

The weight space corresponding to $(k+1, k+q-1)$ has a basis

$$\begin{aligned}\{j_1 &= -lr_1 + (1+e^2-el)r_2 + (1+e^2+l^2)r_3, j_2 = -r_1 - (e+l)r_2 \\ &+ (1+e^2+l^2)r_4, \\ j_3 &= -r_5 + r_6 + lr_7, j_4 = lr_5 + (-1+el-l^2)r_7 + r_8\},\end{aligned}$$

where

$$\begin{aligned}r_1 =& eu_0 \otimes g_6(k-1,0,q-2) + 2e^3 u_0 \otimes g_8(k-1,0,q-2) \\ &+ eu_0 \otimes g_9(k-2,0,q-2) + u_1 \otimes f_3(k-1,0,q-1), \\ r_2 =& -u_0 \otimes g_6(k-1,0,q-2) - e^2 u_0 \otimes g_8(k-1,0,q-2) \\ &+ eu_1 \otimes f_4(k-1,0,q-1), \\ r_3 =& u_0 \otimes g_7(k-1,0,q-2) + e^2 u_0 \otimes g_8(k-1,0,q-2) + eu_1 \otimes f_5(k-1,0,q-2), \\ r_4 =& eu_0 \otimes g_8(k-1,0,q-2) + u_1 \otimes f_6(k,0,q-2), \\ r_5 =& eu_0 \otimes f_4(k-1,0,q-1) + eu_0 \otimes f_6(k,0,q-2) + u_1 \otimes g_6(k-1,0,q-2), \\ r_6 =& eu_0 \otimes f_5(k-1,0,q-2) + eu_0 \otimes f_6(k,0,q-2) + u_1 \otimes g_7(k-1,0,q-2), \\ r_7 =& -u_0 \otimes f_6(k,0,q-2) + eu_1 \otimes g_8(k-1,0,q-2), \\ r_8 =& -u_0 \otimes f_3(k-1,0,q-1) - e^2 u_0 \otimes f_4(k-1,0,q-1) + e^2 u_0 \otimes f_6(k,0,q-2) \\ &+ eu_1 \otimes g_9(k-2,0,q-2).\end{aligned}$$

Next, I will express $s_1 e_\beta$, $s_2 e_\beta$, $s_1 e'_\beta$ and $s_2 s'_\beta$ as linear combinations of j_1, j_2, j_3, j_4. We compute

$$\begin{aligned}&s_1 e_\beta = \\ &u_0 \otimes [-qf_3(k-1,0,q-1) + 2k(q-1)f_6(k,0,q-2) \\ &+ (l-e)e(-q+1)f_5(k-1,0,q-2) + (l-e)e(q-1)f_6(k,0,q-2) \\ &+ (l-e)ef_4(k-1,0,q-1)] + u_1 \otimes [eqg_6(k-1,0,q-2) + 2ekg_8(k-1,0,q-2) \\ &+ eqg_9(k-2,0,q-2) + (l-e)g_6(k-1,0,q-2) + (l-e)(q-1)g_7(k-1,0,q-2) \\ &+ (l-e)kg_8(k-1,0,q-2)] = u_0 \otimes [(e^2+l^2)f_3(k-1,0,q-1) \\ &+ (l-e)ef_4(k-1,0,q-1) - (l-e)e(e^2+l^2+1)f_5(k-1,0,q-2) \\ &- (l+e)e(e^2+l^2+1)f_6(k,0,q-2)] + u_1 \otimes [(-e^3-el^2+l-e)g_6(k-1,0,q-2) \\ &- (l-e)(e^2+l^2+1)g_7(k-1,0,q-2) + (l+e)e^2 g_8(k-1,0,q-2) \\ &- e(e^2+l^2)g_9(k-2,0,q-2)]\end{aligned}$$

and

$$\begin{aligned}
&(-e^3-el^2+l-e)r_5-(l-e)(e^2+l^2+1)r_6+e(l+e)r_7-(e^2+l^2)r_8=\\
&u_1\otimes[(-e^3-el^2+l-e)g_6(k-1,0,q-2)-(l-e)(e^2+l^2+1)g_7(k-1,0,q-2)\\
&+(l+e)e^2g_8(k-1,0,q-2)-e(e^2+l^2)g_9(k-2,0,q-2)]\\
&+u_0\otimes\{(e^2+l^2)f_3(k-1,0,q-1)+[e(-e^3-el^2+l-e)\\
&+e^2(e^2+l^2)]f_4(k-1,0,q-1)-e(l-e)(e^2+l^2+1)f_5(k-1,0,q-2)\\
&+[e(e^3-el^2+l-e)-e(l-e)(e^2+l^2+1)-e(l+e)-e^2(e^2+l^2)]f_6(k,0,q-2)\}.
\end{aligned}$$

Since

$$e(-e^3-el^2+l-e)+e^2(e^2+l^2)=e(l-e)$$

and

$$\begin{aligned}
&e(-e^3-el^2+l-e)-e(l-e)(e^2+l^2+1)-e(l+e)-e^2(e^2+l^2)\\
&=-e^4-e^2-e^2l^2-e^3l-el^3-el=-(l+e)e(e^2+l^2+1),
\end{aligned}$$

we have

$$s_1e_\beta=(-e^3-el^2+l-e)r_5-(l-e)(e^2+l^2+1)r_6+e(l+e)r_7-(e^2+l^2)r_8.$$

Since

$$\begin{aligned}
&-(l-e)(e^2+l^2+1)j_3-(e^2+l^2)j_4=\\
&[(l-e)(e^2+l^2+1)-l(e^2+l^2)]r_5-(l-e)(e^2+l^2+1)r_6\\
&+[-l(l-e)(e^2+l^2+1)-(e^2+l^2)(-1+el-l^2)]r_7-(e^2+l^2)r_8
\end{aligned}$$

and

$$(l-e)(e^2+l^2+1)-l(e^2+l^2)=l-e-e^3-el^2,$$

$$-l(l-e)(e^2+l^2+1)-(e^2+l^2)(-1+el-l^2)=(e^2+l^2)(-l^2+le+1-el+l^2)-l^2+el=e^2+el,$$

we derive that

$$s_1e_\beta=(e-l)(e^2+j^2+1)j_3-(e^2+l^2)j_4.$$

Moreover, we have

$$\begin{aligned}
&s_2e_\beta=\\
&u_0\otimes[-(e+l)g_6(k-1,0,q-2)-(e+l)(q-1)g_7(k-1,0,q-2)\\
&-(e+l)kg_8(k-2,0,q-2)+eqg_6(k-1,0,q-2)+2ekg_8(k-1,0,q-2)\\
&+eqg_9(k-2,0,q-2)+u_1\otimes[+e(e+l)f_4(k-1,0,q-1)\\
&-e(e+l)(q-1)f_5(k-1,0,q-2)+e(e+l)(q-1)f_6(k,0,q-2)\\
&+qf_3(k-1,0,q-1)-2k(q-1)f_6(k,0,q-2)]\\
&=u_0\otimes[(-e^3-el^2-e-l)g_6(k-1,0,q-2)+(e+l)(e^2+l^2+1)g_7(k-1,0,q-2)\\
&+e^2(e-l)g_8(k-1,0,q-2)+(-e^3-el^2)g_9(k-2,0,q-2)]\\
&+u_1\otimes[-(e^2+l^2)f_3(k-1,0,q-1)+e(e+l)f_4(k-1,0,q-1)\\
&+e(e+l)(e^2+l^2+1)f_5(k-1,0,q-2)+(e^2-el)(e^2+l^2+1)f_6(k,0,q-2)]
\end{aligned}$$

and

$$\begin{aligned}
&-(e^2+l^2)r_1+(e+l)r_2+(e+l)(e^2+l^2+1)r_3+(e^2-el)(e^2+l^2+1)r_4\\
&=u_1\otimes[-(e^2+l^2)f_3(k-1,0,q-1)+e(e+l)f_4(k-1,0,q-1)\\
&+e(e+l)(e^2+l^2+1)f_5(k-1,0,q-2)+(e^2-el)(e^2+l^2+1)f_6(k,0,q-2)]\\
&+u_0\otimes\{[-e^3-l^2e-e-l]g_6(k-1,0,q-2)+(e+l)(e^2+l^2+1)g_7(k-1,0,q-2)\\
&+[(-e^2-l^2)2e^3-(e+l)e^2+e^2(e+l)(e^2+l^2+1)+e(-e^2+el)(e^2+l^2+1)]\\
&\times g_8(k-1,0,q-2)-e(e^2+l^2)g_9(k-2,0,q-2)
\end{aligned}$$

Since

$$\begin{aligned}
&(-e^2-l^2)2e^3-(e+l)e^2+e^2(e+l)(e^2+l^2+1)+e(e^2-el)(e^2+l^2+1)]\\
&=-2e^3(e^2+l^2)-e^3-e^2l+2e^3(e^2+l^2+1)=e^3-e^2l,
\end{aligned}$$

we have

$$s_2e_\beta=-(e^2+l^2)r_1+(e+l)r_2+(e+l)(e^2+l^2+1)r_3+(e^2-el)(e^2+l^2+1)r_4.$$

Using

$$\begin{aligned}
&(e+l)j_1+(e^2-el)j_2=\\
&[-el-l^2-e^2+el]r_1+[(e+l)(1+e^2-el)-(e+l)(e^2-el)]r_2\\
&+(e+l)(1+e^2+l^2)r_3+(e^2-el)(1+e^2+l^2)r_4,
\end{aligned}$$

we establish that

$$s_2e_\beta=(e+l)j_1+(e^2-el)j_2.$$

Further, we have

$$\begin{aligned}
&s_1e'_\beta=\\
&u_0\otimes[(2k+q)g_6(k-1,0,q-2)+(2k+q)kg_8(k-1,0,q-2)\\
&(l-e)eg_6(k-1,0,q-2)+(l-e)e(2k+q-1)g_8(k-1,0,q-2)\\
&+(l-e)eg_9(k-2,0,q-2)]+u_1\otimes[-e(2k+q)f_4(k-1,0,q-1)\\
&+(l-e)f_3(k-1,0,q-1)+(l-e)(q-1)f_6(k,0,q-2)]\\
&=u_0\otimes[(el-l^2)g_6(k-1,0,q-2)+(le^3-el^3-el+e^2)g_8(k-1,0,q-2)\\
&+(l-e)eg_9(k-2,0,q-2)]+u_1\otimes[-e(e^2-l^2)f_4(k-1,0,q-1)\\
&+(l-e)f_3(k-1,0,q-1)-(l-e)(e^2+l^2+1)f_6(k,0,q-2)]
\end{aligned}$$

and

$$\begin{aligned}
&(l-e)r_1+(l^2-e^2)r_2-(l-e)(e^2+l^2+1)r_4=\\
&u_0\otimes\{[e(l-e)-l^2+e^2]g_6(k-1,0,q-2)+[2e^3(l-e)-e^2(l^2-e^2)\\
&-e(l-e)(e^2+l^2+1)]g_8(k-1,0,q-2)+e(l-e)g_9(k-2,0,q-2)\}\\
&+u_1\otimes[(l-e)f_3(k-1,0,q-1)+el^2f_4(k-1,0,q-1)\\
&-(l-e)(e^2+l^2+1)f_6(k,0,q-2)].
\end{aligned}$$

The equality

$$2e^3(l-e)-e^2(l^2-e^2)-e(l-e)(e^2+l^2+1)=e^3l-el^3-el+e^2$$

implies

$$s_1e'_\beta=(l-e)r_1+(l^2-e^2)r_2-(l-e)(e^2+l^2+1)r_4=(e-l)j_2.$$

Finally,

$$
\begin{aligned}
&s_2e'_\beta = \\
&u_0\otimes[-(e+l)f_3(k-1,0,q-1)-(e+l)(q-1)f_6(k,0,q-2)\\
&-e(2k+q)f_4(k-1,0,q-1)]+u_1\otimes[e(e+l)g_6(k-1,0,q-2)\\
&+e(e+l)(2k+q-1)g_8(k-1,0,q-2)+e(e+l)g_9(k-2,0,q-2)\\
&-k(2k+q)g_8(k-1,0,q-2)-(2k+q)g_6(k-1,0,q-2)]\\
&=u_0\otimes[(-e-l)f_3(k-1,0,q-1)-e(e^2-l^2)f_4(k-1,0,q-1)\\
&+(e+l)(e^2+l^2+1)f_6(k,0,q-2)+u_1\otimes[(el+l^2)g_6(k-1,0,q-2)\\
&+[el(e^2-l^2)-e(e+l)]g_8(k-1,0,q-2)+e(e+l)g_9(k-2,0,q-2)]
\end{aligned}
$$

and

$$
\begin{aligned}
&(el+l^2)r_5+(l(e^2-l^2)-e-l)r_7+(e+l)r_8 = \\
&u_0\otimes\{-(e+l)f_3(k-1,0,q-1)+[e(el+l^2)-e^2(e+l)]f_4(k-1,0,q-1)\\
&=[e(el+l^2)-l(e^2-l^2)+e+l+e^2(e+l)]f_6(k,0,q-2)\}\\
&+u_1\otimes[(el+l^2)g_6(k-1,0,q-2)+[el(e^2-l^2)-e-l]g_8(k-1,0,q-2)\\
&+e(e+l)g_9(k-2,0,q-2)]
\end{aligned}
$$

together with

$$
\begin{aligned}
e(el+l^2)-e^2(e+l) &= -e^3+el^2\\
e(el+l^2)-l(e^2-l^2)+e+l+e^2(e+l) &= (e+l)(e^2+l^2+1)
\end{aligned}
$$

gives

$$s_2e'_\beta=(el+l^2)r_5+(l(e^2-l^2)-e-l)r_7+(e+l)r_8=(e+l)j_4.$$

Next, we determine the dimension of the spaces spanned by the elements

$$\{s_2e_\beta=(e+l)j_1+(e^2-el)j_2, s_1e'_\beta=(e-l)j_2\}$$

and

$$\{s_1e_\beta=(e-l)(e^2+l^2+1)j_3-(e^2+l^2)j_4, s_2e'_\beta=(e+l)j_4\}.$$

If $(e+l)(e-l)(e^2+l^2+1)$ is not a multiple of p, then both spaces are two-dimensional. Note that $e^2+l^2+1=-q+1$ is not a multiple of p since $q\not\equiv 1 \pmod p$.

Thus, if $(e+l)(e-l)=e^2-l^2=2k+q$ is not divisible by p, then both spaces are two-dimensional, which implies that the induced module $H^0(k,k+q)$ is irreducible.

Assume now that $2k+q$ is divisible by p. Then, $e\equiv \pm l \pmod p$.

If $e\equiv l \pmod p$, then $e+l$ is not divisible by p and the first space is spanned by j_1 while the second space is spanned by j_4.

If $e\equiv -l \pmod p$, then $e-l$ is not divisible by p and the first space is spanned by j_2 while the second space is spanned by s_1e_β.

Therefore, we conclude that if p does not divide $2k+q$, then the induced supermodule $H^0(k,k+q)$ is irreducible. If p divides $2k+q$, then it is not irreducible, and the weights $(k,k+q)$ and $(k+1,k+q-1)$ are linked.

PROPOSITION 9.3. *Assume p does not divide $k, k+q$ and $0\leq q\leq p$.*

If $q=0$ and $p\nmid k$, then the induced supermodule $H^0(k,k)=0$.

If $q=1$, then the induced supermodule $H^0(k,k+1)$ is irreducible of the character $2x_1^kx_2^{k+1}+2x_1^{k+1}x_2^k$.

If $q > 1$ and $2k + q$ is not a multiple of p, then the induced supermodule $H^0(k, k + q)$ is irreducible of character

$$2x_1^k x_2^{k+q} + 4\sum_{v=1}^{q-1} x_1^{k+v} x_2^{k+q-v} + 2x_1^{k+q} x_2^k.$$

If $q > 1$ and p divides $2k + q$, then the simple supermodule $L(k, k + q)$ has the character

$$2\sum_{v=0}^{q} x_1^{k+v} x_2^{k+q-v}.$$

If $q > 1$ and p divides $2k + q$, then the induced supermodule $H^0(k, k + q)$ has the simple composition factors $L(k, k + q)$ and $L(k + 1, k + q - 1)$.

PROOF. If $q = 0$, the statement follows from Theorem 7.1.

If $q = 1$, then $[H^0(k, k + 1)] = 2[L_{GL(2)}(k, k + 1)]$.

Assume $1 < q < p$. In this case,

$$[H^0(k, k + q)] = 2[L_{GL(2)}(k, k + q)] + 2[L_{GL(2)}(k + 1, k + q - 1)].$$

If $2k + q$ is not divisible by p, then the above computations in 9.4 show that $H^0(k, k + q)$ is irreducible.

If $2k+q$ is a multiple of p, then by the above computations in 9.4, the induced module $H^0(k, k + q)$ is not irreducible. Its simple socle $L(k, k + q)$ satisfies

$$L(k, k + q)] = 2[L_{GL(2)}(k, k + q)],$$

and its simple top $L(k + 1, k + q - 1)$ satisfies

$$[L(k + 1, k + q - 1)] = 2[L_{GL(2)}(k + 1, k + q - 1)].$$

The character of $L(k, k + q)$ equals $2\sum_{v=0}^{q} x_1^{k+v} x_2^{k+q-v}$.

Finally, assume $q = p$. In this case,

$$[H^0(k, k + q)] = 2[L_{GL(2)}(k, k + p)] + 4[L_{GL(2)}(k + 1, k + q - 1)].$$

If p does not divide k, then the above computation in 9.4 shows that the weight-space of $L(k, k+q)_{(k+1,k+q-1)}$ is four-dimensional, which implies that $H^0(k, k+q) = L(k, k + q)$ is irreducible.

If p divides k, then the above computation in 9.4 shows that the weight-space of $L(k, k + p)_{(k+1,k+p-1)}$ is two-dimensional. This implies

$$[L(k, k + p)] = 2[L_{GL(2)}(k, k + p)] + 2[L_{SL(2)}(k + 1, k + p - 1)].$$

The supermodule $L(k, k + p)$ is the socle of $H^0(k, k + p)$ such that

$$[L(k, k + p)] = 2[L_{GL(2)}(k, k + p)],$$

and its top is $L(k + 1, k + p - 1)$, which satisfies

$$[L(k + 1, k + p - 1)] = 2[L_{GL(2)}(k + 1, k + p - 1)].$$

□

10. A composition series for induced supermodules

At this time, we switch from the Borel supergroup $\mathbf{B}^+$ and anti-dominant weights $(k, k+q)$ to Borel supergroup $\mathbf{B}^-$ and dominant weights. We will change the notation and write dominant weights as $(q+k, q)$ instead of $(k+q, k)$ to conform to the paper [**6**] we will be citing. Also, we will denote by $L_-(\lambda)$ the simple socle of $H^0_-(\lambda)$ of the highest dominant weight λ.

As it has been observed in [**2**, Example 5.2] that $H^0_-(-w_0\lambda)^*$, where w_0 is the longest element of the Weyl group $W \simeq S_2$, is the universal highest weight supermodule, whose top is isomorphic to $L_-(\lambda)$. Thus $L_-(\lambda)^* \simeq L_-(-w_0\lambda)$. On the other hand, the simple supermodule $L_-(\lambda)^*$ has $-\lambda$ as the largest weight with respect to the antidominant order, hence $L_-(\lambda) \simeq L(w_0\lambda)$.

This duality allows us to translate earlier results formulated for the induced supermodules $H^0(\lambda)$ for the Borel supergroup $\mathbf{B}^+$, and simple supermodules $L(\lambda)$ for antidominant weights λ to induced supermodules $H^0_-(\kappa)$ for the Borel supergoup $\mathbf{B}^-$, and simple supermodules $L_-(\kappa)$ for dominant weights κ.

We reformulate Corollary 7.2 for dominant weight $\lambda = (q+k, q)$ as follows.

COROLLARY 10.1. *The induced $\mathbf{Q}(2)$-supermodule $H^0_-(q,q)$ is irreducible and one-dimensional if p divides q, and vanishes otherwise.*

The induced $\mathbf{Q}(2)$-supermodule $H^0_-(q+1,q)$ is irreducible and has the character $2x_1^q x_2^{q+1} + 2x_1^{q+1}x_2^q$. Also,

$$[H^0_-(q+1,q)] = 2[H^0_{GL(2)}(q+1,q)].$$

Assume $k > 1$. If $p \nmid q$ or $p \nmid q+k$, then the induced $\mathbf{Q}(2)$-supermodule $H^0(q+k,q)$ has the character $2x_1^q x_2^{k+q} + 4\sum_{v=1}^{q-1} x_1^{k+v}x_2^{k+q-v} + 2x_1^{k+q}x_2^k$. Also,

$$[H^0_-(q+k,q)] = 2[H^0_{GL(2)}(q+k,q)] + 2[H^0_{GL(2)}(q+k-1,q+1)].$$

If $p|q$ and $p|k$, then the induced $\mathbf{Q}(2)$-supermodule $H^0(q+k,q)$ has the character $x_1^q x_2^{k+q} + 4\sum_{v=1}^{q-1} x_1^{k+v}x_2^{k+q-v} + x_1^{k+q}x_2^k$. Also,

$$[H^0_-(q+k,q)] = [H^0_{GL(2)}(q+k,k)] + 3[H^0_{GL(2)}(q+k-1,q+1)].$$

Recall from [**3**] that λ is p-restricted, if $0 \leq \lambda_1 - \lambda_2 < p$ if p divides λ_1 and $0 \leq \lambda_1 - \lambda_2 \leq p$ if p does not divide λ_1.

The Steinberg tensor product theorem for $\mathbf{Q}(n)$ was proved in Theorem 9.9 of [**3**]. It states that

$$L(\lambda + p\mu) \simeq L_-(\lambda) \otimes F^* L_{ev}(\mu),$$

where λ, μ are dominant weights, λ is p-restricted and F^* is the Frobenius twist functor.

We will apply the above Steinberg tensor product theorem and work with characters of irreducible supermodules for p-restricted weights, together with the characters of the induced supermodule, to determine the composition factors of induced supermodules and thus establish the strong linkage principle for $\mathbf{Q}(2)$.

LEMMA 10.2. *We have $[L_-(q,q)] = [L_{GL(2)}(q,q)]$ if $p|q$ and $[L_-(q,q)] = 0$ if $p \nmid q$. Assume $k > 0$ and $k \equiv k_0 \pmod{p}$, where $0 \leq k_0 < p$.*

If $k_0 = 1$, then $[L_-(q+k,q)] = 2[L_{GL(2)}(q+k,q)]$.

If $k_0 > 1$ and $p \nmid 2q+k$, then

$$[L_-(q+k,q)] = 2[L_{GL(2)}(q+k,q)] + 2[L_{GL(2)}(q+k-1,q+1)].$$

If $k_0 > 1$ *and* $p|2q+k$, *then* $[L_-(q+k,q) = 2[L_{GL(2)}(q+k,k)]$.
If $k_0 = 0$ *and* $p|q$, *then* $[L_-(q+k,q)] = [L_{GL(2)}(q+k,q)]$.

PROOF. We have already proved the corresponding statements for p-restricted anti-dominant weights $(q, q+k)$. Using the above duality, the statement for p-restricted dominant weights $(q+k,q)$ follows. The general case follows from the application of the Steinberg tensor product theorem. □

Compare to Lemma 6.4 of [**6**].

Denote by $[M]_{SL(2)}$ the image of a $SL(2)$-module in the Grothendieck group for $SL(2)$.

LEMMA 10.3. *If* $k = p$ *and* $p \nmid q$, *then*

$$[L_-(q+k,q)]_{SL(2)} = 2[L_{SL(2)}(p)]_{SL(2)} + 4[L_{SL(2)}(p-2)]_{SL(2)}.$$

If $k = pt$, *where* $t > 1$, *and* $p \nmid q$, *then*

$$[L_-(q+k,q)]_{SL(2)} = 2[F^*(L_{SL(2)}(1) \otimes L_{SL(2)}(t-1))]_{SL(2)} + 4[L_{SL(2)}(pt-2)]_{SL(2)}.$$

PROOF. The statement for $t = 1$ follows from the proof of Proposition 9.3. Therefore, we assume $t > 1$.

We have

$$[L_-(q+p,q)] = 2[L_{GL(2)}(q+p,q)] + 4[L_{GL(2)}(q+p-1,q+1)],$$

and by the Steinberg tensor product theorem

$$L_-(q+pt,q) \simeq L_-(q+p,q) \otimes F^*(L_{SL(2)}(t-1)).$$

Considering the above tensor product as a $SL(2)$-module, it is filtered by 2 copies of

$$L_{SL(2)}(p) \otimes F^*(L(_{SL(2)}(t-1)) \simeq F^*(L_{SL(2)}(1) \otimes L_{SL(2)}(t-1))$$

and four copies of

$$L_{SL(2)}(p-2) \otimes F^*(L_{SL(2)}(t-1)) \simeq L_{SL(2)}(tp-2).$$

The sum of the weightspaces $L_-(q+k,q)_\mu$, where $p \nmid (\mu_1 - \mu_2)$ corresponds to the sum of μ-weightspaces in the 4 copies of $L_{GL(2)}(q+k-1,q+1)$.

The remaining weightspaces $L_-(q+k,q)_\mu$, where $p|(\mu_1 - \mu_2)$ correspond to the elements in the 2 copies of $F^*(L_{SL(2)}(1) \otimes L_{SL(2)}(t-1))$. □

Note that since $L_{SL(2)}(1) = H^0_{SL(2)}(1)$, by Pierri's rule, there is an equality in the Grothendieck group of $SL(2)$-modules

$$[L_{SL(2)}(1) \otimes H^0_{SL(2)}(t-1)]_{SL(2)} = [H^0_{SL(2)}(t)]_{SL(2)} + [H^0_{SL(2)}(t-2)]_{SL(2)}.$$

The reader is asked to review Section 6 of [**6**]. For the convenience of the reader, we copy the notation and relevant statement from [**6**] here.

Assume that the p-adic expansion of $k+1$ is given as $k+1 = \sum_{i=0}^u a_i p^i$, where each $0 \le a_i < p$.

We will consider specific words $w = w_0 \dots w_u$ (to be defined recursively later) of length $u+1$ consisting of symbols $<$, $\le$, $\ge$ and $>$. To each such word w, we assign a subset $S(w)$ of the set $\{0, \dots, k\}$ consisting of those numbers s with the p-adic expansion $s = \sum_{i=0}^u s_i p^i$ such that

$$0 \le s_i \le a_i - 1 \text{ if } w_i = \text{“}<\text{”}, \qquad 0 \le s_i \le a_i \text{ if } w_i = \text{“}\le\text{”},$$
$$a_i \le s_i < p \text{ if } w_i = \text{“}\ge\text{” and } \quad a_i + 1 \le s_i < p \text{ if } w_i = \text{“}>\text{”}.$$

To each word w we assign a weight $\ell(w)$ by

$$\ell_k(w) = k - \sum_{w_i=\text{“}\geq\text{”}} 2a_ip^i - \sum_{w_i=\text{“}>\text{”}} 2(a_i+1)p^i.$$

Due to the property $\binom{k}{i} = \binom{k}{k-i}$, we note that the character of a simple module $L_{SL(2)}(\ell)$ is uniquely determined by the set $S(\ell)$ of its nonnegative weights. We will see that if $L_{SL(2)}(\ell)$ is a composition factor of $H^0_{SL(2)}(k)$, then $\ell = \ell_k(w)$ for some word w as above.

We will define the ordered sets $W^s = \cup_{j=-1}^{s-1} W_j$ and $W = \cup_{j=-1}^{\infty} W_j$ recursively as follows.

First define W_{-1} to consist of a single word $w =$ “ $<\leq \ldots \leq$ ”. Then define W_0 to consist of a single word $w =$ “ $\geq<\leq \ldots \leq$ ”. Assume the ordered sets $W_j = \{w^j_1, \ldots, w^j_{2^j}\}$ for $0 \leq j \leq s-1$ are defined, define $W_s = \{w^s_1, \ldots w^s_{2^s}\}$ as follows.

For $j = 1, \ldots 2^{s-1}$, w^s_j is obtained from w^{s-1}_j by replacing its s-th symbol $<$ by $\geq$ and its $(s+1)$-st symbol $\leq$ by $<$. The remaining elements w^s_j for $j = 2^{s-1}+1, \ldots, 2^s$ are obtained by taking elements of W_{-1} through W_{s-2} in the order they are listed and changing their s-th symbols $\leq$ to $>$ and their $(s+1)$-st symbols $\leq$ to $<$.

EXAMPLE 10.4. Let $k+1 = a_0 + a_1p + a_2p^2 + a_3p^3 + a_4p^4$, where $a_4 \neq 0$. Then the ordered set W^4 consists of the following words w (listed together with $\ell_k(w)$)

$$\begin{array}{llllll}
<\leq\leq\leq\leq & \text{of weight } k & & & & \\
\geq<\leq\leq\leq & \text{of weight } k & -2a_0 & & & \\
\geq\geq<\leq\leq & \text{of weight } k & -2a_0 - 2a_1p & & & \\
<><\leq\leq & \text{of weight } k & - 2(a_1+1)p & & & \\
\geq\geq\geq<\leq & \text{of weight } k & -2a_0 - 2a_1p & -2a_2p^2 & & \\
<>\geq<\leq & \text{of weight } k & - 2(a_1+1)p & -2a_2p^2 & & \\
<\leq><\leq & \text{of weight } k & & -2(a_2+1)p^2 & & \\
\geq<><\leq & \text{of weight } k & -2a_0 & -2(a_2+1)p^2 & & \\
\geq\geq\geq\geq< & \text{of weight } k & -2a_0 - 2a_1p & -2a_2p^2 & - 2a_3p^3 & \\
<>\geq\geq< & \text{of weight } k & - 2(a_1+1)p & -2a_2p^2 & - 2a_3p^3 & \\
<\leq>\geq< & \text{of weight } k & & -2(a_2+1)p^2 & - 2a_3p^3 & \\
\geq<>\geq< & \text{of weight } k & -2a_0 & -2(a_2+1)p^2 & - 2a_3p^3 & \\
<\leq\leq>< & \text{of weight } k & & & - 2(a_3+1)p^3 & \\
\geq<\leq>< & \text{of weight } k & -2a_0 & & - 2(a_3+1)p^3 & \\
\geq\geq<>< & \text{of weight } k & -2a_0 - 2a_1p & & - 2(a_3+1)p^3 & \\
<><>< & \text{of weight } k & - 2(a_1+1)p & & - 2(a_3+1)p^3 &
\end{array}$$

Denote by $D(k)$ the set of highest weights of simple composition factors in $H^0_{SL(2)}(k)$. The set $D(k)$ consists of dominant weights of type $\ell_k(w)$ for $w \in W$, and is thoroughly described in the following statement.

PROPOSITION 10.5. *(Proposition 6.2 of* [**6**]*) Let* $k + 1 = \sum_{i=0}^{u} a_i p^i$, *where* $0 \leq a_i < p$.

If k *is such that every* $a_i \neq 0, p-1$, *then the simple composition factors of* $H^0_{SL(2)}(k)$ *are exactly those* $L_{SL(2)}(\ell_k(w))$, *where* $w \in W^u$ *and* $\ell_k(w) \geq 0$.

For each i *such that* $a_i = p-1$, *the weights* $\ell_k(w)$, *corresponding to* w *such that* $w_i =$ "$>$", *are removed from this list. For each* i *such that* $a_i = 0$, *we remove all weights* $\ell_k(w)$, *corresponding to* w *such that* $w_i =$ "$<$". *After removing these weights and duplicate listings of the remaining weights, we obtain the list of all simple composition factors* $L_{SL(2)}(\ell)$ *of* $H^0_{SL(2)}(k)$.

Consider a word $w \in W \backslash W_{-1}$ and the corresponding weight $\ell_k(w)$. We say that w and $\ell_k(w)$ are of the *first kind* if $w_0 =$ "$\geq$" (this is equivalent to $\ell_k(w) \equiv k - 2a_0 \pmod{2p}$). We say that w and $\ell_k(w)$ are of the *second kind* if $w_0 =$ "$<$" (this is equivalent to $a_0 \neq 0$ and $\ell_k(w) \equiv k \pmod{2p}$).

Now we can describe a composition series for induced supermodules $H^0_-(q+k, q)$.

Let $\alpha = (1, -1)$. There is a bijective correspondence between the weights of the induced supermodule $H^0_-(q+k,q)$ and the corresponding $SL(2)$-weights. The $SL(2)$-weight $\ell_k(w)$ corresponds to the $GL(2)$ or **Q**(2)-weight

$$s_k(w) = (q + k - r_k(w), q + r_k(w)) = (q + k, q) - r_k(w)\alpha,$$

where

$$r_k(w) = \frac{k - \ell_k(w)}{2} = \sum_{w_i = \text{"}\geq\text{"}} a_i p^i + \sum_{w_i = \text{"}>\text{"}} (a_i + 1) p^i.$$

It is essential to observe that the length of all weights in $H^0_-(q+k,q)$, which include weights $s_k(w)$, $s_k(w) - \alpha$ for all $w \in D(k)$ and $(q+k,q)$, equals $2q+k$.

We exclude cases $k = 0, k = 1$ since the corresponding induced supermodules $H^0_-(q+k,q)$ are irreducible. Thus, we assume $k > 1$.

For an integer k, denote by k_0 the number $0 \leq k_0 < p$ such that $k \equiv k_0 \pmod{p}$.

PROPOSITION 10.6. *Assume* $p \nmid 2q + k$.

If $1 < k_0 < p-1$, *then the simple composition factors of* $H^0_-(q+k,q)$ *are* $L_-(s_k(w))$ *for* $w \in D(k)$ *of the first kind and* $L_-(s_{k-2}(w'))$ *for* $w' \in D(k-2)$ *of the second kind, all with simple multiplicities.*

If $k_0 = 1$, *then the simple composition factors of* $H^0_-(q+k,q)$ *are* $L_-(s_k(w))$ *for* $w \in D(k)$ *of the first kind and* $L_-(s_{k-2}(w'))$ *for* $w' \in D(k-2)$ *of the first kind, all with simple multiplicities.*

If $k_0 = p-1$, *then the simple composition factors of* $H^0_-(q+k,q)$ *are* $L_-(s_k(w))$ *for* $w \in D(k)$ *of the first kind and* $L_-(s_{k-2}(w'))$ *for* $w' \in D(k-2)$ *of the second kind, all with simple multiplicities.*

Assume $p | 2q + k$ *but* $p \nmid k$.

Then the composition factors of $H^0_-(q+k,q)$ *are* $L_-(s_k(w))$ *for* $w \in D(k)$ *and* $L_-(s_k(w'))$ *for* $w' \in D(k-2)$, *all with simple multiplicities.*

Assume $p | q$ *and* $p | k$ *and write* $q = pq'$, $k = pk'$.

Then $H^0_-(q+k,q)$ *has simple composition factors* $F^*(L_{SL(2)}(s_{k'}(w'))$ *for* $w' \in D(k')$, *with simple multiplicities,* $F^*(L_{SL(2)}(s_{k'-2}(w')))$ *for* $w' \in D(k'-2)$, *with multiplicity three,and* $L_-(s_{k-2}(w))$ *for* $w \in D(k-2)$ *of the first kind, with multiplicity four.*

PROOF. By Corollary 10.1,

$$[H^0_-(q+k,q)] = 2[H^0_{GL(2)}(q+k,q)] + 2[H^0_{SL(2)}(q+k-1,q+1)].$$

The sum of characters of all simple modules $L_{GL(2)}(s_k(w))$ for $w \in D(k)$ equals the character of $H^0_{GL(2)}(q+k,k)$. Since $k > 1$, then the sum of characters of all simple modules $L_{GL(2)}(s_{k-2}(w))$ for $w \in D(k-2)$ equals the character of $H^0_{GL(2)}(q+k-1,k+1)$. If $1 < k_0$, then the sum of characters of all simple modules $L_{GL(2)}(s_k(w)-\alpha)$ for $w \in D(k)$ equals the character of $H^0_{GL(2)}(q+k-1,k+1)$.

Assume $p \nmid 2q+k$.

If $1 < k_0 < p-1$, then all $1 < (\ell_k(w))_0 < p-1$ for $w \in D(k)$. Then by Lemma 10.2, each $L_-(s_k(w))$ for $w \in D(k)$, satisfies

$$[L_-(s_k(w))] = 2[L_{GL(2)}(s_k(w))] + 2[L_{GL(2)}(s_k(w)-\alpha)],$$

and each $L_-(s_{k-2}(w'))$ for $w' \in D(k-2)$, satisfies

$$[L_-(s_{k-2}(w'))] = 2[L_{GL(2)}(s_{k-2}(w'))] + 2[L_{GL(2)}(s_{k-2}(w')-\alpha)].$$

Consider the weight $s_k(w)$ corresponding to $w \in D(k)$ of the first kind. Then the weight $s_k(w)-\alpha$ corresponds to $w' \in D(k-2)$ of the first kind. If $w' \in D(k-2)$ is of the second kind, then the weight $s_{k-2}(w')-\alpha$ corresponds to some $w \in D(k)$ of the second kind. Therefore, we can describe these weights for $w' \in D(k-2)$ as $s_k(w)+\alpha$ for $w \in D(k)$ of the second kind.

The comparison of characters using Proposition 10.5 gives that the composition factors of $H^0_-(q+k,q)$ are $L_-(s_k(w))$ for $w \in D(k)$ of the first kind and $L_-(s_{k-2}(w'))$ for $w' \in D(k-2)$ of the second kind (equivalently, $L_-(s_k(w)+\alpha)$ for $w \in D(k)$ of the second kind), all with simple multiplicities.

If $k_0 = 1$, then $(\ell_k(w))_0 = 1$ and $[L_-(s_k(w))] = 2[L_{GL(2)}(s_k(w))]$ for $w \in D(k)$ of the first kind, and $(\ell_k(w))_0 = p-1$ and

$$[L_-(s_{k-2}(w'))] = 2[L_{GL(2)}(s_{k-2}(w'))] + 2[L_{GL(2)}(s_{k-2}(w')-\alpha)]$$

for $w' \in D(k-2)$. Note that in this case, all $w' \in D(k-2)$ are of the first kind. The weights $\ell_{k-2}(w')-\alpha = \ell_k(w)$ for $w' \in D(k-2)$ correspond to $w \in D(k)$ of the second kind. Therefore, we can describe these weights for $w' \in D(k-2)$ as $s_k(w)+\alpha$ for $w \in D(k)$ of the second kind. The comparison of characters using Proposition 10.5 gives that the composition factors of $H^0_-(q+k,q)$ are $L_-(s_k(w))$ for $w \in D(k)$ of the first kind and $L_-(s_{k-2}(w'))$ for $w' \in D(k-2)$ of the first kind (equivalently, $L_-(s_k(w)+\alpha)$ for $w \in D(k)$ of the second kind), all with simple multiplicities.

If $k_0 = p-1$, then $(\ell_k(w))_0 = p-1$ and

$$[L_-(s_k(w))] = 2[L_{GL(2)}(s_k(w))] + 2[L_{GL(2)}(s_k(w)-\alpha)],$$

for $w \in D(k)$, and $(\ell_k(w'))_0 = 1$ and $[L_-(s_k(w'))] = 2[L_{GL(2)}(s_{k-2}(w'))]$ for $w' \in D(k-2)$ of the second kind. Note that in this case, all $w \in D(k)$ are of the first kind. The weights $\ell_k(w)-\alpha = \ell_{k-2}(w')$ for $w \in D(k)$ correspond to $w' \in D(k-2)$ of the first kind. The comparison of characters using Proposition 10.5 gives that the composition factors of $H^0_-(q+k,q)$ are $L_-(s_k(w))$ for $w \in D(k)$ of the first kind and $L_-(s_{k-2}(w'))$ for $w' \in D(k-2)$ of the second kind, all with simple multiplicities.

Assume $p|2q+k$. Then by Lemma 10.2, $[L_-(s_k(w))] = 2[L_{GL(2)}(s_k(w))]$ and $[L_-(s_k(w)-\alpha)] = 2[L_{GL(2)}(s_k(w)-\alpha)]$. The situation is much simpler than in the previous case when $p \nmid 2q+k$, because we are essentially dealing with $SL(2)$-modules.

If $p \nmid k$, then the comparison of characters using Proposition 10.5 gives that the composition factors of $H^0_-(q+k,q)$ are $L_-(s_k(w))$ for $w \in D(k)$ and $L_-(s_k(w'))$ for $w' \in D(k-2)$, all with simple multiplicities.

If $p|k$, then write $q = pq'$, $k = pk'$, and we consider the $GL(2)$-composition factors of $H^0_-(q'+k',q')$. They are $L_{GL(2)}(s_{k'}(w'))$ for $w' \in D(k')$. Applying the Frobenius transform F^*, we get $F^*(L_{GL(2)}(s_{k'}(w')))$ for $w' \in D(k')$ and these composition factors of $H^0_-(q+k,q)$ constitute one-dimensional subspaces of weight spaces of $H^0_-(q+k,q)_\mu$, where both components μ_1, μ_2 are divisible by p. This covers the weight spaces for $\mu = (q+k,q)$ and $\mu = (q,q+k)$ exactly. To cover the outstanding three dimensions in remaining weight spaces $H^0_0(q+k,q)_\mu$, where $p|\mu_1,\mu_2$, we use composition factors $F^*(L_{GL(2)}(s_{k'-2}(w')))$ for $w' \in D(k'-2)$, each with multiplicity three.

The remaining weight spaces $H^0_-(q+k,q)_\mu$, where both components μ_1,μ_2 are not divisible by p, are covered by four copies of simple supermodules $L_-(s_{k-2}(w))$ for $w \in D(k-2)$ of the first kind. The comparison of characters using Proposition 10.5 gives that the composition factors of $H^0_-(q+k,q)$ are $F^*(L_{GL(2)}(s_{k'}(w'))$ for $w' \in D(k')$, with simple multiplicities, $F^*(L_{GL(2)}(s_{k'-2}(w')))$ for $w' \in D(k'-2)$, with multiplicity three, and $L_-(s_{k-2}(w))$ for $w \in D(k-2)$ of the first kind, with multiplicities four. □

REMARK 10.7. It follows from the linkage principle of Theorem 8.10 of [**3**] and the proof of Lemma 8.9 of [**3**] that, if λ and μ are linked, then $cont(\lambda) = cont(\mu)$, where $cont(\lambda) = (a_r(\lambda) - b_r(\lambda)|r = 0,\dots,p-1)$,

$$a_r(\lambda) = |\{i = 1,2|(\lambda_i+1)^2 - (\lambda_i+1) \equiv r \pmod p\}|$$

and

$$b_r(\lambda) = |\{i = 1,2|(\lambda_i)^2 - (\lambda_i) \equiv r \pmod p\}|$$

for each $r = 0,\dots,p-1$.

We point out that the content of the weights appearing in the composition series is the same, conforming to the linkage principle of [**3**].

Additionally, in the case $p|2q+k$, all appearing weights have the same content, so in this case the linkage principle of [**3**] does not give sufficient information to restrict the weights strongly linked to λ.

See also Section 11 and Remark 11.4 for additional information.

Note the "dot action" like setup in the following proposition. We move from the $\mathbf{Q}(2)$-weight λ to $\lambda - p\epsilon_1$ and the $SL(2)$-weight $\lambda_1 - \lambda_2 - p = p(t-1)$. Then we determine the $SL(2)$-composition factors $L(ps)$ of $H^0_{SL(2)}(p(t-1))$, move to the corresponding $\mathbf{Q}(2)$-weight μ such that $\mu_1 - \mu_2 = ps$, and finally, add $p\epsilon_1$ to $p\mu + p\epsilon_1$.

PROPOSITION 10.8. *Assume $k = pt$, where $t > 1$ and p does not divide q. Then the simple composition factors of $H^0_-(q+k,q)$ are $L_-(ps_{t-1}(w) + p\epsilon_1)$ for $w \in D(t-1)$.*

PROOF. By Corollary 10.1 we have

$$[H^0_-(q+k,q)] = 2[H^0_{GL(2)}(q+k,q)] + 2[H^0_{GL(2)}(q+k-1,q+1)],$$

while by Lemma 10.3, if $k = pt$, where $t > 1$, and $p \nmid q$, then

$[L_-(q+k, q)]_{SL(2)} = 2[F^*(L_{SL(2)}(1) \otimes L_{SL(2)}(t-1))]_{SL(2)} + 4[L_{SL(2)}(pt-2)]_{SL(2)}$, where the $SL(2)$-module $L_{SL(2)}(pt-2)$ corresponds to the $GL(2)$-module $L_{GL(2)}(q+k-1, q+1)$.

The module $H^0_{GL(2)}(q+k, q)$ has composition factors $L_{GL(2)}(s_k(w))$ for $w \in D(k)$. Those simple modules corresponding to $w \in D(k)$ of the second kind cover exactly the weight spaces $H^0_{GL(2)}(q+k, q)_\mu$, where $p \nmid \mu_1 - \mu_2$.

The module $H^0_{GL(2)}(q+k-1, q+1)$ has composition factors $L_{GL(2)}(s_{k-2}(w'))$ for $w' \in D(k-2)$. Those simple modules corresponding to $w' \in D(k-2)$ of the first kind cover exactly the weight spaces $H^0_{GL(2)}(q+k-1, q+1)_\mu$, where $p \nmid \mu_1 - \mu_2$.

Since there is a one-to-one correspondance between $w \in D(k)$ of the second kind and $w' \in D(k-2)$ of the first kind, we can combine these to four copies of $L_{GL(2)}(s_{k-2}(w'))$, where $w' \in D(k-2)$ is of the first kind. There is another description of the weights $s_k(w')$ for $w' \in D(k-2)$ of the first kind. There is a one-to-one correspondence between $w' \in D(k-2)$ of the first kind and $w'' \in D(t-1)$ given by $\ell_{k-2}(w') = (p-2) + p\ell_{t-1}(w'')$ for $w'' \in D(t-1)$.

This implies that simple supermodules $L_-(ps_{t-1}(w'') + p\epsilon_1)$ for $w'' \in D(t-1)$ are composition factors of $H^0_-(q+k, q)$. The sum of the dimensions of their weightspaces at μ such that $p \nmid \mu_1 - \mu_2$ coincides with the dimension of the weightspaces $H^0_-(q+k, q)_\mu$, where $p \nmid \mu_1 - \mu_2$.

Since the $SL(2)$-characters of $L_{SL(2)}(\ell(w''))$ for $w'' \in D(t-1)$ add up to the character of $H^0_{SL(2)}(t-1)$, we have

$$\sum_{w'' \in D(t-1)} [L_{SL(2)}(1) \otimes L_{SL(2)}(\ell(w''))]_{SL(2)} = [H^0_{SL(2)}(1) \otimes H^0_{SL(2)}(t-1)]_{SL(2)}$$
$$= [H^0_{SL(2)}(t)]_{SL(2)} + [H^0_{SL(2)}(t-2)]_{SL(2)}.$$

The sum of the dimensions of the weightspaces $L_-(ps_{t-1}(w'') + p\epsilon_1)_\mu$ for $w'' \in D(t-1)$ and $p|\mu_1 - \mu_2$ is given by adding up dimensions of the corresponding $\mu_1 - \mu_2$-weightspaces in $F^*(L_{SL(2)}(1) \otimes L_{SL(2)}(\ell(w'')))$ for $w'' \in D(t-1)$. Therefore, the sum of the dimensions of the weightspaces $L_-(ps_{t-1}(w'') + p\epsilon_1)_\mu$ for $w'' \in D(t-1)$ such that $p|\mu_1 - \mu_2$ coincides with the dimension of the weightspaces $H^0_-(q+k, q)_\mu$, where $p|\mu_1 - \mu_2$. This proves that $H^0_-(q+k, q)$ has a filtration by simple supermodules $L_-(ps_{t-1}(w'') + p\epsilon_1)$ for $w'' \in D(t-1)$. □

11. Blocks of $\mathbf{Q}(2)$

We start with the following statement, which relates to the linkage and strong linkage of weights.

LEMMA 11.1. *Assume $\lambda \neq \mu$ and $Ext^1_{\mathbf{Q}(2)}(L_-(\mu), L_-(\lambda)) \neq 0$. Then $L_-(\mu)$ is a composition factor of $H^0_-(\lambda)$, hence λ and μ are strongly linked.*

PROOF. Let $\lambda \neq \mu$ and

$$0 \to L_-(\lambda) \to M \to L_-(\mu) \to 0$$

be a non-split exact sequence of $\mathbf{Q}(2)$-supermodules. Denote by V a subspace of the weightspace M_μ that is mapped isomorphically onto the weightspace $L_-(\mu)_\mu$. If $\mathbf{U}^+V = 0$, then the supermodule $\mathbf{Q}(2)V$ generated by V is a supersubmodule of M that is mapped isomorphically onto $L_-(\mu)$, or $\mathbf{Q}(2)V = M$. In the first case, the above exact sequence splits, which is a contradiction. If $\mathbf{Q}(2)V = M$, i.e. M is generated by "primitive" subspace V, then by the universality of the Weyl supermodule $V_-(\mu)$ it follows that M is an epimorphic image of $V_-(\mu)$. Since the formal characters of $V_-(\mu)$ and $H^0_-(\mu)$ are the same, we obtain that $L_-(\lambda)$ is a composition factor of $H^0_-(\mu)$.

If $\mathbf{U}^+V \neq 0$, then $\mathbf{U}^+$ acts nontrivially on V, which implies $\mu < \lambda$. By Frobenius reciprocity,

$$Hom_{\mathbf{Q}(2)}(M, H^0_-(\lambda)) \simeq Hom_{\mathbf{B}^-}(M, \mathfrak{u}(\lambda)) \simeq Hom_{\mathbf{B}^-}(M/Dist(\mathbf{U}^-)^+M, \mathfrak{u}(\lambda)).$$

Since all weights of $Dist(\mathbf{U}^-)^+M$ are strictly less than λ, the $\mathbf{H}(2)$-supermodule $M/Dist(\mathbf{U}^-)^+M$ has a unique composition factor $\mathfrak{u}(\lambda)$. On the other hand, we have seen in subsection 2.2 that every indecomposable injective $\mathbf{H}(2)$-supermodule has composition factors isomorphic to the simple $\mathfrak{u}(\pi)$ or its parity shift. In particular, $\pi \neq \pi'$ implies $Ext^1_{\mathbf{H}(2)}(\mathfrak{u}(\pi), \mathfrak{u}(\pi')) = 0$. Therefore, $\mathfrak{u}(\lambda)$ appears as a summand of the top of $M/Dist(\mathbf{U}^-)^+M$. This shows

$$Hom_{\mathbf{B}^-}(M/Dist(\mathbf{U}^-)^+M, \mathfrak{u}(\lambda)) \neq 0,$$

and it implies that M is embedded into $H^0_-(\lambda)$. □

The length of a weight λ is defined as $|\lambda| = \lambda_1 + \lambda_2$. Denote by Λ^+_t the set of dominant weights of $\mathbf{Q}(2)$ of length t. Since the action of the Lie superalgebra on weights preserves their lengths, every block of $\mathbf{Q}(2)$ is a subset of Λ^+_t for some t. Within Λ^+_t, the weights $\lambda = (\lambda_1, \lambda_2)$ are uniquely identified with their $SL(2)$-weight $\lambda_1 - \lambda_2$.

Assume t is not divisible by p. Denote by C_t the set of all weights $\mu \in \Lambda^+_t$ such that $p|\mu_1 - \mu_2$.

Let $\lambda, \mu \in C_t$ be such that $\lambda_1 - \lambda_2 = pa$ and $\mu_1 - \mu_2 = pb$. Denote by $\tilde{B}_\lambda$ the set of all weights μ such that $|\mu| = |\lambda|$ and $b - 1 \sim_{ev} a - 1$.

Denote by B_λ the block of $\mathbf{Q}(2)$ containing the weight λ.

Proposition 11.2. *Assume t is not divisible by p and $\lambda \in C_t$. Then $\tilde{B}_\lambda = B_\lambda$.*

Proof. We can rephrase Proposition 10.8 on the level of corresponding $SL(2)$-weights before the application of the Frobenius twist functor as follows. For $\kappa = s$, we look at the weights ν strongly $SL(2)$-linked to $\kappa - 1 = s - 1$, and then strongly link κ to each $\nu + 1$. This shows that the block B_λ of C_t contains $\tilde{B}_\lambda$. Lemma 11.1 proves the opposite inclusion. □

If t is even, then for $0 < i \leq \frac{p-1}{2}$ denote by

$$\begin{aligned} B_{t,i} = &\{(\frac{t}{2} + i + ps, \frac{t}{2} - i - ps)|s \geq 0\} \cup \\ &\{(\frac{t}{2} + p - i + ps, \frac{t}{2} - p + i - ps)|s \geq 0\} \end{aligned}$$

a subset of the set Λ^+_t. Then $\cup_{i=1}^{\frac{p-1}{2}} B_{t,i} = \Lambda^+_t \setminus C_t$. In this case, C_t consists of weights $(\frac{t}{2} + ps, \frac{t}{2} - ps)$ for $s \geq 0$.

If t is odd, then for $0 \leq i < \frac{p-1}{2}$ denote by

$$\begin{aligned} B_{t,i} = &\{(\frac{t+1}{2} + i + ps, \frac{t-1}{2} - i - ps)|s \geq 0\} \cup \\ &\{(\frac{t+1}{2} + p - i + ps, \frac{t-1}{2} - p + i - ps)|s \geq 0\} \end{aligned}$$

a subset of the set Λ_t^+. Then $\cup_{i=0}^{\frac{p-1}{2}-1} B_{t,i} = \Lambda_t^+ \setminus C_t$. In this case, C_t consists of weights $(\frac{t+1}{2} + \frac{p-1}{2} + ps, \frac{t-1}{2} - \frac{p-1}{2} - ps)$ for $s \geq 0$.

PROPOSITION 11.3. *If t is a multiple of p, then all weights of Λ_t^+ are linked. If p does not divide t, then $\Lambda_t^+ \setminus C_t = \cup_{j\in J_t} B_{t,j}$ as above, and all weights in each $B_{t,j}$ for $j \in J_t$ are linked.*

PROOF. Write $\lambda = (q+k, q)$, where $t = 2q+k$ and $t \equiv b_0 \pmod p$, where $0 \leq b_0 < p$.

If t is a multiple of p, which means $b_0 = 0$, then by Proposition 10.6 we have $(q+k, q) \sim (q+k-1, q+1)$ for each k. Therefore, all weights in Λ_t are linked.

Assume t is not divisible by p, which means $b_0 \neq 0$. We proceed similarly to the proof of Proposition 6.7 of [**6**].

The induced supermodule $H^0_-(q+2p-b_0+2ps, q)$ has simple composition factors $L_-(q+p+2ps, q+p-b_0)$ and $L_-(q+2p-b_0+2ps, q)$, hence $(q+p+2ps, q+p-b_0) \sim (q+2p-b_0+2ps, q)$.

The induced supermodule $H^0_-(q+p+2pt, q+p-b_0)$ has simple composition factors $L_-(q+p-b_0+2pt, q+p)$ and $L_-(q+p+2pt, q+p-b_0)$, hence $(q+p+2pt, q+p-b_0) \sim (q+p-b_0+2pt, q+p)$.

Therefore, $(q+2p-b_0+2ps, q)$, $(q+p+2ps, q+p-b_0)$ and $(q+p-b_0+2ps, q+p)$ are all linked. This implies that all weights in $B_{t,j}$ for $j \in J$ are linked. The equality $\cup_j B_{t,j} = \Lambda_t^+ \setminus C_t$ is clear. □

REMARK 11.4. If $\mu = (\lambda_1+pt, \lambda_2-pt)$ or $\mu = (\lambda_2+pt, \lambda_1-pt)$, then $cont(\lambda) = cont(\mu)$.

Note that every weight (a, b) such that $a+b \equiv 0 \pmod p$ satisfies $cont(a, b) = 0$ because $a^2 - a = (b+1)^2 - (b+1)$ and $b^2 - b = (a+1)^2 - (a+1)$. Additionally, it is clear that if $c \equiv b \pmod p$ and $d \equiv a \pmod p$, that is $(c, d) \equiv (b, a) \pmod p$, then $cont(c, d) = cont(b, a)$.

Therefore, the content of weights $(q+k, q)$ and $(q+k-1, q+1)$ is identical provided $2q+k \equiv 0 \pmod p$. Also, the contents of weights $(q+2p-b_0+2ps, q)$, $(q+p+2ps, q+p-b_0)$, and $(q+p-b_0+2ps, q+p)$ appearing in the proof of the above proposition are also identical.

Note also that all weights inside C_t have identical content.

PROPOSITION 11.5. *Assume t is not divisible by p. If t is even, then the content of weights $\mu_i = (\frac{t}{2}+i, \frac{t}{2}-i)$ for $i = 0, \ldots, \frac{p-1}{2}$ are mutually distint. If t is odd, then the content of weights $\mu_i = (\frac{t+1}{2}+i, \frac{t-1}{2}-i)$, where $i = 0, \ldots, \frac{p-1}{2}$ are mutually distinct.*

PROOF. If $x^2 + x \equiv y^2 + y \pmod p$, then $(x-y)(x+y+1) \equiv 0 \pmod p$, and either $x \equiv y \pmod p$ or $x+y+1 \equiv 0 \pmod p$.

If $x^2 - x \equiv y^2 - y \pmod p$, then $(x-y)(x+y-1) \equiv 0 \pmod p$, and either $x \equiv y \pmod p$ or $x+y-1 \equiv 0 \pmod p$.

If $x^2 + x \equiv y^2 - y \pmod p$, then $(x+y)(x-y+1) \equiv 0 \pmod p$, and either $x+y \equiv 0 \pmod p$ or $x-y+1 \equiv 0 \pmod p$.

Denote $\mu_i = (a_i, b_i)$ and $\mu_j = (a_j, b_j)$.

First, assume that <u>t is even</u>.

Assume that weights μ_i and μ_j, for $i \neq j$, have the same content, which contains *two entries equal to 1 or an entry equal to 2.* Since $0 \leq i \neq j \leq \frac{p-1}{2}$, we have $a_i = \frac{t}{2}+i \not\equiv a_j = \frac{t}{2}+j \pmod p$, $a_i = \frac{t}{2}+i \not\equiv b_j = \frac{t}{2}-j \pmod p$, $b_i = \frac{t}{2}-i \not\equiv a_j = \frac{t}{2}+j$

(mod p), and $b_i = \frac{t}{2} - i \not\equiv b_j = \frac{t}{2} - j \pmod p$. Therefore, we have two possibilities, either

$a_i + a_j + 1 = \frac{t}{2} + i + \frac{t}{2} + j + 1 \equiv 0 \pmod p$ and $b_i + b_j + 1 = \frac{t}{2} - i + \frac{t}{2} - j + 1 \equiv 0 \pmod p$,

or

$a_i + b_j + 1 = \frac{t}{2} + i + \frac{t}{2} - j + 1 \equiv 0 \pmod p$ and $b_i + a_j + 1 = \frac{t}{2} - i + \frac{t}{2} + j + 1 \equiv 0 \pmod p$.

In the first case, we obtain $i + j \equiv 0 \pmod p$, which is not possible, while in the second case, we obtain $i - j \equiv 0 \pmod p$, which is not possible.

If the content of μ_i contains only *one or no entry equal to 1*, then we get one of the following possibilities

- $a_i + a_i = \frac{t}{2} + i + \frac{t}{2} + i \equiv 0 \pmod p$,
- $a_i + b_i = \frac{t}{2} + i + \frac{t}{2} - i = t \equiv 0 \pmod p$ which is not possible,
- $b_i + b_i = \frac{t}{2} - i + \frac{t}{2} - i \equiv 0 \pmod p$,
- $a_i - a_i + 1 = 1 \equiv 0 \pmod p$ which is not possible,
- $a_i - b_i + 1 = \frac{t}{2} + i - \frac{t}{2} + i + 1 \equiv 0 \pmod p$,
- $b_i - b_i + 1 = 1 \equiv 0 \pmod p$ which is not possible, or
- $b_i - a_i + 1 = \frac{t}{2} - i - \frac{t}{2} - i + 1 \equiv 0 \pmod p$, which is not possible.

Therefore, $t + 2i \equiv 0 \pmod p$, $t - 2i \equiv 0 \pmod p$, or $i = \frac{p-1}{2}$.

Possible solutions are $i_1 = \frac{p-1}{2}$, $i_2 \equiv \frac{t}{2} \pmod p$ and $i_3 \equiv -\frac{t}{2} \pmod p$.

Since only one of i_2, i_3 can fall within the range $0 \leq i_2, i_3 \leq \frac{p-1}{2}$, we need to verify that if $0 \leq i_2 \leq \frac{p-1}{2}$ and $i_1 \neq i_2$, then the contents of μ_{i_1} and μ_{i_2} are different, and if $0 \leq i_3 \leq \frac{p-1}{2}$ and $i_1 \neq i_3$, then the contents of μ_{i_1} and μ_{i_3} are different.

If $p = 3$, there are only two possible values of $i = 0, 1$. Since $i_1 = 1$, we must have $i_2 = 0$ or $i_3 = 0$, which implies t is a multiple of 3, a contradiction. Therefore, we can assume $p > 3$.

For the content of μ_{i_1}, consider the congruence classes modulo p of $(\frac{t}{2}+i_1)(\frac{t}{2}+i_1+1)$, $(\frac{t}{2}+i_1)(\frac{t}{2}+i_1-1)$, $(\frac{t}{2}-i_1)(\frac{t}{2}-i_1+1)$, $(\frac{t}{2}-i_1)(\frac{t}{2}-i_1-1)$. The contributions of $(\frac{t}{2}+i_1)(\frac{t}{2}+i_1+1)$ and $(\frac{t}{2}-i_1)(\frac{t}{2}-i_1-1)$ to the content are offsetting. We are left with

$(\frac{t}{2}+i_1)(\frac{t}{2}+i_1-1) = \frac{t^2}{4} + ti_1 + i_1^2 - \frac{t}{2} - i_1 \equiv \frac{t^2}{4} - t + \frac{3}{4} \pmod p$ and

$(\frac{t}{2}-i_1)(\frac{t}{2}-i_1+1) = \frac{t^2}{4} - ti_1 + i_1^2 + \frac{t}{2} - i_1 \equiv \frac{t^2}{4} + t + \frac{3}{4} \pmod p$.

The congruence classes of the last two expressions differ because $t \not\equiv 0 \pmod p$. This implies that the content of μ_{i_1} has exactly one entry 1 and -1.

For the content of μ_{i_2}, consider the congruence classes modulo p of $(\frac{t}{2}+i_2)(\frac{t}{2}+i_2+1)$, $(\frac{t}{2}+i_2)(\frac{t}{2}+i_2-1)$, $(\frac{t}{2}-i_2)(\frac{t}{2}-i_2+1)$, $(\frac{t}{2}-i_2)(\frac{t}{2}-i_2-1)$. The contributions of $(\frac{t}{2}-i_2)(\frac{t}{2}-i_2+1)$ and $(\frac{t}{2}-i_2)(\frac{t}{2}-i_2-1)$ are offsetting. We are left with

$(\frac{t}{2}+i_2)(\frac{t}{2}+i_2+1) = \frac{t^2}{4} + ti_2 + i_2^2 + \frac{t}{2} + i_2 \equiv t^2 + t \pmod p$ and

$(\frac{t}{2}+i_2)(\frac{t}{2}+i_2-1) = \frac{t^2}{4} + ti_2 + i_2^2 - \frac{t}{2} - i_2 \equiv t^2 - t \pmod p$.

The congruence classes of the last two expressions differ because $t \not\equiv 0 \pmod p$. This implies that the content of μ_{i_2} has exactly one entry 1 and -1.

For the content of μ_{i_3}, consider the congruence classes modulo p of $(\frac{t}{2}+i_3)(\frac{t}{2}+i_3+1)$, $(\frac{t}{2}+i_3)(\frac{t}{2}+i_3-1)$, $(\frac{t}{2}-i_3)(\frac{t}{2}-i_3+1)$, $(\frac{t}{2}-i_3)(\frac{t}{2}-i_3-1)$. The contributions of $(\frac{t}{2}+i_3)(\frac{t}{2}+i_3+1)$ and $(\frac{t}{2}+i_3)(\frac{t}{2}+i_3-1)$ are offsetting. We are left with

$(\frac{t}{2}-i_3)(\frac{t}{2}-i_3+1) = \frac{t^2}{4} - ti_3 + i_3^3 + \frac{t}{2} - i_3 \equiv t^2 + t \pmod p$ and

$(\frac{t}{2}-i_3)(\frac{t}{2}-i_3-1)=\frac{t^2}{4}-ti_3+i_3^2-\frac{t}{2}+i_3\equiv t^2-t \pmod p$.

The congruence classes of the last two expressions differ because $t\not\equiv 0 \pmod p$. This implies that the content of μ_{i_3} has exactly one entry, 1 and -1.

The congruence $\frac{t^2}{4}+t+\frac{3}{4}\equiv t^2+t \pmod p$ is equivalent to $3t^2-3\equiv 0 \pmod p$. The congruence $\frac{t^2}{4}-t+\frac{3}{4}\equiv t^2-t \pmod p$ is equivalent to $3t^2-3\equiv 0 \pmod p$. Since $p\neq 3$, we obtain that $t\equiv \pm 1 \pmod p$. However, in this case the condition $1\leq i_2,i_3\leq \frac{p-1}{2}$ implies $i_1=i_2$ or $i_1=i_3$, which is a contradiction.

Now assume that <u>t is odd</u>.

Assume that weights μ_i and μ_j, for $i\neq j$, have the same content, which contains *two entries equal to 1 or an entry equal to 2*. Since $0\leq i\neq j\leq \frac{p-1}{2}$, we have $a_i=\frac{t+1}{2}+i\not\equiv a_j=\frac{t+1}{2}+j \pmod p$, $a_i=\frac{t+1}{2}+i\not\equiv b_j=\frac{t-1}{2}-j \pmod p$, $b_i=\frac{t-1}{2}-i\not\equiv a_j=\frac{t+1}{2}+j \pmod p$, and $b_i=\frac{t-1}{2}-i\not\equiv b_j=\frac{t-1}{2}-j \pmod p$. Therefore, we have two possibilities, either

$a_i+a_j+1=\frac{t+1}{2}+i+\frac{t-1}{2}+j+1\equiv 0 \pmod p$ and $b_i+b_j+1=\frac{t-1}{2}-i+\frac{t-1}{2}-j+1\equiv 0 \pmod p$,

or

$a_i+b_j+1=\frac{t+1}{2}+i+\frac{t-1}{2}-j+1\equiv 0 \pmod p$ and $b_i+a_j+1=\frac{t-1}{2}-i+\frac{t+1}{2}+j+1\equiv 0 \pmod p$.

In the first case, we obtain $i+j\equiv 0 \pmod p$, while in the second case, we obtain $i-j\equiv 0 \pmod p$, which is not possible.

If the content of μ_i contains only *one or no entry equal to 1*, then we get one of the following possibilities

- $a_i+a_i=\frac{t+1}{2}+i+\frac{t+1}{2}+=t+1+2i\equiv 0 \pmod p$,
- $a_i+b_i=\frac{t+1}{2}+i+\frac{t-1}{2}-i=t\equiv 0 \pmod p$ which is not possible,
- $b_i+b_i=\frac{t-1}{2}-i+\frac{t-1}{2}-i=t-1-2i\equiv 0 \pmod p$,
- $a_i-a_i+1=1\equiv 0 \pmod p$ which is not possible,
- $a_i-b_i+1=\frac{t+1}{2}+i-\frac{t-1}{2}+i+1=2i+2\equiv 0 \pmod p$, which is not possible,
- $b_i-b_i+1=1\equiv 0 \pmod p$, which is not possible, or
- $b_i-a_i+1=\frac{t-1}{2}-i-\frac{t+1}{2}-i+1=-2i\equiv 0 \pmod p$.

Therefore, $t+1+2i\equiv 0 \pmod p$, $t-1-2i\equiv 0 \pmod p$, or $i\equiv 0 \pmod p$.

Possible solutions are $i_1=0$, $i_2\equiv\frac{t-1}{2} \pmod p$ and $i_3\equiv -\frac{t+1}{2} \pmod p$.

If $p=3$, there are only two possible values of $i=0,1$. Since $i_1=0$, we get that either $i_2=1$ or $i_3=1$, which implies that t is a multiple of 3, a contradiction. Therefore, we can assume $p>3$.

Since we cannot have $0\leq i_2\neq i_3\leq\frac{p-1}{2}$, we need to verify that if $0\leq i_2\leq\frac{p-1}{2}$ and $i_1\neq i_2$, then the contents of μ_{i_1} and μ_{i_2} are different, and if $0\leq i_3\leq \frac{p-1}{2}$ and $i_1\neq i_3$, then the contents of μ_{i_1} and μ_{i_3} are different.

For the content of μ_{i_1}, consider the congruence classes modulo p of $(\frac{t+1}{2})(\frac{t+1}{2}+1)$, $(\frac{t+1}{2})(\frac{t+1}{2}-1)$, $(\frac{t-1}{2})(\frac{t-1}{2}+1)$, $(\frac{t-1}{2})(\frac{t-1}{2}-1)$. The contributions of $(\frac{t-1}{2})(\frac{t-1}{2}+1)$ and $(\frac{t+1}{2})(\frac{t+1}{2}-1)$ are offsetting. We are left with

$(\frac{t+1}{2})(\frac{t+1}{2}+1)=\frac{(t+1)^2}{4}+\frac{t+1}{2}\equiv\frac{t^2+4t+3}{4} \pmod p$ and

$(\frac{t-1}{2})(\frac{t-1}{2}-1)=\frac{(t-1)^2}{4}-\frac{t-1}{2}\equiv\frac{t^2-4t+3}{4} \pmod p$.

The congruence classes of the last two expressions differ because $t\not\equiv 0 \pmod p$. This implies that the content of μ_{i_1} has exactly one entry 1 and -1.

For the content of μ_{i_2}, consider the congruence classes modulo p of $(\frac{t+1}{2}+i_2)(\frac{t+1}{2}+i_2+1)$, $(\frac{t+1}{2}+i_2)(\frac{t+1}{2}+i_2-1)$, $(\frac{t-1}{2}-i_2)(\frac{t-1}{2}-i_2+1)$, $(\frac{t-1}{2}-i_2)(\frac{t-1}{2}-i_2-1)$. The contributions of $(\frac{t-1}{2}-i_2)(\frac{t-1}{2}-i_2+1)$ and $(\frac{t-1}{2}-i_2)(\frac{t-1}{2}-i_2-1)$ are offsetting. We are left with

$(\frac{t+1}{2}+i_2)(\frac{t+1}{2}+i_2+1) \equiv t(t+1) \pmod p$ and

$(\frac{t+1}{2}+i_2)(\frac{t+1}{2}+i_2-1) \equiv t(t-1) \pmod p$.

The congruence classes of the last two expressions differ because $t \not\equiv 0 \pmod p$. This implies that the content of μ_{i_2} has exactly one entry 1 and -1.

For the content of μ_{i_3}, consider the congruence classes modulo p of $(\frac{t+1}{2}+i_3)(\frac{t+1}{2}+i_3+1)$, $(\frac{t+1}{2}+i_3)(\frac{t+1}{2}+i_3-1)$, $(\frac{t-1}{2}-i_3)(\frac{t-1}{2}-i_3+1)$, $(\frac{t-1}{2}-i_3)(\frac{t-1}{2}-i_3-1)$. The contributions of $(\frac{t+1}{2}+i_3)(\frac{t+1}{2}+i_3+1)$ and $(\frac{t+1}{2}+i_3)(\frac{t+1}{2}+i_3-1)$ are offsetting. We are left with

$(\frac{t-1}{2}-i_3)(\frac{t-1}{2}-i_3+1) \equiv t(t+1) \pmod p$ and

$(\frac{t-1}{2}-i_3)(\frac{t-1}{2}-i_3-1) \equiv t(t-1) \pmod p$.

The congruence classes of the last two expressions differ because $t \not\equiv 0 \pmod p$. This implies that the content of μ_{i_3} has exactly one entry, 1 and -1.

The congruence $\frac{t^2+4t+3}{4} \equiv t^2+t \pmod p$ is equivalent to $3t^2-3 \equiv 0 \pmod p$. The congruence $\frac{t^2-4t+3}{4} \equiv t^2-t \pmod p$ is equivalent to $3t^2-3 \equiv 0 \pmod p$. Therefore, $t \equiv \pm 1 \pmod p$. If $t \equiv 1 \pmod p$, then $i_2 = 0$ and $i_3 \equiv -1 \pmod p$, which is a contradiction with $i_2 \neq i_1$ and $0 \leq i_3 \leq \frac{p-1}{2}$. If $t \equiv -1 \pmod p$, then $i_3 = 0$ and $i_2 \equiv -1 \pmod p$, which is a contrradiction with $i_3 \neq i_1$ and $0 \leq i_2 \leq \frac{p-1}{2}$. □

Consequently, if $\lambda \in C_t$ and $\mu \notin C_t$, then $\lambda \not\sim \mu$.

THEOREM 11.6. *Every block of* $\mathbf{Q}(2)$ *is of the form* Λ_t^+ *for* $p|t$, $B_{t,j}$ *for* $j \in J_t$ *for* $p \nmid t$, *or* $\ddot{B}_\lambda$ *for some* $\lambda \in C_t$ *such that* $p \nmid t$.

PROOF. The statement follows from Propositions 11.2, 11.3, 11.5, and the linkage principle of [**3**]. □

12. Examples

Let $k = a_0 + a_1 p + \ldots + a_s p^s$ be the p-adic expansion of k.

EXAMPLE 12.1. Assume $k = a_0 + p$, where $a_0 \neq 0$, and p does not divide $2q+k$. Then the character of $L_-(q+k,q)$ is

$$[2x_1^{q+k}x_2^q + 4\sum_{v=1}^{a_0-1} x_1^{q+k-v}x_2^{q+v} + 2x_1^{q+k-a_0}x_2^{q+a_0}](1+x_1^p x_2^{-p})$$

and we get additional composition factor of $H^0(q+k,q)$ of the highest weight $\mu = (q+k-a_0, q+a_0)$. The character of $L_-(\mu)$ is

$$2x_1^{q+k-a_0}x_2^{q+a_0} + 4\sum_{v=1}^{p-1-a_0} x_1^{q+k-a_0-v}x_2^{q+a_0+v} + 2x_1^{q+k-p}x_2^{q+p}.$$

Thus, $H^0(q+a_0+p,q)$ has the composition series with factors $L_-(q+a_0+p,q)$ and $L_-(q+p,q+a_0)$. Therefore, $(q+a_0+p,q) \sim (q+p,q+a_0)$.

The following example illustrates the case when $p|2q+k$ but $p \nmid q,k$.

EXAMPLE 12.2. Let $0 < q < \frac{p}{2}$ and $k = p - 2q$ so that $2q + k = p$. The character of $L_{-}(q+k, q)$ is $2\sum_{i=0}^{k} x_1^{q+k-i} x_2^{q+i}$, and the character of $L_{-}(q+k-1, q+1)$ is $2\sum_{j=0}^{k-2} x_1^{q+k-1-j} x_2^{q+1+j}$. The induced supermodule $H^0(q+k, q)$ has the composition series with factors $L_{-}(q+k, q)$ and $L_{-}(q+k-1, q+1)$. Therefore, $(q+k, q) \sim (q+k-1, q+1)$.

EXAMPLE 12.3. Let $q = 0$ and $k = p$. Then the character of $L_{-}(p, 0)$ is $x_1^p + x_2^p$. The composition series of $H_{-}^0(p, 0)$ consists of $L_{-}(p, 0)$ and four copies of $L_{-}(p-1, 1)$. Therefore, $(p, 0) \sim (p-1, 0)$.

EXAMPLE 12.4. Let $\lambda = (1, 4)$, $\iota = \sqrt{-1}$, $k = 1$, $q = 3$, $e = 1$, $l = -2\iota$. The following element I belongs to $H^0(1, 4)$:

$$\begin{aligned} I = u_0 \otimes [&- s_{22}^3 s'_{11} s'_{21} - s_{21}^2 s_{22} s'_{12} s'_{22} + s_{21} s_{22}^2 s'_{11} s'_{22} + s_{21} s_{ijj}^2 s'_{12} s'_{21} \\ &+ 2\iota s_{11} s_{22} s_{22} s'_{21} s'_{22} - 2\iota s_{12} s_{21} s_{22} s'_{21} s'_{22}] \\ +u_1 \otimes [&- s_{11} s_{22}^3 s'_{21} + s_{12} s_{21} s_{22}^2 s'_{21} + s_{11} s_{21} s_{22}^2 s'_{22} - s_{12} s_{21}^2 s_{22} s'_{22} \\ &- 2\iota s_{22}^2 s'_{11} s'_{21} s'_{22} + 2\iota s_{21} s_{22} s'_{12} s'_{21} s'_{22}]. \end{aligned}$$

We have

$$\begin{aligned} e_\alpha(I) =& u_0 \otimes [-s_{22}^3 (s'_{21})^2 + s_{21}^2 s_{22} (s'_{22})^2 - s_{22}^2 s_{21} (s'_{21} s'_{22} + s'_{22} s'_{21})] \\ &+ u_1 \otimes [2\iota s_{22}^2 (s'_{21})^2 s'_{22} + 2\iota s_{21} s_{22} s'_{21} (s'_{22})^2] = 0, \\ e'_\alpha(I) =& u_0 \otimes [s_{21} s_{22}^3 s'_{21} + s_{21}^2 s_{22}^2 s'_{22} - s_{21}^2 s_{22}^2 s'_{22} - s_{21} s_{22}^3 s'_{21}] \\ &+ u_1 \otimes [(-s_{22} s'_{21} + s_{21} s'_{22})(s_{22}^2 s'_{21} - s_{21} s_{22} s'_{22}) \\ &+ \iota (2 s_{21} s_{22}^2 s'_{21} s'_{22} - 2 s_{21} s_{22}^2 s'_{21} s'_{22})] = 0, \\ h'_1(I) =& u_1 \otimes [-s_{22}^3 s'_{11} s'_{21} - s_{21}^2 s_{22} s'_{12} s'_{22} + s_{21} s_{22}^2 s'_{11} s'_{22} + s_{21} s_{22}^2 s'_{12} s'_{21} \\ &+ 2\iota s_{11} s_{22}^2 s'_{21} s'_{22} - 2\iota s_{12} s_{21} s_{22} s'_{21} s'_{22}] \\ &+ u_0 \otimes [-s_{11} s_{22}^3 s'_{21} + s_{12} s_{21} s_{22}^2 s'_{21} + s_{11} s_{21} s_{22} s'_{22} - s_{12} s_{21}^2 s'_{22} \\ &- 2\iota s_{22}^2 s'_{11} s'_{21} s'_{22} + 2\iota s_{21} s_{22} s'_{12} s'_{21} s'_{22}] \\ &+ u_0 \otimes [s_{11} s_{22}^3 s'_{21} + s_{12} s_{21}^2 s'_{22} - s_{11} s_{21} s_{22} s'_{22} - s_{12} s_{21} s_{22}^2 s'_{21} \\ &+ 2\iota s_{22}^2 s'_{11} s'_{21} s'_{22} - 2\iota s_{21} s_{22} s'_{12} s'_{21} s'_{22}] \\ &- u_1 \otimes [-s_{22}^3 s'_{11} s'_{21} + s_{21} s_{22}^2 s'_{12} s'_{21} + s_{21} s_{22}^2 s'_{11} s'_{22} - s_{21}^2 s_{22} s'_{12} s'_{22} \\ &+ 2\iota s_{11} s_{22}^2 s'_{21} s'_{22} - 2\iota s_{12} s_{21} s_{22} s'_{21} s'_{22}] = 0, \end{aligned}$$

$$\begin{aligned}
h_2'(I) =& -2\iota u_1 \otimes [-s_{22}^3 s_{11}' s_{21}' - s_{21}^2 s_{22} s_{12}' s_{22}' + s_{21} s_{22}^2 s_{11}' s_{22}' + s_{21} s_{22}^2 s_{12}' s_{21}' \\
&+2\iota s_{11} s_{22}^2 s_{21}' s_{22}' - 2\iota s_{12} s_{21} s_{22} s_{21}' s_{22}'] \\
&+ 2\iota u_0 \otimes [-s_{11} s_{22}^3 s_{21}' + s_{12} s_{21} s_{22}^2 s_{21}' + s_{11} s_{21} s_{22}^2 s_{22}' - s_{12} s_{21}^2 s_{22} s_{22}' \\
&-2\iota s_{22}^2 s_{11}' s_{21}' s_{22}' + 2\iota s_{21} s_{22} s_{12}' s_{21}' s_{22}'] \\
&+ u_0 \otimes [-3 s_{22}^2 s_{11}' s_{21}' s_{22}' - s_{21} s_{22}^3 s_{11}' + 2 s_{21} s_{22} s_{12}' s_{21}' s_{22}' - s_{21}^2 s_{22}^2 s_{12}' \\
&+ 2 s_{21} s_{22} s_{12}' s_{21}' s_{22}' - s_{22}^2 s_{11}' s_{21}' s_{22}' + s_{21} s_{22}^3 s_{11}' + s_{21}^2 s_{22}^2 s_{12}' \\
&-2\iota s_{11} s_{21} s_{22}^2 s_{22}' + 2\iota s_{12} s_{21}^2 s_{22} s_{22}' + 2\iota s_{11} s_{22}^3 s_{21}' - 2\iota s_{12} s_{21} s_{22}^2 s_{21}'] \\
&- u_1 \otimes [s_{11} s_{22}^2 s_{21}' s_{22}' - s_{12} s_{21} s_{22} s_{21}' s_{22}' + 2 s_{11} s_{22}^2 s_{21}' s_{22}' + s_{11} s_{21} s_{22}^3 \\
&-2 s_{12} s_{21} s_{22} s_{21}' s_{22}' - s_{12} s_{21}^2 s_{22}^2 + s_{11} s_{22}^2 s_{21}' s_{22}' - s_{12} s_{21} s_{22} s_{21}' s_{22}' \\
&- s_{11} s_{21} s_{22}^3 + s_{12} s_{21}^2 s_{22}^2 - 2\iota s_{21} s_{22}^2 s_{11}' s_{22}' + 2\iota s_{22}^3 s_{11}' s_{21}' \\
&+2\iota s_{21}^2 s_{22} s_{12}' s_{22}' - 2\iota s_{21} s_{22}^2 s_{12}' s_{21}'] = 0.
\end{aligned}$$

The above equalities are absolute, and they are also valid modulo any prime p. Computations from The weightspaces of $H^0(1,4)$ are $(1,4), (2,3), (3,2)$ and $(4,1)$. The above example shows that the weightspace of $H^0(1,4)_{(2,3)}$ is nontrivial. If the characteristic of the ground field is $p=3$, then the weight $(1,4)$ is 3-restricted and the induced supermodule $H^0_-(4,1)$ is irreducible.

13. Appendix

In this section, we give detailed proof of the selected statements from Sections 5 through 8. Most of the proofs given here are standard, but they are tedious. We moved them here to free up the space in the main text. The reader might decide to skip this appendix at first read.

THE PROOF OF PROPOSITION 5.1. We compute

$$\begin{aligned}
e_\alpha'(x_1) =& aB(a-1,b,c)(s_{22}s_{21}' - s_{21}s_{22}') = aB(a-1,b,c+1)s_{21}' \\
&- aB(a-1,b+1,c)s_{22}' = aw_1(a-1,b,c+1) - aw_2(a-1,b+1,c); \\
e_\alpha'(w_1) =& aB(a-1,b+1,c)s_{21}'s_{22}' = av_1(a-1,b+1,c); \\
e_\alpha'(w_2) =& aB(a-1,b,c+1)s_{21}'s_{22}' = av_1(a-1,b,c+1); \\
e_\alpha'(w_3) =& aB(a-1,b+1,c)s_{11}s_{21}'s_{22}' + aB(a-1,b+1,c+1)s_{11}'s_{21}' \\
&- aB(a-1,b+2,c)s_{11}'s_{22}' + B(a,b+2,c) = av_5(a-1,b+1,c) \\
&+ av_3(a-1,b+1,c+1) + x_1(a,b+2,c); \\
e_\alpha'(w_4) =& aB(a-1,b,c+1)s_{11}s_{21}'s_{22}' + aB(a-1,b+1,c+1)s_{12}'s_{21}' \\
&- aB(a-1,b+2,c)s_{12}'s_{22}' + B(a,b,c)s_{21}'s_{22}' + B(a,b+1,c+1) \\
&= -av_6(a-1,b,c+1) - av_2(a-1,b+2,c) + v_1(a,b,c) \\
&\quad + x_1(a,b+1,c+1);
\end{aligned}$$

$$
\begin{aligned}
e'_\alpha(w_5) =& aB(a-1,1,c)s_{12}s'_{21}s'_{22} + aB(a-1,0,c+2)s'_{11}s'_{21} - aB(a-1,1,c+1)s'_{11}s'_{22}\\
&- B(a,0,c)s'_{21}s'_{22} + B(a,1,c+1) = av_9(a-1,0,c)s_{21} + av_3(a-1,0,c+2)\\
&- v_1(a,0,c) + x_1(a,1,c+1) = -(a+1)v_1(a,0,c) + av_5(a-1,0,c+1)\\
&+ av_3(a-1,0,c+2) + x_1(a,1,c+1);\\
e'_\alpha(w_6) =& aB(a-1,0,c+1)s_{12}s'_{21}s'_{22} + aB(a-1,0,c+2)s'_{12}s'_{21}\\
&- aB(a-1,1,c+1)s'_{12}s'_{22} + B(a,0,c+2) = av_{10}(a-1,0,c+1)\\
&- av_2(a-1,1,c+1) + x_1(a,0,c+2);\\
e'_\alpha(v_1) =& 0;\\
e'_\alpha(v_2) =& -aB(a-1,b,c+1)s'_{12}s'_{21}s'_{22} - B(a,b,c+1)s'_{22} = -az_2(a-1,b,c+1)\\
&- w_2(a,b,c+1);\\
e'_\alpha(v_3) =& -aB(a-1,b+1,c)s'_{11}s'_{21}s'_{22} - B(a,b+1,c)s'_{21} = -az_1(a-1,b+1,c)\\
&- w_1(a,b+1,c);\\
e'_\alpha(v_4) =& aB(a-1,b,c)[s_{11}s_{21}s_{22}s'_{11}s'_{21}s'_{22} + s_{21}^2s_{22}s'_{11}s'_{12}s'_{21} - s_{11}s_{21}^2s'_{12}s'_{21}s'_{22}\\
&- s_{21}^3s'_{11}s'_{12}s'_{22}] + B(a,b,c)[s_{21}s'_{11}s'_{21}s'_{22} + s_{11}s_{21}^2s'_{22} - s_{11}s_{21}s_{22}s'_{21}\\
&- s_{21}^3s'_{12} + s_{21}^2s_{22}s'_{11}] = az_3(a-1,b,c+1) - az_4(a-1,b+1,c)\\
&+ z_1(a,b+1,c) + w_4(a,b+2,c) - w_3(a,b+1,c+1);\\
e'_\alpha(v_5) =& aB(a-1,b+1,c+1)s'_{11}s'_{21}s'_{22} + B(a,b+2,c)s'_{22}\\
&= az_1(a-1,b+1,c+1) + w_2(a,b+2,c);\\
e'_\alpha(v_6) =& aB(a-1,b+2,c)s'_{12}s'_{21}s'_{22} + B(a,b+1,c+1)s'_{21}\\
&= az_2(a-1,b+2,c) + w_1(a,b+1,c+1);\\
e'_\alpha(v_7) =& aB(a-1,0,c)[s_{12}s_{21}s_{22}s'_{11}s'_{21}s'_{22} + s_{21}s_{22}^2s'_{11}s'_{12}s'_{21} - s_{11}s_{21}s_{22}s'_{12}s'_{21}s'_{22}\\
&- s_{21}^2s_{22}s'_{11}s'_{12}s'_{22}] + B(a,0,c)[s_{12}s_{21}^2s'_{22} - s_{11}s_{22}^2s'_{21} - s_{21}^2s_{22}s'_{12}\\
&+ s_{12}s_{22}^2s'_{11}] = az_5(a-1,0,c+1) - az_4(a-1,0,c+1) - w_2(a+1,1,c)\\
&+ w_4(a,1,c+1) - w_3(a,0,c+2);\\
e'_\alpha(v_8) =& aB(a-1,0,c)[s_{12}s_{22}^2s'_{11}s'_{21}s'_{22} + s_{22}^3s'_{11}s'_{12}s'_{21} - s_{12}s_{21}s_{22}s'_{12}s'_{21}s'_{22}\\
&- s_{21}s_{22}^2s'_{11}s'_{12}s'_{22}] + B(a,0,c)[s_{22}s'_{12}s'_{21}s'_{22} + s_{12}s_{21}s_{22}s'_{22} - s_{12}s_{22}^2s'_{21}\\
&- s_{21}s_{22}^2s'_{12} + s_{22}^3s'_{11}] = az_7(a-1,0,c+1) - az_6(a-1,0,c+1)\\
&+ az_2(a,0,c+1) + z_2(a,0,c+1) + w_4(a,0,c+2) - w_2(a+1,0,c+1)\\
&- w_5(a,0,c+2);\\
e'_\alpha(v_9) =& aB(a-1,0,c+2)s'_{11}s'_{21}s'_{22} + B(a,0,c+2)s'_{22} = az_1(a-1,0,c+2)\\
&+ w_2(a,1,c+1);\\
e'_\alpha(v_{10}) =& aB(a-1,1,c+1)s'_{12}s'_{21}s'_{22} + B(a,0,c+2)s'_{21} = az_2(a-1,1,c+1)\\
&+ w_1(a,0,c+2);\\
e'_\alpha(v_{11}) =& aB(a-1,0,1)s'_{11}s'_{21}s'_{22} - aB(a-1,1,0)s'_{12}s'_{21}s'_{22} - B(a,1,0)s'_{22}\\
&- B(a,0,1)s'_{21} = az_1(a-1,0,1) - az_2(a-1,1,0) - w_2(a,1,0)\\
&- w_1(a,0,1);
\end{aligned}
$$

$$
\begin{aligned}
e'_\alpha(z_1) =& - B(a,b+1,c)s'_{21}s'_{22} = -v_1(a,b+1,c);\\
e'_\alpha(z_2) =& - B(a,b,c+1)s'_{21}s'_{22} = -v_1(a,b,c+1);\\
e'_\alpha(z_3) =& aB(a-1,b+3,c)s'_{11}s'_{12}s'_{21}s'_{22} + B(a,b+2,c)(-s_{11}s'_{21}s'_{22} - s_{21}s'_{12}s'_{22})\\
&+ B(a,b+2,c+1)s'_{11}s'_{22} = ay_1(a-1,b+3,c) + v_6(a,b+2,c)\\
&+ v_3(a,b+2,c+1);\\
e'_\alpha(z_4) =& aB(a-1,b+2,c+1)s'_{11}s'_{12}s'_{21}s'_{22}\\
&- B(a,b+1,c+1)(s_{11}s'_{21}s'_{22} - s_{21}s'_{11}s'_{22}) - B(a,b+3,c)s'_{12}s'_{22}\\
&= ay_1(a-1,b+2,c+1) - v_5(a,b+1,c+1) - v_2(a,b+3,c);\\
e'_\alpha(z_5) =& aB(a-1,2,c+1)s'_{11}s'_{12}s'_{21}s'_{22} - B(a,2,c)s_{12}s'_{21}s'_{22} - B(a,2,c+1)s'_{12}s'_{21}\\
&+ B(a,1,c+2)s'_{11}s'_{21} = ay_1(a-1,2,c+1) + v_{10}(a,0,c)s^2_{21}\\
&+ v_3(a,1,c+2) = ay_1(a-1,2,c+1) + v_1(a+1,1,c) + v_6(a,1,c+1)\\
&+ v_3(a,1,c+2);\\
e'_\alpha(z_6) =& aB(a-1,1,c+2)s'_{11}s'_{12}s'_{21}s'_{22} - B(a,0,c+2)(s_{11}s'_{21}s'_{22} - s_{21}s'_{11}s'_{22})\\
&- B(a,2,c+1)s'_{12}s'_{22} = ay_1(a-1,1,c+2) - v_5(a,0,c+2)\\
&- v_2(a,2,c+1);\\
e'_\alpha(z_7) =& aB(a-1,1,c+2)s'_{11}s'_{12}s'_{21}s'_{22} - B(a,1,c+1)s_{12}s'_{21}s'_{22}\\
&- B(a,1,c+2)s'_{12}s'_{21} + B(a,0,c+3)s'_{11}s'_{21} = ay_1(a-1,1,c+2)\\
&+ v_{10}(a,0,c+1)s_{21} + v_3(a,0,c+3) = ay_1(a-1,1,c+2)\\
&+ v_1(a+1,0,c+1) + v_6(a,0,c+2) + v_3(a,0,c+3);\\
e'_\alpha(z_8) =& aB(a-1,0,c+2)s'_{11}s'_{12}s'_{21}s'_{22} - B(a,0,c+2)s_{12}s'_{21}s'_{22}\\
&- B(a,1,c+2)s'_{12}s'_{22} + B(a,0,c+3)s'_{11}s'_{22}\\
=& ay_1(a-1,0,c+3) - v_9(a,0,c+2) - v_2(a,1,c+2);
\end{aligned}
$$

$$
\begin{aligned}
e'_\alpha(z_9) =& aB(a-1,1,1)s'_{11}s'_{12}s'_{21}s'_{22} - B(a,2,0)s'_{12}s'_{22} + B(a,1,1)s'_{11}s'_{22}\\
&- B(a,0,1)s_{11}s'_{21}s'_{22} = ay_1(a-1,1,1) - v_2(a,2,0) + v_5(a,0,1);\\
e'_\alpha(z_{10}) =& aB(a-1,0,2)s'_{11}s'_{12}s'_{21}s'_{22} - B(a,1,1)s'_{12}s'_{22} + B(a,0,2)s'_{11}s'_{22}\\
&- B(a,0,1)s_{12}s'_{21}s'_{22} = ay_1(a-1,0,2) - v_9(a,0,1) - v_2(a,1,1);\\
e'_\alpha(z_{11}) =& - aB(a-1,2,0)s'_{11}s'_{12}s'_{21}s'_{22} + B(a,1,0)s_{11}s'_{21}s'_{22} + B(a,2,0)s'_{12}s'_{21}\\
&- B(a,1,1)s'_{11}s'_{21} = -ay_1(a-1,2,0) - v_6(a,1,0) - v_3(a,1,1);\\
e'_\alpha(z_{12}) =& - aB(a-1,1,1)s'_{11}s'_{12}s'_{21}s'_{22} + B(a,0,1)s_{12}s'_{21}s'_{22} + B(a,1,1)s'_{12}s'_{21}\\
&- B(a,0,2)s'_{11}s'_{21} = -ay_1(a-1,1,1) - v_{10}(a,0,1) - v_3(a,0,2);\\
e'_\alpha(y_1) =& - B(a,b+1,c)s'_{12}s'_{21}s'_{22} + B(a,b,c+1)s'_{11}s'_{21}s'_{22} = -z_2(a,b+1,c)\\
&+ z_1(a,b,c+1).
\end{aligned}
$$

□

THE PROOF OF PROPOSITION 5.2. The lengthy proof is divided into three subsections to organize it better.

1. Relations between the coefficients of η. We derive the relations between the coefficients of η that will describe an $\mathbf{U}^+$-invariant element.

Investigate multiples of x_1. We obtain

$$\sum(c(w_3)x_1(a,b+2,c)+c(w_4)x_1(a,b+1,c+1)+c(w_5)x_1(a,1,c+1)$$
$$+c(w_6)x_1(a,0,c+2))=0.$$

Looking at multiples of $x_1(a,0,c+2)$, we obtain $c(w_6(a,0,c))=0$. After considering all multiples of $x_1(a,1,c+1)$, we obtain

$$c(w_4(a,0,c))+c(w_5(a,0,c))=0.$$

Since there is a unique multiple of $x_1(a,b+2,0)$, we derive that all

$$c(w_3(a,b,0))=0.$$

After considering all multiples of $x_1(a,b+2,c+1)$, we obtain

$$c(w_3(a,b,c+1))+c(w_4(a,b+1,c))=0.$$

Investigate multiples of w_1. We obtain

$$\sum(c(x_1)aw_1(a-1,b,c+1)-c(v_3)w_1(a,b+1,c)+c(v_6)w_1(a,b+1,c+1)$$
$$+c(v_{10})w_1(a,0,c+2)-c(v_{11})w_1(a,0,1))=0.$$

After considering all multiples of $w_1(a,0,1)$, we obtain

$$(a+1)c(x_1(a+1,0,0))-c(v_{11}(a,0,0))=0.$$

After considering all multiples of $w_1(a,0,c+2)$, we obtain

$$(a+1)c(x_1(a+1,0,c+1))+c(v_{10}((a,0,c))=0.$$

After considering all multiples of $w_1(a,b+1,0)$, we obtain

$$-c(v_3(a,b,0))=0.$$

After considering all multiples of $w_1(a,b+1,c+1)$, we obtain

$$(a+1)c(x_1(a+1,b+1,c))-c(v_3(a,b,c+1))+c(v_6(a,b,c))=0.$$

Investigate multiples of w_2. We obtain

$$\sum(-c(x_1)aw_2(a-1,b+1,c)-c(v_2)w_2(a,b,c+1)+c(v_5)w_2(a,b+2,c)$$
$$+c(v_9)w_2(a,0,c+2)-c(v_{11})w_2(a,1,0))=0.$$

Since there is a unique multiple of $w_2(a,0,1)$, we derive that

$$-c(v_2(a,0,0))=0.$$

After considering all multiples of $w_2(a,0,c+2)$, we obtain

$$-c(v_2(a,0,c+1))=0.$$

After considering all multiples of $w_2(a,1,0)$, we obtain

$$(-(a+1)c(x_1(a+1,0,0))-c(v_{11}(a,0,0))=0.$$

After considering all multiples of $w_2(a,b+2,0)$, we obtain

$$-(a+1)c(x_1(a+1,b+1,0))+c(v_5((a,b,0))=0.$$

After considering all multiples of $w_2(a,1,c+1)$, we obtain

$$-(a+1)c(x_1(a+1,0,c+1))-c(v_2((a,1,c))+c(v_9(a,0,c))=0.$$

After considering all multiples of $w_2(a,b+2,c+1)$, we obtain

$$-(a+1)c(x_1(a+1,b+1,c+1))-c(v_2((a,b+2,c))+c(v_5(a,b,c+1))=0.$$

Investigate multiples of w_3. We obtain

$$\sum(- c(v_4)w_3(a,b+1,c+1) - c(v_7)w_3(a,0,c+2) = 0$$

From here we derive $c(v_4(a,b,c)) = 0$ and $c(v_7(a,0,c)) = 0$.

Investigate multiples of w_4. We obtain

$$\sum(-c(v_4)v_4(a,b+2,c) + c(v_7)w_4(a,1,c+1) + c(v_8)w_4(a,0,c+2) = 0$$

We derive $c(v_4(a,b,c)) = 0$, $c(v_7(a,0,c)) = 0$ and $c(v_8(a,0,c)) = 0$.

Investigate multiples of v_1. We obtain

$$\begin{aligned}\sum(c(w_1)av_1(a-1,b+1,c) &+ c(w_2)av_1(a-1,b,c+1) + c(w_4)v_1(a,b,c)\\ &- c(w_5)(a+1)v_1(a,0,c) - c(z_1)v_1(a,b+1,c) - c(z_2)v_1(a,b,c+1)\\ &- c(z_5)v_1(a+1,0,c) + c(z_7)v_1(a+1,0,c+1)) = 0.\end{aligned}$$

After considering all multiples of $v_1(a,0,0)$, we obtain

$$c(w_4)(a,0,0) - (a+1)c(w_5)(a,0,0) = 0.$$

After considering all multiples of $v_1(a+1,0,c+1)$, we obtain

$$\begin{aligned}&(a+2)c(w_2)(a+2,0,c) + c(w_4)(a+1,0,c+1) - (a+2)c(w_5)(a+1,0,c+1)\\ &- c(z_2)(a+1,0,c) + c(z_7)(a,0,c) = 0.\end{aligned}$$

After considering all multiples of $v_1(a+1,1,0)$, we obtain

$$(a+2)c(w_1)(a+2,0,0) + c(w_4)(a+1,1,0) - c(z_1)(a+1,0,0) + c(z_5)(a,0,0) = 0.$$

After considering all multiples of $v_1(a+1,b+2,0)$, we obtain

$$(a+2)c(w_1)(a+2,b+1,0) + c(w_4)(a+1,b+2,0) - c(z_1)(a+1,b+1,0) = 0.$$

After considering all multiples of $v_1(a+1,1,c+1)$, we obtain

$$\begin{aligned}&(a+2)c(w_1)(a+2,0,c+1) + (a+2)c(w_2)(a+2,1,c)\\ &+ c(w_4)(a+1,1,c+1) - c(z_1)(a+1,0,c+1) - c(z_2)(a+1,1,c)\\ &+ c(z_5)(a,0,c+1) = 0.\end{aligned}$$

After considering all multiples of $v_1(a+1,b+2,c+1)$, we obtain

$$\begin{aligned}&(a+2)c(w_1)(a+2,b+1,c+1) + (a+2)c(w_2)(a+2,b+2,c)\\ &+ c(w_4)(a+1,b+2,c+1) - c(z_1)(a+1,b+1,c+1) - c(z_2)(a+1,b+2,c) = 0.\end{aligned}$$

Investigate multiples of v_2. We obtain

$$\begin{aligned}\sum(- c(w_4)av_2(a-1,b+2,c) &- c(z_4)v_2(a,b+3,c) - c(z_6)v_2(a,2,c+1)\\ &- c(z_9)v_2(a,2,0)) = 0.\end{aligned}$$

After considering all multiples of $v_2(a,2,0)$, we obtain

$$-(a+1)c(w_4)(a+1,0,0) - c(z_9)(a,0,0) = 0.$$

After considering all multiples of $v_2(a,2,c+1)$, we obtain

$$-(a+1)c(w_4)(a+1,0,c+1) - c(z_6)(a,0,c) = 0.$$

After considering all multiples of $v_2(a,b+3,c)$, we obtain

$$-(a+1)c(w_4)(a+1,b+1,c) - c(z_4)(a,b,c) = 0.$$

Investigate multiples of v_3. We obtain

$$\sum(c(w_3)av_3(a-1,b+1,c+1)+c(w_5)av_3(a-1,0,c+2)+c(z_3)v_3(a,b+2,c+1) + c(z_5)v_3(a,0,c+2)+c(z_7)v_3(a,0,c+3))-c(z_{12})v_3(a,0,2)=0.$$

After considering all multiples of $v_3(a,0,2)$, we obtain

$$(a+1)c(w_5(a+1,0,0))-c(z_{12}(a,0,0))=0.$$

After considering all multiples of $v_3(a,0,c+3)$, we obtain

$$(a+1)c(w_5)(a+1,0,c+1)+c(z_7)(a,0,c)=0.$$

After considering all multiples of $v_3(a,1,1)$, we obtain

$$(a+1)c(w_3)(a+1,0,0)=0.$$

After considering all multiples of $v_3(a,1,c+2)$, we obtain

$$(a+1)c(w_3)(a+1,0,c+1)+c(z_5)(a,0,c)=0.$$

After considering all multiples of $v_3(a,b+2,c+1)$, we obtain

$$(a+1)c(w_3)(a+1,b+1,c)+c(z_3)(a,b,c)=0.$$

Investigate multiples of v_5. We obtain

$$\sum(c(w_3)av_5(a-1,b+1,c)+c(w_5)av_5(a-1,0,c+1)-c(z_4)v_5(a,b+1,c+1) - c(z_6)v_5(a,0,c+2)+c(z_9)v_5(a,0,1))=0.$$

After considering all multiples of $v_5(a,0,1)$, we obtain

$$(a+1)c(w_5)(a+1,0,0)+c(z_9)(a,0,0)=0.$$

After considering all multiples of $v_5(a,0,c+2)$, we obtain

$$(a+1)c(w_5)(a+1,0,c+1)-c(z_6)(a,0,c)=0.$$

After considering all multiples of $v_5(a,b+1,0)$, we obtain

$$(a+1)c(w_3)(a+1,b,0)=0.$$

After considering all multiples of $v_5(a,b+1,c+1)$, we obtain

$$(a+1)c(w_3)(a+1,b,c+1)-c(z_4)(a,b,c)=0.$$

Investigate multiples of v_6. We obtain

$$\sum(-c(w_4)av_6(a-1,b,c+1)+c(z_3)v_6(a,b+2,c)+c(z_5)v_6(a,1,c+1) + c(z_7)v_6(a,0,c+2))-c(z_{11})v_6(a,1,0)=0.$$

Looking at multiples of $v_6(a,1,0)$, we obtain $c(z_{11}(a,0,0))=0$.

After considering all multiples of $v_6(a,0,1)$, we obtain

$$-(a+1)c(w_4(a+1,0,0))=0$$

After considering all multiples of $v_6(a,0,c+2)$, we obtain

$$-(a+1)c(w_4)(a+1,0,c+1)+c(z_7)(a,0,c)=0.$$

After considering all multiples of $v_6(a,1,c+1)$, we obtain

$$-(a+1)c(w_4)(a+1,1,c)+c(z_5)(a,0,c)=0.$$

After considering all multiples of $v_6(a,b+2,0)$, we obtain

$$c(z_3)(a,b,0)=0.$$

After considering all multiples of $v_6(a, b+2, c+1)$, we obtain

$$-(a+1)c(w_4)(a+1, b+2, c) + c(z_3)(a, b, c+1) = 0.$$

Investigate multiples of v_9. We obtain

$$-c(z_8)v_9(a, 0, c+2) - c(z_{10})v_9(a, 0, 1) = 0.$$

We obtain $c(z_8(a, 0, c)) = 0$ and $c(z_{10}(a, 0, 0)) = 0$.

Investigate multiples of v_{10}. We obtain

$$ac(w_6)(a-1, 0, c+1) - c(z_{12}(a, 0, 0) = 0.$$

Since we already know that $c(w_6)(a, 0, c) = 0$, we get $c(z_{12}(a, 0, 0)) = 0$.

Investigate multiples of z_1. We obtain

$$\sum\big(- c(v_3)az_1(a-1, b+1, c) + c(v_5)az_1(a-1, b+1, c+1) + c(v_9)az_1(a-1, 0, c+2) + c(v_{11})az_1(a-1, 0, 1) + c(y_1)z_1(a, b, c+1)\big) = 0.$$

After considering all multiples of $z_1(a, 0, 1)$, we obtain

$$(a+1)c(v_{11})(a+1, 0, 0) + c(y_1)(a, 0, 0) = 0.$$

After considering all multiples of $z_1(a, 0, c+2)$, we obtain

$$(a+1)c(v_9)(a+1, 0, c) + c(y_1)(a, 0, c+1) = 0.$$

After considering all multiples of $z_1(a, b+1, 0)$, we obtain

$$-(a+1)c(v_3)(a+1, b, 0) = 0.$$

After considering all multiples of $z_1(a, b+1, c+1)$, we obtain

$$-(a+1)c(v_3)(a+1, b, c+1) + (a+1)c(v_5)(a+1, b, c) + c(y_1)(a, b+1, c) = 0.$$

Investigate multiples of z_2. We obtain

$$\sum\big(- c(v_2)az_2(a-1, b, c+1) + c(v_6)az_2(a-1, b+2, c) + c(v_{10})az_2(a-1, 1, c+1) - c(v_{11})az_2(a-1, 1, 0) - c(y_1)z_2(a, b+1, c)\big) = 0.$$

After considering all multiples of $z_2(a, 0, c+1)$, we obtain

$$-(a+1)c(v_2)(a+1, 0, c) = 0.$$

After considering all multiples of $z_2(a, 1, 0)$, we obtain

$$-(a+1)c(v_{11})(a+1, 0, 0) - c(y_1)(a, 0, 0) = 0.$$

After considering all multiples of $z_2(a, 1, c+1)$, we obtain

$$-(a+1)c(v_2)(a+1, 1, c) + (a+1)c(v_{10})(a+1, 0, c) - c(y_1)(a, 0, c+1) = 0.$$

After considering all multiples of $z_2(a, b+2, 0)$, we obtain

$$(a+1)c(v_6)(a+1, b, 0) - c(y_1)(a, b+1, 0) = 0.$$

After considering all multiples of $z_2(a, b+2, c+1)$, we obtain

$$-(a+1)c(v_2)(a+1, b+2, c) + (a+1)c(v_6)(a+1, b, c+1) - c(y_1)(a, b+1, c+1) = 0.$$

Investigate multiples of z_3. We obtain

$$ac(v_4)z_3(a-1, b, c+1) = 0.$$

Investigate multiples of z_4. We obtain

$$-ac(v_4)z_4(a-1, b+1, c) - ac(v_7)z_4(a-1, 0, c+1) = 0.$$

Investigate multiples of z_5. We obtain

$$ac(v_7)z_5(a-1,0,c+1)=0.$$

Investigate multiples of z_6. We obtain

$$-ac(v_8)z_6(a-1,0,c+1)=0.$$

Investigate multiples of z_7. We obtain

$$ac(v_8)z_7(a-1,0,c+1)=0.$$

Investigate multiples of y_1. We obtain

$$\sum(c(z_3)ay_1(a-1,b+3,c)+c(z_4)ay_1(a-1,b+2,c+1)$$
$$+c(z_5)ay_1(a-1,2,c+1)+c(z_6)ay_1(a-1,1,c+2)+c(z_7)ay_1(a-1,1,c+2)$$
$$+c(z_9)ay_1(a-1,1,1)-c(z_{12})ay_1(a-1,1,1)))=0.$$

After considering all multiples of $y_1(a,1,1)$, we obtain

$$(a+1)c(z_9)(a+1,0,0)-(a+1)c(z_{12})(a+1,0,0))=0.$$

After considering all multiples of $y_1(a,1,c+2)$, we obtain

$$(a+1)c(z_6)(a+1,0,c)+(a+1)c(z_7)(a+1,0,c)=0.$$

After considering all multiples of $y_1(a,2,c+1)$, we obtain

$$(a+1)c(z_4)(a+1,0,c)+(a+1)c(z_5)(a+1,0,c)=0.$$

After considering all multiples of $y_1(a,b+3,0)$, we obtain

$$(a+1)c(z_3)(a+1,b,0)=0.$$

After considering all multiples of $y_1(a,b+3,c+1)$, we obtain

$$(a+1)c(z_3)(a+1,b,c+1)+(a+1)c(z_4)(a+1,b+1,c)=0.$$

Next, we describe all $\mathbf{U}^+$-invariant elements of $\Bbbk[\mathbf{Q}(2)]$.

Denote $F_0 = span < x_1 >$, $F_1 = span < w_1,\dots,w_6 >$, $F_2 = span < v_1,\dots,v_{11} >$, $F_3 = span < z_1,\dots,z_{12} >$ and $F_4 = span < y_1 >$.

Since $e'_\alpha(F_0+F_2+F_4)\subset F_1+F_3$ and $e'_\alpha(F_1+F_3)\subset F_0+F_2+F_4$, each $\mathbf{U}^+$-invariant is a sum of an $\mathbf{U}^+$ invariant from $F_0+F_2+F_4$ and an $\mathbf{U}^+$-invariant from F_1+F_3.

2. Invariants from $F_0+F_2+F_4$**.** First, consider an element from $F_0+F_2+F_4$ and analyze relations between coefficients established in the above subsection. We will not explicitly list which relations are considered, but the reader can extract them as needed.

Since there is a unique multiple of $w_4(a,b+2,c)$, we derive that all

$$c(v_4(a,b,c))=0.$$

Since there is a unique multiple of $w_4(a,1,c+1)$, we derive that all

$$c(v_7(a,0,c))=0.$$

Since there is a unique multiple of $w_4(a,0,c+2)$, we derive that all

$$c(v_8(a,0,c))=0.$$

Since there is a unique multiple of $w_2(a,0,1)$ and $w_2(a,0,c+2)$, we derive that all

$$c(v_2(a,0,c))=0.$$

Since there is a unique multiple of $x_1(a, b+2, 0)$, we derive that all

$$c(w_3(a, b, 0)) = 0.$$

The equations

$$(a+1)c(x_1(a+1, 0, 0)) = c(v_{11}(a, 0, 0))$$
$$-(a+1)c(x_1(a+1, 0, 0)) = c(v_{11}(a, 0, 0))$$

imply that

$$c(v_{11}(a, 0, 0)) = 0.$$

The equation

$$(a+1)c(v_{11})(a+1, 0, 0) + c(y_1)(a, 0, 0) = 0$$

then imply that

$$c(y_1(a, 0, 0)) = 0.$$

Next, we determine invariants from $F_0 + F_2 + F_4$ using the following procedure.

We start with a particular coefficient and list all equations in which this coefficient appears. For the coefficients appearing in those equations, we again list all equations involving them, and so on until we reach a point where none of the listed coefficients appear in any additional equations. Analyzing the listed equations, we determine all $\mathbf{U}^+$-invariants that involve only those listed coefficients. From now on, in this section, we write "invariant" short for $\mathbf{U}^+$-invariant and underline new invariants as we construct them.

We already mentioned that $\underline{v_1(a, b, c)}$ are invariant.

Since $e'_\alpha(x_1(pt, b, c)) = 0$, the elements $\underline{x_1(pt, b, c)}$ are invariants for every a, b, t.

Next, we find invariants in which the coefficient $c(x_1)$ is not zero.

Elements $c(x_1(a+1, 0, 0))$ appear only in

$$(a+1)c(x_1(a+1, 0, 0)) = c(v_{11}(a, 0, 0))$$
$$-(a+1)c(x_1(a+1, 0, 0)) = c(v_{11}(a, 0, 0))$$

This implies $c(v_{11}(a, 0, 0)) = 0$ and if $a+1$ is not divisible by p, then $c(x_1(a+1, 0, 0)) = 0$. If p divides $a+1$, then $x_1(a+1, 0, 0)$ is an invariant.

The element $c(v_{11}(a, 0, 0))$ appears in two additional relations

$$ac(v_{11})(a, 0, 0) + c(y_1)(a-1, 0, 0) = 0$$
$$-ac(v_{11})(a, 0, 0) - c(y_1)(a-1, 0, 0) = 0$$

which imply $c(y_1(a-1, 0, 0)) = 0$. No further relations involve $c(y_1(a-1, 0, 0))$. Thus we obtain invariants $x_1(pt, 0, 0)$ for every t.

Elements $c(x_1(a+1, 0, c+1))$ appear only in relations

$$(a+1)c(x_1(a+1, 0, c+1) = -c(v_{10}(a, 0, c))$$
$$-(a+1)c(x_1(a+1, 0, c+1)) - c(v_2((a, 1, c)) + c(v_9(a, 0, c)) = 0.$$

The elements $c(v_{10}(a, 0, c))$ and $c(v_2(a, 1, c))$ appear in only additional relations

$$-ac(v_2(a, 1, c) + ac(v_{10}(a, 0, c)) - c(y_1)(a-1, 0, c+1) = 0$$

if $a \geq 1$ and p does not divide a.

The elements $v_9(a, 0, c)$ and $c(y_1(a-1, 0, c+1))$ appears in only additional relations

$$ac(v_9((a, 0, c)) + c(y_1((a-1, 0, c+1)) = 0.$$

Thus

$$\underline{c_1x_1(a+1,0,c+1)+c_2v_2(a,1,c)-(a+1)c_1v_{10}(a,0,c)}$$
$$\underline{+[(a+1)c_1+c_2]v_9(a,0,c)-[a(a+1)c_1+ac_2]y_1(a-1,0,c+1)}$$

is an invariant for every c_1, c_2 if $a \geq 1$ and p does not divide $a, a+1$.

If p divides a, we use an analogous argument to get the same result, invariant

$$c_1x_1(a+1,0,c+1)+c_2v_2(a,1,c)-c_1v_{10}(a,0,c)+(c_1+c_2)v_9(a,0,c).$$

If p divides $a+1$, then we get invariants

$$c_1x_1(a+1,0,c+1)+c_2v_2(a,1,c)+c_2v_9(a,0,c)-ac_2y_1(a-1,0,c+1).$$

These exceptional cases conform to the general expression above.

The elements $c(x_1(a+1,b+1,0))$ appear only in relations

$$(a+1)c(x_1(a+1,b+1,0)=c(v_3(a,b,1))-c(v_6(a,b,0))$$
$$-(a+1)c(x_1(a+1,b+1,0))=-c(v_5(a,b,0))$$

The element $c(v_3(a,b,1))$ and $c(v_5(a,b,0)$ appear in only additional relations if $a \geq 1$ and p does not divide a

$$-ac(v_3)(a,b,1)+ac(v_5)(a,b,0)+c(y_1(a-1,b+1,0))=0$$

The element $c(v_6(a,b,0))$ appears in only additional relations if $a \geq 1$ and p does not divide a

$$ac(v_6)(a,b,0)-c(y_1((a-1,b+1,0))=0.$$

Those are the only relations involving $c(y_1(a-1,b+1,0))$.

From there, we obtain

$$c(v_6(a,b,0))=c(v_3(a,b,1))-c(v_5(a,b,0))$$

and

$$c(y_1(a-1,b+1,0))=ac(v_6(a,b,0)).$$

The elements

$$\underline{c_1x_1(a+1,b+1,0)+c_3v_3(a,b,1)+(a+1)c_1v_5(a,b,0)+(c_3-(a+1)c_1)v_6(a,b,0)}$$
$$\underline{+a(c_3-(a+1)c_1)y_1(a-1,b+1,0)}$$

are invariants for every c_1, c_3 if $a \geq 1$ and p does not divide a.

If p divides a, we use an analogous argument to get the same result; the invariant elements

$$c_1x_1(a+1,b+1,0)+c_3v_3(a,b,1)+c_1v_5(a,b,0)+c_3v_6(a,b,0)$$

for each c_1, c_3.

If p divides $a+1$, then we get invariants

$$c_1x_1(a+1,b+1,0)+c_3v_3(a,b,1)+c_3v_6(a,b,0)+ac_3y_1(a-1,b+1,0).$$

Elements $c(x_1(a+1,b+1,c+1)$ appear only in relations

$$(a+1)c(x_1(a+1,b+1,c+1))-c(v_3(a,b,c+2))+c(v_6(a,b,c+1))=0$$
$$-(a+1)c(x_1(a+1,b+1,c+1))-c(v_2((a,b+2,c))+c(v_5(a,b,c+1))=0$$

The elements $c(v_3(a,b,c+2))$ and $c(v_5(a,b,c+1))$ appears in only additional relations if $a \geq 1$ and p does not divide a

$$-ac(v_3)(a,b,c+2)+ac(v_5)(a,b,c+1)+c(y_1)(a-1,b+1,c+1)=0.$$

The elements $c(v_2(a,b+2,c)$ and $c(v_6(a,b,c+1)$ appears in only additional relations if $a \geq 1$ and p does not divide a

$$-ac(v_2)(a,b+2,c) + ac(v_6)(a,b,c+1) - c(y_1)(a-1,b+1,c+1) = 0.$$

Those are the only relations involving $c(y_1(a-1,b+1,c+1))$.

From here, we obtain

$$c(v_3(a,b,c+2)) - c(v_6(a,b,c+1)) = -c(v_2((a,b+2,c)) + c(v_5(a,b,c+1))$$

and invariants

$$\underline{c_1x_1(a+1,b+1,c+1) + c_2v_2(a,b+2,c) + c_3v_3(a,b,c+2)}$$
$$\underline{+((a+1)c_1 + c_2)v_5(a,b,c+1) + (-(a+1)c_1 + c_3)v_6(a,b,c+1)}$$
$$\underline{+a(-c_2 - (a+1)c_1 + c_3)y_1(a-1,b+1,c+1)}$$

for each c_1, c_2, c_3 if $a \geq 1$ and p does not divide a.

If p divides a, we use an analogous argument to get the same result; the invariant elements

$$c_1x_1(a+1,b+1,c+1) + c_2v_2(a,b+2,c) + c_3v_3(a,b,c+2)$$
$$+ (c_1 + c_2)v_5(a,b,c+1) + (-c_1 + c_3)v_6(a,b,c+1)$$

for each c_1, c_2, c_3.

If p divides $a+1$, then we get the same result, the invariants

$$c_1x_1(a+1,b+1,c+1) + c_2v_2(a,b+2,c) + c_3v_3(a,b,c+2) + c_2v_5(a,b,c+1)$$
$$+ c_3v_6(a,b,c+1) + a(-c_2 + c_3)y_1(a-1,b+1,c+1).$$

Now we consider $c(v_2(a,b,c))$. The cases of $c(v_2(a,1,c))$ and $c(v_2(a,b+2,c))$ were considered earlier and were connected to $c(x_1)$.

Next, we consider $c(v_3(a,b,c))$. The cases of $c(v_3(a,b,1))$ and $c(v_3(a,b,c+2))$ were considered earlier and were connected to $c(x_1)$.

The cases of $v_5(a,b,0)$ and $v_5(a,b,c+1)$ were considered earlier and were connected to $c(x_1)$.

The cases of $v_6(a,b,0)$ and $v_6(a,b,c+1)$ were considered earlier and were connected to $c(x_1)$.

The cases of $v_9(a,0,c)$ were considered earlier and were connected to $c(x_1)$.

The cases of $v_{10}(a,0,c)$ were considered earlier and were connected to $c(x_1)$.

The cases of $y_1(a-1,0,c+1)$, $y_1(a-1,b+1,0)$, $y_1(a-1,b+1,c+1)$ were considered earlier and were connected to $c(x_1)$.

3. Invariants from F_1+F_3. First, we describe coefficients that are vanishing: Since there is a unique multiple of $x_1(a,b+2,0)$, we derive that all

$$c(w_3(a,b,0)) = 0.$$

Since there is a unique multiple of $x_1(a,0,c+2)$, we derive that all

$$c(w_6(a,0,c)) = 0.$$

Since there is a unique multiple of $v_6(a,1,0)$, we derive that all

$$c(z_{11}(a,0,0)) = 0.$$

Since there is a unique multiple of $v_9(a,0,c+2)$, we derive that all

$$c(z_8(a,0,c)) = 0.$$

Since there is a unique multiple of $v_9(a,0,1)$, we derive that all

$$c(z_{10}(a,0,0)) = 0.$$

Since $c(w_6)(a-1,0,c+1) = 0$, we obtain

$$c(z_{12}(a,0,0)) = 0.$$

We will see soon that $c(w_4(a+1,0,0) = 0$, $c(w_5(a+1,0,0)) = 0$ and $c(z_9(a,0,0)) = 0$.

Next, we determine invariants from $F_1 + F_3$, using an analogous approach as for $F_0 + F_2 + F_4$.

Since $e'_\alpha(w_1(pt,b,c)) = e'_\alpha(w_2(pt,b,c)) = 0$, the elements $\underline{w_1(pt,b,c)}$ and $\underline{w_2(pt,b,c)}$ are invariants for each b, c and t.

Investigate elements $c(w_4(a+1,b,c))$.

Elements $c(w_4(a+1,0,0))$ appear only in

$$c(w_4(a+1,0,0)) + c(w_5(a+1,0,0)) = 0$$
$$c(w_4)(a+1,0,0) - (a+2)c(w_5(a+1,0,0) = 0.$$
$$-(a+1)c(w_4)(a+1,0,0) - c(z_9(a,0,0) = 0.$$
$$-(a+1)c(w_4)(a+1,0,0) = 0.$$

If p does not divide $a+1$, then $c(w_4(a+1,0,0)) = c(w_5(a+1,0,0)) = c(z_9(a,0,0)) = 0$. If p divides $a+1$, then $c(w_4(a+1,0,0) = c(w_5(a+1,0,0)) = c(z_9(a,0,0)) = 0$.

The elements $c(w_5(a+1,0,0))$ appear in additional relations

$$(a+1)c(w_5)(a+1,0,0) - c(z_{12}(a,0,0)) = 0$$
$$(a+1)c(w_5)(a+1,0,0) + c(z_9(a,0,0)) = 0$$

which imply $c(z_{12}(a,0,0)) = 0$.

The elements $c(z_9(a,0,0))$ appear in additional relations

$$(a+1)c(z_9)(a+1,0,0) - (a+1)c(z_{12})(a+1,0,0)) = 0.$$

The elements $c(z_{12}(a,0,0)$ appear in additional relations

$$ac(w_6)(a-1,0,c+1) - c(z_{12}(a,0,0)) = 0.$$

Therefore, there is no invariant coming from these elements.

Elements $c(w_4(a+1,0,c+1))$ appears in

$$c(w_4(a+1,0,c+1)) + c(w_5(a+1,0,c+1)) = 0$$
$$(a+2)c(w_2)(a+2,0,c) + c(w_4)(a+1,0,c+1) - (a+2)c(w_5)(a+1,0,c+1)$$
$$- c(z_2)(a+1,0,c) + c(z_7)(a,0,c) = 0$$
$$-(a+1)c(w_4)(a+1,0,c+1) - c(z_6)(a,0,c) = 0$$
$$-(a+1)c(w_4)(a+1,0,c+1) + c(z_7)(a,0,c) = 0$$

The elements $c(w_2((a+2,0,c))$ and $c(z_2((a+1,0,c))$ do not appear in any additional relations.

The elements $c(w_5((a+1,0,c+1))$ appears in additional relation

$$(a+1)c(w_5)(a+1,0,c+1) + c(z_7)(a,0,c) = 0$$
$$(a+1)c(w_5)(a+1,0,c+1) - c(z_6)(a,0,c) = 0$$

Elements $c(z_6(a,0,c))$ and $c(z_7(a,0,c))$ appear in additional relations

$$ac(z_6)(a,0,c) + ac(z_7)(a,0,c) = 0.$$

We obtain the invariants

$$\underline{c_2 w_2(a+2,0,c) + c_4 w_4(a+1,0,c+1) - c_4 w_5(a+1,0,c+1)}$$
$$\underline{+((a+2)c_2 + (2a+4)c_4) z_2(a+1,0,c) - (a+1)c_4 z_6(a,0,c) + (a+1)c_4 z_7(a,0,c)}$$

for each c_2, c_4.

The element $c(w_4(a+1,1,0))$ appears only in

$$c(w_3(a+1,0,1)) + c(w_4(a+1,1,0)) = 0.$$
$$(a+2)c(w_1)(a+2,0,0) + c(w_4)(a+1,1,0) - c(z_1)(a+1,0,0) + c(z_5)(a,0,0) = 0.$$
$$-(a+1)c(w_4)(a+1,1,0) - c(z_4)(a,0,0) = 0.$$
$$-(a+1)c(w_4)(a+1,1,0) + c(z_5)(a,0,0) = 0.$$

The element $c(w_3(a+1,0,1))$ appears in only additional relations

$$(a+1)c(w_3)(a+1,0,1) + c(z_5)(a,0,0) = 0.$$
$$(a+1)c(w_3)(a+1,0,1) - c(z_4)(a,0,0) = 0.$$

The elements $c(w_1(a+2,0,0))$ and $c(z_1(a+1,0,0))$ do not appear in additional relations. The elements $c(z_4(a,0,0))$ and $c(z_5(a,0,0))$ appear only in additional

$$ac(z_4)(a,0,0) + ac(z_5)(a,0,0) = 0.$$

We obtain invariants

$$\underline{c_1 w_1(a+2,0,0) - c_4 w_3(a+1,0,1) + c_4 w_4(a+1,1,0)}$$
$$\underline{+((a+2)c_1 + (a+2)c_4) z_1(a+1,0,0) - (a+1)c_4 z_4(a,0,0) + (a+1)c_4 z_5(a,0,0)}$$

for each c_1, c_4.

The element $c(w_4(a+1,b+2,0))$ appears only in

$$c(w_3(a+1,b+1,1)) + c(w_4(a+1,b+2,0)) = 0.$$
$$(a+2)c(w_1)(a+2,b+1,0) + c(w_4)(a+1,b+2,0) - c(z_1)(a+1,b+1,0) = 0.$$
$$(a+1)c(w_4)(a+1,b+2,0) - c(z_4)(a,b+1,0) = 0.$$
$$-(a+1)c(w_4)(a+1,b+2,0) + c(z_3)(a,b,1) = 0.$$

The elements $c(w_1(a+2,b+1,0))$ and $c(z_1(a+1,b+1,0))$ do not appear in any additional relations. The element $c(w_3(a+1,b+1,1))$ appears only in additional relations

$$(a+1)c(w_3)(a+1,b+1,1) + c(z_3)(a,b,1) = 0.$$
$$(a+1)c(w_3)(a+1,b+1,1) - c(z_4)(a,b+1,0) = 0.$$

The elements $c(z_3(a,b,1))$ and $c(z_4(a,b+1,0))$ appears only in additional relations

$$ac(z_3)(a,b,1) + ac(z_4)(a,b+1,0) = 0.$$

We obtain invariants

$$\underline{c_1 w_1(a+2,b+1,0) - c_4 w_3(a+1,b+1,1) + c_4 w_4(a+1,b+2,0)}$$
$$\underline{+((a+2)c_1 + c_4) z_1(a+1,b+1,0) + (a+1)c_4 z_3(a,b,1) - (a+1)c_4 z_4(a,b+1,0)}$$

for each c_1, c_4.

The element $c(w_4(a+1,1,c+1))$ appears only in

$$c(w_3(a+1,0,c+2)) + c(w_4(a+1,1,c+1)) = 0.$$
$$(a+2)c(w_1)(a+2,0,c+1) + (a+2)c(w_2)(a+2,1,c) + c(w_4)(a+1,1,c+1)$$
$$- c(z_1)(a+1,0,c+1) - c(z_2)(a+1,1,c) + c(z_5)(a,0,c+1) = 0.$$
$$-(a+1)c(w_4)(a+1,1,c+1) - c(z_4)(a,0,c+1) = 0.$$

$$-(a+1)c(w_4)(a+1,1,c+1)+c(z_5)(a,0,c+1)=0.$$

The elements $c(w_1(a+2,0,c+1))$, $c(w_2)(a+2,1,c)$, $c(z_1(a+1,0,c+1))$ and $c(z_2)(a+1,1,c)$ do not appear in any additional relations. The element $c(w_3(a+1,0,c+2))$ appears only in additional relations

$$(a+1)c(w_3)(a+1,0,c+2)+c(z_5)(a,0,c+1)=0.$$

$$(a+1)c(w_3)(a+1,0,c+2)-c(z_4)(a,0,c+1)=0.$$

The elements $c(z_4(a,0,c+1))$ and $c(z_5(a,0,c+1))$ appears only in additional relations

$$ac(z_4)(a,0,c+1)+ac(z_5)(a,0,c+1)=0.$$

We obtain invariants

$$\underline{c_1w_1(a+2,0,c+1)+c_2w_2(a+2,1,c)-c_4w_3(a+1,0,c+2)}$$
$$\underline{+c_4w_4(a+1,1,c+1)+d_1z_1(a+1,0,c+1)+((a+2)c_1+(a+2)c_2}$$
$$\underline{+(a+2)c_4-d_1)z_2(a+1,1,c)-(a+1)c_4z_4(a,0,c+1)+(a+1)c_4z_5(a,0,c+1)}$$

for all c_1, c_2, c_4, d_1.

The element $c(w_4(a+1,b+2,c+1))$ appears only in

$$c(w_3(a+1,b+1,c+2))+c(w_4(a+1,b+2,c+1))=0.$$

$$(a+2)c(w_1)(a+2,b+1,c+1)+(a+2)c(w_2)(a+2,b+2,c)$$
$$+c(w_4)(a+1,b+2,c+1)-c(z_1)(a+1,b+1,c+1)-c(z_2)(a+1,b+2,c)=0.$$

$$-(a+1)c(w_4)(a+1,b+2,c+1)-c(z_4)(a,b+1,c+1)=0.$$

$$-(a+1)c(w_4)(a+1,b+2,c+1)+c(z_3)(a,b,c+2)=0.$$

The elements $c(w_1(a+2,b+1,c+1))$, $c(w_2)(a+2,b+2,c)$, $c(z_1(a+1,b+1,c+1))$ and $c(z_2)(a+1,b+2,c)$ do not appear in any additional relations. The element $c(w_3(a+1,b+1,c+2))$ appears only in additional relations

$$(a+1)c(w_3)(a+1,b+1,c+2)+c(z_3)(a,b,c+2)=0.$$

$$(a+1)c(w_3)(a+1,b+1,c+2)-c(z_4)(a,b+1,c+1)=0.$$

The elements $c(z_3(a,b,c+2))$ and $c(z_4(a,b+1,c+1))$ appears only in additional relations

$$ac(z_3)(a+1,b,c+2)+ac(z_4)(a+1,b+1,c+1)=0.$$

We obtain invariants

$$\underline{c_1w_1(a+2,b+1,c+1)+c_2w_2(a+2,b+2,c)-c_4w_3(a+1,b+1,c+2)}$$
$$\underline{+c_4w_4(a+1,b+2,c+1)+d_1z_1(a+1,b+1,c+1)}$$
$$\underline{+((a+2)c_1+(a+2)c_2+c_4-d_1)z_2(a+1,b+2,c)}$$
$$\underline{+(a+1)c_4z_3(a,b,c+2)-(a+1)c_4z_4(a,b+1,c+1)}$$

for each c_1, c_2, c_4, d_1.

Next, we investigate $c(w_1(a,b,c))$. If p divides a, then $w_1(a,b,c)$ is an invariant.

The cases $w_1(a+2,0,0)$, $w_1(a+2,b+1,0)$, $w_1(a+2,0,c+1)$, $w_1(a+2,b+1,c+1)$ were considered earlier and were connected to $c(w_4)$. No other values were connected in other relations. Therefore, no additional invariants are coming from there.

An analogous situation is for $w_2(a,b,c)$. If p divides a, then $w_2(a,b,c)$ is invariant. The cases $w_2(a+2,0,c)$, $w_2(a+2,1,c)$, $w_2(a+2,b+2,c)$ were considered earlier and were connected to $c(w_4)$. If p divides a, then $w_2(a,b,c)$ is invariant, and there are no other new invariants.

Next, investigate $c(w_3(a,b,c))$. The cases $w_3(a+1,0,1)$, $w_3(a+1,b+1,1)$, $w_3(a+1,0,c+2)$ and $w_3(a+1,b+1,c+2)$, were considered earlier and were connected to $c(w_4)$. Since $c(w_3(a,b,0))=0$, there are no new invariants here.

Next, investigate $c(w_5(a,0,c))$. The cases $w_5(a+1,0,0)$, $w_5(a+1,0,c+1)$ were considered earlier and were connected to $c(w_4)$. Thus, there are no new invariants here.

The cases $z_1(a+1,0,0)$, $z_1(a+1,0,c+1)$, $z_1(a+1,b+1,c+1)$ were investigated earlier and are connected to $c(w_4)$. There are no new invariants. The cases $z_2(a+1,0,c)$, $z_2(a+1,1,c)$, $z_2(a+1,b+2,c)$ were investigated earlier and are connected to $c(w_4)$. There are no new invariants.

Next, investigate $c(z_3(a,b,c))$. The cases $z_3(a,b,1)$ and $z_3(a,b,c+2)$ were investigated earlier and are connected to $c(w_4)$. There are no new invariants.

Next, investigate $c(z_4(a,b,c))$. The cases $z_4(a,0,0)$, $z_4(a,b+1,0)$ and $z_3(a,b+1,c+1)$ were investigated earlier and are connected to $c(w_4)$. There are no new invariants.

Next, investigate $c(z_5(a,0,c))$. The elements $z_5(a,0,0)$ and $z_5(a,0,c+1)$ were investigated earlier and are connected to $c(w_4)$. There are no new invariants.

The cases of $z_6(a,0,c)$ and $z_7(a,0,c)$were investigated earlier and are connected to $c(w_4)$, and there are no new invariants here. □

THE PROOF OF PROPOSITION 6.3. We have

$h_1'(f_1(pt,0,0))=0;$

$h_1'(f_2(a,0,c))=$

$$[(a+1)det^a s_{22}^{c+2}s_{11}' - (a+1)det^a s_{21}^1 s_{22}^{c+1}s_{12}']$$
$$-(a+1)[-adet^{a-1}s_{12}s_{22}^{c+1}s_{11}'s_{21}'s_{22}' + adet^{a-1}s_{12}s_{21}^1 s_{22}^c s_{12}'s_{21}'s_{22}'$$
$$-adet^{a-1}s_{22}^{c+2}s_{11}'s_{12}'s_{21}' + det^a s_{12}s_{22}^{c+1}s_{21}' - det^a s_{22}^c s_{12}'s_{21}'s_{22}']$$
$$+(a+1)[adet^{a-1}s_{12}s_{22}^{c+1}s_{11}'s_{21}'s_{22}' - adet^{a-1}s_{12}s_{21}^1 s_{22}^c s_{12}'s_{21}'s_{22}'$$
$$-adet^{a-1}s_{21}^1 s_{22}^{c+1}s_{11}'s_{12}'s_{22}' + det^a s_{11}s_{22}^{c+1}s_{22}' + det^a s_{22}^c s_{12}'s_{21}'s_{22}']$$
$$-a(a+1)[-det^{a-1}s_{11}s_{22}^{c+1}s_{12}'s_{21}'s_{22}' + det^{a-1}s_{12}s_{22}^{c+1}s_{11}'s_{21}'s_{22}']$$
$$=(a+1)w_4(a,0,c+1) - (a+1)w_5(a,0,c+1) + (2a+2)z_2(a,0,c)$$
$$+a(a+1)z_7(a-1,0,c) + a(a+1)z_2(a,0,c) - a(a+1)z_6(a-1,0,c)$$
$$+a(a+1)z_2(a,0,c) = (a+1)[w_4(a,0,c+1) - w_5(a,0,c+1) + (2a+2)z_2(a,0,c)$$
$$-az_6(a-1,0,c) + az_7(a-1,0,c)] = (a+1)g_2(a,0,c);$$

$h_1'(f_3(a,b,c))=$

$$(a+1)det^a s_{21}^{b+1}s_{22}^{c+1}s_{11}' - (a+1)det^a s_{21}^{b+2}s_{22}^c s_{12}'$$
$$+(a+1)[adet^{a-1}s_{11}s_{21}^b s_{22}^{c+1}s_{11}'s_{21}'s_{22}' - adet^{a-1}s_{11}s_{21}^{b+1}s_{22}^c s_{12}'s_{21}'s_{22}'$$
$$-adet^{a-1}s_{21}^{b+2}s_{22}^c s_{11}'s_{12}'s_{22}' + det^a s_{11}s_{21}^{b+1}s_{22}^c s_{22}' + det^a s_{21}^b s_{22}^c s_{11}'s_{21}'s_{22}']$$
$$-(a+1)[-adet^{a-1}s_{11}s_{21}^b s_{22}^{c+1}s_{11}'s_{21}'s_{22}' + adet^{a-1}s_{11}s_{21}^{b+1}s_{22}^c s_{12}'s_{21}'s_{22}'$$
$$-adet^{a-1}s_{21}^{b+1}s_{22}^{c+1}s_{11}'s_{12}'s_{21}' + det^a s_{12}s_{21}^{b+1}s_{22}^c s_{21}' - det^a s_{21}^b s_{22}^c s_{11}'s_{21}'s_{22}']$$
$$-a(a+1)[-det^{a-1}s_{11}s_{21}^{b+1}s_{22}^c s_{12}'s_{21}'s_{22}' + det^{a-1}s_{12}s_{21}^{b+1}s_{22}^c s_{11}'s_{21}'s_{22}'].$$

If $b > 0$, then

$$\begin{aligned}
&h_1'(f_3(a,b,c) = \\
&(a+1)[w_4(a,b+1,c) - w_3(a,b,c+1) + w_1(a+1,b,c) + 2z_1(a,b,c) \\
&- az_4(a-1,b,c) + az_3(a-1,b-1,c+1) + az_1(a,b,c)]
\end{aligned}$$

If $b, c > 0$, then $w_1(a+1,b,c) = -(a+1)z_2(a,b+1,c-1) + g_6(a,b,c-1)$ and $-z_2(a,b+1,c-1) + z_1(a,b,c) = g_8(a,b,c-1)$. Thus, if $b, c > 0$ then

$$\begin{aligned}
&h_1'(f_3(a,b,c)) = \\
&(a+1)[w_4(a,b+1,c) - w_3(a,b,c+1) + z_2(a,b+1,c-1) - az_4(a-1,b,c) \\
&+ az_3(a-1,b-1,c+1)] + (a+1)g_6(a,b,c-1) \\
&+ (a+1)(a+2)[-z_2(a,b+1,c-1) + z_1(a,b,c)] \\
&= (a+1)g_{10}(a-1,b-1,c-1) + (a+1)g_6(a,b,c-1) \\
&+ (a+1)(a+2)g_8(a,b,c-1).
\end{aligned}$$

If $b = 0$, then

$$\begin{aligned}
&h_1'(f_3(a,0,c) = \\
&(a+1)[w_4(a,1,c) - w_3(a,0,c+1) + w_1(a+1,0,c) + 2z_1(a,0,c) \\
&- az_4(a-1,0,c) + az_5(a-1,0,c) + az_1(a,0,c) + az_1(a,0,c)]
\end{aligned}$$

If $c > 0$ and $b = 0$, then $w_1(a+1,0,c) = -(a+1)z_2(a,1,c-1) + g_6(a,0,c-1)$ and $z_1(a,0,c) = g_8(a,0,c-1) + z_2(a,1,c-1)$ imply

$$\begin{aligned}
&h_1'(f_3(a,0,c) = \\
&(a+1)[w_4(a,1,c) - w_3(a,0,c+1) - (a+1)z_2(a,1,c-1) + (2a+2)z_1(a,0,c) \\
&- az_4(a-1,0,c) + az_5(a-1,0,c)] + (a+1)g_6(a,0,c-1) \\
&= (a+1)[w_4(a,1,c) - w_3(a,0,c+1) + (a+1)z_2(a,1,c-1) - az_4(a-1,0,c) \\
&+ az_5(a-1,0,c)] + (a+1)g_6(a,0,c-1) + (a+1)(2a+2)g_8(a,0,c-1) = \\
&= (a+1)g_9(a-1,0,c-1) + (a+1)g_6(a,0,c-1) + (a+1)(2a+2)g_8(a,0,c-1).
\end{aligned}$$

If $b > 0$ and $c = 0$, then

$$\begin{aligned}
&h_1'(f_3(a,b,0)) = \\
&(a+1)[w_4(a,b+1,0) - w_3(a,b,1) + w_1(a+1,b,0) + (a+2)z_1(a,b,0) \\
&- az_4(a-1,b,0) + az_3(a-1,b-1,1)] \\
&(a+1)[-w_3(a,b,1) + w_4(a,b+1,0) + z_1(a,b,0) + az_3(a-1,b-1,1) \\
&+ az_4(a,b,0)] \\
&+ (a+1)[w_1(a+1,b,0) + (a+1)z_1(a,b,0)] \\
&= (a+1)g_5(a-1,b-1,0) + (a+1)g_3(a,b,0)
\end{aligned}$$

If $b=c=0$, then

$$
\begin{aligned}
&h_1'(f_3(a,0,c)) = \\
&(a+1)[w_4(a,1,0)-w_3(a,0,1)+w_1(a+1,0,0)+(2a+2)z_1(a,0,0) \\
&-az_4(a-1,0,0)+az_5(a-1,0,0)] \\
&(a+1)[-w_3(a,0,1)+w_4(a,1,0)+(a+1)z_1(a,0,0)-az_4(a-1,0,0) \\
&+az_5(a-1,0,0)] \\
&+(a+1)[w_1(a+1,0,0)+(a+1)z_1(a,0,0)] \\
&=(a+1)g_4(a-1,0,0)+(a+1)g_3(a,0,0). \\
&h_1'(f_4(a,b,c)) \\
&=adet^{a-1}s_{21}^{b+1}s_{22}^{c+1}s_{11}'s_{12}'s_{21}'-det^a s_{11}s_{21}^b s_{22}^{c+1}s_{21}' \\
&+[-adet^{a-1}s_{11}s_{21}^b s_{22}^{c+1}s_{11}'s_{21}'s_{22}'+adet^{a-1}s_{11}s_{21}^{b+1}s_{22}^c s_{12}'s_{21}'s_{22}' \\
&-adet^{a-1}s_{21}^{b+1}s_{22}^{c+1}s_{11}'s_{12}'s_{21}'+det^a s_{12}s_{21}^{b+1}s_{22}^c s_{21}'-det^a s_{21}^b s_{22}^c s_{11}'s_{21}'s_{22}'] \\
&+a[-det^{a-1}s_{11}s_{21}^{b+1}s_{22}^c s_{12}'s_{21}'s_{22}'+det^{a-1}s_{12}s_{21}^{b+1}s_{22}^c s_{11}'s_{21}'s_{22}'] \\
&=-w_1(a+1,b,c)-z_1(a,b,c)-az_1(a,b,c) \\
&=-w_1(a+1,b,c)-(a+1)z_1(a,b,c)
\end{aligned}
$$

If $c=0$, this equals $-g_3(a,b,0)$. If $c>0$, it equals

$$
\begin{aligned}
&[-w_1(a+1,b,c)-(a+1)z_2(a,b+1,c-1)]+[(a+1)z_2(a,b+1,c-1) \\
&-(a+1)z_1(a,b,c-1)]=-g_6(a,b,c-1)-(a+1)g_8(a,b,c-1). \\
&h_1'(f_5(a,b,c)) = \\
&adet^{a-1}s_{21}^{b+2}s_{22}^{c+1}s_{11}'s_{12}'s_{22}'-det^a s_{12}s_{21}^{b+2}s_{22}^c s_{22}' \\
&+[adet^{a-1}s_{11}s_{21}^b s_{22}^{c+2}s_{11}'s_{21}'s_{22}'-adet^{a-1}s_{11}s_{21}^{b+1}s_{22}^{c+1}s_{12}'s_{21}'s_{22}' \\
&-adet^{a-1}s_{21}^{b+2}s_{22}^{c+1}s_{11}'s_{12}'s_{22}'+det^a s_{11}s_{21}^{b+1}s_{22}^{c+1}s_{22}'+det^a s_{21}^b s_{22}^{c+1}s_{11}'s_{21}'s_{22}'] \\
&-a[-det^{a-1}s_{11}s_{21}^{b+1}s_{22}^{c+1}s_{12}'s_{21}'s_{22}'+det^{a-1}s_{12}s_{21}^{b+1}s_{22}^{c+1}s_{11}'s_{21}'s_{22}'] \\
&=w_2(a+1,b+1,c)+z_1(a,b,c+1)+az_1(a,b,c+1) \\
&=w_2(a+1,b+1,c)+(a+1)z_1(a,b,c+1) \\
&=[w_2(a+1,b+1,c)+(a+1)z_2(a,b+1,c)] \\
&+(a+1)[-z_2(a,b+1,c)+z_1(a,b,c+1)]=g_7(a,b,c)+(a+1)g_8(a,b,c); \\
&h_1'(f_6(a,b,c)) = \\
&adet^{a-1}s_{21}^b s_{22}^{c+1}s_{11}'s_{21}'s_{22}'-adet^{a-1}s_{21}^{b+1}s_{22}^c s_{12}'s_{21}'s_{22}' \\
&=az_1(a-1,b,c+1)-az_2(a-1,b+1,c)=ag_8(a-1,b,c);
\end{aligned}
$$

$h_1'(f_7(a,0,c) =$
$[adet^{a-1}s_{21}^{1}s_{22}^{c+1}s_{11}'s_{12}'s_{22}' - det^{a}s_{12}s_{21}^{1}s_{22}^{c}s_{22}']$
$+ [adet^{a-1}s_{12}s_{22}^{c+1}s_{11}'s_{21}'s_{22}' - adet^{a-1}s_{12}s_{21}^{1}s_{22}^{c}s_{12}'s_{21}'s_{22}' - adet^{a-1}s_{21}^{1}s_{22}^{c+1}s_{11}'s_{12}'s_{22}'$
$+ det^{a}s_{11}s_{22}^{c+1}s_{22}' + det^{a}s_{22}^{c}s_{12}'s_{21}'s_{22}']$
$- a[-det^{a-1}s_{11}s_{22}^{c+1}s_{12}'s_{21}'s_{22}' + det^{a-1}s_{12}s_{22}^{c+1}s_{11}'s_{21}'s_{22}']$
$= w_2(a+1,0,c) + z_2(a,0,c) + az_2(a,0,c) = w_2(a+1,0,c) + (a+1)z_2(a,0,c)$
$= g_1(a,0,c);$
$h_1'(g_1(a,0,c)) =$
$(a+1)det^{a}s_{22}^{c+1}s_{11}'s_{22}' - (a+1)det^{a}s_{21}^{1}s_{22}^{c}s_{12}'s_{22}'$
$+ (a+1)[adet^{a-1}s_{22}^{c+1}s_{11}'s_{12}'s_{21}'s_{22}' - det^{a}s_{12}s_{22}^{c}s_{21}'s_{22}']$
$= -(a+1)v_9(a,0,c) - (a+1)v_2(a,1,c) + (a+1)ay_1(a-1,0,c+1)$
$= -(a+1)f_7(a,0,c)$
$h_1'(g_2(a,0,c)) =$
$[adet^{a-1}s_{11}s_{22}^{c+2}s_{11}'s_{22}' - adet^{a-1}s_{11}s_{21}^{1}s_{22}^{c+1}s_{12}'s_{22}' - adet^{a-1}s_{21}^{1}s_{22}^{c+2}s_{11}'s_{12}'$
$+ det^{a}s_{12}s_{21}^{1}s_{22}^{c+1} + det^{a}s_{22}^{c+1}s_{11}'s_{22}']$
$- [adet^{a-1}s_{12}s_{22}^{c+2}s_{11}'s_{21}' - adet^{a-1}s_{12}s_{21}^{1}s_{22}^{c+1}s_{12}'s_{21}' - adet^{a-1}s_{21}^{1}s_{22}^{c+2}s_{11}'s_{12}'$
$+ det^{a}s_{11}s_{22}^{c+2} + det^{a}s_{22}^{c+1}s_{12}'s_{21}']$
$+ (2a+2)[adet^{a-1}s_{22}^{c+1}s_{11}'s_{12}'s_{21}'s_{22}' - det^{a}s_{12}s_{22}^{c}s_{21}'s_{22}']$
$- a[(a-1)det^{a-2}s_{11}s_{22}^{c+2}s_{11}'s_{12}'s_{21}'s_{22}' + det^{a-1}s_{12}s_{21}^{1}s_{22}^{c+1}s_{11}'s_{22}'$
$- det^{a-1}s_{11}s_{21}^{1}s_{22}^{c+1}s_{12}'s_{22}' - det^{a-1}s_{11}s_{12}s_{22}^{c+1}s_{21}'s_{22}' + det^{a-1}s_{22}^{c+1}s_{11}'s_{12}'s_{21}'s_{22}']$
$+ a[(a-1)det^{a-2}s_{12}s_{21}^{1}s_{22}^{c+1}s_{11}'s_{12}'s_{21}'s_{22}' + det^{a-1}s_{12}s_{22}^{c+2}s_{11}'s_{21}'$
$- det^{a-1}s_{11}s_{22}^{c+2}s_{12}'s_{21}' - det^{a-1}s_{11}s_{12}s_{22}^{c+1}s_{21}'s_{22}' - det^{a-1}s_{22}^{c+1}s_{11}'s_{12}'s_{21}'s_{22}']$
$= -x_1(a+1,0,c+1) - (a+1)v_9(a,0,c) + (a+1)v_{10}(a,0,c)$
$+ (a+1)ay_1(a-1,0,c+1) - f_2(a,0,c)$

Since

$$det^{a-1}s_{11}s_{22}^{c+2}s_{11}'s_{22}' - det^{a-1}s_{12}s_{21}^{1}s_{22}^{c+1}s_{11}'s_{22}' = det^{a}s_{22}^{c+1}s_{11}'s_{22}'$$

$$det^{a-1}s_{12}s_{21}^{1}s_{22}^{c+1}s_{12}'s_{21}' - det^{a-1}s_{11}s_{22}^{c+2}s_{12}'s_{21}' = -det^{a}s_{22}^{c+1}s_{12}'s_{21}',$$

we have

$$det^{a}s_{22}^{c+1}s_{11}'s_{22}' - det^{a}s_{22}^{c+1}s_{12}'s_{21}' + 2det^{a}s_{22}^{c}s_{12}s_{21}'s_{22}' = -v_9(a,0,c) + v_{10}(a,0,c)$$

and the sum of all unmarked terms is $-(a+1)v_9(a,0,c) + (a+1)v_{10}(a,0,c)$.

$h_1'(g_3(a,b,0)) =$
$(a+1)det^{a}s_{21}^{b}s_{22}^{1}s_{11}'s_{21}' - (a+1)det^{a}s_{21}^{b+1}s_{12}'s_{21}'$
$+ (a+1)[adet^{a-1}s_{21}^{b+1}s_{11}'s_{12}'s_{21}'s_{22}' - det^{a}s_{11}s_{21}^{b}s_{21}'s_{22}']$
$= (a+1)[v_3(a,b,1) + v_6(a,b,0) + ay_1(a-1,b+1,0)]$
$= (a+1)f_4(a,b,0);$

$$
\begin{aligned}
&h_1'(g_4(a,0,0)) = \\
&-[(a+1)det^a s_{11}s_{22}^2 s_{11}'s_{21}' - (a+1)det^a s_{11}s_{21}^1 s_{22}^1 s_{12}'s_{21}' - (a+1)det^a s_{21}^2 s_{22}^1 s_{11}'s_{12}' \\
&+ det^{a+1}s_{11}s_{21}^1 s_{22}^1 + det^{a+1}s_{22}^1 s_{11}'s_{21}'] \\
&+[(a+1)det^a s_{11}s_{21}^1 s_{22}^1 s_{11}'s_{22}' - (a+1)det^a s_{11}s_{21}^2 s_{12}'s_{22}' - (a+1)det^a s_{21}^2 s_{22}^1 s_{11}'s_{12}' \\
&+ det^{a+1}s_{12}s_{21}^2 + det^{a+1}s_{21}^1 s_{11}'s_{22}'] \\
&(a+2)[(a+1)det^a s_{21}^1 s_{11}'s_{12}'s_{21}'s_{22}' - det^{a+1}s_{11}s_{21}'s_{22}'] \\
&-(a+1)[a det^{a-1}s_{11}s_{21}^1 s_{22}^1 s_{11}'s_{12}'s_{21}'s_{22}' + det^a s_{12}s_{21}^2 s_{11}'s_{22}' \\
&- det^a s_{11}s_{21}^2 s_{12}'s_{22}' - det^a s_{11}s_{12}s_{21}^1 s_{21}'s_{22}' + det^a s_{21}^1 s_{11}'s_{12}'s_{21}'s_{22}'] \\
&+(a+1)[a det^{a-1}s_{12}s_{21}^2 s_{11}'s_{12}'s_{21}'s_{22}' + det^a s_{12}s_{21}^1 s_{22}^1 s_{11}'s_{21}' \\
&- det^a s_{11}s_{21}^1 s_{22}^1 s_{12}'s_{21}' - det^a s_{11}s_{12}s_{21}^1 s_{21}'s_{22}' - det^a s_{21}^1 s_{11}'s_{12}'s_{21}'s_{22}'] \\
&= -x_1(a+2,1,0) - (a+2)v_3(a+1,0,1) - (a+2)v_5(a,0,0) \\
&= -f_3(a+1,0,0) - (a+2)f_4(a+1,0,0)
\end{aligned}
$$

Here,

$$
\begin{aligned}
&-det^{a+1}s_{22}^1 s_{11}'s_{21}' + det^{a+1}s_{21}s_{11}'s_{22}' - det^{a+1}s_{11}s_{21}'s_{22}' \\
&= -v_3(a+1,0,1) - v_5(a+1,0,0)
\end{aligned}
$$

and

$$
\begin{aligned}
&[-det^a s_{11}s_{22}^2 s_{11}'s_{21}' + det^a s_{12}s_{21}^1 s_{22}^1 s_{11}'s_{21}'] + [det^a s_{11}s_{21}^1 s_{22}^1 s_{11}'s_{22}' - det^a s_{12}s_{21}^2 s_{11}'s_{22}'] \\
&- det^{a+1}s_{11}s_{21}'s_{22}' = -det^{a+1}s_{22}s_{11}'s_{21}' + det^{a+1}s_{21}s_{11}'s_{22}' - det^{a+1}s_{11}s_{21}'s_{22}' \\
&= -v_3(a+1,0,1) - v_5(a+1,0,0)
\end{aligned}
$$

$$
\begin{aligned}
&h_1'(g_5(a,b,0)) = \\
&-[(a+1)det^a s_{11}s_{21}^{b+1}s_{22}^2 s_{11}'s_{21}' - (a+1)det^a s_{11}s_{21}^{b+2}s_{22}^1 s_{12}'s_{21}' \\
&-(a+1)det^a s_{21}^{b+3}s_{22}^1 s_{11}'s_{12}' + det^{a+1}s_{11}s_{21}^{b+2}s_{22}^1 + det^{a+1}s_{21}^{b+1}s_{22}^1 s_{11}'s_{21}'] \\
&+[(a+1)det^a s_{11}s_{21}^{b+2}s_{22}^1 s_{11}'s_{22}' - (a+1)det^a s_{11}s_{21}^{b+3}s_{12}'s_{22}' \\
&-(a+1)det^a s_{21}^{b+3}s_{22}^1 s_{11}'s_{12}' + det^{a+1}s_{12}s_{21}^{b+3} + det^{a+1}s_{21}^{b+2}s_{11}'s_{22}'] \\
&+[(a+1)det^a s_{21}^{b+2}s_{11}'s_{12}'s_{21}'s_{22}' - det^{a+1}s_{11}s_{21}^{b+1}s_{21}'s_{22}'] \\
&+(a+1)[a det^{a-1}s_{11}s_{21}^{b+2}s_{22}^1 s_{11}'s_{12}'s_{21}'s_{22}' + det^a s_{12}s_{21}^{b+2}s_{22}^1 s_{11}'s_{21}' \\
&- det^a s_{11}s_{21}^{b+2}s_{22}^1 s_{12}'s_{21}' - det^a s_{11}^2 s_{21}^{b+1}s_{22}^1 s_{21}'s_{22}'] \\
&-(a+1)[a det^{a-1}s_{11}s_{21}^{b+2}s_{22}^1 s_{11}'s_{12}'s_{21}'s_{22}' + det^a s_{12}s_{21}^{b+3}s_{11}'s_{22}' \\
&- det^a s_{11}s_{21}^{b+3}s_{12}'s_{22}' - det^a s_{11}s_{12}s_{21}^{b+2}s_{21}'s_{22}' + det^a s_{21}^{b+2}s_{11}'s_{12}'s_{21}'s_{22}'] \\
&= -x_1(a+2,b+2,0) - v_3(a+1,b+1,1) - v_5(a+1,b+1,0) \\
&-(a+1)det^{a+1}s_{21}^{b+1}s_{22}s_{11}'s_{21}' + (a+1)det^{a+1}s_{21}^{b+2}s_{11}'s_{22}' - (a+1)det^{a+1}s_{11}s_{21}^{b+1}s_{21}'s_{22}' \\
&= -x_1(a+2,b+2,0) - (a+2)v_3(a+1,b+1,1) - (a+2)v_5(a+1,b+1,0) \\
&= -f_3(a+1,b+1,0) - (a+2)f_4(a+1,b+1,0)
\end{aligned}
$$

Here,

$$
\begin{aligned}
&-det^{a+1}s_{21}^{b+1}s_{22}^1 s_{11}'s_{21}' + det^{a+1}s_{21}^{b+2}s_{11}'s_{22}' - det^{a+1}s_{11}s_{21}^{b+1}s_{21}'s_{22}' \\
&= -v_3(a+1,b+1,1) - v_5(a+1,b+1,0).
\end{aligned}
$$

$$h'_1(g_6(a,b,c)) =$$
$$[(a+1)det^a s_{21}^b s_{22}^{c+2} s'_{11}s'_{21} - (a+1)det^a s_{21}^{b+1} s_{22}^{c+1} s'_{12}s'_{21}]$$
$$+ (a+1)[adet^{a-1} s_{21}^{b+1} s_{22}^{c+1} s'_{11}s'_{12}s'_{21}s'_{22} - det^a s_{12}s_{21}^{b+1} s_{22}^{c} s'_{21}s'_{22}]$$
$$= (a+1)v_6(a,b,c+1) + (a+1)v_1(a+1,b,c) + (a+1)v_3(a,b,c+2)$$
$$+ a(a+1)y_1(a-1,b+1,c+1) = (a+1)f_4(a,b,c+1) + (a+1)f_6(a+1,b,c);$$
$$h'_1(g_7(a,b,c)) =$$
$$[(a+1)det^a s_{21}^{b+1} s_{22}^{c+1} s'_{11}s'_{22} - (a+1)det^a s_{21}^{b+2} s_{22}^{c} s'_{12}s'_{22}]$$
$$+ (a+1)[adet^{a-1} s_{21}^{b+1} s_{22}^{c+1} s'_{11}s'_{12}s'_{21}s'_{22} - det^a s_{12}s_{21}^{b+1} s_{22}^{c} s'_{21}s'_{22}]$$
$$= (a+1)[v_1(a+1,b,c) - v_5(a,b,c+1) + ay_1(a-1,b+1,c+1) - v_2(a,b+2,c)]$$
$$= -(a+1)f_5(a,b,c) + (a+1)f_6(a+1,b,c);$$
$$h'_1(g_8(a,b,c)) =$$
$$[adet^{a-1} s_{21}^{b+1} s_{22}^{c+1} s'_{11}s'_{12}s'_{21}s'_{22} - det^a s_{11}s_{21}^{b} s_{22}^{c+1} s'_{21}s'_{22}]$$
$$- [adet^{a-1} s_{21}^{b+1} s_{22}^{c+1} s'_{11}s'_{12}s'_{21}s'_{22} - det^a s_{12}s_{21}^{b+1} s_{22}^{c} s'_{21}s'_{22}]$$
$$= -v_1(a+1,b,c) = -f_6(a+1,b,c);$$
$$h'_1(g_9(a,0,c)) =$$
$$- [(a+1)det^a s_{11}s_{22}^{c+3} s'_{11}s'_{21} - (a+1)det^a s_{11}s_{21}^{1} s_{22}^{c+2} s'_{12}s'_{21}$$
$$- (a+1)det^a s_{21}^{2} s_{22}^{c+2} s'_{11}s'_{12} + det^{a+1} s_{11}s_{21}^{1} s_{22}^{c+2} + det^{a+1} s_{22}^{c+2} s'_{11}s'_{21}]$$
$$+ [(a+1)det^a s_{11}s_{21}^{1} s_{22}^{c+2} s'_{11}s'_{22} - (a+1)det^a s_{11}s_{21}^{2} s_{22}^{c+1} s'_{12}s'_{22}$$
$$(a+1)det^a s_{21}^{2} s_{22}^{c+2} s'_{11}s'_{12} + det^{a+1} s_{12}s_{21}^{2} s_{22}^{c+1} + det^{a+1} s_{21}^{1} s_{22}^{c+1} s'_{11}s'_{22}]$$
$$+(a+2)[(a+1)det^a s_{21}^{1} s_{22}^{c+1} s'_{11}s'_{12}s'_{21}s'_{22} - det^{a+1} s_{12}s_{21}^{1} s_{22}^{c} s'_{21}s'_{22}]$$
$$- (a+1)[adet^{a-1} s_{11}s_{21}^{1} s_{22}^{c+2} s'_{11}s'_{12}s'_{21}s'_{22} + det^a s_{12}s_{21}^{2} s_{22}^{c+1} s'_{11}s'_{22}$$
$$- det^a s_{11}s_{21}^{2} s_{22}^{c+1} s'_{12}s'_{22} - det^a s_{11}s_{12}s_{21}^{1} s_{22}^{c+1} s'_{21}s'_{22} + det^a s_{21}^{1} s_{22}^{c+1} s'_{11}s'_{12}s'_{21}s'_{22}]$$
$$+ (a+1)[adet^{a-1} s_{12}s_{21}^{2} s_{22}^{c+1} s'_{11}s'_{12}s'_{21}s'_{22} + det^a s_{12}s_{21}^{1} s_{22}^{c+2} s'_{11}s'_{21}$$
$$- det^a s_{11}s_{21}^{1} s_{22}^{c+2} s'_{12}s'_{21} - det^a s_{11}s_{12}s_{21}^{1} s_{22}^{c+1} s'_{21}s'_{22} - det^a s_{21}^{1} s_{22}^{c+1} s'_{11}s'_{12}s'_{21}s'_{22}]$$
$$= -x_1(a+2,1,c+1) - (a+1)v_3(a+1,0,c+2)$$
$$- v_3(a+1,0,c+2) - (a+2)v_5(a+1,0,c+1) + av_1(a+2,0,c)$$
$$= -[x_1(a+2,1,c+1) + (a+2)v_3(a+1,0,c+2) + (a+2)v_5(a+1,0,c+1)]$$
$$+ (a+2)v_1(a+2,0,c)$$
$$= -f_3(a+1,0,c+1) - (a+2)f_4(a+1,0,c+1) + (a+2)f_6(a+2,0,c)$$

Here

$$- det^{a+1} s_{22}^{c+2} s'_{11}s'_{21} + det^{a+1} s_{21}^{1} s_{22}^{c+1} s'_{11}s'_{22} - det^{a+1} s_{12}s_{21}^{1} s_{22}^{c} s'_{21}s'_{22}$$
$$= -v_3(a+1,0,c+2) - v_5(a+1,0,c+1) + v_1(a+2,0,c)$$

and

$$det^a s_{11}s_{21}^{1} s_{22}^{c+2} s'_{11}s'_{22} - det^{a+1} s_{12}s_{21}^{1} s_{22}^{c} s'_{21}s'_{22} - det^a s_{12}s_{21}^{2} s_{22}^{c+1} s'_{11}s'_{22}$$
$$= det^{a+1} s_{21}^{1} s_{22}^{c+1} s'_{11}s'_{22} - det^{a+1} s_{12}s_{21}^{1} s_{22}^{c} s'_{21}s'_{22}$$
$$= -v_5(a+1,0,c+1) + v_1(a+2,0,c)$$

$$
\begin{aligned}
&h_1'(g_{10}(a,b,c)) = \\
&-[(a+1)det^a s_{11}s_{21}^{b+1}s_{22}^{c+3}s_{11}'s_{21}' - (a+1)det^a s_{11}s_{21}^{b+2}s_{22}^{c+2}s_{12}'s_{21}' \\
&-(a+1)det^a s_{21}^{b+3}s_{22}^{c+2}s_{11}'s_{12}' + det^{a+1}s_{11}s_{21}^{b+2}s_{22}^{c+2} + det^{a+1}s_{21}^{b+1}s_{22}^{c+2}s_{11}'s_{21}'] \\
&+[(a+1)det^a s_{11}s_{21}^{b+2}s_{22}^{c+2}s_{11}'s_{22}' - (a+1)det^a s_{11}s_{21}^{b+3}s_{22}^{c+1}s_{12}'s_{22}' \\
&-(a+1)det^a s_{21}^{b+3}s_{22}^{c+2}s_{11}'s_{12}' + det^{a+1}s_{12}s_{21}^{b+3}s_{22}^{c+1} + det^{a+1}s_{21}^{b+2}s_{22}^{c+1}s_{11}'s_{22}'] \\
&+[(a+1)det^a s_{21}^{b+2}s_{22}^{c+1}s_{11}'s_{12}'s_{21}'s_{22}' - det^{a+1}s_{12}s_{21}^{b+2}s_{22}^{c}s_{21}'s_{22}'] \\
&+(a+1)[a\,det^{a-1}s_{11}s_{21}^{b+2}s_{22}^{c+2}s_{11}'s_{12}'s_{21}'s_{22}' + det^a s_{12}s_{21}^{b+2}s_{22}^{c+2}s_{11}'s_{21}' \\
&-det^a s_{11}s_{21}^{b+2}s_{22}^{c+2}s_{12}'s_{21}' - det^a s_{11}^2 s_{21}^{b+1}s_{22}^{c+2}s_{21}'s_{22}'] \\
&-(a+1)[a\,det^{a-1}s_{11}s_{21}^{b+2}s_{22}^{c+2}s_{11}'s_{12}'s_{21}'s_{22}' + det^a s_{12}s_{21}^{b+3}s_{22}^{c+1}s_{11}'s_{22}' \\
&-det^a s_{11}s_{21}^{b+3}s_{22}^{c+1}s_{12}'s_{22}' - det^a s_{11}s_{12}s_{21}^{b+2}s_{22}^{c+1}s_{21}'s_{22}' + det^a s_{21}^{b+2}s_{22}^{c+1}s_{11}'s_{12}'s_{21}'s_{22}'] \\
&= -x_1(a+2,b+2,c+1) + v_1(a+2,b+1,c) - v_5(a+1,b+1,c+1) \\
&- v_3(a+1,b+1,c+2) - (a+1)v_3(a+1,b+1,c+2) \\
&-(a+1)v_5(a+1,b+1,c+1) = -[x_1(a+2,b+2,c+1) \\
&+(a+2)v_3(a+1,b+1,c+2) + (a+2)v_5(a+1,b+1,c+1)] + v_1(a+2,b+1,c) \\
&= -f_3(a+1,b+1,c+1) - (a+2)f_4(a+1,b+1,c+1) + f_6(a+2,b+1,c)
\end{aligned}
$$

Here,

$$det^a s_{11}s_{21}^{b+2}s_{22}^{c+2}s_{11}'s_{22}' - det^a s_{12}s_{21}^{b+3}s_{22}^{c+1}s_{11}'s_{22}' - det^{a+1}s_{21}^{b+2}s_{22}^{c+1}s_{11}'s_{22}',$$

$$-det^a s_{11}^2 s_{21}^{b+1}s_{22}^{c+2}s_{21}'s_{22}' + det^a s_{11}s_{12}s_{21}^{b+2}s_{22}^{c+1}s_{21}'s_{22}' = -det^{a+1}s_{11}s_{21}^{b+1}s_{22}^{c+1}s_{21}'s_{22}'$$

and their sum equals $-v_5(a+1,b+1,c+1)$. □

THE PROOF OF PROPOSITION 6.4. We have

$$h_2'(f_1(pt,0,0)) = 0;$$

$$
\begin{aligned}
&h_2'(f_2(a,0,c)) = \\
&[(a+1)det^a s_{11}s_{22}^{c+1}s_{22}' - (a+1)det^a s_{12}s_{22}^{c+1}s_{21}' + (c+1)det^{a+1}s_{22}^{c}s_{22}'] \\
&-(a+1)[-a\,det^{a-1}s_{11}s_{22}^{c+1}s_{12}'s_{21}'s_{22}' - c\,det^a s_{22}^{c}s_{12}'s_{21}'s_{22}' - det^a s_{21}^{1}s_{22}^{c+1}s_{12}' \\
&-det^a s_{12}s_{22}^{c+1}s_{21}' + det^a s_{12}s_{21}^{1}s_{22}^{c}s_{22}' - det^a s_{22}^{c}s_{12}'s_{21}'s_{22}'] \\
&+(a+1)[-a\,det^{a-1}s_{12}s_{22}^{c+1}s_{11}'s_{21}'s_{22}' - det^a s_{22}^{c+2}s_{11}' + det^a s_{12}s_{22}^{c+1}s_{21}' \\
&-det^a s_{12}s_{21}^{1}s_{22}^{c}s_{22}'] - a(a+1)[-det^{a-1}s_{21}^{1}s_{22}^{c+1}s_{11}'s_{12}'s_{22}' + det^{a-1}s_{22}^{c+2}s_{11}'s_{12}'s_{21}'] \\
&= (a+c+2)w_2(a+1,0,c) + (a+1)w_5(a,0,c+1) + (a+1)w_2(a+1,0,c) \\
&-(a+1)w_4(a,0,c+1) + (a+1)c\,z_2(a,0,c) + a(a+1)z_6(a,0,c) - a(a+1)z_7(a,0,c) \\
&+(a+1)z_2(a,0,c) = -(a+1)[w_4(a,0,c+1) - w_5(a,0,c+1) + (2a+2)z_2(a,0,c) \\
&-az_6(a-1,0,c) + az_7(a-1,0,c)] + 2(a+1)[w_2(a+1,0,c) + (a+1)z_2(a,0,c)] \\
&+(c+1)[w_2(a+1,0,c) + (a+1)z_2(a,0,c)] \\
&= -(a+1)g_2(a,0,c) + (2a+c+3)g_1(a,0,c);
\end{aligned}
$$

$$
\begin{aligned}
&h_2'(f_3(a,b,c)) = \\
&[(a+1)det^a s_{11}s_{21}^{b+1}s_{22}^c s_{22}' - (a+1)det^a s_{12}s_{21}^{b+1}s_{22}^c s_{21}' \\
&+ (b+1)det^{a+1}s_{21}^b s_{22}^c s_{21}' + c\,det^{a+1}s_{21}^{b+1}s_{22}^{c-1}s_{22}'] \\
&+ (a+1)[-a\,det^{a-1}s_{12}s_{21}^{b+1}s_{22}^c s_{11}'s_{21}'s_{22}' + b\,det^a s_{21}^b s_{22}^c s_{11}'s_{21}'s_{22}' - det^a s_{21}^{b+1}s_{22}^{c+1}s_{11}' \\
&+ det^a s_{11}s_{21}^b s_{22}^{c+1}s_{21}' - det^a s_{11}s_{21}^{b+1}s_{22}^c s_{22}' + det^a s_{21}^b s_{22}^c s_{11}'s_{21}'s_{22}'] \\
&- (a+1)[-a\,det^{a-1}s_{11}s_{21}^{b+1}s_{22}^c s_{12}'s_{21}'s_{22}' - c\,det^a s_{21}^{b+1}s_{22}^{c-1}s_{12}'s_{21}'s_{22}' \\
&- det^a s_{21}^{b+2}s_{22}^c s_{12}' - det^a s_{11}s_{21}^b s_{22}^{c+1}s_{21}' + det^a s_{11}s_{21}^{b+1}s_{22}^c s_{22}'] \\
&- a(a+1)[-det^{a-1}s_{21}^{b+2}s_{22}^c s_{11}'s_{12}'s_{22}' + det^{a-1}s_{21}^{b+1}s_{22}^{c+1}s_{11}'s_{12}'s_{21}'] \\
&= -(a+1)w_4(a,b+1,c) + (a+1)w_3(a,b,c+1) + (a+b+2)w_1(a+1,b,c) \\
&+ cw_2(a+1,b+1,c-1) + (a+1)bz_1(a,b,c) + (a+1)cz_2(a,b+1,c-1) \\
&+ (a+1)az_4(a-1,b,c) + (a+1)z_1(a,b,c) \\
&- a(a+1)det^{a-1}s_{12}s_{21}^{b+1}s_{22}^c s_{11}'s_{21}'s_{22}' - a(a+1)det^{a-1}s_{21}^{b+1}s_{22}^{c+1}s_{11}'s_{12}'s_{21}'.
\end{aligned}
$$

If $b,c>0$, then

$$
\begin{aligned}
&- det^{a-1}s_{12}s_{21}^{b+1}s_{22}^c s_{11}'s_{21}'s_{22}' - det^{a-1}s_{21}^{b+1}s_{22}^{c+1}s_{11}'s_{12}'s_{21}' \\
&= z_1(a,b,c) - z_3(a-1,b-1,c+1)
\end{aligned}
$$

and

$$
\begin{aligned}
&h_2'(f_3(a,b,c) = \\
&-(a+1)[-w_3(a,b,c+1) + w_4(a,b+1,c) + z_2(a,b+1,c-1) \\
&+ az_3(a-1,b-1,c+1) - az_4(a-1,b,c)] + (a+1)(a+b)[z_1(a,b,c) \\
&- z_2(a,b+1,c-1)] + (a+b+2)[w_1(a+1,b,c) + (a+1)z_2(a,b+1,c-1)] \\
&+ c[w_2(a+1,b+1,c-1) + (a+1)z_2(a,b+1,c-1)] \\
&+ (a+1)[z_1(a,b,c) - z_2(a,b+1,c-1)] \\
&= -(a+1)g_{10}(a-1,b-1,c-1) + (a+b+2)g_6(a,b,c-1) \\
&+ cg_7(a,b,c-1) + (a+1)(a+b+1)g_8(a,b,c-1).
\end{aligned}
$$

If $b>0$, then

$$
\begin{aligned}
&h_2'(f_3(a,b,0)) = \\
&-(a+1)w_4(a,b+1,0) + (a+1)w_3(a,b,1) + (a+b+2)w_1(a+1,b,0) \\
&+ (a+1)bz_1(a,b,0) + (a+1)az_4(a-1,b,0) + (a+1)az_1(a,b,0) \\
&- (a+1)az_3(a-1,b-1,1) + (a+1)z_1(a,b,0) \\
&= -(a+1)[-w_3(a,b,1) + w_4(a,b+1,0) + z_1(a,b,0) + az_3(a-1,b-1,1) \\
&- az_4(a-1,b,0)] + (a+b+2)[w_1(a+1,b,0) + (a+1)z_1(a,b,0)] \\
&= -(a+1)g_5(a-1,b-1,0) + (a+b+2)g_3(a,b,0).
\end{aligned}
$$

If $c > 0$, then

$$
\begin{aligned}
&h_2'(f_3(a,0,c)) = \\
&-(a+1)w_4(a,1,c) + (a+1)w_3(a,0,c+1) + (a+2)w_1(a+1,0,c) \\
&+ cw_2(a+1,1,c-1) + (a+1)cz_2(a,1,c-1) \\
&+ (a+1)az_4(a-1,0,c) - (a+1)az_5(a-1,0,c) + (a+1)z_1(a,0,c) \\
&= -(a+1)[-w_3(a,0,c+1) + w_4(a,1,c) + (a+1)z_2(a,1,c-1) - az_4(a-1,0,c) \\
&+ az_5(a-1,0,c)] + (a+2)[w_1(a+1,0,c) + (a+1)z_2(a,1,c-1)] \\
&+ c[w_2(a+1,1,c-1) + (a+1)z_2(a,1,c-1)] + (a+1)[z_1(a,0,c) - z_2(a,1,c-1)] \\
&= -(a+1)g_9(a-1,0,c-1) + (a+2)g_6(a,0,c-1) + cg_7(a,0,c-1) \\
&+ (a+1)g_8(a,0,c-1); \\
&h_2'(f_3(a,0,0)) = \\
&-(a+1)w_4(a,1,0) + (a+1)w_3(a,0,1) + (a+2)w_1(a+1,0,0) \\
&+ (a+1)az_4(a-1,0,0) + (a+1)z_1(a,0,0) - (a+1)az_5(a-1,0,0) \\
&= -(a+1)[-w_3(a,0,1) + w_4(a,1,0) + (a+1)z_1(a,0,0) - az_4(a-1,0,0) \\
&+ az_5(a-1,0,0)] + (a+2)[w_1(a+1,0,0) + (a+1)z_1(a,0,0)] \\
&= -(a+1)g_4(a-1,0,0) + (a+2)g_3(a,0,0); \\
&h_2'(f_4(a,b,c)) = \\
&[adet^{a-1}s_{11}s_{21}^{b}s_{22}^{c+1}s_{11}'s_{21}'s_{22}' + (c+1)det^{a}s_{21}^{b}s_{22}^{c}s_{11}'s_{21}'s_{22}' + det^{a}s_{21}^{b+1}s_{22}^{c+1}s_{11}'] \\
&+ [-adet^{a-1}s_{11}s_{21}^{b+1}s_{22}^{c}s_{12}'s_{21}'s_{22}' - cdet^{a}s_{21}^{b+1}s_{22}^{c-1}s_{12}'s_{21}'s_{22}' \\
&- det^{a}s_{21}^{b+2}s_{22}^{c}s_{12}' - det^{a}s_{11}s_{21}^{b}s_{22}^{c+1}s_{21}' + det^{a}s_{11}s_{21}^{b+1}s_{22}^{c}s_{22}'] \\
&+ a[-det^{a-1}s_{21}^{b+2}s_{22}^{c}s_{11}'s_{12}'s_{22}' + det^{a-1}s_{21}^{b+1}s_{22}^{c+1}s_{11}'s_{12}'s_{21}'] \\
&= -w_3(a,b,c+1) + w_4(a,b+1,c) + (c+1)z_1(a,b,c) - cz_2(a,b+1,c-1) \\
&- az_4(a-1,b,c) + adet^{a-1}s_{11}s_{21}^{b}s_{22}^{c+1}s_{11}'s_{21}'s_{22}' + adet^{a-1}s_{21}^{b+1}s_{22}^{c+1}s_{11}'s_{12}'s_{21}'.
\end{aligned}
$$

Here

$$
\begin{aligned}
&adet^{a-1}s_{11}s_{21}^{b}s_{22}^{c+1}s_{11}'s_{21}'s_{22}' + adet^{a-1}s_{21}^{b+1}s_{22}^{c+1}s_{11}'s_{12}'s_{21}' \\
&= az_3(a-1,b-1,c+1)
\end{aligned}
$$

if $b > 0$. It equals $az_5(a-1,0,c) + az_1(a,0,c)$ if $b = 0$.

If $b, c > 0$, then

$$
\begin{aligned}
&h_2'(f_4(a,b,c)) = \\
&-w_3(a,b,c+1) + w_4(a,b+1,c) + (c+1)z_1(a,b,c) - cz_2(a,b+1,c-1) \\
&- az_4(a-1,b,c) + az_3(a-1,b-1,c+1) \\
&= [-w_3(a,b,c+1) + w_4(a,b+1,c) + z_2(a,b+1,c-1) + az_3(a-1,b-1,c+1) \\
&- az_4(a-1,b,c)] + (c+1)[z_1(a,b,c) - z_2(a,b+1,c-1)] \\
&= g_{10}(a-1,b-1,c-1) + (c+1)g_8(a,b,c-1).
\end{aligned}
$$

If $b > 0$, then

$$\begin{aligned}
&h'_2(f_4(a,b,0)) = \\
&- w_3(a,b,1) + w_4(a,b+1,0) + z_1(a,b,0) - az_4(a-1,b,0) + az_3(a-1,b-1,1) \\
&= g_5(a-1,b-1,0).
\end{aligned}$$

If $c > 0$, then

$$\begin{aligned}
&h'_2(f_4(a,0,c)) = \\
&- w_3(a,0,c+1) + w_4(a,1,c) + (c+1)z_1(a,0,c) - cz_2(a,1,c-1) \\
&- az_4(a-1,0,c) + az_5(a-1,0,c) + az_1(a,0,c) = \\
&[-w_3(a,0,c+1) + w_4(a,1,c) + (a+1)z_2(a,1,c-1) - az_4(a-1,0,c) \\
&+ az_5(a-1,0,c)] + (c+a+1)[z_1(a,0,c) - z_2(a,1,c-1)] \\
&= g_9(a-1,0,c-1) + (c+a+1)g_8(a,0,c-1); \\
&h'_2(f_4(a,0,0)) = \\
&- w_3(a,0,1) + w_4(a,1,0) + (a+1)z_1(a,0,0) - az_4(a-1,0,0) + az_5(a-1,0,0) \\
&+ az_1(a,0,0) = g_4(a-1,0,0); \\
&h'_2(f_5(a,b,c)) = \\
&[a det^{a-1} s_{12}s_{21}^{b+2}s_{22}^{c}s'_{12}s'_{21}s'_{22} - (b+2)det^a s_{21}^{b+1}s_{22}^{c}s'_{12}s'_{21}s'_{22} + det^a s_{21}^{b+2}s_{22}^{c+1}s'_{12}] \\
&+ [-a det^{a-1} s_{12}s_{21}^{b+1}s_{22}^{c+1}s'_{11}s'_{21}s'_{22} + b det^a s_{21}^{b}s_{22}^{c+1}s'_{11}s'_{21}s'_{22} - det^a s_{21}^{b+1}s_{22}^{c+2}s'_{11} \\
&+ det^a s_{11}s_{21}^{b}s_{22}^{c+2}s'_{21} - det^a s_{11}s_{21}^{b+1}s_{22}^{c+1}s'_{22} + det^a s_{21}^{b}s_{22}^{c+1}s'_{11}s'_{21}s'_{22}] \\
&- a[-det^{a-1} s_{21}^{b+2}s_{22}^{c+1}s'_{11}s'_{12}s'_{22} + det^{a-1} s_{21}^{b+1}s_{22}^{c+2}s'_{11}s'_{12}s'_{21}] \\
&= w_3(a,b,c+2) - w_4(a,b+1,c+1) + (b+1)z_1(a,b,c+1) - (b+2)z_2(a,b+1,c) \\
&- az_2(a,b+1,c) + az_4(a-1,b,c+1) \\
&- a[det^{a-1} s_{12}s_{21}^{b+1}s_{22}^{c+1}s'_{11}s'_{21}s'_{22} + det^{a-1} s_{21}^{b+1}s_{22}^{c+2}s'_{11}s'_{12}s'_{21}].
\end{aligned}$$

Here

$$\begin{aligned}
&- a det^{a-1} s_{12}s_{21}^{b+1}s_{22}^{c+1}s'_{11}s'_{21}s'_{22} - a det^{a-1} s_{21}^{b+1}s_{22}^{c+2}s'_{11}s'_{12}s'_{21} \\
&= az_1(a,b,c+1) - az_3(a-1,b-1,c+2)
\end{aligned}$$

if $b > 0$. It equals $-az_5(a-1,0,c+1)$ if $b = 0$.

Hence, if $b > 0$, then

$$h_2'(f_5(a,b,c)) =$$
$$w_3(a,b,c+2) - w_4(a,b+1,c+1) + (b+1+a)z_1(a,b,c+1)$$
$$-(a+b+2)z_2(a,b+1,c) - az_3(a-1,b-1,c+2) + az_4(a-1,b,c+1)$$
$$= -[-w_3(a+1,b+1,c+2) + w_4(a+1,b+2,c+1) + z_2(a,b+1,c)$$
$$+ az_3(a-1,b-1,c+2) - az_4(a-1,b,c)] + (b+a+1)[z_1(a,b,c+1)$$
$$- z_2(a,b+1,c)] = -g_{10}(a-1,b-1,c) + (b+a+1)g_8(a,b,c);$$
$$h_2'(f_5((a,0,c)) =$$
$$w_3(a,0,c+2) - w_4(a,1,c+1) + z_1(a,0,c+1) - 2z_2(a,1,c)$$
$$- az_2(a,1,c) + az_4(a-1,0,c+1) - az_5(a-1,0,c+1)$$
$$= -[-w_3(a,0,c+2) + w_4(a,1,c+1) + (a+1)z_2(a,1,c) - az_4(a-1,0,c+1)$$
$$+ az_5(a-1,0,c+1)] + [z_1(a,0,c+1) - z_2(a,1,c)]$$
$$- g_9(a-1,0,c) + g_8(a,0,c);$$
$$h_2'(f_6(a,b,c)) =$$
$$det^a s_{21}^b s_{22}^{c+1} s_{21}' - det^a s_{21}^{b+1} s_{22}^c s_{22}' = w_1(a,b,c+1) - w_2(a,b+1,c)$$
$$= g_6(a-1,b,c) - g_7(a-1,b,c);$$
$$h_2'(f_7(a,0,c)) =$$
$$[a det^{a-1} s_{12} s_{21}^1 s_{22}^c s_{12}' s_{21}' s_{22}' - det^a s_{22}^c s_{12}' s_{21}' s_{22}' + det^a s_{21}^1 s_{22}^{c+1} s_{12}']$$
$$+ [-a det^{a-1} s_{12} s_{22}^{c+1} s_{11}' s_{21}' s_{22}' - det^a s_{22}^{c+2} s_{11}' + det^a s_{12} s_{22}^{c+1} s_{21}' - det^a s_{12} s_{21}^1 s_{22}^c s_{22}']$$
$$- a[-det^{a-1} s_{21}^1 s_{22}^{c+1} s_{11}' s_{12}' s_{22}' + det^{a-1} s_{22}^{c+2} s_{11}' s_{12}' s_{21}']$$
$$= +w_5(a,0,c+1) + w_2(a+1,0,c) - w_4(a,0,c+1) - z_2(a,0,c)$$
$$+ az_6(a-1,0,c) - az_2(a,0,c) - az_7(a-1,0,c)$$
$$-[w_5(a,0,c+1) - (2a+2)z_2(a,0,c) - w_4(a,0,c+1) + az_6(a-1,0,c)$$
$$- az_7(a-1,0,c)] + g_1(a,0,c) = g_1(a,0,c) - g_2(a,0,c);$$
$$h_2'(g_1(a,0,c)) =$$
$$[-(a+1)det^a s_{12} s_{22}^c s_{21}' s_{22}' - det^{a+1} s_{22}^{c+1}] + (a+1)[-det^a s_{22}^{c+1} s_{12}' s_{21}'$$
$$+ det^a s_{21}^1 s_{22}^c s_{12}' s_{22}'] = -x_1(a+1,0,c+1) + (a+1)v_{10}(a,0,c) + (a+1)v_2(a,1,c)$$
$$= -[x_1(a+1,0,c+1) - (a+1)v_{10}(a,0,c) + (a+1)v_9(a,0,c)$$
$$- a(a+1)y_1(a-1,0,c+1))] + (a+1)[v_2(a,1,c) + v_9(a,0,c) - ay_1(a-1,0,c+1)]$$
$$= -f_2(a,0,c) + (a+1)f_7(a,0,c);$$
$$h_2'(g_2(a,0,c)) =$$
$$[-a det^{a-1} s_{11} s_{12} s_{22}^{c+1} s_{21}' s_{22}' + a det^{a-1} s_{11} s_{21}^1 s_{22}^{c+1} s_{12}' s_{22}'$$
$$- a det^{a-1} s_{12} s_{21}^1 s_{22}^{c+1} s_{12}' s_{21}' + (c+1) det^a s_{21}^1 s_{22}^c s_{12}' s_{22}' - det^a s_{11} s_{22}^{c+2}$$
$$+ det^a s_{22}^{c+1} s_{12}' s_{21}'] - [-a det^{a-1} s_{11} s_{12} s_{22}^{c+1} s_{21}' s_{22}' - (c+1) det^a s_{12} s_{22}^c s_{21}' s_{22}'$$
$$+ a det^{a-1} s_{11} s_{22}^{c+2} s_{11}' s_{22}' - a det^{a-1} s_{12} s_{22}^{c+2} s_{11}' s_{21}' + (c+1) det^a s_{22}^{c+1} s_{11}' s_{22}'$$
$$- det^a s_{12} s_{21}^1 s_{22}^{c+1} + det^a s_{22}^{c+1} s_{11}' s_{22}']$$
$$+ (2a+2)[-det^a s_{22}^{c+1} s_{12}' s_{21}' + det^a s_{21}^1 s_{22}^c s_{12}' s_{22}']$$
$$- a[-(a-1) det^{a-2} s_{12} s_{21} s_{22}^{c+1} s_{11}' s_{12}' s_{21}' s_{22}' - det^{a-1} s_{21}^1 s_{22}^{c+2} s_{11}' s_{12}'$$
$$- det^{a-1} s_{11} s_{22}^{c+2} s_{12}' s_{21}' + det^{a-1} s_{11} s_{21}^1 s_{22}^{c+1} s_{12}' s_{22}' + det^{a-1} s_{22}^{c+1} s_{11}' s_{12}' s_{21}' s_{22}']$$

$$+ a[-(a-1)det^{a-2}s_{11}s_{22}^{c+2}s'_{11}s'_{12}s'_{21}s'_{22} - c\,det^{a-1}s_{22}^{c+1}s'_{11}s'_{12}s'_{21}s'_{22}$$
$$- det^{a-1}s_{21}^{1}s_{22}^{c+2}s'_{11}s'_{12} - det^{a-1}s_{12}s_{22}^{c+2}s'_{11}s'_{21} + det^{a-1}s_{12}s_{21}^{1}s_{22}^{c+1}s'_{11}s'_{22}$$
$$- 2det^{a-1}s_{22}^{c+1}s'_{11}s'_{12}s'_{21}s'_{22}] = -x_1(a+1,0,c+1) - a(a+c+2)y_1(a-1,0,c+1)$$
$$+ (c+1)v_9(a,0,c) + (c+1)v_2(a,1,c) - (a+1)det^{a}s_{22}^{c+1}s'_{11}s'_{22}$$
$$- (a+1)det^{a}s_{22}^{c+1}s'_{12}s'_{21} + (2a+2)v_2(a,1,c)$$
$$= (c+1)[v_2(a,1,c) + v_9(a,0,c) - ay_1(a-1,0,c+1)]$$
$$- [x_1(a+1,0,c+1) + (a+1)v_9(a,0,c) - (a+1)v_{10}(a,0,c)$$
$$- a(a+1)y_1(a-1,0,c+1)] + (2a+2)[v_2(a,1,c) + v_9(a,0,c) - ay_1(a-1,0,c+1)]$$
$$= (2a+c+3)f_7(a,0,c) - f_2(a,0,c);$$

$$h'_2(g_3(a,b,0)) =$$
$$[-(a+1)det^{a}s_{11}s_{21}^{b}s'_{21}s'_{22} - det^{a+1}s_{21}^{b+1}]$$
$$+ (a+1)[-det^{a}s_{21}^{b}s_{22}^{1}s'_{11}s'_{21} + det^{a}s_{21}^{b+1}s'_{11}s'_{22}]$$
$$= -x_1(a+1,b+1,0) - (a+1)v_5(a,b,0) - (a+1)v_3(a,b,1)$$
$$= -f_3(a,b,0) - (a+1)f_4(a,b,0);$$

$$h'_2(g_4(a,0,0)) =$$
$$- [-(a+1)det^{a}s_{11}^{2}s_{22}^{1}s'_{21}s'_{22} - det^{a+1}s_{11}s'_{21}s'_{22} + (a+1)det^{a}s_{11}s_{21}^{1}s_{22}^{1}s'_{11}s'_{22}$$
$$- (a+1)det^{a}s_{12}s_{21}^{1}s_{22}^{1}s'_{11}s'_{21} + det^{a+1}s_{21}^{1}s'_{11}s'_{22} - det^{a+1}s_{11}s_{21}^{1}s_{22}^{1} + det^{a+1}s_{22}^{1}s'_{11}s'_{21}]$$
$$+ [-(a+1)det^{a}s_{11}s_{12}s_{21}^{1}s'_{21}s'_{22} + det^{a+1}s_{11}s'_{21}s'_{22} + (a+1)det^{a}s_{11}s_{21}^{2}s'_{12}s'_{22}$$
$$- (a+1)det^{a}s_{12}s_{21}^{2}s'_{12}s'_{21} + det^{a+1}s_{21}^{1}s'_{12}s'_{21} - det^{a+1}s_{11}s_{21}^{1}s_{22}^{1} + det^{a+1}s_{21}^{1}s'_{12}s'_{21}]$$
$$+ (a+2)[-det^{a+1}s_{22}^{1}s'_{11}s'_{21} + det^{a+1}s_{21}^{1}s'_{11}s'_{22}]$$
$$- (a+1)[-a\,det^{a-1}s_{12}s_{21}^{2}s'_{11}s'_{12}s'_{21}s'_{22} - det^{a}s_{21}^{2}s_{22}^{1}s'_{11}s'_{12}$$
$$- det^{a}s_{11}s_{21}^{1}s_{22}^{1}s'_{12}s'_{21} + det^{a}s_{11}s_{21}^{2}s'_{12}s'_{22} + 2det^{a}s_{21}^{1}s'_{11}s'_{12}s'_{21}s'_{22}]$$
$$+ (a+1)[-a\,det^{a-1}s_{11}s_{21}^{1}s_{22}^{1}s'_{11}s'_{12}s'_{21}s'_{22} - det^{a}s_{21}^{2}s_{22}^{1}s'_{11}s'_{12}$$
$$- det^{a}s_{12}s_{21}^{1}s_{22}^{1}s'_{11}s'_{21} + det^{a}s_{12}s_{21}^{2}s'_{11}s'_{22} - det^{a}s_{21}^{1}s'_{11}s'_{12}s'_{21}s'_{22}]$$
$$= -(a+3)(a+1)y_1(a,1,0) + (a+3)det^{a+1}s_{11}s'_{21}s'_{22}$$
$$+ (a+3)det^{a+1}s_{21}s'_{12}s'_{21} - (a+3)v_3(a+1,0,1)$$
$$= -(a+3)[v_3(a+1,0,1) - v_6(a+1,0,0) + (a+1)y_1(a,1,0)]$$
$$= -(a+3)f_4(a+1,0,0);$$

$$h'_2(g_5(a,b,0)) =$$
$$- [-(a+1)det^{a}s_{11}^{2}s_{21}^{b+1}s_{22}^{1}s'_{21}s'_{22} - det^{a+1}s_{11}s_{21}^{b+1}s'_{21}s'_{22}$$
$$+ (a+1)det^{a}s_{11}s_{21}^{b+2}s_{22}^{1}s'_{11}s'_{22} - (a+1)det^{a}s_{12}s_{21}^{b+2}s_{22}^{1}s'_{11}s'_{21}$$
$$+ (b+1)det^{a+1}s_{21}^{b+1}s_{22}^{1}s'_{11}s'_{21} + det^{a+1}s_{21}^{b+2}s'_{11}s'_{22}$$
$$- det^{a+1}s_{11}s_{21}^{b+2}s_{22}^{1} + det^{a+1}s_{21}^{b+1}s_{22}^{1}s'_{11}s'_{21}]$$
$$+ [-(a+1)det^{a}s_{11}s_{12}s_{21}^{b+2}s'_{21}s'_{22} + (b+2)det^{a+1}s_{11}s_{21}^{b+1}s'_{21}s'_{22}$$

$$+ (a+1)det^a s_{11}s_{21}^{b+3}s'_{12}s'_{22} - (a+1)det^a s_{12}s_{21}^{b+3}s'_{12}s'_{21}$$
$$+ (b+2)det^{a+1}s_{21}^{b+2}s'_{12}s'_{21} - det^{a+1}s_{11}s_{21}^{b+2}s_{22}^{1} + det^{a+1}s_{21}^{b+2}s'_{12}s'_{21}]$$
$$+ [-det^{a+1}s_{21}^{b+1}s_{22}^{1}s'_{11}s'_{21} + det^{a+1}s_{21}^{b+2}s'_{11}s'_{22}]$$
$$+ (a+1)[-adet^{a-1}s_{11}s_{21}^{b+2}s_{22}^{1}s'_{11}s'_{12}s'_{21}s'_{22} - det^a s_{21}^{b+2}s'_{11}s'_{12}s'_{21}s'_{22}$$
$$- det^a s_{21}^{b+3}s_{22}^{1}s'_{11}s'_{12} - det^a s_{11}s_{21}^{b+1}s_{22}^{2}s'_{11}s'_{21} + det^a s_{11}s_{21}^{b+2}s_{22}^{1}s'_{11}s'_{22}]$$
$$- (a+1)[-adet^{a-1}s_{12}s_{21}^{b+3}s'_{11}s'_{12}s'_{21}s'_{22} + (b+1)det^a s_{21}^{b+2}s'_{11}s'_{12}s'_{21}s'_{22}$$
$$- det^a s_{21}^{b+3}s_{22}^{1}s'_{11}s'_{12} - det^a s_{11}s_{21}^{b+2}s_{22}^{1}s'_{12}s'_{21}$$
$$+ det^a s_{11}s_{21}^{b+3}s'_{12}s'_{22} + 2det^a s_{21}^{b+2}s'_{11}s'_{12}s'_{21}s'_{22}]$$
$$= -(a+b+4)(a+1)y_1(a,b+2,0) - (b+2)v_3(a+1,b+1,1)$$
$$- (b+2)v_6(a+1,b+1,0) - v_6(a+1,b+1,0) - v_3(a+1,b+1,1)$$
$$+ (a+1)det^{a+1}s_{11}s_{21}^{b+1}s'_{21}s'_{22} - (a+1)det^{a+1}s_{21}^{b+1}s_{22}s'_{11}s'_{21}$$
$$+ (a+1)det^{a+1}s_{21}^{b+2}s'_{12}s'_{21} = -(a+b+4)[v_3(a+1,b+1,1) + v_6(a+1,b+1,0)$$
$$+ (a+1)y_1(a,b+2,0)] = -(a+b+4)f_4(a+1,b+1,0);$$

$$h'_2(g_6(a,b,c)) =$$
$$[-(a+1)det^a s_{11}s_{21}^{b}s_{22}^{c+1}s'_{21}s'_{22} - (c+1)det^{a+1}s_{21}^{b}s_{22}^{c}s'_{21}s'_{22} - det^{a+1}s_{21}^{b+1}s_{22}^{c+1}]$$
$$+ (a+1)[-det^a s_{21}^{b+1}s_{22}^{c+1}s'_{12}s'_{21} + det^a s_{21}^{b+2}s_{22}^{c}s'_{12}s'_{22}]$$
$$= [-x_1(a+1,b+1,c+1) + (a+1)v_6(a,b,c+1) + (a+1)v_2(a,b+2,c)]$$
$$- (c+1)v_1(a+1,b,c)$$
$$= [-f_3(a,b,c+1) + (a+1)f_5(a,b,c)] - (c+1)f_6(a+1,b,c);$$

$$h'_2(g_7(a,b,c)) =$$
$$[-(a+1)det^a s_{12}s_{21}^{b+1}s_{22}^{c}s'_{21}s'_{22} + (b+1)det^{a+1}s_{21}^{b}s_{22}^{c}s'_{21}s'_{22} - det^{a+1}s_{21}^{b+1}s_{22}^{c+1}]$$
$$+ (a+1)[-det^a s_{21}^{b+1}s_{22}^{c+1}s'_{12}s'_{21} + det^a s_{21}^{b+2}s_{22}^{c}s'_{12}s'_{22}]$$
$$= -x_1(a+1,b+1,c+1) + (a+1)[v_1(a+1,b,c) + v_6(a,b,c+1)]$$
$$+ (a+1)v_2(a,b+2,c) + (b+1)v_1(a+1,b,c)$$
$$= [-f_3(a,b,c+1) + (a+1)f_5(a,b,c)] + (a+b+2)f_6(a+1,b,c);$$

$$h'_2(g_8(a,b,c)) =$$
$$[-det^a s_{21}^{b}s_{22}^{c+2}s'_{11}s'_{21} + det^a s_{21}^{b+1}s_{22}^{c+1}s'_{11}s'_{22}]$$
$$- [-det^a s_{21}^{b+1}s_{22}^{c+1}s'_{12}s'_{21} + det^a s_{21}^{b+2}s_{22}^{c}s'_{12}s'_{22}]$$
$$= [-v_5(a,b,c+1) - v_6(a,b,c+1) - v_3(a,b,c+2) - v_2(a,b+2,c)$$
$$= -f_4(a,b,c+1) - f_5(a,b,c);$$

$$h'_2(g_9(a,0,c)) =$$
$$- [-(a+1)det^a s_{11}^{2}s_{22}^{c+2}s'_{21}s'_{22} - (c+2)det^{a+1}s_{11}s_{22}^{c+1}s'_{21}s'_{22}$$
$$+ (a+1)det^a s_{11}s_{21}^{1}s_{22}^{c+2}s'_{11}s'_{22} - (a+1)det^a s_{12}s_{21}^{1}s_{22}^{c+2}s'_{11}s'_{21}$$
$$+ (c+2)det^{a+1}s_{21}^{1}s_{22}^{c+1}s'_{11}s'_{22} - det^{a+1}s_{11}s_{21}^{1}s_{22}^{c+2} + det^{a+1}s_{22}^{c+2}s'_{11}s'_{21}]$$
$$+ [-(a+1)det^a s_{11}s_{12}s_{21}^{1}s_{22}^{c+1}s'_{21}s'_{22} + det^{a+1}s_{11}s_{22}^{c+1}s'_{21}s'_{22}$$
$$+ (a+1)det^a s_{11}s_{21}^{2}s_{22}^{c+1}s'_{12}s'_{22} - (a+1)det^a s_{12}s_{21}^{2}s_{22}^{c+1}s'_{12}s'_{21}$$
$$+ det^{a+1}s_{21}^{1}s_{22}^{c+1}s'_{12}s'_{21} + (c+1)det^{a+1}s_{21}^{2}s_{22}^{c}s'_{12}s'_{22}$$

$$- det^{a+1}s_{11}s_{21}^{1}s_{22}^{c+2} + det^{a+1}s_{21}^{1}s_{22}^{c+1}s'_{12}s'_{21}]$$
$$+ (a+2)[-det^{a+1}s_{21}^{1}s_{22}^{c+1}s'_{12}s'_{21} + det^{a+1}s_{21}^{2}s_{22}^{c}s'_{12}s'_{22}]$$
$$- (a+1)[-adet^{a-1}s_{12}s_{21}^{2}s_{22}^{c+1}s'_{11}s'_{12}s'_{21}s'_{22}$$
$$- det^{a}s_{21}^{2}s_{22}^{c+2}s'_{11}s'_{12} - det^{a}s_{11}s_{21}^{1}s_{22}^{c+2}s'_{12}s'_{21}$$
$$+ det^{a}s_{11}s_{21}^{2}s_{22}^{c+1}s'_{12}s'_{22} + 2det^{a}s_{21}^{1}s_{22}^{c+1}s'_{11}s'_{12}s'_{21}s'_{22}]$$
$$+ (a+1)[-adet^{a-1}s_{11}s_{21}^{1}s_{22}^{c+2}s'_{11}s'_{12}s'_{21}s'_{22} - (c+1)det^{a}s_{21}^{1}s_{22}^{c+1}s'_{11}s'_{12}s'_{21}s'_{22}$$
$$- det^{a}s_{21}^{2}s_{22}^{c+2}s'_{11}s'_{12} - det^{a}s_{12}s_{21}^{1}s_{22}^{c+2}s'_{11}s'_{21}$$
$$+ det^{a}s_{12}s_{21}^{2}s_{22}^{c+1}s'_{11}s'_{22} - det^{a}s_{21}^{1}s_{22}^{c+1}s'_{11}s'_{12}s'_{21}s'_{22}]$$
$$= (a+1)(-a-c-4)y_1(a,1,c+1) + (c+2)v_5(a+1,0,c+1) - v_6(a+1,0,c+1)$$
$$+ (c+1)v_2(a+1,2,c) - v_3(a+1,0,c+2) - (a+1)det^{a+1}s_{21}s_{22}^{c+1}s'_{11}s'_{22}$$
$$+ (a+1)det^{a+1}s_{11}s_{22}^{c+1}s'_{21}s'_{22} + (a+2)v_2(a+1,2,c)$$
$$= +(c+2)[v_2(a+1,2,c) + v_5(a+1,0,c+1) - (a+1)y_1(a,1,c+1)]$$
$$+ (a+1)[v_2(a+1,2,c) + v_5(a+1,0,c+1) - (a+1)y_1(a,1,c+1)]$$
$$- [v_3(a+1,0,c+2) + v_6(a+1,0,c+1) + (a+1)y_1(a,1,c+1)]$$
$$= (a+c+3)f_5(a+1,0,c) - f_4(a+1,0,c+1);$$

$$h'_2(g_{10}(a,b,c)) =$$
$$- [-(a+1)det^{a}s_{11}^{2}s_{21}^{b+1}s_{22}^{c+2}s'_{21}s'_{22} - (c+2)det^{a+1}s_{11}s_{21}^{b+1}s_{22}^{c+1}s'_{21}s'_{22}$$
$$+ (a+1)det^{a}s_{11}s_{21}^{b+2}s_{22}^{c+2}s'_{11}s'_{22} - (a+1)det^{a}s_{12}s_{21}^{b+2}s_{22}^{c+2}s'_{11}s'_{21}$$
$$+ (b+1)det^{a+1}s_{21}^{b+1}s_{22}^{c+2}s'_{11}s'_{21} + (c+2)det^{a+1}s_{21}^{b+2}s_{22}^{c+1}s'_{11}s'_{22}$$
$$- det^{a+1}s_{11}s_{21}^{b+2}s_{22}^{c+2} + det^{a+1}s_{21}^{b+1}s_{22}^{c+2}s'_{11}s'_{21}]$$
$$+ [-(a+1)det^{a}s_{11}s_{12}s_{21}^{b+2}s_{22}^{c+1}s'_{21}s'_{22} + (b+2)det^{a+1}s_{11}s_{21}^{b+1}s_{22}^{c+1}s'_{21}s'_{22}$$
$$+ (a+1)det^{a}s_{11}s_{21}^{b+3}s_{22}^{c+1}s'_{12}s'_{22} - (a+1)det^{a}s_{12}s_{21}^{b+3}s_{22}^{c+1}s'_{12}s'_{21}$$
$$+ (b+2)det^{a+1}s_{21}^{b+2}s_{22}^{c+1}s'_{12}s'_{21} + (c+1)det^{a+1}s_{21}^{b+3}s_{22}^{c}s'_{12}s'_{22}$$
$$- det^{a+1}s_{11}s_{21}^{b+2}s_{22}^{c+2} + det^{a+1}s_{21}^{b+2}s_{22}^{c+1}s'_{12}s'_{21}]$$
$$+ [-det^{a+1}s_{21}^{b+2}s_{22}^{c+1}s'_{12}s'_{21} + det^{a+1}s_{21}^{b+3}s_{22}^{c}s'_{12}s'_{22}]$$
$$+ (a+1)[-adet^{a-1}s_{11}s_{21}^{b+2}s_{22}^{c+2}s'_{11}s'_{12}s'_{21}s'_{22} - (c+2)det^{a}s_{21}^{b+2}s_{22}^{c+1}s'_{11}s'_{12}s'_{21}s'_{22}$$
$$- det^{a}s_{21}^{b+3}s_{22}^{c+2}s'_{11}s'_{12} - det^{a}s_{11}s_{21}^{b+1}s_{22}^{c+3}s'_{11}s'_{21} + det^{a}s_{11}s_{21}^{b+2}s_{22}^{c+2}s'_{11}s'_{22}]$$
$$- (a+1)[-adet^{a-1}s_{12}s_{21}^{b+3}s_{22}^{c+1}s'_{11}s'_{12}s'_{21}s'_{22} + (b+1)det^{a}s_{21}^{b+2}s_{22}^{c+1}s'_{11}s'_{12}s'_{21}s'_{22}$$
$$- det^{a}s_{21}^{b+3}s_{22}^{c+2}s'_{11}s'_{12} - det^{a}s_{11}s_{21}^{b+2}s_{22}^{c+2}s'_{12}s'_{21} + det^{a}s_{11}s_{21}^{b+3}s_{22}^{c+1}s'_{12}s'_{22}$$
$$+ 2det^{a}s_{21}^{b+2}s_{22}^{c+1}s'_{11}s'_{12}s'_{21}s'_{22}] = -(a+1)(a+b+c+5)y_1(a,b+2,c+1)$$
$$- (a+1)v_6(a+1,b+1,c+1) - (a+1)v_3(a+1,b+1,c+2)$$
$$+ (c+2)v_5(a+1,b+1,c+1) + (c+1)v_2(a+1,b+3,c)$$
$$- (b+2)v_6(a+1,b+1,c+1) - (b+2)v_3(a+1,b+1,c+2) + v_2(a+1,b+3,c)$$
$$= (c+2)[v_2(a+1,b+3,c) + v_5(a+1,b+1,c+1) - (a+1)y_1(a,b+2,c+1)]$$
$$- (b+2)[v_3(a+1,b+1,c+2) + v_6(a+1,b+1,c+1) + (a+1)y_1(a,b+2,c+1)]$$
$$- (a+1)[v_3(a+1,b+1,c+2) + v_6(a+1,b+1,c+1) + (a+1)y_1(a,b+2,c+1)]$$
$$= (c+2)f_5(a+1,b+1,c) - (a+b+3)f_4(a+1,b+1,c+1). \quad \square$$

THE PROOF OF THEOREM 7.1. We divide this lengthy computational proof into six subsections.

1. $\mathbf{q = 0}$.

Case $\lambda_1 = \lambda_2 = k$ and $p|k$

Basis of $\mathbf{U}^+$-invariants with the weight (k,k) consists of $\{f_1(k,0,0)\}$.

Since $h_1'(f_1(k,0,0)) = 0 = h_2'(f_1(k,0,0))$, the element $f_1(k,0,0) = det^k$ is the primitive element of weight (k,k).

In this case $\mathfrak{u}(\lambda) = K.1$, ${}_\lambda\Bbbk[\mathbf{Q}(2)]^{\mathbf{U}^+} = Kf_1(0,0,0) = K.1$ and $\mathfrak{u}(\lambda) \otimes_\lambda \Bbbk[\mathbf{Q}(2)]^{\mathbf{U}^+} = K(1 \otimes 1)$

The induced supermodule $H^0(k,k)$ is a one-dimensional simple supermodule $L(k,k)$.

Case $\lambda_1 = \lambda_2 = k$ and $p \nmid k$

In this case there are no $\mathbf{U}^+$-invariants of weight (k,k) and $H^0(k,k) = 0$.

2. $\mathbf{q = 1}$.

$\lambda = (k, k+1)$ We have

$$\begin{aligned}
h_1'(f_2(k-1,0,0)) &= kg_2(k-1,0,0);\\
h_1'(f_3(k-1,0,0)) &= kg_3(k-1,0,0) + kg_4(k-2,0,0);\\
h_1'(f_4(k-1,0,0)) &= -g_3(k-1,0,0);\\
h_1'(f_7(k-1,0,0) &= g_1(k-1,0,0);\\
h_1'(g_1(k-1,0,c)) &= -kf_7(k-1,0,0);\\
h_1'(g_2(k-1,0,0)) &= -f_2(k-1,0,0);\\
h_1'(g_3(k-1,0,0)) &= kf_4(k-1,0,0);\\
h_1'(g_4(k-2,0,0)) &= -f_3(k-1,0,0) - kf_4(k-1,0,0).
\end{aligned}$$

The matrix of the action of h_1' on $D_{(k,k+1)}$ is

$$\begin{pmatrix}
0 & 0 & 0 & 0 & 0 & -1 & 0 & 0\\
0 & 0 & 0 & 0 & 0 & 0 & 0 & -1\\
0 & 0 & 0 & 0 & 0 & 0 & k & -k\\
0 & 0 & 0 & 0 & -k & 0 & 0 & 0\\
0 & 0 & 0 & 1 & 0 & 0 & 0 & 0\\
k & 0 & 0 & 0 & 0 & 0 & 0 & 0\\
0 & k & -1 & 0 & 0 & 0 & 0 & 0\\
0 & k & 0 & 0 & 0 & 0 & 0 & 0
\end{pmatrix}.$$

We have

$$\begin{aligned}
h_2'(f_2(k-1,0,0)) &= (2k+1)g_1(k-1,0,0) - kg_2(k-1,0,0);\\
h_2'(f_3(-2,0,0)) &= (k+1)g_3(k-1,0,0) - kg_4(k-2,0,0);\\
h_2'(f_4(k-1,0,0)) &= g_4(k-2,0,0);\\
h_2'(f_7(k-1,0,0) &= g_1(k-1,0,0) - g_2(k-1,0,0);\\
h_2'(g_1(k-1,0,0)) &= -f_2(k-1,0,0) + kf_7(k-1,0,0);\\
h_2'(g_2(k-1,0,0)) &= (2k+1)f_7(k-1,0,0) - f_2(k-1,0,0);\\
h_2'(g_3(k-1,0,0)) &= -f_3(k-1,0,0) - kf_4(k-1,0,0);\\
h_2'(g_4(k-2,0,0)) &= -(k+1)f_4(k-1,0,0).
\end{aligned}$$

The matrix of the action of h_2' on $B_{(k,k+1)}$ is

$$\begin{pmatrix} 0 & 0 & 0 & 0 & -1 & -1 & 0 & 0 \\ 0 & 0 & 0 & 0 & 0 & 0 & -1 & 0 \\ 0 & 0 & 0 & 0 & 0 & 0 & -k & -k-1 \\ 0 & 0 & 0 & 0 & k & 2k+1 & 0 & 0 \\ 2k+1 & 0 & 0 & 1 & 0 & 0 & 0 & 0 \\ -k & 0 & 0 & -1 & 0 & 0 & 0 & 0 \\ 0 & k+1 & 0 & 0 & 0 & 0 & 0 & 0 \\ 0 & -k & 1 & 0 & 0 & 0 & 0 & 0 \end{pmatrix}.$$

Case $p|k$, $p \nmid k+1$

In this case, $\mathfrak{u}(\lambda) = Kt_0 + Kt_{(01)}$, where

$$h_1'(t_0) = 0, h_1'(t_{(01)}) = 0, h_2'(t_0) = \lambda_2 t_{(01)}, h_2'(t_{(01)}) = t_0.$$

The matrix of the action of h_1' on $U = \{u_0, u_1\}$ is

$$\begin{pmatrix} 0 & 0 \\ 0 & 0 \end{pmatrix}$$

and the matrix of action of h_2' on $U = \{u_0, u_1\}$ is

$$\begin{pmatrix} 0 & 1 \\ 1 & 0 \end{pmatrix}.$$

The matrix of the action of h_1' on $B_{(k,k+1)}$ is

$$\begin{pmatrix} 0 & 0 & 0 & 0 & 0 & -1 & 0 & 0 \\ 0 & 0 & 0 & 0 & 0 & 0 & 0 & -1 \\ 0 & 0 & 0 & 0 & 0 & 0 & 0 & 0 \\ 0 & 0 & 0 & 0 & 0 & 0 & 0 & 0 \\ 0 & 0 & 0 & 1 & 0 & 0 & 0 & 0 \\ 0 & 0 & 0 & 0 & 0 & 0 & 0 & 0 \\ 0 & 0 & -1 & 0 & 0 & 0 & 0 & 0 \\ 0 & 0 & 0 & 0 & 0 & 0 & 0 & 0 \end{pmatrix}.$$

The matrix of action of h_1' on $U \otimes B_{k,k+1} = \{u_0 \otimes B_{(k,k+1)}, u_1 \otimes B_{(k,k+1)}\}$ is

$$\begin{pmatrix} 0 & 0 & 0 & 0 & 0 & -1 & 0 & 0 & 0 & 0 & 0 & 0 & 0 & 0 & 0 & 0 \\ 0 & 0 & 0 & 0 & 0 & 0 & 0 & -1 & 0 & 0 & 0 & 0 & 0 & 0 & 0 & 0 \\ 0 & 0 & 0 & 0 & 0 & 0 & 0 & 0 & 0 & 0 & 0 & 0 & 0 & 0 & 0 & 0 \\ 0 & 0 & 0 & 0 & 0 & 0 & 0 & 0 & 0 & 0 & 0 & 0 & 0 & 0 & 0 & 0 \\ 0 & 0 & 0 & 1 & 0 & 0 & 0 & 0 & 0 & 0 & 0 & 0 & 0 & 0 & 0 & 0 \\ 0 & 0 & 0 & 0 & 0 & 0 & 0 & 0 & 0 & 0 & 0 & 0 & 0 & 0 & 0 & 0 \\ 0 & 0 & -1 & 0 & 0 & 0 & 0 & 0 & 0 & 0 & 0 & 0 & 0 & 0 & 0 & 0 \\ 0 & 0 & 0 & 0 & 0 & 0 & 0 & 0 & 0 & 0 & 0 & 0 & 0 & 0 & 0 & 0 \\ 0 & 0 & 0 & 0 & 0 & 0 & 0 & 0 & 0 & 0 & 0 & 0 & 0 & 1 & 0 & 0 \\ 0 & 0 & 0 & 0 & 0 & 0 & 0 & 0 & 0 & 0 & 0 & 0 & 0 & 0 & 0 & 1 \\ 0 & 0 & 0 & 0 & 0 & 0 & 0 & 0 & 0 & 0 & 0 & 0 & 0 & 0 & 0 & 0 \\ 0 & 0 & 0 & 0 & 0 & 0 & 0 & 0 & 0 & 0 & 0 & 0 & 0 & 0 & 0 & 0 \\ 0 & 0 & 0 & 0 & 0 & 0 & 0 & 0 & 0 & 0 & 0 & -1 & 0 & 0 & 0 & 0 \\ 0 & 0 & 0 & 0 & 0 & 0 & 0 & 0 & 0 & 0 & 0 & 0 & 0 & 0 & 0 & 0 \\ 0 & 0 & 0 & 0 & 0 & 0 & 0 & 0 & 0 & 0 & 1 & 0 & 0 & 0 & 0 & 0 \\ 0 & 0 & 0 & 0 & 0 & 0 & 0 & 0 & 0 & 0 & 0 & 0 & 0 & 0 & 0 & 0 \end{pmatrix}.$$

Hence, the basis of elements annihilated by h_1' is

$$\begin{aligned} B' = \{ & r_1 = u_0 \otimes f_2(k-1,0,0), r_2 = u_0 \otimes f_3(k-1,0,0), \\ & r_3 = u_0 \otimes g_1(k-1,0,0), r_4 = u_0 \otimes g_3(k-1,0,0), \\ & r_5 = u_1 \otimes f_2(k-1,0,0), r_6 = u_1 \otimes f_3(k-1,0,0), \\ & r_7 = u_1 \otimes g_1(k-1,0,0), r_8 = u_1 \otimes g_3(k-1,0,0)\}, \end{aligned}$$

constructed using elements of

$$B'' = \{f_2(k-1,0,0), f_3(k-1,0,0), g_1(k-1,0,0), g_3(k-1,0,0)\}.$$

The corresponding matrix of the action of h_2' on B'' is

$$\begin{pmatrix} 0 & 0 & -1 & 0 \\ 0 & 0 & 0 & -1 \\ 1 & 0 & 0 & 0 \\ 0 & 1 & 0 & 0 \end{pmatrix}.$$

The matrix of action of h_2' on $B' = U \otimes B'' = \{u_0 \otimes B'', u_1 \otimes B''\}$ is

$$\begin{pmatrix} 0 & 0 & -1 & 0 & 1 & 0 & 0 & 0 \\ 0 & 0 & 0 & -1 & 0 & 1 & 0 & 0 \\ 1 & 0 & 0 & 0 & 0 & 0 & 1 & 0 \\ 0 & 1 & 0 & 0 & 0 & 0 & 0 & 1 \\ 1 & 0 & 0 & 0 & 0 & 0 & 1 & 0 \\ 0 & 1 & 0 & 0 & 0 & 0 & 0 & 1 \\ 0 & 0 & 1 & 0 & 1 & 0 & 0 & 0 \\ 0 & 0 & 0 & 1 & 0 & -1 & 0 & 0 \end{pmatrix}.$$

The basis of $H^0(k, k+1)$ consists of elements

$$\begin{aligned} & -r_2 + r_8 = -u_0 \otimes f_3(k-1,0,0) + u_1 \otimes g_3(k-1,0,0), \\ & -r_1 + r_7 = -u_0 \otimes f_2(k-1,0,0) + u_1 \otimes g_1(k-1,0,0), \\ & r_4 + r_6 = u_0 \otimes g_3(k-1,0,0) + u_1 \otimes f_3(k-1,0,0), \\ & r_3 + r_5 = u_0 \otimes g_1(k-1,0,0) + u_1 \otimes f_2(k-1,0,0). \end{aligned}$$

The vectors of right weight $(k+1, k)$ are

$$-r_2 + r_8 = -u_0 \otimes f_3(k-1,0,0) + u_1 \otimes g_3(k-1,0,0)$$

and

$$r_4 + r_6 = u_0 \otimes g_3(k-1,0,0) + u_1 \otimes f_3(k-1,0,0).$$

The vectors

$$-r_1 + r_7 = -u_0 \otimes f_2(k-1,0,0) + u_1 \otimes g_1(k-1,0,0)$$

and

$$r_3 + r_5 = u_0 \otimes g_1(k-1,0,0) + u_1 \otimes f_2(k-1,0,0)$$

have right weight $(k, k+1)$. The character of $H^0(k, k+1)$ is $2x_1^k x_2^{k+1} + 2x_1^{k+1} x_2^k$, and the induced supermodule $H^0(k, k+1)$ is the simple supermodule $L(k, k+1)$.

Case $p|k+1$, $p \nmid k$

In this case, $\mathfrak{u}(\lambda) = Kt_0 + Kt_{(10)}$, where

$$h_1'(t_0) = t_{(10)}, h_1'(t_{(10)}) = -t_0, h_2'(t_0) = 0, h_2'(t_{(10)}) = 0.$$

The matrix of action of h'_1 on $U = \{u_0, u_1\}$ is

$$\begin{pmatrix} 0 & 1 \\ -1 & 0 \end{pmatrix}$$

and the matrix of the action of h'_2 on $U = \{u_0, u_1\}$ is

$$\begin{pmatrix} 0 & 0 \\ 0 & 0 \end{pmatrix}.$$

The matrix of the action of h'_2 on $B_{(-1,0)}$ is

$$\begin{pmatrix} 0 & 0 & 0 & 0 & -1 & -1 & 0 & 0 \\ 0 & 0 & 0 & 0 & 0 & 0 & -1 & 0 \\ 0 & 0 & 0 & 0 & 0 & 0 & 1 & 0 \\ 0 & 0 & 0 & 0 & -1 & -1 & 0 & 0 \\ -1 & 0 & 0 & 1 & 0 & 0 & 0 & 0 \\ 1 & 0 & 0 & -1 & 0 & 0 & 0 & 0 \\ 0 & 0 & 0 & 0 & 0 & 0 & 0 & 0 \\ 0 & 1 & 1 & 0 & 0 & 0 & 0 & 0 \end{pmatrix}.$$

Hence, the basis of elements annihilated by h'_2 is

$$\begin{aligned} B' = \{r_1 &= u_0 \otimes (f_2(k-1,0,0) + f_7(k-1,0,0)), \\ r_2 &= u_0 \otimes (f_3(k-1,0,0) - f_4(k-1,0,0)), \\ r_3 &= u_0 \otimes (g_1(k-1,0,0) - g_2(k-1,0,0)), r_4 = u_0 \otimes g_4(k-2,0,0), \\ r_5 &= u_1 \otimes (f_2(k-1,0,0) + f_7(k-1,0,0)), \\ r_6 &= u_1 \otimes (f_3(k-1,0,0) - f_4(k-1,0,0)), \\ r_7 &= u_1 \otimes (g_1(k-1,0,0) - g_2(k-1,0,0)), r_8 = u_0 \otimes g_4(k-2,0,0)\}. \end{aligned}$$

constructed using elements from

$$\begin{aligned} B'' = \{&f_2(k-1,0,0) + f_7(k-1,0,0), f_3(k-1,0,0) - f_4(k-1,0,0), \\ &g_1(k-1,0,0) - g_2(k-1,0,0), g_4(k-2,0,0)\}. \end{aligned}$$

Since

$$\begin{aligned} &h'_1(f_2(k-1,0,0) + f_7(k-1,0,0)) = g_1(k-1,0,0) - g_2(k-1,0,0), \\ &h'_1(f_3(k-1,0,0) - f_4(k-1,0,0)) = -g_4(k-2,0,0), \\ &h'_1(g_1(k-1,0,0) - g_2(k-1,0,0)) = f_7(k-1,0,0) + f_2(k-1,0,0), \\ &h'_1(g_4(k-2,0,0)) = -f_3(k-1,0,0) + f_4(k-1,0,0), \end{aligned}$$

the corresponding matrix of the action of h'_1 on B'' is

$$\begin{pmatrix} 0 & 0 & 1 & 0 \\ 0 & 0 & 0 & -1 \\ 1 & 0 & 0 & 0 \\ 0 & -1 & 0 & 0 \end{pmatrix}.$$

The matrix of action of h_1' on $U \otimes B''$ is

$$\begin{pmatrix} 0 & 0 & 1 & 0 & 1 & 0 & 0 & 0 \\ 0 & 0 & 0 & -1 & 0 & 1 & 0 & 0 \\ 1 & 0 & 0 & 0 & 0 & 0 & 1 & 0 \\ 0 & -1 & 0 & 0 & 0 & 0 & 0 & 1 \\ -1 & 0 & 0 & 0 & 0 & 0 & -1 & 0 \\ 0 & -1 & 0 & 0 & 0 & 0 & 0 & 1 \\ 0 & 0 & -1 & 0 & -1 & 0 & 0 & 0 \\ 0 & 0 & 0 & -1 & 0 & 1 & 0 & 0 \end{pmatrix}.$$

The basis of $H^0(k, k+1)$ consists of elements

$$\begin{aligned} &\{u_0 \otimes (f_3(k-1,0,0) - f_4(k-1,0,0)) + u_1 \otimes g_4(k-2,0,0), \\ &u_0 \otimes g_4(k-2,0,0) + u_1 \otimes (f_3(k-1,0,0) - f_4(k-1,0,0)), \\ &u_0 \otimes (f_2(k-1,0,0) + f_7(k-1,0,0)) - u_1 \otimes (g_1(k-1,0,0) - g_2(k-1,0,0)), \\ &u_0 \otimes (g_1(k-1,0,0) - g_2(k-1,0,0) - u_1 \otimes (f_2(k-1,0,0) + f_7(k-1,0,0))\}. \end{aligned}$$

The character of $H^0(k, k+1)$ is $2x_1^k x_2^{k+1} + 2x_1^{k+1} x_2^k$, and the induced supermodule $H^0(k, k+1)$ is the simple supermodule $L(k, k+1)$.

Case $p \nmid k, k+1$

Since $p \nmid \lambda_1, p \nmid \lambda_2$, we set $e = \sqrt{\lambda_1}$, $f = \sqrt{\lambda_2}$, $\iota = \sqrt{-1}$,

$$w_0 = \frac{t_0}{f} + e\iota t_{(11)}, w_1 = \frac{et_{(10)}}{f} + \iota t_{(01)}, u_0 = \frac{t_0}{f} - e\iota t_{(11)} \text{ and } u_1 = \frac{et_{(10)}}{f} - \iota t_{(01)}.$$

Then

$$h_1'(w_0) = ew_1, h_1'(w_1) = ew_0, h_2'(w_0) = -f\iota w_1, h_2'(w_1) = f\iota w_0,$$

$$h_1'(u_0) = eu_1, h_1'(u_1) = eu_0, h_2'(u_0) = f\iota u_1 \text{ and } h_2'(u_1) = -f\iota u_0,$$

which shows that $M_1 = \Bbbk u_0 + \Bbbk u_1$ and $M_2 = \Bbbk w_0 + \Bbbk w_1$ are irreducible $\mathbf{H}(2)$-supermodules. The formal characters are equal to $2x^\lambda$. Since both t_0 and $t_{(11)}$ are even, $\mathfrak{u}(\lambda) \simeq M_1$ is absolutely irreducible.

The matrix of action of h_1' on $U = \{u_0, u_1\}$ is

$$\begin{pmatrix} 0 & e \\ e & 0 \end{pmatrix},$$

where $k = e^2$.

The matrix of action of h_2' on $U = \{u_0, u_1\}$ is

$$\begin{pmatrix} 0 & -f\iota \\ f\iota & 0 \end{pmatrix},$$

where $f^2 = k+1$ and $k = -(f\iota)^2 - 1$. Denote $l = f\iota$. Then $k = -l^2 - 1$.

The action of h_1' on $U \otimes B_{(k,k+1)}$ is given by the matrix

$$\begin{pmatrix}
0 & 0 & 0 & 0 & 0 & -1 & 0 & 0 & e & 0 & 0 & 0 & 0 & 0 & 0 & 0\\
0 & 0 & 0 & 0 & 0 & 0 & 0 & -1 & 0 & e & 0 & 0 & 0 & 0 & 0 & 0\\
0 & 0 & 0 & 0 & 0 & 0 & e^2 & -e^2 & 0 & 0 & e & 0 & 0 & 0 & 0 & 0\\
0 & 0 & 0 & 0 & -e^2 & 0 & 0 & 0 & 0 & 0 & 0 & e & 0 & 0 & 0 & 0\\
0 & 0 & 0 & 1 & 0 & 0 & 0 & 0 & 0 & 0 & 0 & 0 & e & 0 & 0 & 0\\
e^2 & 0 & 0 & 0 & 0 & 0 & 0 & 0 & 0 & 0 & 0 & 0 & 0 & e & 0 & 0\\
0 & e^2 & -1 & 0 & 0 & 0 & 0 & 0 & 0 & 0 & 0 & 0 & 0 & 0 & e & 0\\
0 & e^2 & 0 & 0 & 0 & 0 & 0 & 0 & 0 & 0 & 0 & 0 & 0 & 0 & 0 & e\\
e & 0 & 0 & 0 & 0 & 0 & 0 & 0 & 0 & 0 & 0 & 0 & 0 & 1 & 0 & 0\\
0 & e & 0 & 0 & 0 & 0 & 0 & 0 & 0 & 0 & 0 & 0 & 0 & 0 & 0 & 1\\
0 & 0 & e & 0 & 0 & 0 & 0 & 0 & 0 & 0 & 0 & 0 & 0 & 0 & -e^2 & e^2\\
0 & 0 & 0 & e & 0 & 0 & 0 & 0 & 0 & 0 & 0 & 0 & e^2 & 0 & 0 & 0\\
0 & 0 & 0 & 0 & e & 0 & 0 & 0 & 0 & 0 & 0 & -1 & 0 & 0 & 0 & 0\\
0 & 0 & 0 & 0 & 0 & e & 0 & 0 & -e^2 & 0 & 0 & 0 & 0 & 0 & 0 & 0\\
0 & 0 & 0 & 0 & 0 & 0 & e & 0 & 0 & -e^2 & 1 & 0 & 0 & 0 & 0 & 0\\
0 & 0 & 0 & 0 & 0 & 0 & 0 & e & 0 & -e^2 & 0 & 0 & 0 & 0 & 0 & 0
\end{pmatrix}.$$

Hence, the basis of elements annihilated by h_1' is

$$\begin{aligned}
B' = \{r_1 &= eu_0 \otimes g_2(k-1,0,0) + u_1 \otimes f_2(k-1,0,0),\\
r_2 &= -u_0 \otimes f_2(k-1,0,0) + eu_1 \otimes g_2(k-1,0,0),\\
r_3 &= -u_0 \otimes g_3(k-1,0,0) + eu_1 \otimes f_4(k-1,0,0),\\
r_4 &= eu_0 \otimes f_4(k-1,0,0)) + u_1 \otimes g_3(k-1,0,0),\\
r_5 &= u_0 \otimes g_1(k-1,0,0) + eu_1 \otimes f_7(k-1,0,0),\\
r_6 &= -eu_0 \otimes f_7(k-1,0,0) + u_1 \otimes g_1(k-1,0,0),\\
r_7 &= eu_0 \otimes (g_3(k-1,0,0) + g_4(k-2,0,0)) + u_1 \otimes f_3(k-1,0,0),\\
r_8 &= -u_0 \otimes f_3(k-1,0,0) + eu_1 \otimes (g_3(k-1,0,0) + g_4(k-2,0,0))\}.
\end{aligned}$$

We have

$$\begin{aligned}
h_2'(r_1) =& elu_1 \otimes g_2(k-1,0,0) - eu_0 \otimes f_2(k-1,0,0) + e(2k+1)u_0 \otimes f_7(k-1,0,0)\\
&- lu_0 \otimes f_2(k-1,0,0) - (2k+1)u_1 \otimes g_1(k-1,0,0) + ku_1 \otimes g_2(k-1,0,0)\\
=& u_0 \otimes [(-e-l)f_2(k-1,0,0) + (2k+1)ef_7(k-1,0,0)]\\
&+ u_1 \otimes [-(2k+1)g_1(k-1,0,0) + (k+el)g_2(k-1,0,0)];\\
h_2'(r_2) =& -lu_1 \otimes f_2(k-1,0,0) - (2k+1)u_0 \otimes g_1(k-1,0,0) + ku_0 \otimes g_2(k-1,0,0)\\
&- elu_0 \otimes g_2(k-1,0,0) - u_1 \otimes f_2(k-1,0,0) - (2k+1)u_1 \otimes f_7(k-1,0,0)\\
=& u_0 \otimes [(2k+1)g_1(k-1,0,0) + (k-el)eg_2(k-1,0,0)]\\
&+ u_1 \otimes [(e-l)f_2(k-1,0,0) - (2k+1)ef_7(k-1,0,0)];
\end{aligned}$$

$$\begin{aligned}
h_2'(r_3) =& -lu_1 \otimes g_3(k-1,0,0) + u_0 \otimes f_3(k-1,0,0) + ku_0 \otimes f_4(k-1,0,0) \\
& - elu_0 \otimes f_4(k-1,0,0) - eu_1 \otimes g_4(k-2,0,0) \\
=& u_0 \otimes [f_3(k-1,0,0) + (k-el)f_4(k-1,0,0)] \\
& + u_1 \otimes [-lg_3(k-1,0,0) - eg_4(k-2,0,0)]; \\
h_2'(r_4) =& elu_1 \otimes f_4(k-1,0,0) + eu_0 \otimes g_4(k-2,0,0) \\
& - lu_0 \otimes g_3(k-1,0,0) + u_1 \otimes f_3(k-1,0,0) + ku_1 \otimes f_4(k-1,0,0) \\
=& u_0 \otimes [-lg_3(k-1,0,0) + eg_4(k-1,0,0)] \\
& + u_1 \otimes [f_3(k-1,0,0) + (k+el)f_4(k-1,0,0)]; \\
h_2'(r_5) =& lu_1 \otimes g_1(k-1,0,0) - u_0 \otimes f_2(k-1,0,0) + ku_0 \otimes f_7(k-1,0,0) \\
& - elu_0 \otimes f_7(k-1,0,0) - eu_1 \otimes g_1(k-1,0,0) + eu_1 \otimes g_2(k-1,0,0) \\
=& u_0 \otimes [-f_2(k-1,0,0) + (k-el)f_7(k-1,0,0)] \\
& + u_1 \otimes [(l-e)g_1(k-1,0,0) + eg_2(k-1,0,0)]; \\
h_2'(r_6) =& -elu_1 \otimes f_7(k-1,0,0) - eu_0 \otimes g_1(k-1,0,0) + eu_0 \otimes g_2(k-1,0,0) \\
& - lu_0 \otimes g_1(k-1,0,0) + u_1 \otimes f_2(k-1,0,0) - ku_1 \otimes f_7(k-1,0,0) \\
=& u_0 \otimes [(-e-l)g_1(k-1,0,0) + eg_2(k-1,0,0)] \\
& + u_1 \otimes [f_2(k-1,0,0) + (-k-le)f_7(k-1,0,0)];
\end{aligned}$$

$$\begin{aligned}
h_2'(r_7) =& elu_1 \otimes g_3(k-1,0,0) + elu_1 \otimes g_4(k-2,0,0) - eu_0 \otimes f_3(k-1,0,0) \\
& - (2k+1)u_0 \otimes f_4(k-1,0,0) - lu_0 \otimes f_3(k-1,0,0) - (k+1)u_1 \otimes g_3(k-1,0,0) \\
& + ku_1 \otimes g_4(k-2,0,0) = u_0 \otimes [(-e-l)f_3(k-1,0,0) + (2k+1)f_4(k-1,0,0)] \\
& + u_1 \otimes [(-e-k-1)g_3(k-1,0,0) + (k+el)g_4(k-2,0,0)]; \\
h_2'(r_8) =& -lu_1 \otimes f_3(k-1,0,0) - (k+1)u_0 \otimes g_3(k-1,0,0) + ku_0 \otimes g_4(k-2,0,0) \\
& - elu_0 \otimes g_3(k-1,0,0) - elu_0 \otimes g_4(k-2,0,0) + eu_1 \otimes f_3(k-1,0,0) \\
& + (2k+1)u_1 \otimes f_4(k-1,0,0) \\
=& u_0 \otimes [(-el-k-1)g_3(k-1,0,0) + (k-el)g_4(k-2,0,0)] \\
& + u_1 \otimes [(-l+e)f_3(k-1,0,0) + e(2k+1)f_4(k-1,0,0)].
\end{aligned}$$

To get a vector annihilated by h_2', we need to consider linear combinations of $\{r_1, r_5\}, \{r_2, r_6\}, \{r-3, r_7\}$ and $\{r_4, r_8\}$ separately.

The solutions are $r_1 + (e+l)r_5$, $r_2 + (l-e)r_6$, $(e+l)r_3 + r_7$ and $(l-e)r_4 + r_8$. Thus, the character of $H^0(k, k+1)$ is $2x_1^k x_2^{k+1} + 2x_1^{k+1} x_2^k$ and $H^0(k, k+1)$ is simple $L(k, k+1)$.

3. $S_{k,k+q}$ **for** $q \geq 2$.

The weightspace of right weight $(k, k+q)$ for $q \geq 2$ has a basis

$$S_{k,k+q} = \{f_2(k-1,0,q-1), f_7(k-1,0,q-1), g_1(k-1,0,q-1), g_2(k-1,0,q-1)\}.$$

The matrix of h_1' on this set is

$$\begin{pmatrix} 0 & 0 & 0 & -1 \\ 0 & 0 & -k & 0 \\ 0 & 1 & 0 & 0 \\ k & 0 & 0 & 0 \end{pmatrix}.$$

The matrix of h_2' on this set is

$$\begin{pmatrix} 0 & 0 & -1 & -1 \\ 0 & 0 & k & 2k+q \\ 2k+q & 1 & 0 & 0 \\ -k & -1 & 0 & 0 \end{pmatrix}.$$

Case $p|k$ and $p|k+q$

The basis of elements annihilated by h_1' is $\{f_2(k-1,0,q-1), g_1(k-1,0,q-1)\}$. Out of this space, it is $f_2(k-1,0,q-1)$ that is annihilated by h_2'.

Case $p|k$ and $p\nmid k+q$

The action of h_1' on U is given by

$$\begin{pmatrix} 0 & 0 \\ 0 & 0 \end{pmatrix},$$

and the action of h_2' on U is given by

$$\begin{pmatrix} 0 & 1 \\ q & 0 \end{pmatrix}.$$

The basis of elements annihilated by h_1' is

$$\begin{aligned} B' = \{&u_0 \otimes f_2(k-1,0,q-1), u_0 \otimes g_1(k-1,0,q-1), \\ &u_1 \otimes f_2(k-1,0,q-1), u_1 \otimes g_1(k-1,0,q-1)\}. \end{aligned}$$

The matrix of action of h_2' on B' is

$$\begin{pmatrix} 0 & -1 & 1 & 0 \\ q & 0 & 0 & 1 \\ q & 0 & 0 & 1 \\ 0 & q & -q & 0 \end{pmatrix}.$$

Therefore, we obtain two elements

$$\begin{aligned} r_1 =& u_0 \otimes g_1(k-1,0,q-1) + u_1 \otimes f_2(k-1,0,q-1); \\ r_2 =& u_0 \otimes f_2(k-1,0,q-1) - q u_1 \otimes g_1(k-1,0,q-1). \end{aligned}$$

Case $p\nmid k$ and $p|k+q$

The action of h_2' on U is given by

$$\begin{pmatrix} 0 & 0 \\ 0 & 0 \end{pmatrix}.$$

The action of h_1' on U is given by

$$\begin{pmatrix} 0 & 1 \\ k & 0 \end{pmatrix}.$$

The basis of elements annihilated by h_2' is

$$\begin{aligned} B' = \{&u_0 \otimes (f_2(k-1,0,q-1) - k f_7(k-1,0,q-1)), \\ &u_0 \otimes (g_1(k-1,0,q-1) - g_2(k-1,0,q-1)), \\ &u_1 \otimes (f_2(k-1,0,q-1) - k f_7(k-1,0,q-1)), \\ &u_1 \otimes (g_1(k-1,0,q-1) - g_2(k-1,0,q-1))\}. \end{aligned}$$

From the matrix of action of h_1' on B'

$$\begin{pmatrix} 0 & 1 & 1 & 0 \\ -k & 0 & 0 & 1 \\ k & 0 & 0 & -1 \\ 0 & k & k & 0 \end{pmatrix},$$

we obtain two elements

$$r_1 = u_0 \otimes (f_2(k-1,0,q-1) - kf_7(k-1,0,q-1)) \\ + ku_1 \otimes (g_1(k-1,0,q-1) - g_2(k-1,0,q-1));$$

$$r_2 = u_0 \otimes (g_1(k-1,0,q-1) - g_2(k-1,0,q-1)) \\ - u_1 \otimes (f_2(k-1,0,q-1) - kf_7(k-1,0,q-1)).$$

Case $p \nmid k$, $p \nmid k+q$.
The action of h_1' on U is given by

$$\begin{pmatrix} 0 & e \\ e & 0 \end{pmatrix},$$

where $k = e^2$. The action of h_2' on U is given by

$$\begin{pmatrix} 0 & -l \\ l & 0 \end{pmatrix},$$

where $l^2 = -k - q$.
The matrix of h_1' on $U \otimes S_{k,k+q}$ is

$$\begin{pmatrix} 0 & 0 & 0 & -1 & e & 0 & 0 & 0 \\ 0 & 0 & -k & 0 & 0 & e & 0 & 0 \\ 0 & 1 & 0 & 0 & 0 & 0 & e & 0 \\ k & 0 & 0 & 0 & 0 & 0 & 0 & e \\ e & 0 & 0 & 0 & 0 & 0 & 0 & 1 \\ 0 & e & 0 & 0 & 0 & 0 & k & 0 \\ 0 & 0 & e & 0 & 0 & -1 & 0 & 0 \\ 0 & 0 & 0 & e & -k & 0 & 0 & 0 \end{pmatrix}.$$

The basis of elements annihilated by h_1' consists of

$$B' = \begin{matrix} \{r_1 = u_0 \otimes f_2(k-1,0,q-1) - eu_1 \otimes g_2(k-1,0,q-1); \\ r_2 = eu_0 \otimes f_7(k-1,0,q-1) - u_1 \otimes g_1(k-1,0,q-1); \\ r_3 = u_0 \otimes g_1(k-1,0,q-1) + eu_1 \otimes f_7(k-1,0,q-1); \\ r_4 = eu_0 \otimes g_2(k-1,0,q-1) + u_1 \otimes f_2(k-1,0,q-1)\}. \end{matrix}$$

We have

$$\begin{aligned}
h_2'(r_1) =&lu_1 \otimes f_2(k-1,0,q-1) + (2k+q)u_0 \otimes g_1(k-1,0,q-1)\\
&- ku_0 \otimes g_2(k-1,0,q-1) + elu_0 \otimes g_2(k-1,0,q-1)\\
&- eu_1 \otimes f_2(k-1,0,q-1) + e(2k+q)u_1 \otimes f_7(k-1,0,q-1)\\
&= (2k+q)u_0 \otimes g_1(k-1,0,q-1) + (el-k)u_0 \otimes g_2(k-1,0,q-1)\\
&+ (-e+l)u_1 \otimes f_2(k-1,0,q-1) + e(2k+q)u_1 \otimes f_7(k-1,0,q-1);\\
h_2'(r_2) =&elu_1 \otimes f_7(k-1,0,q-1) + eu_0 \otimes g_1(k-1,0,q-1)\\
&- eu_0 \otimes g_2(k-1,0,q-1) + lu_0 \otimes g_1(k-1,0,q-1)\\
&- u_1 \otimes f_2(k-1,0,q-1) + ku_1 \otimes f_7(k-1,0,q-1)\\
&= (e+l)u_0 \otimes g_1(k-1,0,q-1) - eu_0 \otimes g_2(k-1,0,q-1)\\
&- u_1 \otimes f_2(k-1,0,q-1) + (k+el)u_1 \otimes f_7(k-1,0,q-1);\\
h_2'(r_3) =&lu_1 \otimes g_1(k-1,0,q-1) - u_0 \otimes f_2(k-a,0,q-1)\\
&+ ku_0 \otimes f_7(k-1,0,q-1) - elu_0 \otimes f_7(k-1,0,q-1)\\
&- eu_1 \otimes g_1(k-1,0,q-1) + eu_1 \otimes g_2(k-1,0,q-1)\\
&= -u_0 \otimes f_2(k-1,0,q-1) + (k-el)u_0 \otimes f_7(k-1,0,q-1)\\
&+ (l-e)u_1 \otimes g_1(k-1,0,q-1) + eu_1 \otimes g_2(k-1,0,q-1);\\
h_2'(r_4) =&elu_1 \otimes g_2(k-1,0,q-1) - eu_0 \otimes f_2(k-1,0,q-1)\\
&+ e(2k+q)u_0 \otimes f_7(k-1,0,q-1) - lu_0 \otimes f_2(k-1,0,q-1)\\
&- (2k+q)u_1 \otimes g_1(k-1,0,q-1) + ku_1 \otimes g_2(k-1,0,q-1)\\
&= (-e-l)u_0 \otimes f_2(k-1,0,q-1) + e(2k+q)u_0 \otimes f_7(k-1,0,q-1)\\
&- (2k+q)u_1 \otimes g_1(k-1,0,q-1) + (k+el)u_1 \otimes g_2(k-1,0,q-1).
\end{aligned}$$

Using identities

$$\begin{aligned}
&(e+l)(-e+l) = -e^2 + l^2 = -k-k-q = -2k-q;\\
&-e(-e+l) = e^2 - el = k - el;\\
&(k+el)(-e+l) = -ke - e^2l + kl + el^2 = -ke - kl + kl - ek - eq = -2ke - eq
\end{aligned}$$

and

$$\begin{aligned}
&(e+l)(k-el) = ek + kl - e^2l - el^2 = ek + kl - kl + ek + eq = 2ek + eq;\\
&(l-e)(e+l) = l^2 - e^2 = -k-q-k = -2k-q;\\
&e(e+l) = e^2 + el = k + el
\end{aligned}$$

we obtain that the elements annihilated by h_2' are

$$r_1 = r_1 + (-e+l)r_2; (e+l)r_3 - r_4.$$

4. $S_{k+q,k}$ **for** $q \geq 2$.

The weightspace of right weight $(k+q,k)$ for $q \geq 2$ has a basis

$$S_{k+q,k} = \{f_3(k-1,q-1,0), f_4(k-1,q-1,0), g_3(k-1,q-1,0), g_5(k-2,q-2,0))\}.$$

The matrix of h_1' on this set is

$$\begin{pmatrix} 0 & 0 & 0 & -1 \\ 0 & 0 & k & -k \\ k & -1 & 0 & 0 \\ k & 0 & 0 & 0 \end{pmatrix}.$$

The matrix of h_2' on this set is

$$\begin{pmatrix} 0 & 0 & -1 & 0 \\ 0 & 0 & -k & -k-q \\ k+q & 0 & 0 & 0 \\ -k & 1 & 0 & 0 \end{pmatrix}.$$

Case $p|k$ and $p|k+q$

The basis of elements annihilated by h_1' is $\{f_3(k-1,q-1,0), g_3(k-1,q-1,0)\}$. Out of this space, it is $f_3(k-1,q-1,0)$ that is annihilated by h_2'.

Case $p|k$ and $p\nmid k+q$

The action of h_1' on U is given by

$$\begin{pmatrix} 0 & 0 \\ 0 & 0 \end{pmatrix},$$

and the action of h_2' on U is given by

$$\begin{pmatrix} 0 & 1 \\ q & 0 \end{pmatrix}.$$

The basis of elements annihilated by h_1' is

$$B' = \{u_0 \otimes f_3(k-1,q-1,0), u_0 \otimes g_3(k-1,q-1,0),$$
$$u_1 \otimes f_3(k-1,q-1,0), u_1 \otimes g_3(k-1,q-1,0)\}.$$

The matrix of action of h_2' on B' is

$$\begin{pmatrix} 0 & -1 & 1 & 0 \\ q & 0 & 0 & 1 \\ q & 0 & 0 & 1 \\ 0 & q & -q & 0 \end{pmatrix}.$$

Therefore, we obtain two elements

$$r_1 = u_0 \otimes f_3(k-1,0,q-1) - qu_1 \otimes g_3(k-1,0,q-1);$$
$$r_2 = u_0 \otimes g_3(k-1,0,q-1) + qu_1 \otimes f_3(k-1,0,q-1).$$

Case $p\nmid k$ and $p|k+q$

The action of h_2' on U is given by

$$\begin{pmatrix} 0 & 0 \\ 0 & 0 \end{pmatrix},$$

and the action of h_1' on U is given by

$$\begin{pmatrix} 0 & 1 \\ k & 0 \end{pmatrix}.$$

The basis of elements annihilated by h_2' is

$$B' = \{u_0 \otimes (f_3(k-1,q-1,0) + kf_4(k-1,q-1,0)), u_0 \otimes g_5(k-2,q-2,0),$$
$$u_1 \otimes (f_3(k-1,q-1,0) + kf_4(k-1,q-1,0)), u_1 \otimes g_5(k-2,q-2,0)\}.$$

The matrix of action of h_1' on B' is

$$\begin{pmatrix} 0 & -1 & 1 & 0 \\ k & 0 & 0 & 1 \\ k & 0 & 0 & 1 \\ 0 & k & -k & 0 \end{pmatrix}.$$

Therefore, we obtain two elements

$$r_1 = u_0 \otimes (f_2(k-1,0,q-1) - kf_7(k-1,0,q-1)) - ku_1 \otimes g_5(k-2,q-2,0);$$

$$r_2 = u_0 \otimes g_5(k-2,q-2,0) + u_1 \otimes (f_3(k-1,q-1,0) + kf_4(k-1,q-1,0)).$$

Case $p \nmid k$, $p \nmid k+q$.
The action of h_1' on U is given by

$$\begin{pmatrix} 0 & e \\ e & 0 \end{pmatrix},$$

where $k = e^2$. The action of h_2' on U is given by

$$\begin{pmatrix} 0 & -l \\ l & 0 \end{pmatrix},$$

where $l^2 = -k-q$.

The matrix of h_1' on $U \otimes S_{k,k+q}$ is

$$\begin{pmatrix} 0 & 0 & 0 & -1 & e & 0 & 0 & 0 \\ 0 & 0 & k & -k & 0 & e & 0 & 0 \\ k & -1 & 0 & 0 & 0 & 0 & e & 0 \\ k & 0 & 0 & 0 & 0 & 0 & 0 & e \\ e & 0 & 0 & 0 & 0 & 0 & 0 & 1 \\ 0 & e & 0 & 0 & 0 & 0 & -k & k \\ 0 & 0 & e & 0 & -k & 1 & 0 & 0 \\ 0 & 0 & 0 & e & -k & 0 & 0 & 0 \end{pmatrix}.$$

The basis B' of elements annihilated by h_1' consists of

$$\begin{aligned} \{r_1 &= u_0 \otimes f_3(k-1,q-1,0) + e^2 u_0 \otimes f_4(k-1,q-1,0) - eu_1 \otimes g_5(k-2,q-1,0); \\ r_2 &= eu_0 \otimes f_4(k-1,q-1,0) + u_1 \otimes g_3(k-1,q-1,0); \\ r_3 &= u_0 \otimes g_3(k-1,q-1,0) - eu_1 \otimes f_4(k-1,q-1,0); \\ r_4 &= eu_0 \otimes g_3(k-1,q-1,0) + eu_0 \otimes g_5(k-2,q-1,0) + u_1 \otimes f_3(k-1,q-1,0)\}. \end{aligned}$$

We have

$$
\begin{aligned}
h_2'(r_1) =& lu_1 \otimes f_3(k-1,q-1,0) + (k+q)u_0 \otimes g_3(k-1,q-1,0) \\
& - ku_0 \otimes g_5(k-2,q-2,0) + e^2 lu_1 \otimes f_4(k-1,q-1,0) \\
& + e^2 u_0 \otimes g_5(k-2,q-2,0) + elu_0 \otimes g_5(k-2,q-2,0) \\
& - e(k+q)u_1 \otimes f_4(k-1,q-1,0) \\
& = (k+q)u_0 \otimes g_3(k-1,q-1,0) + elu_0 \otimes g_5(k-2,q-2,0) \\
& + lu_1 \otimes f_3(k-1,q-1,0) + (-ek-eq+e^2 l)u_1 \otimes f_4(k-1,q-1,0); \\
h_2'(r_2) =& elu_1 \otimes f_4(k-1,q-1,0) + eu_0 \otimes g_5(k-2,q-2,0) \\
& - lu_0 \otimes g_3(k-1,q-1,0) + u_1 \otimes f_3(k-1,q-1,0) \\
& + ku_1 \otimes f_4(k-1,q-1,0) \\
& = -lu_0 \otimes g_3(k-1,q-1,0) + eu_0 \otimes g_5(k-2,q-2,0) \\
& + u_1 \otimes f_3(k-1,q-1,0) + (k+el)u_1 \otimes f_4(k-1,q-1,0); \\
h_2'(r_3) =& lu_1 \otimes g_3(k-1,q-1,0) - u_0 \otimes f_3(k-1,q-1,0) \\
& - ku_0 \otimes f_4(k-1,q-1,0) + elu_0 \otimes f_4(k-1,q-1,0) \\
& + eu_1 \otimes g_5(k-2,q-2,0) \\
& = -u_0 \otimes f_3(k-1,q-1,0) + (-k+el)u_0 \otimes f_4(k-1,q-1,0) \\
& + lu_1 \otimes g_3(k-1,q-1,0) + eu_1 \otimes g_5(k-2,q-2,0); \\
h_2'(r_4) =& elu_1 \otimes g_3(k-1,q-1,0) - eu_0 \otimes f_3(k-1,q-1,0) \\
& - eku_0 \otimes f_4(k-1,q-1,0) + elu_1 \otimes g_5(k-2,q-1,0) \\
& - (k+q)eu_0 \otimes f_4(k-1,q-1,0) - lu_0 \otimes f_3(k-1,q-1,0) \\
& - (k+q)u_1 \otimes g_3(k-1,q-1,0) + ku_1 \otimes g_5(k-2,q-2,0) \\
& = (-e-l)u_0 \otimes f_3(k-1,q-1,0) - e(2k+q)u_0 \otimes f_4(k-1,q-1,0) \\
& (-k-q+el)u_1 \otimes g_3(k-1,q-1,0) + (k+el)u_1 \otimes g_5(k-2,q-2,0).
\end{aligned}
$$

Using identities

$$
\begin{aligned}
&(-l)(-l) = l^2 = -k-q; \\
&(-l)e = -el; \\
&(-l)(k+el) = -kl - el^2 = -kl + ek + eq = -e^2 l + ek + eq
\end{aligned}
$$

and

$$
\begin{aligned}
&(e+l)(-k+el) = -ke + kl - kl + el^2 = -2ke - eq; \\
&(l+e)l = l^2 + el = -k - q + el; \\
&(e+l)e = k + el
\end{aligned}
$$

we obtain that the elements annihilated by h_2' are

$$r_1 = r_1 - lr_2; (e+l)r_3 - r_4.$$

Remark 13.1. The matrix of h_2' on $U \otimes S_{k,k+q}$ is

$$\begin{pmatrix} 0 & 0 & -1 & 0 & -l & 0 & 0 & 0 \\ 0 & 0 & -k & -k-q & 0 & -l & 0 & 0 \\ k+q & 0 & 0 & 0 & 0 & 0 & -l & 0 \\ -k & 1 & 0 & 0 & 0 & 0 & 0 & -l \\ l & 0 & 0 & 0 & 0 & 0 & 1 & 0 \\ 0 & l & 0 & 0 & 0 & 0 & k & k+q \\ 0 & 0 & l & 0 & -k-q & 0 & 0 & 0 \\ 0 & 0 & 0 & l & k & -1 & 0 & 0 \end{pmatrix}.$$

The basis of elements annihilated by h_1' consists of

$$B' = \begin{array}{l} \{r_1 = u_0 \otimes f_3(k-1, q-1, 0) - lu_1 \otimes g_3(k-1, q-1, 0); \\ r_2 = lu_0 \otimes f_4(k-1, q-1, 0) + u_1 \otimes g_5(k-2, q-2, 0); \\ r_3 = lu_0 \otimes g_3(k-1, q-1, 0) - u_1 \otimes f_3(k-1, q-1, 0); \\ r_4 = u_0 \otimes g_5(k-2, q-1, 0) + lu_1 \otimes f_4(k-1, q-1, 0)\}. \end{array}$$

5. $S_{(k+1,k+q-1)}$ **for** $q \geq 2$.

The weightspace of right weight $(k+1, k+q-1)$ for $q \geq 2$ has a basis $S_{k+1,k+q-1} =$

$$\begin{array}{l} \{f_3(k-1, 0, q-1), f_4(k-1, 0, q-1), f_5(k-1, 0, q-2), f_6(k, 0, q-2), \\ g_6(k-1, 0, q-2), g_7(k-1, 0, q-2), g_8(k-1, 0, q-2), g_9(k-2, 0, q-2)\}. \end{array}$$

We compute

$$\begin{aligned} h_1'(f_3(a,0,c)) =& (a+1)g_9(a-1,0,c-1) + (a+1)g_6(a,0,c-1) \\ & + (a+1)(2a+2)g_8(a,0,c-1), \\ h_1'(f_4(a,b,c)) =& - g_6(a,b,c-1) - (a+1)g_8(a,b,c-1), \\ h_1'(f_5(a,b,c)) =& g_7(a,b,c) + (a+1)g_8(a,b,c), \\ h_1'(f_6(a,b,c)) =& ag_8(a-1,b,c), \\ h_1'(g_6(a,b,c)) =& (a+1)f_4(a,b,c+1) + (a+1)f_6(a+1,b,c), \\ h_1'(g_7(a,b,c)) =& - (a+1)f_5(a,b,c) + (a+1)f_6(a+1,b,c), \\ h_1'(g_8(a,b,c)) =& - f_6(a+1,b,c), \\ h_1'(g_9(a,0,c)) =& - f_3(a+1,0,c+1) - (a+2)f_4(a+1,0,c+1) \\ & + (a+2)f_6(a+2,0,c). \end{aligned}$$

The matrix of h_1' on $S_{k+1,k+q-1}$ is

$$\begin{pmatrix} 0 & 0 & 0 & 0 & 0 & 0 & 0 & -1 \\ 0 & 0 & 0 & 0 & k & 0 & 0 & -k \\ 0 & 0 & 0 & 0 & 0 & -k & 0 & 0 \\ 0 & 0 & 0 & 0 & k & k & -1 & k \\ k & -1 & 0 & 0 & 0 & 0 & 0 & 0 \\ 0 & 0 & 1 & 0 & 0 & 0 & 0 & 0 \\ 2k^2 & -k & k & k & 0 & 0 & 0 & 0 \\ k & 0 & 0 & 0 & 0 & 0 & 0 & 0 \end{pmatrix}.$$

If $c > 0$, then

$$\begin{aligned} h_2'(f_3(a,0,c) =& -(a+1)g_9(a-1,0,c-1) + (a+2)g_6(a,0,c-1) \\ &+ cg_7(a,0,c-1) + (a+1)g_8(a,0,c-1). \end{aligned}$$

If $c > 0$, then

$$h_2'(f_4(a,0,c)) = g_9(a-1,0,c-1) + (c+a+1)g_8(a,0,c-1),$$

$$\begin{aligned} h_2'(f_5(a,0,c)) =& -g_9(a-1,0,c) + g_8(a,0,c,) \\ h_2'(f_6(a,b,c)) =& g_6(a-1,b,c) - g_7(a-1,b,c), \\ h_2'(g_6(a,b,c)) =& [-f_3(a,b,c+1) + (a+1)f_5(a,b,c)] - (c+1)f_6(a+1,b,c), \\ h_2'(g_7(a,b,c)) =& [-f_3(a,b,c+1) + (a+1)f_5(a,b,c)] + (a+b+2)f_6(a+1,b,c), \\ h_2'(g_8(a,b,c)) =& -f_4(a,b,c+1) - f_5(a,b,c), \\ h_2'(g_9(a,0,c)) =& (a+c+3)f_5(a+1,0,c) - f_4(a+1,0,c+1). \end{aligned}$$

The matrix of h_2' on $S_{k+1,k+q-1}$ is

$$\begin{pmatrix} 0 & 0 & 0 & 0 & -1 & -1 & 0 & 0 \\ 0 & 0 & 0 & 0 & 0 & 0 & -1 & -1 \\ 0 & 0 & 0 & 0 & k & k & -1 & k+q-1 \\ 0 & 0 & 0 & 0 & -q+1 & k+1 & 0 & 0 \\ k+1 & 0 & 0 & 1 & 0 & 0 & 0 & 0 \\ q-1 & 0 & 0 & -1 & 0 & 0 & 0 & 0 \\ k & k+q-1 & 1 & 0 & 0 & 0 & 0 & 0 \\ -k & 1 & -1 & 0 & 0 & 0 & 0 & 0 \end{pmatrix}.$$

<u>Case $p|k$ and $p|k+q$</u>

The basis of elements in the span of $S_{k+1,k+q-1}$ annihilated by h_1' is

$\{f_3(k-1,0,q-1), f_6(k,0,q-2), g_6(k-1,0,q-2), g_7(k-1,0,q-2)\}$.

In this space, there are two basis elements

$$f_3(k-1,0,q-1) - f_6(k,0,q-2), g_6(k-1,0,q-2) - g_7(k-1,0,q-2)$$

of the space annihilated by h_2'.

In $U \otimes S_{k+1,k+q-1}$, we have elements

$$\begin{aligned} &\{u_0 \otimes [f_3(k-1,0,q-1) - f_6(k,0,q-2)], \\ &u_0 \otimes [g_6(k-1,0,q-2) - g_7(k-1,0,q-2)], \\ &u_1 \otimes [f_3(k-1,0,q-1) - f_6(k,0,q-2)], \\ &u_1 \otimes [g_6(k-1,0,q-2) - g_7(k-1,0,q-2)]\} \end{aligned}$$

that form a basis of the solutions.

<u>Case $p|k$ and $p \nmid k+q$</u>

The action of h_1' on U is given by

$$\begin{pmatrix} 0 & 0 \\ 0 & 0 \end{pmatrix},$$

and the action of h_2' on U is given by

$$\begin{pmatrix} 0 & 1 \\ q & 0 \end{pmatrix}.$$

The basis of elements in $U \otimes S_{k+1,k+q-1}$ annihilated by h_1' is

$$\begin{aligned} B' = \{&u_0 \otimes f_3(k-1,0,q-1), u_0 \otimes f_6(k,0,q-2),\\ &u_0 \otimes g_6(k-1,0,q-2), u_0 \otimes g_7(k-1,0,q-2),\\ &u_1 \otimes f_3(k-1,0,q-1), u_1 \otimes f_6(k,0,q-2),\\ &u_1 \otimes g_6(k-1,0,q-2), u_1 \otimes g_7(k-1,0,q-2)\}.\end{aligned}$$

The matrix of action of h_2' on $S_{k+1,k+q-1}$ is

$$\begin{pmatrix}
0 & 0 & 0 & 0 & -1 & -1 & 0 & 0 & 1 & 0 & 0 & 0 & 0 & 0 & 0 & 0\\
0 & 0 & 0 & 0 & 0 & 0 & -1 & -1 & 0 & 1 & 0 & 0 & 0 & 0 & 0 & 0\\
0 & 0 & 0 & 0 & 0 & 0 & -1 & q-1 & 0 & 0 & 1 & 0 & 0 & 0 & 0 & 0\\
0 & 0 & 0 & 0 & -q+1 & 1 & 0 & 0 & 0 & 0 & 0 & 1 & 0 & 0 & 0 & 0\\
1 & 0 & 0 & 1 & 0 & 0 & 0 & 0 & 0 & 0 & 0 & 0 & 1 & 0 & 0 & 0\\
q-1 & 0 & 0 & -1 & 0 & 0 & 0 & 0 & 0 & 0 & 0 & 0 & 0 & 1 & 0 & 0\\
0 & q-1 & 1 & 0 & 0 & 0 & 0 & 0 & 0 & 0 & 0 & 0 & 0 & 0 & 1 & 0\\
0 & 1 & -1 & 0 & 0 & 0 & 0 & 0 & 0 & 0 & 0 & 0 & 0 & 0 & 0 & 1\\
q & 0 & 0 & 0 & 0 & 0 & 0 & 0 & 0 & 0 & 0 & 0 & 1 & 1 & 0 & 0\\
0 & q & 0 & 0 & 0 & 0 & 0 & 0 & 0 & 0 & 0 & 0 & 0 & 0 & 1 & 1\\
0 & 0 & q & 0 & 0 & 0 & 0 & 0 & 0 & 0 & 0 & 0 & 0 & 0 & 1 & -q+1\\
0 & 0 & 0 & q & 0 & 0 & 0 & 0 & 0 & 0 & 0 & 0 & q-1 & -1 & 0 & 0\\
0 & 0 & 0 & 0 & q & 0 & 0 & 0 & -1 & 0 & 0 & -1 & 0 & 0 & 0 & 0\\
0 & 0 & 0 & 0 & 0 & q & 0 & 0 & -q+1 & 0 & 0 & 1 & 0 & 0 & 0 & 0\\
0 & 0 & 0 & 0 & 0 & 0 & q & 0 & 0 & -q+1 & -1 & 0 & 0 & 0 & 0 & 0\\
0 & 0 & 0 & 0 & 0 & 0 & 0 & q & 0 & -1 & 1 & 0 & 0 & 0 & 0 & 0
\end{pmatrix}.$$

The basis of elements in $U \otimes S_{k+1,k+q-1}$ annihilated by h_2' consists of

$r_1 = -u_0 \otimes g_6(k-1,0,q-2) + (1-q)u_0 \otimes g_7(k-1,0,q-2) - qu_1 \otimes f_3(k-1,0,q-1)$,

$r_2 = (1-q)u_0 \otimes g_8(k-1,0,q-2) - u_0 \otimes g_9(k-2,0,q-2) - qu_1 \otimes f_4(k-1,0,q-1)$,

$r_3 = -u_0 \otimes g_8(k-1,0,q-2) + u_0 \otimes g_9(k-2,0,q-2) - qu_1 \otimes f_5(k-1,0,q-2)$,

$r_4 = -u_0 \otimes g_6(k-1,0,q-2) + u_0 \otimes g_7(k-2,0,q-2) - qu_1 \otimes f_6(k,0,q-2)$,

$r_5 = u_0 \otimes f_3(k-1,0,q-1) + (-1+q)u_0 \otimes f_6(k,0,q-2) - qu_1 \otimes g_6(k-1,0,q-2)$,

$r_6 = u_0 \otimes f_3(k-1,0,q-1) - u_0 \otimes f_6(k,0,q-2) - qu_1 \otimes g_7(k-1,0,q-2)$,

$r_7 = u_0 \otimes f_4(k-1,0,q-1) + u_0 \otimes f_5(k-1,0,q-2) - qu_1 \otimes g_8(k-1,0,q-2)$,

$r_8 = u_0 \otimes f_4(k-1,0,q-1) + (1-q)u_0 \otimes f_5(k-1,0,q-2) - qu_1 \otimes g_9(k-2,0,q-2)$.

Compute

$$\begin{aligned}
h_1'(r_1) =&0;\\
h_1'(r_2) =&-(1-q)u_0 \otimes f_6(k,0,q-2) + u_0 \otimes f_3(k-1,0,q-1)\\
&- qu_1 \otimes g_6(k-1,0,q-2);\\
h_1'(r_3) =&u_0 \otimes f_6(k,0,q-2) - u_0 \otimes f_3(k-1,0,q-1)\\
&+ qu_1 \otimes g_7(k-1,0,q-2);\\
h_1'(r_4) =&0;\\
h_1'(r_5) =&0;\\
h_1'(r_6) =&0;\\
h_1'(r_7) =&-u_0 \otimes g_6(k-1,0,q-2) + u_0 \otimes g_7(k-1,0,q-2)\\
&- qu_1 \otimes f_3(k-1,0,q-1);\\
h_1'(r_8) =&-u_0 \otimes g_6(k-1,0,q-2) + (1-q)u_0 \otimes g_7(k-1,0,q-2)\\
&- qu_1 \otimes f_3(k-1,0,q-1).
\end{aligned}$$

The basis of the solutions is

$$\{r_1, r_4, r_5, r_6\}.$$

Case $p \nmid k$ and $p | k+q$

The action of h'_2 on U is given by

$$\begin{pmatrix} 0 & 0 \\ 0 & 0 \end{pmatrix},$$

and the action of h'_1 on U is given by

$$\begin{pmatrix} 0 & 1 \\ k & 0 \end{pmatrix}.$$

The matrix of action of h'_1 on $U \otimes S_{k+1,k+q-1}$ is

$$\begin{pmatrix}
0 & 0 & 0 & 0 & 0 & 0 & 0 & -1 & 1 & 0 & 0 & 0 & 0 & 0 & 0 & 0 \\
0 & 0 & 0 & 0 & k & 0 & 0 & -k & 0 & 1 & 0 & 0 & 0 & 0 & 0 & 0 \\
0 & 0 & 0 & 0 & 0 & -k & 0 & 0 & 0 & 0 & 1 & 0 & 0 & 0 & 0 & 0 \\
0 & 0 & 0 & 0 & k & k & -1 & k & 0 & 0 & 0 & 1 & 0 & 0 & 0 & 0 \\
k & -1 & 0 & 0 & 0 & 0 & 0 & 0 & 0 & 0 & 0 & 0 & 1 & 0 & 0 & 0 \\
0 & 0 & 1 & 0 & 0 & 0 & 0 & 0 & 0 & 0 & 0 & 0 & 0 & 1 & 0 & 0 \\
2k^2 & -k & k & k & 0 & 0 & 0 & 0 & 0 & 0 & 0 & 0 & 0 & 0 & 1 & 0 \\
k & 0 & 0 & 0 & 0 & 0 & 0 & 0 & 0 & 0 & 0 & 0 & 0 & 0 & 0 & 1 \\
k & 0 & 0 & 0 & 0 & 0 & 0 & 0 & 0 & 0 & 0 & 0 & 0 & 0 & 0 & 1 \\
0 & k & 0 & 0 & 0 & 0 & 0 & 0 & 0 & 0 & 0 & 0 & -k & 0 & 0 & k \\
0 & 0 & k & 0 & 0 & 0 & 0 & 0 & 0 & 0 & 0 & 0 & 0 & k & 0 & 0 \\
0 & 0 & 0 & k & 0 & 0 & 0 & 0 & 0 & 0 & 0 & 0 & -k & -k & 1 & -k \\
0 & 0 & 0 & 0 & k & 0 & 0 & 0 & -k & 1 & 0 & 0 & 0 & 0 & 0 & 0 \\
0 & 0 & 0 & 0 & 0 & k & 0 & 0 & 0 & 0 & 1 & 0 & 0 & 0 & 0 & 0 \\
0 & 0 & 0 & 0 & 0 & 0 & k & 0 & -2k^2 & k & -k & -k & 0 & 0 & 0 & 0 \\
0 & 0 & 0 & 0 & 0 & 0 & 0 & k & k & 0 & 0 & 0 & 0 & 0 & 0 & 0
\end{pmatrix}.$$

The basis of elements annihilated by h'_1 on $U \otimes S_{k+1,k+q-1}$ is

$$\begin{aligned}
\{r_1 =& - u_0 \otimes g_6(k-1,0,q-2) - 2ku_0 \otimes g_8(k-1,0,q-2) - u_0 \otimes g_9(k-2,0,q-2) \\
&- u_1 \otimes f_3(k-1,0,q-1), \\
r_2 =& u_0 \otimes g_6(k-1,0,q-2) + ku_0 \otimes g_8(k-1,0,q-2) - ku_1 \otimes f_4(k-1,0,q-1), \\
r_3 =& - u_0 \otimes g_7(k-1,0,q-2) - ku_0 \otimes g_8(k-1,0,q-2) - ku_1 \otimes f_5(k-1,0,q-2), \\
r_4 =& - u_0 \otimes g_8(k-1,0,q-2) - u_1 \otimes f_6(k,0,q-2), \\
r_5 =& - u_0 \otimes f_4(k-1,0,q-1) - u_0 \otimes f_6(k,0,q-2) - u_1 \otimes g_6(k-1,0,q-2), \\
r_6 =& u_0 \otimes f_5(k-1,0,q-2) - u_0 \otimes f_6(k,0,q-2) - u_1 \otimes g_7(k-1,0,q-2), \\
r_7 =& u_0 \otimes f_6(k,0,q-2) - ku_1 \otimes g_8(k-1,0,q-2), \\
r_8 =& u_0 \otimes f_3(k-1,0,q-1) + ku_0 \otimes f_4(k-1,0,q-1) - ku_0 \otimes f_6(k,0,q-2) \\
&- ku_1 \otimes g_9(k-2,0,q-2)\}.
\end{aligned}$$

Compute

$$\begin{aligned}
h_2'(r_1) &= -u_0 \otimes [-f_3(k-1,0,q-1) + kf_5(q-1,0,a-2) + (k+1)f_6(k,0,q-2)] \\
&\quad - 2ku_0 \otimes [-f_4(k-1,0,q-1) - f_5(k-1,0,q-2)] \\
&\quad - u_0 \otimes [-f_4(k-1,0,q-1) - f_5(k-1,0,q-2)] \\
&\quad + u_1 \otimes [(k+1)g_6(q-1,0,a-2) + (-k-1)g_7(k-1,0,q-2) \\
&\quad + kg_8(q-1,0,q-2) - kg_9(k-2,0,q-2)]; \\
h_2'(r_2) &= u_0 \otimes [-f_3(k-1,0,q-1) + kf_5(q-1,0,q-2) + (k+1)f_6(k,0,q-2)] \\
&\quad + ku_0 \otimes [-f_4(k-1,0,q-1) - f_5(k-1,0,q-2)] \\
&\quad + ku_1 \otimes [-g_8(q-1,0,q-2) + g_9(k,0,q-2)]; \\
h_2'(r_3) &= -u_0 \otimes [-f_3(k-1,0,q-1) + kf_5(q-1,0,q-2) + (k+1)f_6(k,0,q-2)] \\
&\quad - ku_0 \otimes [-f_4(k-1,0,q-1) - f_5(k-1,0,q-2)] \\
&\quad + ku_1 \otimes [g_8(q-1,0,q-2) - g_9(k,0,q-2)]; \\
h_2'(r_4) &= -u_0 \otimes [-f_4(k-1,0,q-1) - f_5(k-1,0,q-2)] \\
&\quad + u_1 \otimes [g_6(k-1,0,q-2) - g_7(k-1,0,q-2)]; \\
h_2'(r_5) &= -u_0 \otimes [-g_8(k-1,0,q-2 + g_9(k-2,0,q-2)] \\
&\quad - u_0 \otimes [g_6(k-1,0,q-2) - g_7(k-1,0,q-2)] \\
&\quad + u_1 \otimes [-f_4(k-1,0,q-1) + kf_5(k-1,0,q-2) + (k+1)f_6(k,0,q-2)]; \\
h_2'(r_6) &= u_0 \otimes [g_8(k-1,0,q-2 - g_9(k-2,0,q-2)] \\
&\quad - u_0 \otimes [g_6(k-1,0,q-2) - g_7(k-1,0,q-2)] \\
&\quad + u_1 \otimes [-f_4(k-1,0,q-1) + kf_5(k-1,0,q-2) + (k+1)f_6(k,0,q-2)]; \\
h_2'(r_7) &= u_0 \otimes [g_6(k-1,0,q-2) - g_7(k-1,0,q-2)] \\
&\quad + ku_1 \otimes [-f_4(k-1,0,q-1) - f_5(k-1,0,q-2)]; \\
h_2'(r_8) &= u_0 \otimes [(k+1)g_6(k-1,0,q-2) + (-k-1)g_7(k-1,0,q-2) \\
&\quad + kg_8(k-1,0,q-2) - kg_9(k-2,q-2)] \\
&\quad + ku_0 \otimes [-g_8(k-1,0,q-2) + g_9(k-1,0,q-2)] \\
&\quad - ku_0 \otimes [g_6(k-1,0,q-2) - g_7(k-1,0,q-2)] \\
&\quad ku_1 \otimes [-f_4(k-1,0,q-1) - f_5(k-1,0,q-2)].
\end{aligned}$$

On $\{r_1, r_2, r_3, r_4\}$ we get a matrix

$$\begin{pmatrix} -1 & 1 & -1 & 0 \\ -2k-1 & k & -k & -1 \\ k+1 & 0 & 0 & 1 \\ -k & k & -k & 0 \end{pmatrix}.$$

The solutions are $r_2 + r_3, r_1 + r_2 - (k+1)r_4$.

On $\{r_5, r_6, r_7, r_8\}$ we get a matrix

$$\begin{pmatrix} 1 & 1 & 0 & 0 \\ 0 & 0 & k & k \\ -1 & -1 & 1 & 1 \\ 1 & 1 & 0 & 0 \end{pmatrix}.$$

The solutions are $r_5 - r_6, r_7 - r_8$.

The basis of the solutions is

$$\{r_2 + r_3, r_1 + r_2 - (k+1)r_4, r_5 - r_6, r_7 - r_8\}.$$

<u>Case $p \nmid k$, $p \nmid k+q$.</u>
The action of h_1' on U is given by

$$\begin{pmatrix} 0 & e \\ e & 0 \end{pmatrix},$$

where $k = e^2$. The action of h_2' on U is given by

$$\begin{pmatrix} 0 & -l \\ l & 0 \end{pmatrix},$$

where $l^2 = -k - q$.

The matrix of h_1' on $U \otimes S_{k+1,k+q-1}$ is

$$\begin{pmatrix}
0 & 0 & 0 & 0 & 0 & 0 & 0 & -1 & e & 0 & 0 & 0 & 0 & 0 & 0 & 0 \\
0 & 0 & 0 & 0 & k & 0 & 0 & -k & 0 & e & 0 & 0 & 0 & 0 & 0 & 0 \\
0 & 0 & 0 & 0 & 0 & -k & 0 & 0 & 0 & 0 & e & 0 & 0 & 0 & 0 & 0 \\
0 & 0 & 0 & 0 & k & k & -1 & k & 0 & 0 & 0 & e & 0 & 0 & 0 & 0 \\
k & -1 & 0 & 0 & 0 & 0 & 0 & 0 & 0 & 0 & 0 & 0 & e & 0 & 0 & 0 \\
0 & 0 & 1 & 0 & 0 & 0 & 0 & 0 & 0 & 0 & 0 & 0 & 0 & e & 0 & 0 \\
2k^2 & -k & k & k & 0 & 0 & 0 & 0 & 0 & 0 & 0 & 0 & 0 & 0 & e & 0 \\
k & 0 & 0 & 0 & 0 & 0 & 0 & 0 & 0 & 0 & 0 & 0 & 0 & 0 & 0 & e \\
e & 0 & 0 & 0 & 0 & 0 & 0 & 0 & 0 & 0 & 0 & 0 & 0 & 0 & 0 & 1 \\
0 & e & 0 & 0 & 0 & 0 & 0 & 0 & 0 & 0 & 0 & 0 & -k & 0 & 0 & k \\
0 & 0 & e & 0 & 0 & 0 & 0 & 0 & 0 & 0 & 0 & 0 & 0 & k & 0 & 0 \\
0 & 0 & 0 & e & 0 & 0 & 0 & 0 & 0 & 0 & 0 & 0 & -k & -k & 1 & -k \\
0 & 0 & 0 & 0 & e & 0 & 0 & 0 & -k & 1 & 0 & 0 & 0 & 0 & 0 & 0 \\
0 & 0 & 0 & 0 & 0 & e & 0 & 0 & 0 & 0 & -1 & 0 & 0 & 0 & 0 & 0 \\
0 & 0 & 0 & 0 & 0 & 0 & e & 0 & 2k^2 & k & k & k & 0 & 0 & 0 & 0 \\
0 & 0 & 0 & 0 & 0 & 0 & 0 & e & -k & 0 & 0 & 0 & 0 & 0 & 0 & 0
\end{pmatrix}$$

The basis of elements annihilated by h_1' on $U \otimes S_{k+1,k+q-1}$ consists of

$$\begin{aligned}
r_1 =& eu_0 \otimes g_6(k-1,0,q-2) + 2e^3 u_0 \otimes g_8(k-1,0,q-2) + eu_0 \otimes g_9(k-2,0,q-2) \\
&+ u_1 \otimes f_3(k-1,0,q-1), \\
r_2 =& -u_0 \otimes g_6(k-1,0,q-2) - e^2 u_0 \otimes g_8(k-1,0,q-2) + eu_1 \otimes f_4(k-1,0,q-1), \\
r_3 =& u_0 \otimes g_7(k-1,0,q-2) + e^2 u_0 \otimes g_8(k-1,0,q-2) + eu_1 \otimes f_5(k-1,0,q-2), \\
r_4 =& eu_0 \otimes g_8(k-1,0,q-2) + u_1 \otimes f_6(k,0,q-2), \\
r_5 =& eu_0 \otimes f_4(k-1,0,q-1) + eu_0 \otimes f_6(k,0,q-2) + u_1 \otimes g_6(k-1,0,q-2), \\
r_6 =& -eu_0 \otimes f_5(k-1,0,q-2) + eu_0 \otimes f_6(k,0,q-2) + u_1 \otimes g_7(k-1,0,q-2), \\
r_7 =& -u_0 \otimes f_6(k,0,q-2) + eu_1 \otimes g_8(k-1,0,q-2), \\
r_8 =& -u_0 \otimes f_3(k-1,0,q-1) - e^2 u_0 \otimes f_4(k-1,0,q-1) + e^2 u_0 \otimes f_6(k,0,q-2) \\
&+ eu_1 \otimes g_9(k-2,0,q-2).
\end{aligned}$$

We have

$$\begin{aligned}
h_2'(r_1) =& elu_1 \otimes g_6(k-1,0,q-2) + 2e^3 lu_1 \otimes g_8(k-1,0,q-2) \\
&+ elu_1 \otimes g_9(k-1,0,q-2) \\
&- lu_0 \otimes f_3(k-1,0,q-1) + eu_0 \otimes [-f_3(k-1,0,q-1) + kf_5(k-1,0,q-2) \\
&+ (-q+1)f_6(k,0,q-1)] + 2e^3 u_0 \otimes [-f_4(k-1,0,q-1) - f_5(k-1,0,q-2)] \\
&+ eu_0 \otimes [-f_4(k-1,0,q-1) + (k+q-1)f_5(k-1,0,q-2)] \\
&- u_1 \otimes [(k+1)g_6(k-1,0,q-2) + (q-1)g_7(k-1,0,q-2) \\
&+ kg_8(k-1,0,q-2) \\
&- kg_9(k-2,0,q-2)] = u_0 \otimes [(-e-l)f_3(k-1,0,q-1) \\
&+ (-2e^3 - e)f_4(k-1,0,q-1) \\
&+ (ek - 2e^3 + e(k+q-1))f_5(k-1,0,q-2) + e(-q+1)f_6(k,0,q-2)] \\
&+ u_1 \otimes [(el-k-1)g_6(k-1,0,q-2) + (1-q)g_7(k-1,0,q-2) \\
&+ (2e^3 l - k)g_8(k-1,0,q-2) + (el+k)g_9(k-2,0,q-2)],
\end{aligned}$$

$$\begin{aligned}
h_2'(r_2) =& - lu_1 \otimes g_6(k-1,0,q-2) - e^2 lu_1 \otimes g_8(k-1,0,q-2) \\
&- elu_0 \otimes f_4(k-1,0,q-1) \\
&- u_0 \otimes [-f_3(k-1,0,q-1) + kf_5(k,0,q-2) + (-q+1)f_6(k,0,q-2)] \\
&- e^2 u_0 \otimes [-f_4(k-1,0,q-1) - f_5(k-1,0,q-2)] \\
&- eu_1 \otimes [(k+q-1)g_8(k-1,0,q-2) + g_9(k-2,0,q-2)] \\
&= u_0 \otimes [f_3(k-1,0,q-1) + (-el+e^2)f_4(k-1,0,q-1) \\
&+ (-k+e^2)f_5(k-1,0,q-2) + (q-1)f_6(k,0,q-2)] \\
&+ u_1 \otimes [(-l)g_6(k-1,0,q-2) + (-e^2 l - ke - qe + e)g_8(k-1,0,q-2) \\
&- eg_9(k-2,0,q-2)],
\end{aligned}$$

$$\begin{aligned}
h_2'(r_3) =& lu_1 \otimes g_7(k-1,0,q-2) + e^2 lu_1 \otimes g_8(k-1,0,q-2) - elu_0 \otimes f_5(k-1,0,q-2) \\
&+ u_0 \otimes [-f_3(k-1,0,q-1) + kf_5(k-1,0,q-2) + (k+1)f_6(k-1,0,q-2)] \\
&+ e^2 u_0 \otimes [-f_4(k-1,0,q-1) - f_5(k-1,0,q-2)] \\
&- eu_1 \otimes [g_8(k-1,0,q-2) - g_9(k-2,0,q-2)] \\
&= u_0 \otimes [-f_3(k-1,0,q-1) - e^2 f_4(k-1,0,q-1) \\
&+ (-el+k-e^2)f_5(k-1,0,q-2) + (k+1)f_6(k,0,q-2)] \\
&+ u_1 \otimes [lg_7(k-1,0,q-2) + (e^2 l - e)g_8(k-1,0,q-2) + eg_9(k-2,0,q-2)]
\end{aligned}$$

$$\begin{aligned}
h_2'(r_4) =& elu_1 \otimes g_8(k-1,0,q-2) - lu_0 \otimes f_6(k,0,q-2) \\
&+ eu_0 \otimes [-f_4(k-1,0,q-1) - f_5(k-1,0,q-2)] \\
&- u_1 \otimes [g_6(k-1,0,q-2) - g_7(k-1,0,q-2)] \\
&= u_0 \otimes [-ef_4(k-1,0,q-1) - ef_5(k-1,0,q-2) - lf_6(k,0,q-2)] \\
&+ u_1 \otimes [-g_6(k-1,0,q-2) + g_7(k-1,0,q-2) + elg_8(k-1,0,q-2)],
\end{aligned}$$

$$\begin{aligned}
h_2'(r_5) =& elu_1 \otimes f_4(k-1,0,q-1) + elu_1 \otimes f_6(k,0,q-2) - lu_0 \otimes g_6(k-1,0,q-2) \\
&+ eu_0 \otimes [(k+q-1)g_8(k-1,0,q-2) + g_9(k-2,0,q-2)] \\
&+ eu_0 \otimes [g_6(k-1,0,q-2) - g_7(k-1,0,q-2)] \\
&- u_1 \otimes [-f_3(k-1,0,q-1) + kf_5(k-1,0,q-2) + (-q+1)f_6(k,0,q-2)] \\
&= u_0 \otimes [(-l+e)g_6(k-1,0,q-2) - eg_7(k-1,0,q-2) \\
&+ (ek+eq-e)g_8(k-1,0,q-2) + eg_9(k-2,0,q-2)] \\
&+ u_1 \otimes [f_3(k-1,0,q-1) + elf_4(k-1,0,q-1) - kf_5(k-1,0,q-2) \\
&+ (el+q-1)f_6(k,0,q-2)],
\end{aligned}$$

$$\begin{aligned}
h_2'(r_6) =& - elu_1 \otimes f_5(k-1,0,q-2) + elu_1 \otimes f_6(k,0,q-2) - lu_0 \otimes g_7(k-1,0,q-2) \\
&- eu_0 \otimes [g_8(k-1,0,q-2) - g_9(k-2,0,q-2)] \\
&+ eu_0 \otimes [g_6(k-1,0,q-2) - g_7(k-1,0,q-2)] \\
&- u_1 \otimes [-f_3(k-1,0,q-1) + kf_5(k-1,0,q-2) + (k+1)f_6(k,0,q-2)] \\
&= u_0 \otimes [eg_6(k-1,0,q-2) + (-l-e)g_7(k-1,0,q-2) \\
&- eg_8(k-1,0,q-2) + eg_9(k-2,0,q-2)] \\
&+ u_1 \otimes [f_3(k-1,0,q-1) + (-el-k)f_5(k-1,0,q-2) \\
&+ (el-k-1)f_6(k,0,q-2)],
\end{aligned}$$

$$\begin{aligned}
h_2'(r_7) =& - lu_1 \otimes f_6(k,0,q-2) - elu_0 \otimes g_8(k-1,0,q-2) \\
&- u_0 \otimes [g_6(k-1,0,q-2) - g_7(k-1,0,q-2)] \\
&- eu_1 \otimes [-f_4(k-1,0,q-1) - f_5(k-1,0,q-2)] \\
&= u_0 \otimes [-g_6(k-1,0,q-2) + g_7(k-1,0,q-2) - elg_8(k-1,0,q-2)] \\
&+ u_1 \otimes [+ef_4(k-1,0,q-1) + ef_5(k-1,0,q-2) - lf_6(k,0,q-2)],
\end{aligned}$$

$$\begin{aligned}
h_2'(r_8) =& - lu_1 \otimes f_3(k-1,0,q-1) - e^2lu_1 \otimes f_4(k-1,0,q-1) \\
&+ e^2lu_1 \otimes f_6(k,0,q-2) - elu_0 \otimes g_9(k-2,0,q-2) \\
&- u_0 \otimes [(k+1)g_6(k-1,0,q-2) + (q-1)g_7(k-1,0,q-2) \\
&+ kg_8(k-1,0,q-2) - kg_9(k-2,0,q-2)] \\
&- e^2u_0 \otimes [(k+q-1)g_8(k-1,0,q-2) + g_9(k-2,0,q-2)] \\
&+ e^2u_0 \otimes [g_6(k-1,0,q-2) - g_7(k-1,0,q-2)] \\
&- eu_1 \otimes [-f_4(k-1,0,q-1) + (k+q-1)f_5(k-1,0,q-1)] \\
&= u_0 \otimes [(-k-1+e^2)g_6(k-1,0,q-2) + (-q+1-e^2)g_7(k-1,0,q-2) \\
&+ (-k-e^2k-e^2q+e^2)g_8(k-1,0,q-2) - elg_9(k-2,0,q-2)] \\
&+ u_1 \otimes [-lf_3(k-1,0,q-1) + (-e^2l+e)f_4(k-1,0,q-1) \\
&- e(k+q-1)f_5(k-1,0,q-2) + e^2lf_6(k,0,q-2)],
\end{aligned}$$

On the set $\{r_1, r_2, r_3, r_4\}$, we get the matrix

$$\begin{pmatrix} -e-l & 1 & -1 & 0 \\ -2e^3-e & -el+e^2 & -e^2 & -e \\ -e^3-el^2-e & 0 & -el & -e \\ e^3+el^2+e & -e^2-l^2-1 & e^2+1 & -l \\ el-e^2-1 & -l & 0 & -1 \\ 1+e^2+l^2 & 0 & l & 1 \\ 2e^3l-e^2 & -e^2l+l^2e+e & e^2l-e & el \\ el+e^2 & -e & e & 0 \end{pmatrix},$$

which gives solutions $-lr_1+(1+e^2-el)r_2+(1+e^2+l^2)r_3$ and $-r_1-(e+l)r_2+(1+e^2+l^2)r_4$.

On the set $\{r_5, r_6, r_7, r_8\}$, we get the matrix

$$\begin{pmatrix} -l+e & e & -1 & -1 \\ -e & -l-e & 1 & l^2+1 \\ -el^2-e & -e & -el & e^2l^2 \\ e & e & 0 & -el \\ 1 & 1 & 0 & -l \\ el & 0 & e & -e^2l+e \\ -e^2 & -el-e^2 & e & e(l^2+1) \\ el-e^2-l^2-1 & el-e^2-1 & -l & e^2l \end{pmatrix},$$

which gives solutions $-r_5+r_6+lr_7$ and $lr_5+(-1+el-l^2)r_7+r_8$.

The basis of the solutions is

$$\{-lr_1+(1+e^2-el)r_2+(1+e^2+l^2)r_3, -r_1-(e+l)r_2+(1+e^2+l^2)r_4,$$
$$-r_5+r_6+lr_7, lr_5+(-1+el-l^2)r_7+r_8\}.$$

6. $S_{(k+q-v,k+v)}$ **for** $0<v<q-1$.

The weightspace of right weight $(k+q-v,k+v)$ for $0<v<q-1$ has a basis $S_{(k+q-v,k+v)}=$

$$\{f_3(k-1,q-1-v,v), f_4(k-1,q-1-v,v), f_5(k-1,q-1-v,v-1),$$
$$f_6(k,q-v-1,v-1), g_6(k-1,q-1-v,v-1), g_7(k-1,q-1-v,v-1),$$
$$g_8(k-1,q-1-v,v-1), g_{10}(k-2,q-v-2,v-1)\}.$$

The matrix of h_1' on this set is

$$\begin{pmatrix} 0 & 0 & 0 & 0 & 0 & 0 & 0 & -1 \\ 0 & 0 & 0 & 0 & k & 0 & 0 & -k \\ 0 & 0 & 0 & 0 & 0 & -k & 0 & 0 \\ 0 & 0 & 0 & 0 & k & k & -1 & 1 \\ k & -1 & 0 & 0 & 0 & 0 & 0 & 0 \\ 0 & 0 & 1 & 0 & 0 & 0 & 0 & 0 \\ k^2+k & -k & k & k & 0 & 0 & 0 & 0 \\ k & 0 & 0 & 0 & 0 & 0 & 0 & 0 \end{pmatrix},$$

and the matrix of h_2' on this set is

$$\begin{pmatrix} 0 & 0 & 0 & 0 & -1 & -1 & 0 & 0 \\ 0 & 0 & 0 & 0 & 0 & 0 & -1 & -k-q+v+1 \\ 0 & 0 & 0 & 0 & k & k & -1 & v+1 \\ 0 & 0 & 0 & 0 & -v & k+q-v & 0 & 0 \\ k+q-v & 0 & 0 & 1 & 0 & 0 & 0 & 0 \\ v & 0 & 0 & -1 & 0 & 0 & 0 & 0 \\ k(k+q-v-1) & v+1 & k+q-v-1 & 0 & 0 & 0 & 0 & 0 \\ -k & 1 & -1 & 0 & 0 & 0 & 0 & 0 \end{pmatrix}.$$

<u>Case $p|k$ and $p|k+q$</u>
The basis of elements in the span of $S_{(k+q-v,k+v)}$ annihilated by h_1' is

$$\{f_3(k-1,0,q-1), f_6(k,0,q-2), g_6(k-1,0,q-2), g_7(k-1,0,q-2)\}.$$

In this space, there are two elements

$$f_3(k-1,0,q-1)+vf_6(k,0,q-2), g_6(k-1,0,q-2)-g_7(k-1,0,q-2)$$

that form a basis of elements annihilated by h_2'.

In $U \otimes S_{(k+q-v,k+v)}$, we have elements

$$\begin{aligned} &\{u_0 \otimes [f_3(k-1,0,q-1)+vf_6(k,0,q-2)], \\ &u_0 \otimes [g_6(k-1,0,q-2)-g_7(k-1,0,q-2)], \\ &u_1 \otimes [f_3(k-1,0,q-1)+vf_6(k,0,q-2)], \\ &u_1 \otimes [g_6(k-1,0,q-2)-g_7(k-1,0,q-2)]\} \end{aligned}$$

forming a basis for the solutions.

<u>Case $p|k$ and $p \nmid k+q$</u>
The action of h_1' on U is given by

$$\begin{pmatrix} 0 & 0 \\ 0 & 0 \end{pmatrix},$$

and the action of h_2' on U is given by

$$\begin{pmatrix} 0 & 1 \\ q & 0 \end{pmatrix}.$$

The basis of elements in $U \otimes S_{(k+q-v,k+v)}$ annihilated by h_1' is

$$\begin{aligned} B' = \{&u_0 \otimes f_3(k-1,0,q-1), u_0 \otimes f_6(k,0,q-2), \\ &u_0 \otimes g_6(k-1,0,q-2), u_0 \otimes g_7(k-1,0,q-2), \\ &u_1 \otimes f_3(k-1,0,q-1), u_1 \otimes f_6(k,0,q-2), \\ &u_1 \otimes g_6(k-1,0,q-2), u_1 \otimes g_7(k-1,0,q-2)\}. \end{aligned}$$

The matrix of action of h_2' is

$$
\begin{pmatrix}
0 & 0 & 0 & 0 & -1 & -1 & 0 & 0 & 1 & 0 & 0 & 0 & 0 & 0 & 0 & 0 \\
0 & 0 & 0 & 0 & 0 & 0 & -1 & -q+v+1 & 0 & 1 & 0 & 0 & 0 & 0 & 0 & 0 \\
0 & 0 & 0 & 0 & 0 & 0 & -1 & v+1 & 0 & 0 & 1 & 0 & 0 & 0 & 0 & 0 \\
0 & 0 & 0 & 0 & -v & q-v & 0 & 0 & 0 & 0 & 0 & 1 & 0 & 0 & 0 & 0 \\
q-v & 0 & 0 & 1 & 0 & 0 & 0 & 0 & 0 & 0 & 0 & 0 & 1 & 0 & 0 & 0 \\
v & 0 & 0 & -1 & 0 & 0 & 0 & 0 & 0 & 0 & 0 & 0 & 0 & 1 & 0 & 0 \\
0 & v+1 & q-v-1 & 0 & 0 & 0 & 0 & 0 & 0 & 0 & 0 & 0 & 0 & 0 & 1 & 0 \\
0 & 1 & -1 & 0 & 0 & 0 & 0 & 0 & 0 & 0 & 0 & 0 & 0 & 0 & 0 & 1 \\
q & 0 & 0 & 0 & 0 & 0 & 0 & 0 & 0 & 0 & 0 & 0 & 1 & 1 & 0 & 0 \\
0 & q & 0 & 0 & 0 & 0 & 0 & 0 & 0 & 0 & 0 & 0 & 0 & 0 & 1 & q-v-1 \\
0 & 0 & q & 0 & 0 & 0 & 0 & 0 & 0 & 0 & 0 & 0 & 0 & 0 & 1 & -v-1 \\
0 & 0 & 0 & q & 0 & 0 & 0 & 0 & 0 & 0 & 0 & 0 & v & -q+v & 0 & 0 \\
0 & 0 & 0 & 0 & q & 0 & 0 & 0 & -q+v & 0 & 0 & -1 & 0 & 0 & 0 & 0 \\
0 & 0 & 0 & 0 & 0 & q & 0 & 0 & -v & 0 & 0 & 1 & 0 & 0 & 0 & 0 \\
0 & 0 & 0 & 0 & 0 & 0 & q & 0 & 0 & -v-1 & -q+v+1 & 0 & 0 & 0 & 0 & 0 \\
0 & 0 & 0 & 0 & 0 & 0 & 0 & q & 0 & -1 & 1 & 0 & 0 & 0 & 0 & 0
\end{pmatrix}.
$$

The basis of elements in $U \otimes S_{(k+q-v,k+v)}$ annihilated by h_2' is

$$\begin{aligned}
r_1 =&(1-q+v)u_0 \otimes f_4(k-1,q-1-v,v) + (1+v)u_0 \otimes f_5(k-1,q-1-v,v-1)\\
&+ qu_1 \otimes g_{10}(k-2,q-v-2,v-1),\\
r_2 =&-u_0 \otimes f_4(k-1,q-1-v,v) - u_0 \otimes f_5(k-1,q-1-v,v-1)\\
&+ qu_1 \otimes g_8(k-1,q-1-v,v-1),\\
r_3 =&-u_0 \otimes f_3(k-1,q-1-v,v) + (q-v)u_0 \otimes f_6(k,q-v-1,v-1)\\
&+ qu_1 \otimes g_7(k-1,q-1-v,v-1),\\
r_4 =&-u_0 \otimes f_3(k-1,q-1-v,v) - vu_0 \otimes f_6(k,q-v-1,v-1)\\
&+ qu_1 \otimes g_6(k-1,q-1-v,v-1),\\
r_5 =&u_0 \otimes g_6(k-1,q-1-v,v-1) - u_0 \otimes g_7(k-1,q-1-v,v-1)\\
&+ qu_1 \otimes f_6(k,q-v-1,v-1),\\
r_6 =&(-1+q-v)u_0 \otimes g_8(k-1,q-1-v,v-1) - u_0 \otimes g_{10}(k-2,q-v-2,v-1)\\
&+ qu_1 \otimes f_5(k-1,q-1-v,v-1),\\
r_7 =&(1+v)u_0 \otimes g_8(k-1,q-1-v,v-1) + u_0 \otimes g_{10}(k-2,q-v-2,v-1)\\
&+ qu_1 \otimes f_4(k-1,q-1-v,v),\\
r_8 =&(q-v)u_0 \otimes g_6(k-1,q-1-v,v-1) + vu_0 \otimes g_7(k-1,q-1-v,v-1)\\
&+ qu_1 \otimes f_3(k-1,q-1-v,v).
\end{aligned}$$

Compute

$$\begin{aligned}
h_1'(r_1) =&-(1-q+v)u_0 \otimes g_6(k-1,q-1-v,v-1)\\
&+ (1+v)u_0 \otimes g_7(k-1,q-1-v,v-1)\\
&- qu_1 \otimes (-f_3(k-1,q-1-v,v) + f_6(k,q-v-1,v-1));\\
h_1'(r_2) =&u_0 \otimes g_6(k-1,q-1-v,v-1) - u_0 \otimes g_7(k-1,q-1-v,v-1)\\
&+ qu_1 \otimes f_6(k,q-v-1,v-1);\\
h_1'(r_3) =&0;\\
h_1'(r_4) =&0;\\
h_1'(r_5) =&0;\\
h_1'(r_6) =&-(-1-q-v)u_0 \otimes f_6(k,q-v-1,v-1)\\
&- u_0 \otimes (-f_3(k-1,q-1-v,v) + f_6(k,q-v-1,v-1))\\
&+ qu_1 \otimes g_6(k-1,q-1-v,v-1);\\
h_1'(r_7) =&-(1+v)u_0 \otimes f_6(k,q-v-1,v-1)\\
&+ u_0 \otimes (-f_3(k-1,q-1-v,v) + f_6(k,q-v-1,v-1))\\
&+ qu_1 \otimes g_6(k-1,q-1-v,v-1);\\
h_1'(r_8) =&0.
\end{aligned}$$

The basis of the solutions is

$$\{r_3, r_4, r_5, r_8\}.$$

Case $p \nmid k$ and $p|k+q$

The action of h'_2 on U is given by

$$\begin{pmatrix} 0 & 0 \\ 0 & 0 \end{pmatrix},$$

and the action of h'_1 on U is given by

$$\begin{pmatrix} 0 & 1 \\ k & 0 \end{pmatrix}.$$

The matrix of action of h'_1 on $U \otimes S_{(k+q-v,k+v)}$ is

$$\begin{pmatrix}
0 & 0 & 0 & 0 & 0 & 0 & 0 & -1 & 1 & 0 & 0 & 0 & 0 & 0 & 0 & 0 \\
0 & 0 & 0 & 0 & k & 0 & 0 & -k & 0 & 1 & 0 & 0 & 0 & 0 & 0 & 0 \\
0 & 0 & 0 & 0 & 0 & -k & 0 & 0 & 0 & 0 & 1 & 0 & 0 & 0 & 0 & 0 \\
0 & 0 & 0 & 0 & k & k & -1 & 1 & 0 & 0 & 0 & 1 & 0 & 0 & 0 & 0 \\
k & -1 & 0 & 0 & 0 & 0 & 0 & 0 & 0 & 0 & 0 & 0 & 1 & 0 & 0 & 0 \\
0 & 0 & 1 & 0 & 0 & 0 & 0 & 0 & 0 & 0 & 0 & 0 & 0 & 1 & 0 & 0 \\
k^2+k & -k & k & k & 0 & 0 & 0 & 0 & 0 & 0 & 0 & 0 & 0 & 0 & 1 & 0 \\
k & 0 & 0 & 0 & 0 & 0 & 0 & 0 & 0 & 0 & 0 & 0 & 0 & 0 & 0 & 1 \\
k & 0 & 0 & 0 & 0 & 0 & 0 & 0 & 0 & 0 & 0 & 0 & 0 & 0 & 0 & 1 \\
0 & k & 0 & 0 & 0 & 0 & 0 & 0 & 0 & 0 & 0 & 0 & -k & 0 & 0 & k \\
0 & 0 & k & 0 & 0 & 0 & 0 & 0 & 0 & 0 & 0 & 0 & 0 & k & 0 & 0 \\
0 & 0 & 0 & k & 0 & 0 & 0 & 0 & 0 & 0 & 0 & 0 & -k & -k & 1 & -1 \\
0 & 0 & 0 & 0 & k & 0 & 0 & 0 & -k & 1 & 0 & 0 & 0 & 0 & 0 & 0 \\
0 & 0 & 0 & 0 & 0 & k & 0 & 0 & 0 & 0 & -1 & 0 & 0 & 0 & 0 & 0 \\
0 & 0 & 0 & 0 & 0 & 0 & k & 0 & -k^2-k & k & -k & -k & 0 & 0 & 0 & 0 \\
0 & 0 & 0 & 0 & 0 & 0 & 0 & k & -k & 0 & 0 & 0 & 0 & 0 & 0 & 0
\end{pmatrix}.$$

The basis of elements annihilated by h'_1 on $U \otimes S_{(k+q-v,k+v)}$ is

$$\begin{aligned}
\{r_1 =& -u_0 \otimes g_6(k-1,q-1-v,v-1) - (1+k)u_0 \otimes g_8(k-1,q-1-v,v-1) \\
&- u_0 \otimes g_{10}(k-2,q-v-2,v-1) - u_1 \otimes f_3(k-1,q-1-v,v-1), \\
r_2 =& u_0 \otimes g_6(k-1,q-1-v,v-1) + ku_0 \otimes g_8(k-1,q-1-v,v-1) \\
&- ku_1 \otimes f_4(k-1,q-1-v,v), \\
r_3 =& -u_0 \otimes g_7(k-1,q-1-v,v-1) - ku_0 \otimes g_8(k-1,q-1-v,v-1) \\
&- ku_1 \otimes f_5(k-1,q-1-v,v-1), \\
r_4 =& -u_0 \otimes g_8(k-1,q-1-v,v-1) - u_1 \otimes f_6(k,q-1-v,v-1), \\
r_5 =& -u_0 \otimes f_4(k-1,q-1-v,v) - u_0 \otimes f_6(k,q-v-1,v-1) \\
&- u_1 \otimes g_6(k-1,q-1-v,v-1), \\
r_6 =& u_0 \otimes f_5(k-1,q-1-v,v-1) - u_0 \otimes f_6(k,q-v-1,v-1) \\
&- u_1 \otimes g_7(k-1,q-1-v,v-1), \\
r_7 =& u_0 \otimes f_6(k,q-v-1,v-1) - ku_1 \otimes g_8(k-1,q-1-v,v-1), \\
r_8 =& u_0 \otimes f_3(k-1,q-1-v,v-1) + ku_0 \otimes f_4(k-1,q-1-v,v) \\
&- u_0 \otimes f_6(k,q-v-1,v-1) - ku_1 \otimes g_{10}(k-2,q-v-2,v-1)\}.
\end{aligned}$$

Compute

$$\begin{aligned}
h_2'(r_1) =& - u_0 \otimes [-f_3(k-1,q-1-v,v) + kf_5(k-1,a-1-v,v-1) \\
&- vf_6(k,q-1-v,v-1)] - (1+k)u_0 \otimes [-f_4(k-1,q-1-v,v) \\
&- f_5(k-1,q-1-v,v-1)] \\
&- u_0 \otimes [(v+1)f_4(k-1,q-1-v,v) + (v+1)f_5(k-1,q-1-v,v-1)] \\
&+ u_1 \otimes [-vg_6(k-1,q-1-v,v-1) + vg_7(k-1,q-1-v,v-1) \\
&+ k(-v-1)g_8(k-1,q-1-v,v-1) - kg_{10}(k-2,q-v-2,v-1)]; \\
h_2'(r_2) =& u_0 \otimes [-f_3(k-1,q-1-v,v) + kf_5(k-1,q-1-v,v-1) \\
&- vf_6(k,q-1-v,v-1)] \\
&+ ku_0 \otimes [-f_4(k-1,q-1-v,v) - f_5(k-1,q-1-v,v-1)] \\
&+ ku_1 \otimes [(v+1)g_8(k-1,q-1-v,v-1) + g_{10}(k-2,q-v-2,v-1)]; \\
h_2'(r_3) =& - u_0 \otimes [-f_3(k-1,q-1-v,v) + kf_5(k-1,q-1-v,v-1) \\
&- vf_6(k,q-1-v,v-1)] \\
&- ku_0 \otimes [-f_4(k-1,q-1-v,v) - f_5(k-1,q-1-v,v-1)] \\
&+ ku_1 \otimes [(-v-1)g_8(k-1,q-1-v,v-1) - g_{10}(k-2,q-v-2,v-1)]; \\
h_2'(r_4) =& - u_0 \otimes [-f_4(k-1,q-1-v,v) - f_5(k-1,q-1-v,v-1)] \\
&+ u_1[g_6(k-1,q-1-v,v-1) - g_7(k-1,q-1-v,v-1)]; \\
h_2'(r_5) =& - u_0 \otimes [(v+1)g_8(k-1,q-1-v,v-1) + g_{10}(k-2,q-2-v,v-1)] \\
&- u_0 \otimes [g_6(k-1,q-1-v,v-1) - g_7(k-1,q-1-v,v-1)] \\
&+ u_1 \otimes [-f_3(k-1,q-1-v,v) + kf_5(k-1,q-1-v,v-1) \\
&- vf_6(k,q-v-1,v-1)]; \\
h_2'(r_6) =& u_0 \otimes [-(v+1)g_8(k-1,q-1-v,v-1) - g_{10}(k-2,q-2-v,v-1)] \\
&- u_0 \otimes [g_6(k-1,q-1-v,v-1) - g_7(k-1,q-1-v,v-1)] \\
&+ u_1 \otimes [-f_3(k-1,q-1-v,v) + kf_4(k-1,q-1-v,v) \\
&- vf_6(k,q-1-v,v-1)]; \\
h_2'(r_7) =& u_0 \otimes [g_6(k-1,q-1-v,v-1) - g_7(k-1,q-1-v,v-1)] \\
&+ ku_1 \otimes [-f_4(k-1,q-1-v,v) - f_5(k-1,q-1-v,v-1)]; \\
h_2'(r_8) =& u_0 \otimes [-vg_6(k-1,q-1-v,v-1) + vg_7(k-1,q-1-v,v-1) \\
&+ k(-v-1)g_8(k-1,q-1-v,v-1) - kg_{10}(k-2,q-2-v,v-1)] \\
&+ ku_0 \otimes [(v+1)g_8(k-1,q-1-v,v-1) + g_{10}(k-2,q-v-2,v-1)] \\
&- u_0 \otimes [g_6(k-1,q-1-v,v-1) - g_7(k-1,q-1-v,v-1)] \\
&+ ku_1 \otimes [(v+1)f_4(k-1,q-1-v,v) + (v+1)f_5(k-1,q-1-v,v-1)].
\end{aligned}$$

On $\{r_1, r_2, r_3, r_4\}$ we get a matrix

$$\begin{pmatrix}
1 & -1 & 1 & 0 \\
k-v & -k & k & 1 \\
-v & 0 & 0 & 1 \\
v & -v & v & 0 \\
-v & 0 & 0 & 1 \\
v & 0 & 0 & -1 \\
k(-v-1) & k(v+1) & -k(v+1) & 0 \\
-k & k & -k & 0
\end{pmatrix}.$$

The solutions are $r_2 + r_3$ and $r_1 + r_2 + vr_4$.
On $\{r_5, r_6, r_7, r_8\}$ we get a matrix

$$\begin{pmatrix} -1 & -1 & 1 & -v-1 \\ 1 & 1 & -1 & v+1 \\ -v-1 & -v-1 & 0 & 0 \\ -1 & -1 & 0 & 0 \\ -1 & -1 & 0 & 0 \\ 0 & 0 & -k & kv+k \\ k & k & -k & kv+k \\ -v & -v & 0 & 0 \end{pmatrix}.$$

The solutions are $r_5 - r_6$ and $(1+v)r_7 + r_8$.
The basis of the solutions is

$$\{r_2 + r_3, r_1 + r_2 + vr_4, r_5 - r_6, (1+v)r_7 + r_8\}.$$

Trying the other approach, for elements of $S_{(k+q-v,k+v)}$ annihilated by h_2', compute the nullspace of the matrix

$$\begin{pmatrix} 0 & 0 & 0 & 0 & -1 & -1 & 0 & 0 \\ 0 & 0 & 0 & 0 & 0 & 0 & -1 & v+1 \\ 0 & 0 & 0 & 0 & k & k & -1 & v+1 \\ 0 & 0 & 0 & 0 & -v & -v & 0 & 0 \\ -v & 0 & 0 & 1 & 0 & 0 & 0 & 0 \\ v & 0 & 0 & -1 & 0 & 0 & 0 & 0 \\ k & v+1 & -v-1 & 0 & 0 & 0 & 0 & 0 \\ -k & 1 & -1 & 0 & 0 & 0 & 0 & 0 \end{pmatrix}.$$

Solutions are

$$f_4 + f_5, g_6 - g_7, (v+1)g_8 + g_{10}.$$

<u>Case $p \nmid k$, $p \nmid k + q$.</u>
The action of h_1' on U is given by

$$\begin{pmatrix} 0 & e \\ e & 0 \end{pmatrix},$$

where $k = e^2$. The action of h_2' on U is given by

$$\begin{pmatrix} 0 & -l \\ l & 0 \end{pmatrix},$$

where $l^2 = -k - q$.

The matrix of h_1' on $U \otimes S_{(k+q-v,k+v)}$ is

$$\begin{pmatrix}
0 & 0 & 0 & 0 & 0 & 0 & 0 & -1 & e & 0 & 0 & 0 & 0 & 0 & 0 & 0 \\
0 & 0 & 0 & 0 & k & 0 & 0 & -k & 0 & e & 0 & 0 & 0 & 0 & 0 & 0 \\
0 & 0 & 0 & 0 & 0 & -k & 0 & 0 & 0 & 0 & e & 0 & 0 & 0 & 0 & 0 \\
0 & 0 & 0 & 0 & k & k & -1 & 1 & 0 & 0 & 0 & e & 0 & 0 & 0 & 0 \\
k & -1 & 0 & 0 & 0 & 0 & 0 & 0 & 0 & 0 & 0 & 0 & e & 0 & 0 & 0 \\
0 & 0 & 1 & 0 & 0 & 0 & 0 & 0 & 0 & 0 & 0 & 0 & 0 & e & 0 & 0 \\
k^2+k & -k & k & k & 0 & 0 & 0 & 0 & 0 & 0 & 0 & 0 & 0 & 0 & e & 0 \\
k & 0 & 0 & 0 & 0 & 0 & 0 & 0 & 0 & 0 & 0 & 0 & 0 & 0 & 0 & e \\
e & 0 & 0 & 0 & 0 & 0 & 0 & 0 & 0 & 0 & 0 & 0 & 0 & 0 & 0 & 1 \\
0 & e & 0 & 0 & 0 & 0 & 0 & 0 & 0 & 0 & 0 & 0 & -k & 0 & 0 & k \\
0 & 0 & e & 0 & 0 & 0 & 0 & 0 & 0 & 0 & 0 & 0 & 0 & k & 0 & 0 \\
0 & 0 & 0 & e & 0 & 0 & 0 & 0 & 0 & 0 & 0 & 0 & -k & -k & 1 & -1 \\
0 & 0 & 0 & 0 & e & 0 & 0 & 0 & -k & 1 & 0 & 0 & 0 & 0 & 0 & 0 \\
0 & 0 & 0 & 0 & 0 & e & 0 & 0 & 0 & 0 & -1 & 0 & 0 & 0 & 0 & 0 \\
0 & 0 & 0 & 0 & 0 & 0 & e & 0 & -k^2-k & k & -k & -k & 0 & 0 & 0 & 0 \\
0 & 0 & 0 & 0 & 0 & 0 & 0 & e & -k & 0 & 0 & 0 & 0 & 0 & 0 & 0
\end{pmatrix}.$$

The basis of elements annihilated by h_1' on $U \otimes S_{(k+q-v,k+v)}$ consists of

$$\begin{aligned}
r_1 =& eu_0 \otimes g_6(k-1, q-1-v, v-1) + (e+e^3)u_0 \otimes g_8(k-1, q-1-v, v-1) \\
&+ eu_0 \otimes g_{10}(k-2, q-v-2, v-1) + u_1 \otimes f_3(k-1, q-1-v, v), \\
r_2 =& -u_0 \otimes g_6(k-1, q-1-v, v-1) - e^2u_0 \otimes g_8(k-1, q-1-v, v-1) \\
&+ eu_1 \otimes f_4(k-1, q-1-v, v), \\
r_3 =& u_0 \otimes g_7(k-1, q-1-v, v-1) + e^2u_0 \otimes g_8(k-1, q-1-v, v-1) \\
&+ eu_1 \otimes f_5(k-1, q-1-v, v-1), \\
r_4 =& eu_0 \otimes g_8(k-1, q-1-v, v-1) + u_1 \otimes f_6(k, q-v-1, v-1), \\
r_5 =& eu_0 \otimes f_4(k-1, q-1-v, v) + eu_0 \otimes f_6(k, q-v-1, v-1) \\
&+ u_1 \otimes g_6(k-1, q-1-v, v-1), \\
r_6 =& -eu_0 \otimes f_5(k-1, q-1-v, v-1) + eu_0 \otimes f_6(k, q-1-v, v-1) \\
&+ u_1 \otimes g_7(k-1, q-1-v, v-1), \\
r_7 =& -u_0 \otimes f_6(k, q-v-1, v-1) + eu_1 \otimes g_8(k-1, q-1-v, v-1), \\
r_8 =& -u_0 \otimes f_3(k-1, q-1-v, v) - e^2u_0 \otimes f_4(k-1, q-1-v, v) \\
&+ u_0 \otimes f_6(k, q-v-1, v-1) + eu_1 \otimes g_{10}(k-2, q-2-v, v-1).
\end{aligned}$$

We have

$$\begin{aligned}
h_2'(r_1) =& elu_1 \otimes g_6(k-1, q-1-v, v) + (e+e^3)lu_1 \otimes g_8(k-1, q-1-v, v-1) \\
&+ elu_1 \otimes g_{10}(k-2, q-2-v, v-1) - lu_0 \otimes f_3(k-1, q-1-v, v) \\
&+ eu_0 \otimes [-f_3(k-1, q-1-v, v) + kf_5(k-1, q-1-v, v-1) \\
&- vf_6(k, q-1-v, v-1)] \\
&+ (e+e^3)u_0 \otimes [-f_4(k-1, q-1-v, v) - f_5(k-1, q-1-v, v-1)] \\
&+ eu_0 \otimes [(-k-q+v+1)f_4(k-1, q-1-v, v) \\
&+ (v+1)f_5(k-1, q-1-v, v-1)]
\end{aligned}$$

$$
\begin{aligned}
&- u_1 \otimes [(k+q-v)g_6(k-1,q-1-v,v-1) + vg_7(k-1,q-1-v,v-1)\\
&+ k(k+q-v-1)g_8(k-1,q-1-v,v-1) - kg_{10}(k-2,q-2-v,v-1)]\\
&= u_0 \otimes [(-e-l)f_3(k-1,q-1-v,v)\\
&+ (-e^3+el^2+ev)f_4(k-1,q-1-v,v)\\
&+ evf_5(k-1,q-1-v,v-1) - evf_6(k,q-1-v,v-1)]\\
&+ u_1 \otimes [(el+l^2+v)g_6(k-1,q-1-v,v-1) - vg_7(k-1,q-1-v,v-1)\\
&+ (el+e^3l-k(k+q-v-1))g_8(k-1,q-1-v,v-1)\\
&+ (el+k)g_{10}(k-2,q-2-v,v-1)],
\end{aligned}
$$

$$
\begin{aligned}
h_2'(r_2) =& - lu_1 \otimes g_6(k-1,q-1-v,v-1) - e^2lu_1 \otimes g_8(k-1,q-1-v,v-1)\\
&- elu_0 \otimes f_4(k-1,q-1-v,v)\\
&- u_0 \otimes [-f_3(k-1,q-1-v,v) + kf_5(k-1,q-1-v,v-1)\\
&- vf_6(k,q-1-v,v-1)]\\
&- e^2u_0 \otimes [-f_4(k-1,q-1-v,v) - f_5(k-1,q-1-v,v-1)]\\
&- eu_1 \otimes [(v+1)g_8(k-1,q-1-v,v-1) + g_{10}(k-2,q-2-v,v-1)]\\
&= u_0 \otimes [f_3(k-1,q-1-v,v) + (-el+e^2)f_4(k-1,q-1-v,v)\\
&+ (-k+e^2)f_5(k-1,q-1-v,v-1) + vf_6(k,q-1-v,v-1)]\\
&+ u_1 \otimes [(-l)g_6(k-1,q-1-v,v-1)\\
&+ (-e^2l-ev-e)g_8(k-1,q-1-v,v-1)\\
&- eg_{10}(k-2,q-2-v,v-1)],
\end{aligned}
$$

$$
\begin{aligned}
h_2'(r_3) =& lu_1 \otimes g_7(k-1,q-1-v,v-1) + e^2lu_1 \otimes g_8(k-1,q-1-v,v-1)\\
&- elu_0 \otimes f_5(k-1,q-1-v,v-1)\\
&+ u_0 \otimes [-f_3(k-1,q-1-v,v) + kf_5(k-1,q-1-v,v-1)\\
&+ (k+q-v)f_6(k-1,q-1-v,v-1)]\\
&+ e^2u_0 \otimes [-f_4(k-1,q-1-v,v) - f_5(k-1,q-1-v,v-1)]\\
&- eu_1 \otimes [(k+q-v-1)g_8(k-1,q-1-v,v-1) - g_{10}(k-2,q-2-v,v-1)]\\
&= u_0 \otimes [-f_3(k-1,q-1-v,v) - e^2f_4(k-1,q-1-v,v)\\
&+ (-el+k-e^2)f_5(k-1,q-1-v,v-1) + (k+q-v)f_6(k,q-1-v,v-1)]\\
&+ u_1 \otimes [lg_7(k-1,q-1-v,v-1)\\
&+ (e^2l-e(k+q-v-1))g_8(k-1,q-1-v,v-1)\\
&+ eg_{10}(k-2,q-2-v,v-1)],
\end{aligned}
$$

$$
\begin{aligned}
h_2'(r_4) =& elu_1 \otimes g_8(k-1,q-1-v,v-1) - lu_0 \otimes f_6(k,q-1-v,v-1)\\
&+ eu_0 \otimes [-f_4(k-1,q-1-v,v) - f_5(k-1,q-1-v,v-1)]\\
&- u_1 \otimes [g_6(k-1,q-1-v,v-1) - g_7(k-1,q-1-v,v-1)]\\
&= u_0 \otimes [-ef_4(k-1,q-1-v,v) - ef_5(k-1,q-1-v,v-1)\\
&- lf_6(k,q-1-v,v-1)]\\
&+ u_1 \otimes [-g_6(k-1,q-1-v,v-1) + g_7(k-1,q-1-v,v-1)\\
&+ elg_8(k-1,q-1-v,v-1)],
\end{aligned}
$$

$$
\begin{aligned}
h_2'(r_5) =& elu_1 \otimes f_4(k-1, q-1-v, v) + elu_1 \otimes f_6(k, q-1-v, v-1) \\
& - lu_0 \otimes g_6(k-1, q-1-v, v-1) \\
& + eu_0 \otimes [(v+1)g_8(k-1, q-1-v, v-1) + g_{10}(k-2, q-2-v, v-1)] \\
& + eu_0 \otimes [g_6(k-1, q-1-v, v) - g_7(k-1, q-1-v, v-1)] \\
& - u_1 \otimes [-f_3(k-1, q-1-v, v) + kf_5(k-1, q-1-v, v-1) \\
& - vf_6(k, q-1-v, v-1)] \\
& = u_0 \otimes [(-l+e)g_6(k-1, q-1-v, v-1) - eg_7(k-1, q-1-v, v-1) \\
& + (ev+e)g_8(k-1, q-1-v, v-1) + eg_{10}(k-2, q-2-v, v-1)] \\
& + u_1 \otimes [f_3(k-1, q-1-v, v) + elf_4(k-1, q-1-v, v) \\
& - kf_5(k-1, q-1-v, v-1) + (el+v)f_6(k, q-1-v, v-1)],
\end{aligned}
$$

$$
\begin{aligned}
h_2'(r_6) =& - elu_1 \otimes f_5(k-1, q-1-v, v-1) + elu_1 \otimes f_6(k, q-1-v, v-1) \\
& - lu_0 \otimes g_7(k-1, q-1-v, v-1) \\
& - eu_0 \otimes [(k+q-v-1)g_8(k-1, q-1-v, v-1) - g_{10}(k-2, q-2-v, v-1)] \\
& + eu_0 \otimes [g_6(k-1, q-1-v, v) - g_7(k-1, q-1-v, v-1)] \\
& - u_1 \otimes [-f_3(k-1, q-1-v, v) + kf_5(k-1, q-1-v, v-1) \\
& + (k+q-v)f_6(k, q-1-v, v-1)] \\
& = u_0 \otimes [eg_6(k-1, q-1-v, v-1) + (-l-e)g_7(k-1, q-1-v, v-1) \\
& + (el^2+ev+e)g_8(k-1, q-1-v, v-1) + eg_{10}(k-2, q-2-v, v-1)] \\
& + u_1 \otimes [f_3(k-1, q-1-v, v) + (-el-k)f_5(k-1, q-1-v, v-1) \\
& + (el-k-q+v)f_6(k, q-1-v, v-1)],
\end{aligned}
$$

$$
\begin{aligned}
h_2'(r_7) =& - lu_1 \otimes f_6(k, q-1-v, v-1) - elu_0 \otimes g_8(k-1, q-1-v, v-1) \\
& u_0 \otimes [g_6(k-1, q-1-v, v) - g_7(k-1, q-1-v, v-1)] \\
& - eu_1 \otimes [-f_4(k-1, q-1-v, v) - f_5(k-1, q-1-v, v-1)] \\
& = u_0 \otimes [-g_6(k-1, q-1-v, v-1) + g_7(k-1, q-1-v, v-1) \\
& - elg_8(k-1, q-1-v, v-1)] \\
& + u_1 \otimes [+ef_4(k-1, q-1-v, v) + ef_5(k-1, q-1-v, v-1) \\
& - lf_6(k, q-1-v, v-1)],
\end{aligned}
$$

$$
\begin{aligned}
h_2'(r_8) =& - lu_1 \otimes f_3(k-1, q-1-v, v) - e^2lu_1 \otimes f_4(k-1, q-1-v, v) \\
& + lu_1 \otimes f_6(k, q-1-v, v-1) \\
& - elu_0 \otimes g_{10}(k-2, q-2-v, v-1) \\
& - u_0 \otimes [(k+q-v)g_6(k-1, q-1-v, v) + vg_7(k-1, q-1-v, v-1) \\
& + k(k+q-v-1)g_8(k-1, q-1-v, v-1) - kg_{10}(k-2, q-2-v, v-1)] \\
& - e^2u_0 \otimes [(v+1)g_8(k-1, q-1-v, v-1) + g_{10}(k-2, q-2-v, v-1)] \\
& + u_0 \otimes [g_6(k-1, q-1-v, v) - g_7(k-1, q-1-v, v-1)] \\
& - eu_1 \otimes [(-k-q+v+1)f_4(k-1, q-1-v, v) \\
& + (v+1)f_5(k-1, q-1-v, v-1)]
\end{aligned}
$$

$$
\begin{aligned}
&= u_0 \otimes [(-k-q+v+1)g_6(k-1,q-1-v,v-1)\\
&+(-v-1)g_7(k-1,q-1-v,v-1)\\
&+(-k(k+q-v-1)-e^2v-e^2)g_8(k-1,q-1-v,v-1)\\
&+(-el+k-e^2)g_{10}(k-2,q-2-v,v-1)]\\
&-u_1 \otimes lf_3(k-1,q-1-v,v)\\
&+(-e^2l-el^2+ek+eq-ev-e)f_4(k-1,q-1-v,v)\\
&-e(v+1)f_5(k-1,q-1-v,v-1)+lf_6(k,q-1-v,v-1)],
\end{aligned}
$$

On the set $\{r_1, r_2, r_3, r_4\}$ we get the matrix

$$
\begin{pmatrix}
-e-l & 1 & -1 & 0\\
-e^3+el^2+ev & -el+e^2 & -e^2 & -e\\
ev & 0 & -el & -e\\
-ev & v & -l^2-v & -l\\
el+l^2+v & -l & 0 & -1\\
-v & 0 & l & 1\\
el+e^3l+e^2l^2+e^2v+e^2 & -e^2l-ev-e & e^2l+el^2+ev+e & el\\
el+e^2 & -e & e & 0
\end{pmatrix},
$$

which has a solution $lr_1+(el+l^2+v)r_2+vr_3$ and $r_2+r_3-lr_4$.

On the set $\{r_5, r_6, r_7, r_8\}$ we get the matrix

$$
\begin{pmatrix}
-l+e & e & -1 & l^2+v+1\\
-e & -l-e & 1 & -v-1\\
ev+e & el^2+e+ev & -el & e^2l^2\\
e & e & 0 & -el\\
1 & 1 & 0 & -l\\
el & 0 & e & -e^2l-el^2-ev-e\\
-e^2 & -el-e^2 & e & -ev-e\\
el+v & el+l^2+v & -l & l
\end{pmatrix},
$$

which has the solution $lr_5+(1+el+v)r_7+r_8$ and $-r_5+r_6+lr_7$.

The basis of the solutions is

$$
\{lr_1+(el+l^2+v)r_2+vr_3, r_2+r_3-lr_4, lr_5+(1+el+v)r_7+r_8, -r_5+r_6+lr_7\}.
$$

□

THE PROOF OF PROPOSITION 8.1. The right action of e_β is given as

$$(s_{22})e_\beta = -s_{21}, (s_{12})e_\beta = -s_{11}, (s_{21})e_\beta = (s_{11})e_\beta = 0,$$

$$(s'_{22})e_\beta = -s'_{21}, (s'_{12})e_\beta = -s'_{11}, (s'_{21})e_\beta = (s'_{11})e_\beta = 0,$$

$$(s_{22})e'_\beta = s'_{21}, (s_{12})e'_\beta = s'_{11}, (s_{21})e'_\beta = (s_{11})e'_\beta = 0,$$

$$(s'_{22})e'_\beta = -s_{21}, (s'_{12})e'_\beta = -s_{11}, (s'_{21})e'_\beta = (s'_{11})e'_\beta = 0.$$

$$
\begin{aligned}
(x_1)e_\beta =& - cB(a,b+1,c-1) = -cx_1(a,b+1,c-1);\\
(w_1)e_\beta =& - cB(a,b+1,c-1)s'_{21} = -cw_1(a,b+1,c-1);\\
(w_2)e_\beta =& - cB(a,b+1,c-1)s'_{22} - B(a,b,c)s'_{21}\\
&= -cw_2(a,b+1,c-1) - w_1(a,b,c);\\
(w_3)e_\beta =& - cB(a,b+1,c-1)(s_{11}s'_{21} - s_{21}s'_{11}) = -cw_3(a,b+1,c-1);\\
(w_4)e_\beta =& - cB(a,b+1,c-1)(s_{11}s'_{22} - s_{21}s'_{12}) - B(a,b,c)(s_{11}s'_{21} - s_{21}s'_{11})\\
&= -cw_4(a,b+1,c-1) - w_3(a,b,c);\\
(w_5)e_\beta =& - cB(a,1,c-1)(s_{12}s'_{21} - s_{22}s'_{11}) - B(a,0,c)(s_{11}s'_{21} - s_{21}s'_{11}) =\\
&= -cw_5(a,1,c-1) - w_3(a,0,c)\\
&= -cw_3(a,0,c) + cw_1(a+1,0,c-1) - w_3(a,0,c);\\
&(because\ s_{21}(s_{12}s'_{21} - s_{22}s'_{11}) = s_{22}(s_{11}s'_{21} - s_{21}s'_{11}) - dets'_{21})\\
(v_1)e_\beta =& - cB(a,b+1,c-1)s'_{21}s'_{22} = -cv_1(a,b+1,c-1);\\
(v_2)e_\beta =& - cB(a,b+1,c-1)s'_{12}s'_{22} - B(a,b,c)s'_{11}s'_{22} - B(a,b,c)s'_{12}s'_{21}\\
&= -cv_2(a,b+1,c-1) + v_6(a,b-1,c) + v_5(a,b-1,c) \text{ if } b > 0;\\
&(because\ (s'_{11}s'_{22} + s'_{12}s'_{21})s_{21}\\
&= -(s_{11}s'_{22} - s'_{12}s_{21})s'_{21} - (s_{11}s'_{21} - s'_{11}s_{21})s'_{22})\\
&= -cv_2(a,b+1,c-1) - v_9(a,0,c-1) - v_{10}(a,0,c-1) \text{ if } b = 0,\ c > 0;\\
&= -v_{11}(a,0,0) \text{ if } b = c = 0;\\
(v_3)e_\beta =& - cB(a,b+1,c-1)s'_{11}s'_{21} = -cv_3(a,b+1,c-1);\\
(v_5)e_\beta =& - cB(a,b+1,c-1)(s_{11}s'_{21}s'_{22} - s_{21}s'_{11}s'_{22}) - B(a,b,c)(-s_{21}s'_{11}s'_{21})\\
&= -cv_5(a,b+1,c-1) + v_3(a,b+1,c);\\
(v_6)e_\beta =& - cB(a,b+1,c-1)(s_{11}s'_{21}s'_{22} - s_{21}s'_{12}s'_{21}) - B(a,b,c)(-s_{21}s'_{11}s'_{21})\\
&= -cv_6(a,b+1,c-1) + v_3(a,b+1,c);\\
(v_9)e_\beta =& - cv_9(a,1,c-1) - B(a,0,c)[s_{11}s'_{21}s'_{22} - s_{21}s'_{11}s'_{22} - s_{22}s'_{11}s'_{21}]\\
&= -cv_9(a,1,c-1) - v_5(a,0,c) + v_3(a,0,c+1);\\
&= -cv_5(a,0,c) + cv_1(a+1,0,c-1) - v_5(a,0,c) + v_3(a,0,c+1);\\
(v_{10})e_\beta =& - cv_{10}(a,1,c-1) - B(a,0,c)[\ s_{11}s'_{21}s'_{22} \quad s_{21}s'_{12}s'_{21} \quad s_{22}s'_{11}s'_{21}]\\
&- cv_{10}(a,1,c-1) - v_6(a,0,c) + v_3(a,0,c+1);\\
&= -cv_6(a,0,c) - cv_1(a+1,0,c-1) - v_6(a,0,c) + v_3(a,0,c+1);\\
(z_1)e_\beta =& - cz_1(a,b+1,c-1);\\
(z_2)e_\beta =& - cz_2(a,b+1,c-1) - B(a,b,c)s'_{11}s'_{21}s'_{22}\\
&= -cz_2(a,b+1,c-1) - z_1(a,b,c);\\
(z_3)e_\beta =& - cz_3(a,b+1,c-1);\\
(z_4)e_\beta =& - cz_4(a,b+1,c-1) - B(a,b,c)[s_{11}s_{21}s'_{11}s'_{21}s'_{22} + s_{21}^2s'_{11}s'_{12}s'_{21}]\\
&= -cz_4(a,b+1,c-1) - z_3(a,b,c);\\
(z_5)e_\beta =& - cz_5(a,1,c-1) - B(a,0,c)[s_{11}s_{21}s'_{11}s'_{21}s'_{22} + s_{21}^2s'_{11}s'_{12}s'_{21}];\\
&= -cz_5(a,1,c-1) - z_3(a,0,c)\\
&= -cz_3(a,0,c) + cz_1(a+1,1,c-1) - z_3(a,0,c);
\end{aligned}
$$

$$\begin{aligned}(z_6)e_\beta =& - cz_6(a,1,c-1) - B(a,0,c)[s_{11}s_{21}s'_{12}s'_{21}s'_{22} + s_{11}s_{22}s'_{11}s'_{21}s'_{22}\\ &+ s_{21}^2s'_{11}s'_{12}s'_{22} + s_{21}s_{22}s'_{11}s'_{12}s'_{21}]\\ &= -cz_6(a,1,c-1) - z_4(a,0,c) - z_5(a,0,c) - z_1(a+1,0,c)\\ &= -cz_4(a,0,c) - z_4(a,0,c) - z_5(a,0,c) - z_1(a+1,0,c);\\ (z_7)e_\beta =& - cz_7(a,1,c-1) - B(a,0,c)[s_{11}s_{22}s'_{11}s'_{21}s'_{22} + s_{12}s_{21}s'_{11}s'_{21}s'_{22};\\ &+ 2s_{22}s_{21}s'_{11}s'_{12}s'_{21}] = -cz_7(a,1,c-1) - 2z_5(a,0,c) - z_1(a+1,0,c)\\ &= -cz_5(a,0,c) - 2z_5(a,0,c) - z_1(a+1,0,c);\\ (y_1)e_\beta =& - cy_1(a,b+1,c-1).\end{aligned}$$

We have used the formulae

$$\begin{aligned}v_9(a,1,c-1) =& v_5(a,0,c) - v_1(a+1,0,c-1)\\ &= B(a,0,c-1)[s_{11}s_{22}s'_{21}s'_{22} - s_{21}s_{22}s'_{11}s'_{22} - s_{11}s_{22}s'_{21}s'_{22}\\ &+ s_{12}s_{21}s'_{21}s'_{22}],\\ v_{10}(a,1,c-1) =& v_6(a,0,c) + v_1(a+1,0,c-1)\\ &= B(a,0,c-1)[-s_{11}s_{22}s'_{21}s'_{22} - s_{21}s_{22}s'_{12}s'_{21} + s_{11}s_{22}s'_{21}s'_{22}\\ &- s_{12}s_{21}s'_{21}s'_{22}],\\ z_5(a,1,c-1) =& z_3(a,0,c) - z_1(a+1,1,c-1)\\ =& B(a,0,c-1)[s_{11}s_{21}s_{22}s'_{11}s'_{21}s'_{22} + s_{21}^2s_{22}s'_{11}s'_{12}s'_{21}\\ &- s_{11}s_{22}s_{21}s'_{11}s'_{21}s'_{22} + s_{12}s_{21}^2s'_{11}s'_{21}s'_{22}],\\ z_6(a,1,c-1) =& z_4(a,0,c)\\ =& B(a,0,c-1)[s_{11}s_{21}s_{22}s'_{12}s'_{21}s'_{22} + s_{21}^2s_{22}s'_{11}s'_{12}s'_{22}],\\ z_7(a,1,c-1) =& z_5(a,0,c)\\ =& B(a,0,c-1)[s_{12}s_{21}s_{22}s'_{11}s'_{21}s'_{22} + s_{21}s_{22}^2s'_{11}s'_{12}s'_{21}].\end{aligned}$$

Based on these formulas, we now compute the right action by e_β as follows:

$$\begin{aligned}f_1(pt,0,0)e_\beta =& 0;\\ f_2(a,0,c)e_\beta =& - (c+1)x_1(a+1,1,c)\\ &+ (a+1)cv_6(a,0,c) + (a+1)cv_1(a+1,0,c-1) + (a+1)v_6(a,0,c)\\ &- (a+1)v_3(a,0,c+1)\\ &- (a+1)cv_5(a,0,c) + (a+1)cv_1(a+1,0,c-1) - (a+1)v_5(a,0,c)\\ &+ (a+1)v_3(a,0,c+1)\\ &+ a(a+1)(c+1)y_1(a-1,1,c)\\ &= -(c+1)f_3(a,0,c) + 2(a+1)cf_6(a+1,0,c-1);\\ f_3(a,b,c)e_\beta =& - cx_1(a+1,b+2,c-1)\\ &- (a+1)cv_5(a,b+1,c-1) + (a+1)v_3(a,b+1,c)\\ &+ (a+1)cv_6(a,b+1,c-1) - (a+1)v_3(a,b+1,c)\\ &+ a(a+1)cy_1(a-1,b+2,c-1)\\ &= -cf_3(a,b+1,c-1);\end{aligned}$$

$$
\begin{aligned}
f_4(a,b,c)e_\beta =& -(c+1)v_3(a,b+1,c)\\
& - cv_6(a,b+1,c-1) + v_3(a,b+1,c)\\
& - acy_1(a-1,b+2,c-1)\\
& = -cf_4(a,b+1,c-1);
\end{aligned}
$$

$$
\begin{aligned}
f_5(a,b,c)e_\beta =& - cv_2(a,b+3,c-1) + v_6(a,b+1,c) + v_5(a,b+1,c)\\
& - (c+1)v_5(a,b+1,c) + v_3(a,b+1,c+1)\\
& - a(c+1)y_1(a-1,b+2,c)\\
& = -cf_5(a,b+1,c-1) + f_4(a,b+1,c);\\
f_6(a,b,c)e_\beta =& - cv_1(a,b+1,c-1) = -cf_6(a,b+1,c-1);\\
f_7(a,b,c)e_\beta =& - cv_2(a,2,c-1) + v_6(a,0,c) + v_5(a,0,c)\\
& - cv_5(a,0,c) + cv_1(a+1,0,c-1) - v_5(a,0,c) + v_3(a,0,c+1)\\
& + (c+1)ay_1(a-1,1,c)\\
& = -cf_5(a,0,c-1) + cf_6(a+1,0,c-1) + f_4(a,0,c);\\
g_1(a,0,c)e_\beta =& - cw_2(a+1,1,c-1) - w_1(a+1,0,c) - (a+1)cz_2(a,1,c-1)\\
& - (a+1)z_1(a,0,c)\\
& = -c[w_2(a+1,1,c-1) + (a+1)z_2(a,1,c-1)] - [w_1(a+1,0,c)\\
& + (a+1)z_2(a,1,c-1)]\\
& - (a+1)[z_1(a,0,c) - z_2(a,1,c-1)]\\
& = -cg_7(a,0,c-1) - g_6(a,0,c-1) - (a+1)g_8(a,0,c-1) \text{ if } c>0;\\
& - -g_3(a,0,0) \text{ if } c=0;\\
g_2(a,0,c)e_\beta =& - (c+1)w_4(a,1,c) - w_3(a,0,c+1)\\
& + (c+1)w_3(a,0,c+1) - (c+1)w_1(a+1,0,c) + w_3(a,0,c+1)\\
& - (2a+2)cz_2(a,1,c-1) - (2a+2)z_1(a,0,c)\\
& + acz_4(a-1,0,c) + az_4(a-1,0,c) + az_5(a-1,0,c) + az_1(a,0,c)\\
& - acz_5(a-1,0,c) - 2az_5(a-1,0,c) - az_1(a,0,c)\\
& = -(c+1)g_9(a-1,0,c-1) - (c+1)g_6(a,0,c-1)\\
& - (2a+2)g_8(a,0,c-1) \text{ if } c>0;\\
& = -g_4(a-1,0,0) - g_3(a,0,0) \text{ if } c=0;\\
g_3(a,b,0)e_\beta =& 0;\\
g_4(a,0,0)e_\beta =& - w_3(a+1,1,0)\\
& - w_3(a+1,1,0)\\
& + 0\\
& + (a+1)z_3(a,0,0)\\
& - (a+1)z_3(a,0,0)\\
& = 0;
\end{aligned}
$$

$$\begin{aligned}
g_5(a,b,0)e_\beta =& w_3(a+1,b+2,0) \\
& - w_3(a+1,b+2,0) \\
& + 0 \\
& - (a+1)z_3(a,b+1,0) \\
& + (a+1)z_3(a,b+1,0) \\
& = 0;
\end{aligned}$$

$$\begin{aligned}
g_6(a,b,c)e_\beta =& -(c+1)w_1(a+1,b+1,c) \\
& - (a+1)cz_2(a,b+2,c-1) - (a+1)z_1(a,b+1,c) \\
& = -cg_6(a,b+1,c-1) - g_6(a,b+1,c-1) \\
& + (a+1)g_8(a,b+1,c-1) \text{ if } c > 0; \\
& = -g_3(a,b+1,0) \text{ if } c = 0; \\
g_7(a,b,c)e_\beta =& -cw_2(a+1,b+2,c-1) - w_1(a+1,b+1,c) \\
& - (a+1)cz_2(a,b+2,c-1) - (a+1)z_1(a,b+1,c) \\
& = -cg_7(a,b+1,c-1) - g_6(a,b+1,c-1) \\
& + (a+1)g_8(a,b+1,c-1) \text{ if } c > 0; \\
& = -g_3(a,b+1,0) \text{ if } c = 0; \\
g_8(a,b,c)e_\beta =& -(c+1)z_1(a,b+1,c) \\
& + cz_2(a,b+2,c-1) + z_1(a,b+1,c) \\
& = -cg_8(a,b+1,c-1); \\
g_9(a,0,c)e_\beta =& (c+2)w_3(a+1,1,c+1) \\
& - (c+1)w_4(a+1,2,c) - w_3(a+1,1,c+1) \\
& - (a+2)cz_2(a+1,2,c-1) - (a+2)z_1(a+1,1,c) \\
& + (a+1)(c+1)z_4(a,1,c) + (a+1)z_3(a,0,c+1) \\
& - (a+1)(c+1)z_3(a,0,c+1) + (a+1)(c+1)z_1(a+1,1,c) \\
& - (a+1)z_3(a,0,c+1) \\
& = -cg_{10}(a,0,c-1) + (a+1)cg_8(a+1,1,c-1) + [w_3(a+1,1,c+1) \\
& - w_4(a+1,2,c) \\
& - z_2(a+1,2,c-1) + (a+1)z_4(a,1,c) - (a+1)z_3(a,0,c+1)] \\
& - [z_1(a+1,1,c) - z_2(a+1,2,c-1)] \\
& = -cg_{10}(a,0,c-1) + (a+1)cg_8(a+1,1,c-1) - g_{10}(a,0,c-1) \\
& - g_8(a+1,1,c-1) \text{ if } c > 0; \\
& = w_3(a+1,1,1) - w_4(a+1,2,0) - z_1(a+1,1,0) + (a+1)z_4(a,1,0) \\
& - (a+1)z_3(a,0,1) \\
& = -g_5(a,0,0) \text{ if } c = 0;
\end{aligned}$$

$$
\begin{aligned}
g_{10}(a,b,c)e_\beta =&(c+2)w_3(a+1,b+2,c+1)\\
&-(c+1)w_4(a+1,b+3,c)-w_3(a+1,b+2,c+1)\\
&-cz_2(a+1,b+3,c-1)-z_1(a+1,b+2,c)\\
&-(a+1)(c+2)z_3(a,b+1,c+1)\\
&+(a+1)(c+1)z_4(a,b+2,c)+(a+1)z_3(a,b+1,c+1)\\
&=-cg_{10}(a,b+1,c-1)-g_{10}(a,b+1,c-1)\\
&+g_8(a+1,b+1,c-1) \text{ if } c>0;\\
&=-g_5(a,b+1,0) \text{ if } c=0; \quad \square
\end{aligned}
$$

THE PROOF OF PROPOSITION 8.2. The right action of e'_β is given as

$$
\begin{aligned}
&(s_{22})e'_\beta = s'_{21}, (s_{12})e'_\beta = s'_{11}, (s_{21})e'_\beta = (s_{11})e'_\beta = 0,\\
&(s'_{22})e'_\beta = -s_{21}, (s'_{12})e'_\beta = -s_{11}, (s'_{21})e'_\beta = (s'_{11})e'_\beta = 0.
\end{aligned}
$$

$$
\begin{aligned}
(x_1)e'_\beta =&aB(a-1,b,c)(s_{11}s'_{21}-s_{21}s'_{11})+cB(a,b,c-1)s'_{21}\\
&=aw_3(a-1,b,c)+cw_1(a,b,c-1);\\
(w_1)e'_\beta =&-aB(a-1,b,c)(-s_{21}s'_{11})s'_{21}=av_3(a-1,b+1,c);\\
(w_2)e'_\beta =&-B(a,b,c)s_{21}-[aB(a-1,b,c)(s_{11}s'_{21}-s_{21}s'_{11})+cB(a,b,c-1)s'_{21}]s'_{22}\\
&=-x_1(a,b+1,c)-av_5(a-1,b,c)-cv_1(a,b,c-1);\\
(w_3)e'_\beta =&-cB(a,b,c-1)s'_{21}(s_{11}s'_{21}-s_{21}s'_{11})=-cB(a,b,c-1)s_{21}s'_{11}s'_{21}\\
&=-cv_3(a,b+1,c-1);\\
(w_4)e'_\beta =&-[aB(a-1,b,c)(s_{11}s'_{21}-s_{21}s'_{11})+cB(a,b,c-1)s'_{21}](s_{11}s'_{22}-s_{21}s'_{12})\\
&+B(a,b,c)(-s_{11}s_{21}+s_{11}s_{21})=aB(a-1,b,c)[-s_{11}^2s'_{21}s'_{22}\\
&+s_{11}s_{21}s'_{11}s'_{22}-s_{11}s_{21}s'_{12}s'_{21}-s_{21}^2s'_{11}s'_{12}]\\
&+cB(a,b,c-1)[-s_{11}s'_{21}s'_{22}-s_{21}s'_{12}s'_{21}]=-av_4(a-1,b,c)+cv_6(a,b,c-1);\\
(w_5)e'_\beta =&-B(a,0,c)(-s'_{11}s'_{21}+s'_{21}s'_{11})-[aB(a-1,0,c)(s_{11}s'_{21}-s_{21}s'_{11})\\
&+cB(a,0,c-1)s'_{21}](s_{12}s'_{21}-s_{22}s'_{11})\\
&-2v_3(a,0,c)-aB(a-1,0,c)(s_{11}s_{22}-s_{12}s_{21})s'_{11}s'_{21}-cB(a,0,c-1)s_{22}s'_{11}s'_{21}\\
&=(-a-c-2)v_3(a,0,c);\\
(v_1)e'_\beta =&B(a,b,c)(-s_{21}s'_{21})+[aB(a-1,b,c)(s_{11}s'_{21}-s_{21}s'_{11})\\
&+cB(a,b,c-1)s'_{21}]s'_{21}s'_{22}\\
&=-w_1(a,b+1,c)-az_1(a-1,b+1,c);\\
(v_2)e'_\beta =&B(a,b,c)[-s_{21}s'_{12}+s_{11}s'_{22}]+[aB(a-1,b,c)(s_{11}s'_{21}-s_{21}s'_{11})\\
&+cB(a,b,c-1)s'_{21}]s'_{12}s'_{22}\\
&=w_4(a,b,c)+aB(a-1,b,c)[-s_{11}s'_{12}s'_{21}s'_{22}-s_{21}s'_{11}s'_{12}s'_{22}]\\
&-cB(a,b,c-1)s'_{12}s'_{21}s'_{22}\\
&=w_4(a,b,c)-az_4(a-1,b-1,c)-cz_2(a,b,c-1) \text{ if } b>0,\\
&=w_4(a,0,c)-az_6(a-1,0,c)-cz_2(a,0,c-1) \text{ if } b=0,c>0,\\
&=w_4(a,0,0)-az_9(a-1,0,0) \text{ if } b=c=0;
\end{aligned}
$$

$$
\begin{aligned}
(v_3)e'_\beta =&[aB(a-1,b,c)(s_{11}s'_{21}-s_{21}s'_{11})+cB(a,b,c-1)s'_{21}]s'_{11}s'_{21}=0;\\
(v_5)e'_\beta =&[aB(a-1,b,c)(s_{11}s'_{21}-s_{21}s'_{11})+cB(a,b,c-1)s'_{21}](s_{11}s'_{21}s'_{22}-s_{21}s'_{11}s'_{22})\\
&+B(a,b,c)(-s_{11}s_{21}s'_{21}+s_{21}^2s'_{11})\\
&=aB(a-1,b,c)(-s_{11}s_{21}s'_{11}s'_{21}s'_{22}+s_{11}s_{21}s'_{11}s'_{21}s'_{22})\\
&+cB(a,b,c-1)s_{21}s'_{11}s'_{21}s'_{22}-w_3(a,b+1,c)\\
&=cz_1(a,b+1,c-1)-w_3(a,b+1,c);\\
(v_6)e'_\beta =&[aB(a-1,b,c)(s_{11}s'_{21}-s_{21}s'_{11})+cB(a,b,c-1)s'_{21}](-s_{11}s'_{21}s'_{22}-s_{21}s'_{12}s'_{21})\\
&+B(a,b,c)(+s_{11}s_{21}s'_{21}-s_{11}s_{21}s'_{21})\\
&=aB(a-1,b,c)(s_{11}s_{21}s'_{11}s'_{21}s'_{22}+s_{21}^2s'_{11}s'_{12}s'_{21})=az_3(a-1,b,c);\\
(v_9)e'_\beta =&B(a,0,c)[-s_{12}s_{21}s'_{21}+s'_{11}s'_{21}s'_{22}+s_{21}s_{22}s'_{11}-s'_{21}s'_{11}s'_{22}]\\
&+[aB(a-1,0,c)(s_{11}s'_{21}-s_{21}s'_{11})+cB(a,0,c-1)s'_{21}](s_{12}s'_{21}s'_{22}-s_{22}s'_{11}s'_{22}]\\
&=-w_3(a,0,c+1)+w_1(a+1,0,c)+2z_1(a,0,c)+az_1(a,0,c)+cz_1(a,0,c);\\
(v_{10})e'_\beta =&B(a,0,c)[+s_{12}s_{21}s'_{21}-s'_{11}s'_{21}s'_{22}-s_{11}s_{22}s'_{21}]\\
&[aB(a-1,0,c)(s_{11}s'_{21}-s_{21}s'_{11})+cB(a,0,c-1)s'_{21}](-s_{12}s'_{21}s'_{22}-s_{22}s'_{12}s'_{21})\\
&=-w_1(a+1,0,c)-z_1(a,0,c)\\
&+aB(a-1,0,c)[s_{12}s_{21}s'_{11}s'_{21}s'_{22}+s_{21}s_{22}s'_{11}s'_{12}s'_{21}]\\
&=-w_1(a+1,0,c)-z_1(a,0,c)+az_5(a-1,0,c);\\
(z_1)e'_\beta =&-[aB(a-1,b,c)(s_{11}s'_{21}-s_{21}s'_{11})+cB(a,b,c-1)s'_{21}]s'_{11}s'_{21}s'_{22}\\
&-B(a,b,c)s_{21}s'_{11}s'_{21}=-v_3(a,b+1,c);\\
(z_2)e'_\beta =&B(a,b,c)[-s_{21}s'_{12}s'_{21}-s_{11}s'_{21}s'_{22}]\\
&-[aB(a-1,b,c)(s_{11}s'_{21}-s_{21}s'_{11})+cB(a,b,c-1)s'_{21}]s'_{12}s'_{21}s'_{22}\\
&=v_6(a,b,c)+ay_1(a-1,b+1,c);\\
(z_3)e'_\beta =&-[aB(a-1,b,c)(s_{11}s'_{21}-s_{21}s'_{11})+cB(a,b,c-1)s'_{21}](s_{11}s_{21}s'_{11}s'_{21}s'_{22}\\
&+s_{21}^2s'_{11}s'_{12}s'_{21})+B(a,b,c)(-s_{11}s_{21}^2s'_{11}s'_{21}+s_{11}s_{21}^2s'_{11}s'_{21})=0;\\
(z_4)e'_\beta =&-[aB(a-1,b,c)(s_{11}s'_{21}-s_{21}s'_{11})+cB(a,b,c-1)s'_{21}](s_{11}s_{21}s'_{12}s'_{21}s'_{22}\\
&+s_{21}^2s'_{11}s'_{12}s'_{22})\\
&+B(a,b,c)(-s_{11}^2s_{21}s'_{21}s'_{22}-s_{11}s_{21}^2s'_{12}s'_{21}+s_{11}s_{21}^2s'_{11}s'_{22}-s_{21}^3s'_{11}s'_{12})\\
&=aB(a-1,b,c)[-s_{11}s_{21}^2s'_{11}s'_{12}s'_{21}s'_{22}+s_{11}s_{21}^2s'_{11}s'_{12}s'_{21}s'_{22}]\\
&-cB(a,b,c-1)(s_{21}^2s'_{11}s'_{12}s'_{21}s'_{22})+v_4(a,b+1,c)\\
&=-cy_1(a,b+2,c-1)-v_4(a,b+1,c);\\
(z_5)e'_\beta =&-[aB(a-1,0,c)(s_{11}s'_{21}-s_{21}s'_{11})+cB(a,0,c-1)s'_{21}](s_{12}s_{21}s'_{11}s'_{21}s'_{22}\\
&+s_{21}s_{22}s'_{11}s'_{12}s'_{21})+B(a,0,c)(-s_{12}s_{21}^2s'_{11}s'_{21}+s_{11}s_{21}s_{22}s'_{11}s'_{21})\\
&=v_3(a+1,1,c);
\end{aligned}
$$

$$
\begin{aligned}
(z_6)e'_\beta =& B(a,0,c)[-s_{11}s_{21}s_{22}s'_{12}s'_{21} - s_{11}^2 s_{22}s'_{21}s'_{22} - s_{21}^2 s_{22}s'_{11}s'_{12} + s_{11}s_{21}s_{22}s'_{11}s'_{22} \\
& - s_{21}s'_{21}s'_{11}s'_{12}s'_{22}] \\
& - [aB(a-1,0,c)(s_{11}s'_{21} - s_{21}s'_{11}) + cB(a,0,c-1)s'_{21}](s_{11}s_{22}s'_{12}s'_{21}s'_{22} \\
& + s_{21}s_{22}s'_{11}s'_{12}s'_{22}) \\
& = -v_4(a,0,c+1) - y_1(a,1,c) + aB(a-1,0,c)[-s_{11}s_{21}s_{22}s'_{11}s'_{12}s'_{21}s'_{22} \\
& + s_{11}s_{21}s_{22}s'_{11}s'_{12}s'_{21}s'_{22}] - cB(a,1,c)s'_{11}s'_{12}s'_{21}s'_{22} \\
& = -v_4(a,0,c+1) - (c+1)y_1(a,0,c+1); \\
(z_7)e'_\beta =& B(a,0,c)[-s_{12}s_{21}s_{22}s'_{11}s'_{21} + s_{11}s_{22}^2 s'_{11}s'_{21}] \\
& - [aB(a-1,0,c)(s_{11}s'_{21} - s_{21}s'_{11}) + cB(a,0,c-1)s'_{21}](s_{12}s_{22}s'_{11}s'_{21}s'_{22} \\
& + s_{22}^2 s'_{11}s'_{12}s'_{21}) = v_3(a+1,0,c+1); \\
(y_1)e'_\beta =& [aB(a-1,b,c)(s_{11}s'_{21} - s_{21}s'_{11}) + cB(a,b,c-1)s'_{21}]s'_{11}s'_{12}s'_{21}s'_{22} \\
& + B(a,b,c)[-s_{11}s'_{11}s'_{21}s'_{22} - s_{21}s'_{11}s'_{12}s'_{21}] = -z_3(a,b-1,c) \text{ if } b>0; \\
& = -z_5(a,0,c-1) - z_1(a+1,0,c-1) \text{ if } b=0, c>0; \\
& = z_{11}(a,0,0) \text{ if } b=c=0;
\end{aligned}
$$

Based on this, we compute

$$
\begin{aligned}
f_1(pt,0,0)e'_\beta =& 0; \\
f_2(a,0,c)e'_\beta =& (a+1)w_3(a,0,c+1) + (c+1)w_1(a+1,0,c) + (a+1)w_1(a+1,0,c) \\
& + (a+1)z_1(a,0,c) - (a+1)az_5(a-1,0,c) - (a+1)w_5(a,0,c+1) \\
& + (a+1)w_1(a+1,0,c) + (a+1)(a+c+2)z_1(a,0,c) \\
& + a(a+1)z_5(a-1,0,c) + a(a+1)z_1(a,0,c) \\
& = (2a+c+3)w_1(a+1,0,c) + (a+1)(2a+c+3)z_1(a,0,c) \\
& = [(2a+c+3)g_6(a,0,c-1) + (a+1)(2a+c+3)g_8(a,0,c-1)] \text{ if } c>0; \\
& = (2a+3)g_3(a,0,0) \text{ if } c=0; \\
f_3(a,b,c)e'_\beta =& (a+1)w_3(a,b+1,c) + cw_1(a+1,b+1,c-1) \\
& + (a+1)cz_1(a,b+1,c-1) - (a+1)w_3(a,b+1,c) \\
& - (a+1)az_3(a-1,b,c) + a(a+1)z_3(a-1,b,c) \\
& = c[w_1(a+1,b+1,c-1) + (a+1)z_2(a,b+2,c-2)] \\
& + (a+1)c[z_1(a,b+1,c-1) - z_2(a,b+2,c-2)] \\
& = cg_6(a,b+1,c-2) + (a+1)g_8(a,b+1,c-2) \text{ if } c>1; \\
& = g_3(a,b+1,0) \text{ if } c=1; \\
& = 0 \text{ if } c=0; \\
f_4(a,b,c)e'_\beta =& 0 + az_3(a-1,b,c) - az_3(a-1,b,c) = 0; \\
f_5(a,b,c)e'_\beta =& w_4(a,b+2,c) - az_4(a-1,b+1,c) - cz_2(a,b+2,c-1) \\
& + (c+1)z_1(a,b+1,c) - w_3(a,b+1,c+1) + az_3(a-1,b,c+1) \\
& = [-w_3(a,b+1,c+1 + w_4(a,b+2,c) + z_2(a,b+2.c-1)
\end{aligned}
$$

$$
\begin{aligned}
&+ az_3(a-1,b,c+1) - az_4(a-1,b_1,c)]\\
&+ (c+1)[z_1(a,b+1,c) - z_2(a,b+2,c-1)]\\
&= g_{10}(a-1,b,c-1) + (c+1)g_8(a,b+1,c-1) \text{ if } c>0;\\
&= g_5(a-1,b,0) \text{ if } c=0;
\end{aligned}
$$

$$
\begin{aligned}
f_6(a,b,c)e'_\beta =& - w_1(a,b+1,c) - az_1(a-1,b+1,c)\\
&= -[w_1(a,b+1,c) + az_2(a-1,b+2,c-1)]\\
&- a[z_1(a-1,b+1,c) - z_2(a-1,b+2,c-1)]\\
&= g_6(a-1,b+1,c-1) - ag_8(a-1,b+1,c-1) \text{ if } c>0;\\
&= -g_3(a-1,b+1,0) \text{ if } c=0;
\end{aligned}
$$

$$
\begin{aligned}
f_7(a,0,c)e'_\beta =& w_4(a,1,c) - az_4(a-1,0,c) - cz_2(a,1,c-1) - w_3(a,0,c+1)\\
&+ w_1(a+1,0,c) + (a+c+2)z_1(a,0,c) + az_5(a-1,0,c) + az_1(a,0,c)\\
&= [-w_3(a,0,c+1) + w_4(a,1,c) + (a+1)z_2(a,1,c-1)\\
&- az_4(a-1,0,c) + az_5(a-1,0,c)] + (2a+c+2)[z_1(a,0,c)\\
&- z_2(a,1,c-1)] + [w_1(a+1,0,c) + (a+1)z_2(a,1,c-1)]\\
&= g_9(a-1,0,c-1) + (2a+c+2)g_8(a,0,c-1) + g_6(a,0,c-1) \text{ if } c>0;\\
&= [-w_3(a,0,1) + w_4(a,1,0) + (a+1)z_1(a,0,0)\\
&- az_4(a-1,0,0) + az_5(a-1,0,0)]\\
&+ [w_1(a+1,0,0) + (a+1)z_1(a,0,0)] = g_4(a-1,0,0) + g_3(a,0,0) \text{ if } c=0;\\
g_1(a,0,c)e'_\beta =& - x_1(a+1,1,c) - (a+1)v_5(a,0,c) - cv_1(a+1,0,c-1)\\
&+ (a+1)v_6(a,0,c) + a(a+1)y_1(a-1,1,c)\\
&- [x_1(a+1,1,c) + (a+1)v_5(a-1,0,c) - (a+1)v_6(a,0,c)\\
&- a(a+1)y_1(a-1,1,c)]\\
&- cv_1(a+1,0,c-1) = -f_3(a,0,c) - cf_6(a+1,0,c-1);\\
g_2(a,0,c)e'_\beta =& - av_4(a-1,0,c+1) + (c+1)v_6(a,0,c) + (a+c+3)v_3(a,0,c+1)\\
&+ (2a+2)v_6(a,0,c) + a(2a+2)y_1(a-1,1,c)\\
&+ av_4(a-1,0,c+1) + a(c+1)y_1(a-1,1,c) + av_3(a,0,c+1)\\
&= (2a+c+3)[v_3(a,0,c+1) + v_6(a,0,c) + ay_1(a-1,1,c)]\\
&= (2a+c+3)f_4(a,0,c);\\
g_3(a,b,0)e'_\beta =& (a+1)v_3(a,b+1,0) - (a+1)v_3(a,b+1,0) = 0;\\
g_4(a,0,0)e'_\beta =& v_3(a+1,1,0) - (a+1)v_4(a,1,0) - (a+2)v_3(a+1,1,0)\\
&+ (a+1)v_4(a,1,0) + (a+1)v_3(a+1,1,0) = 0;
\end{aligned}
$$

$$
\begin{aligned}
g_5(a,b,0)e'_\beta =& v_3(a+1,b+2,0) - (a+1)v_4(a,b+2,0) - v_3(a+1,b+2,0)\\
&+ (a+1)v_4(a,b+2,0) = 0;
\end{aligned}
$$

$$
\begin{aligned}
g_6(a,b,c)e'_\beta =&(a+1)v_3(a,b+1,c+1)+(a+1)v_6(a,b+1,c)\\
&+a(a+1)y_1(a-1,b+2,c)=(a+1)f_4(a,b+1,c);\\
g_7(a,b,c)e'_\beta =&-x_1(a+1,b+2,c)-(a+1)v_5(a,b+1,c)\\
&-cv_1(a+1,b+1,c-1)+(a+1)v_6(a,b+1,c)\\
&+a(a+1)y_1(a-1,b+2,c)\\
&=-f_3(a,b+1,c)-cf_6(a+1,b+1,c-1);\\
g_8(a,b,c)e'_\beta =&-v_3(a,b+1,c+1)-v_6(a,b+1,c)-ay_1(a-1,b+2,c)\\
&=-f_4(a,b+1,c);\\
g_9(a,0,c)e'_\beta =&(c+2)v_3(a+1,1,c+1)-(a+1)v_4(a,1,c+1)\\
&+(c+1)v_6(a+1,1,c)+(a+2)v_6(a+1,1,c)\\
&+(a+2)(a+1)y_1(a,2,c)+(a+1)(c+1)y_1(a,2,c)\\
&+(a+1)v_4(a,1,c+1)+(a+1)v_3(a+1,1,c+1)\\
&=(a+c+3)[v_3(a+1,1,c+1)+v_6(a+1,1,c)+(a+1)y_1(a,2,c)]\\
&=(a+c+3)f_4(a+1,1,c);
\end{aligned}
$$

$$
\begin{aligned}
g_{10}(a,b,c)e'_\beta =&+(c+2)v_3(a+1,b+2,c+1)-(a+1)v_4(a,b+2,c+1)\\
&+(c+1)v_6(a+1,b+2,c)+v_6(a+1,b+2,c)+(a+1)y_1(a,b+3,c)\\
&+(a+1)(c+1)y_1(a,b+3,c)+(a+1)v_4(a,b+2,c+1)\\
&=(c+2)[v_3(a+1,b+2,c+1)+v_6(a+1,b+2,c)\\
&+(a+1)y_1(a,b+3,c)]=(c+2)f_4(a+1,b+2,c);
\end{aligned}
$$

□

References

[1] Jonathan Brundan, *Kazhdan-Lusztig polynomials and character formulae for the Lie superalgebra* $\mathfrak{q}(n)$, Adv. Math. **182** (2004), no. 1, 28–77, DOI 10.1016/S0001-8708(03)00073-2. MR2028496

[2] Jonathan Brundan, *Modular representations of the supergroup* $Q(n)$*. II*, Pacific J. Math. **224** (2006), no. 1, 65–90, DOI 10.2140/pjm.2006.224.65. MR2231652

[3] Jonathan Brundan and Alexander Kleshchev, *Modular representations of the supergroup* $Q(n)$*. I*, J. Algebra **260** (2003), no. 1, 64–98, DOI 10.1016/S0021-8693(02)00620-8. Special issue celebrating the 80th birthday of Robert Steinberg. MR1973576

[4] Jonathan Brundan and Alexander Kleshchev, *Projective representations of symmetric groups via Sergeev duality*, Math. Z. **239** (2002), no. 1, 27–68, DOI 10.1007/s002090100282. MR1879328

[5] Jens Carsten Jantzen, *Representations of algebraic groups*, 2nd ed., Mathematical Surveys and Monographs, vol. 107, American Mathematical Society, Providence, RI, 2003. MR2015057

[6] František Marko and Alexandr N. Zubkov, *Linkage principle for ortho-symplectic supergroups*, J. Algebra **493** (2018), 444–482, DOI 10.1016/j.jalgebra.2017.09.010. MR3715221

[7] Akira Masuoka and Alexandr N. Zubkov, *Solvability and nilpotency for algebraic supergroups*, J. Pure Appl. Algebra **221** (2017), no. 2, 339–365, DOI 10.1016/j.jpaa.2016.06.012. MR3545265

[8] Klaus Pommerening, *Ordered sets with the standardizing property and straightening laws for algebras of invariants*, Adv. in Math. **63** (1987), no. 3, 271–290, DOI 10.1016/0001-8708(87)90057-0. MR877787

[9] Taiki Shibata, *Borel-Weil theorem for algebraic supergroups*, J. Algebra **547** (2020), 179–219, DOI 10.1016/j.jalgebra.2019.11.019. MR4039427

[10] A. N. Zubkov, *Some properties of general linear supergroups and of Schur superalgebras* (Russian, with Russian summary), Algebra Logika **45** (2006), no. 3, 257–299, 375, DOI 10.1007/s10469-006-0014-7; English transl., Algebra Logic **45** (2006), no. 3, 147–171. MR2289090

The Pennsylvania State University, 76 University Drive, Hazleton, 18202 Pennsylvania

Email address: fxm13@psu.edu

Sobolev Institute of Mathematics, Omsk Branch, Pevtzova 13, 644043 Omsk, Russian Federation

Email address: a.zubkov@yahoo.com

Contemporary Mathematics
Volume **845**, 2026
https://doi.org/10.1090/conm/845/16958

An introduction to representation-theoretic canonical bases of cluster algebras

Fan Qin

Abstract. This is an introduction to cluster algebras and their common triangular bases. These bases are Kazhdan-Lusztig-type and serve as the canonical bases of cluster algebras from the representation-theoretic point of view. We review seeds associated with signed words, cluster operations (freezing and base change), the extension technique from finite to infinite rank, which are convenient tools to study these algebras and their bases. We present recent results on the topic and provide examples. An appendix collects detailed examples of some cluster algebras from Lie theory, including those from Lie groups and representations of quantum affine algebras.

1. Introduction

1.1. An informal introduction to cluster algebras. Let us begin with an informal but intuitive introduction to cluster algebras, originally invented by Fomin and Zelevinsky [**FZ02**]. Recall that an m-dimensional smooth real manifold takes the form $\mathcal{M} = \cup_{\mathbf{s}} V_{\mathbf{s}}$, such that $V_{\mathbf{s}}$ are open subsets, $\mathbf{s}$ denote their indices, and for each $V_{\mathbf{s}}$, we have a homeomorphism:

$$(x_1, \ldots, x_m) : V_{\mathbf{s}} \simeq \mathbb{R}^m.$$

The pair $(V_{\mathbf{s}}, (x_1, \ldots, x_m))$ is called a chart. We might call $(x_1, \ldots, x_m)$ its cluster of local coordinates. Furthermore, we require that the transition maps (change of local coordinates) are smooth. The collection of all charts is called the atlas of $\mathcal{M}$.

In mathematics and physics (for example, in algebraic geometry and in quantum mechanics), instead of working directly with a geometric object, it is often useful to consider the algebra of functions on it. For a smooth manifold $\mathcal{M}$, we may consider the algebra $C^\infty(\mathcal{M})$ of smooth functions instead. This viewpoint also leads to a quantization: a commutative algebra $C^\infty(\mathcal{M})$ endowed with a Poisson bracket could be replaced by a non-commutative one.

Cluster algebras can be understood in a similar spirit. A rank-m cluster algebra A is an algebra endowed with clusters of distinguished elements called cluster variables:

(1) For any chosen cluster $(x_1, \ldots, x_m)$, every element $f \in \mathsf{A}$ could be expressed as a rational function in $x_1, \ldots, x_m$.
(2) We impose additional requirements:

2020 *Mathematics Subject Classification.* Primary 13F60.

(2a) There is an initial cluster, and all other clusters are obtained recursively from it.
(2b) The transition maps between adjacent clusters (in the recursive process) have a specific form (see (2.5)).

Here we work with rational functions instead of smooth functions because we are in an algebraic-geometric setting. The recursive construction (2a) was an essential part of the original definition by Fomin and Zelevinsky [**FZ03**]. It is natural to ask whether these algebras can be generalized by allowing more general forms of transition maps. This leads to the notion of generalized cluster algebras; see [**CS14**][**GSV21**].

In brief, a cluster algebra is an algebra endowed with an atlas (a collection of clusters) satisfying certain additional requirements. From this point of view, cluster algebras are very fundamental objects. And unsurprisingly they appear in many, sometimes unexpected, areas of mathematics. Their atlas might come from coordinate systems, wall-chamber structures, t-structures of triangulated categories, and so on.

We say that an algebra has a cluster structure if it can be endowed with an atlas making it into a cluster algebra. Not all algebras admit cluster structures. And while all cluster algebras have rich structures, some possess stronger properties. Many fascinating results have been obtained for cluster algebras arising from quiver representations, Lie theory, and (higher) Teichmüller theory; see [**Fom10**] [**Kel08**] [**Kel11**] [**Rei10**] [**Lec10**] [**Thu14**] [**Mul16b**] [**FWZ16**, **FWZ17**, **FWZ20**, **FWZ21**] for an incomplete list of surveys and accessible papers on these topics.

Cluster algebras are interesting objects for at least the following reasons:

- They possess rich and beautiful structures.
- They provide a common platform for connecting and comparing theories form different areas.
- They allow the discovery of new phenomenon even in well-studied mathematical objects.

While cluster algebras were not invented for having applications in other fields, they have proven useful in solving open problems. For example, they have been used to verify the periodicity conjecture of T-systems (certain discrete dynamical systems)[**KNS11**][**IIK$^+$10**][**Kel13**], to show the existence of infinite Lagrangian fillings [**CG22**], and to prove conjectural properties of the (q,t)-characters of simple modules of quantum affine algebras [**FHOO22**].

1.2. Canonical bases of cluster algebras. Cluster algebras were invented by Fomin and Zelevinsky [**FZ02**] as a framework to study the total positivity (see [**Lus94**]) and the dual canonical bases of quantized enveloping algebras (see [**Lus90**, **Lus91**][**Kas91**]). These algebras admit natural quantizations [**BZ05**]. They possess distinguished elements called cluster monomials, namely monomials in $x_1, \ldots, x_m$ from the same cluster, which can be computed recursively.

Fomin and Zelevinsky have the following ambitious expectations [**FZ02**]:

- For many varieties $\mathscr{A}$ arising from Lie theory, the coordinate ring $\Bbbk[\mathscr{A}]$ is a cluster algebra (we might take $\Bbbk = \mathbb{C}$ in the classical case and $\Bbbk = \mathbb{Q}(v)$ or $\mathbb{Q}[v^{\pm}]$ in the quantum case).

- The cluster algebra has a basis that contains all cluster monomials and serves as an analog of the dual canonical basis $\mathbf{B}^*$ of the quantized enveloping algebra $\Bbbk[N]$.

The progress on the second expectation was slow for a long time. The cluster structure on the quantum unipotent subgroup $\Bbbk[N(w)]$ was developed in [**GLS13**] [**GY16, GY21**] (recall that $N = N(w_0)$ when the associated Lie algebra $\mathfrak{g}$ is semisimple). For the quantized coordinate ring $\Bbbk[N(w)]$ of a unipotent subgroup $N(w)$ associated with any Kac-Moody algebra $\mathfrak{g}$, the original dual canonical basis $\mathbf{B}^*$ is known to meet this expectation [**Qin20b**] (see also [**Qin17**][**KKKO18**] [**McN21**][1]).

A key ingredient in the approach of [**Qin17, Qin20b**] is the definition of the common triangular basis $\mathbf{L}$ for a large family of quantum cluster algebras. This basis is a Kazhdan-Lusztig-type basis for cluster algebras (see Remark 6.1 for an earlier construction by Berenstein and Zelevinsky). [**Qin20b**] proved that the dual canonical basis $\mathbf{B}^*$ of any $\Bbbk[N(w)]$ is the common triangular basis of $\Bbbk[N(w)]$ (Theorem 6.7). This result suggests that the common triangular basis $\mathbf{L}$ provides a natural analogue of the dual canonical basis $\mathbf{B}^*$ in the framework of cluster theory. Recently, [**Qin24a**] showed that $\mathbf{L}$ indeed meets the second expectation for almost all known cluster algebras from Lie theory (Theorem 6.11). Moreover, when $\mathfrak{g}$ is a symmetric Kac-Moody algebra, these bases admit (quasi-)categorifications, i.e., they correspond to simple objects in an appropriate monoidal category (i.e, tensor category). See Theorem 6.16, where we use a weaker version of the categorification in [**HL10**] (Section 6.5).

In representation theory, we know several families of canonical bases, including the canonical bases $\mathbf{C}$ for Hecke algebras, the canonical bases $\mathbf{B}$ for quantized enveloping algebras $U_q(\mathfrak{n})$, and their dual bases $\mathbf{B}^*$ (dual canonical bases) for the quantized coordinate rings $\Bbbk[N]$. These bases are of Kazhdan-Lusztig type (Section 5), and they often correspond to *minimal* objects in an appropriate module tensor category (i.e., the simple modules or the indecomposable projective modules). From the above discussion, we see that the common triangular basis $\mathbf{L}$ of a cluster algebra shares similar properties. Therefore, we might call $\mathbf{L}$ the *representation-theoretic canonical basis* of the cluster algebra.

REMARK 1.1 (Different canonical bases). The word "canonical" in "canonical basis" does not seem to have a precise definition. We might interpret it as "the most natural", "the best", or, from a more aesthetic perspective, "the most beautiful".

On the one hand, from the representation-theoretic point of view, the dual canonical basis $\mathbf{B}^*$ of $\Bbbk[N(w)]$ (equivalently, the common triangular basis $\mathbf{L}$ of $\Bbbk[N(w)]$) is its canonical basis, at least when $\mathfrak{g}$ is of symmetric type.[2]

[1] [**Qin17**] proved the ADE case and the case where w has an $\underline{c}$-adapted word and $\mathfrak{g}$ is symmetric Kac-Moody (Section A.2), and [**Qin20b**] treated all cases. Based on a completely different approach, [**KKKO18**] proved the symmetric Kac–Moody case, and [**McN21**] generalized their results to all cases.

[2] For non-symmetric $\mathfrak{g}$, the simple modules of quiver Hecke algebras seem to provide a better basis [**KL09, KL11**][**Rou08**]. Though our knowledge of this basis is still limited and its relation with cluster theory is unknown.

On the other hand, there is yet another canonical basis for cluster algebras: the theta basis ϑ proposed by [**GHKK18**][3]. This basis, constructed from geometric considerations, has exceptionally strong properties, and we might call it the *geometric canonical basis.*

Examples suggest that the following phenomena might hold in general:

- The two canonical bases differ when the cluster algebra has infinitely many cluster variables.
- Each element of $\mathbf{L}$ is a nonnegative linear combination of the basis elements of ϑ (see [**Qin23**, Section 3]), when the initial seed of the cluster algebra is associated with a skew-symmetric matrix (Section 2).

At present, there is insufficient evidence to determine whether one basis could be regarded as more canonical than the other.

In addition, there is a mysterious phenomenon. On the one hand, the elements of $\mathbf{L}$ correspond to *minimal* objects in a module category—either as established theorems or as guiding principles. On the other hand, they admit nonnegative decompositions into the theta basis elements, which are the *atoms* for a cluster algebra (see the atomic basis in [**LLZ14**][**LLRZ14**][**CGM$^+$17**][**Man17**]). It is desirable to understand and interpret the decomposition.

1.3. Contents. In Section 2, we give basic definitions and properties of cluster algebras.

In Section 3, we introduce seeds associated with signed words. They often appear in cluster algebras from Lie theory. We also provide examples.

In Section 4, we introduce operations on cluster algebras (freezing and base change). This part provides the key tools for studying bases of cluster algebras from Lie theory in [**Qin24a**]. It is technical, and the reader could skip it without loss of continuity in the subsequent sections.

In Section 5, we review basics of Kazhdan-Lusztig-type bases and their categorifications.

In Section 6, we introduce the common triangular bases $\mathbf{L}$ for cluster algebras and present known results: their existence (Theorems 6.9 6.7 6.11), quasi-categorifications (Theorem 6.16), and in the special cases, standard bases (Theorem 6.17) and categorifications (Theorem 6.18).

In Section 7, we present the technique of extending finite rank cluster algebras to infinite rank ones. In Example 7.5, we apply it to obtain the quantization matrices of an infinite rank cluster algebra.

In Appendix A, we review some cluster algebras arising from Lie theory: those from Lie group G, double Bruhat cells $G^{u,w}$, and representations of quantum affine algebras. We also provide examples in details.

1.4. Convention. All vectors will be column vectors unless otherwise specified.

Let σ be a permutation of an index set I. For any $m = (m_i)_{i\in I} \in \mathbb{R}^I$, we define $\sigma m \in \mathbb{R}^I$ such that $(\sigma m)_{\sigma i} := m_i$. Similarly, let $B = (b_{ij})_{i,j\in I}$ be a matrix. We define $\sigma B = (b'_{ij})_{i,j\in I}$ such that $b'_{\sigma i,\sigma j} := b_{ij}$. We use $\mathrm{col}_k B$ to denote the k-th column vector of B. For $I_1, I_2 \subset I$, we denote the submatrix $B_{I_1,I_2} := (b_{ij})_{i\in I_1, j\in I_2}$.

[3]For cluster algebras arsing from Teichmüller theory, the canonical basis introduced in [**FG06**] (or the bracelets basis in [**MSW13**]) is the theta basis ϑ in [**GHKK18**]; see [**MQ23**]. Also, see [**DM21**] for the quantum theta basis.

We use Id_n to denote a rank n identity matrix, and Id_I to denote the identity matrix on the index set I.

For $I' \subset I$, we use $\mathrm{pr}_{I'}$ to denote the natural projection from $\mathbb{R}^I$ to $\mathbb{R}^{I'}$.

For any vector $m = (m_i)_{i\in I} \in \mathbb{R}^I$, its support is defined to be $\mathrm{supp}\, m := \{i \in I | m_i \neq 0\}$.

2. Cluster algebras

2.1. Definition of cluster algebras.

Seeds. Let us define cluster algebras following the convention in [**Qin24c**]. We choose and fix a countable set I,[4] called the set of vertices. We further choose a partition $I = I_{\mathrm{uf}} \sqcup I_{\mathrm{f}}$, where I_{uf} denotes its unfrozen part and I_{f} the frozen part. Choose and fix strictly positive integers d_i, $i \in I$.

Assume $(b_{ij})_{i,j\in I}$ is an $I \times I$ $\frac{\mathbb{Z}}{2}$-matrix such that the following conditions hold:

(1) $\mathsf{d}_i b_{ij} = -\mathsf{d}_j b_{ji}$ for any i, j;
(2) $b_{ik} \in \mathbb{Z}$ if $k \in I_{\mathrm{uf}}$;
(3) It is locally finite, i.e., for any i, only finitely many b_{ij} are nonzero and only finitely many b_{ji} are nonzero.

We will denote $\widetilde{B} := (b_{ik})_{i\in I, k\in I_{\mathrm{uf}}}$. $B := \widetilde{B}_{I_{\mathrm{uf}}\times I_{\mathrm{uf}}}$ is called the principal part of $\widetilde{B}$.

A (classical) seed $\mathbf{s}$ is a collection $(I, I_{\mathrm{uf}}, (\mathsf{d}_i)_{i\in I}, \widetilde{B}, (x_i)_{i\in I})$, where $\widetilde{B}$ is called its B-matrix, and x_i are indeterminates called its cluster variables. $(x_i)_{i\in I}$ is called the cluster. We will call x_j, $j \in I_{\mathrm{f}}$ its frozen variables. The rank of $\mathbf{s}$ is defined to be the cardinality $|I|$.

REMARK 2.1. For our purpose, it is sufficient and convenient to define the B-matrix of $\mathbf{s}$ to be $\widetilde{B}$, like most references on cluster algebras. However, one could replace $\widetilde{B}$ by $(b_{ij})_{i,j\in I}$ in the definition of a seed $\mathbf{s}$. This latter definition will be important if we want to consider the corresponding cluster Poisson algebra; see [**GHKK18**][**KQW23**] for examples.

In examples, we will provide a full matrix $(b_{ij})_{i,j\in I}$ for a given seed $\mathbf{s}$. The choice of entries $(b_{ij})_{i,j\in I_{\mathrm{f}}}$ is not unique and will be unimportant for our purpose.

By a quiver, we mean an oriented graph, whose oriented edges will be called arrows. We always assume it to be locally finite, i.e., each vertex is an end of finitely many arrows. It is often convenient to draw a weighted (locally finite) quiver $\widetilde{Q}$, called an ice quiver, to represent the matrix $(b_{ij})_{i,j\in I}$:

- The set of vertices of $\widetilde{Q}$ is I. The full subquiver on I_{uf} is denote by Q. We often call $\widetilde{Q}$ an ice quiver and Q its principal part.
- For any $i, j \in I$, if $b_{ij} > 0$, we draw arrows with positive weights from i to j, such that their total weight is b_{ij}.

Note that $b_{ji} = -\frac{\mathsf{d}_i}{\mathsf{d}_j} b_{ij}$. Only the total weight of arrows matters for us: if two quivers have the same total weight of arrows for any given pair (i, j), we view them as the same. Conversely, any such a weighted quiver determines a matrix (b_{ij}).

In our figures, solid (resp. dashed) arrows represent arrows of weight 1 (resp. $\frac{1}{2}$), and circular (resp. rectangular) nodes represent unfrozen (resp. frozen) vertices.

[4]See [**Gra15**] for general infinite rank cluster algebras and [**Gra15**, Remark 2.2 Remark 3.18] for a discussion on the countability.

FIGURE 2.1. The ice quiver for a seed of $\mathbb{C}[SL_3]$

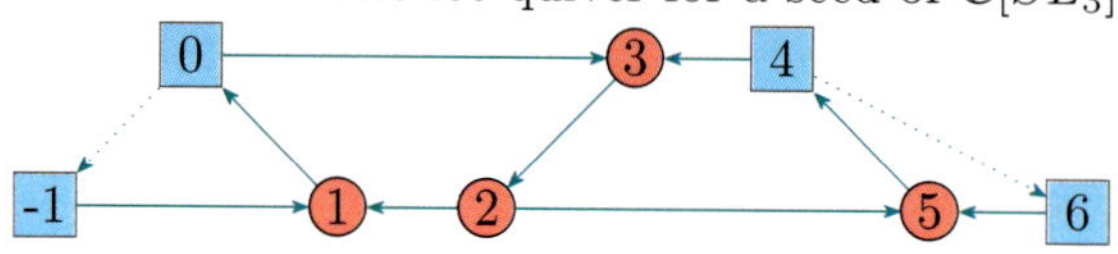

FIGURE 2.2. The quiver for an infinite seed of type A_1 in [**GHL24**]

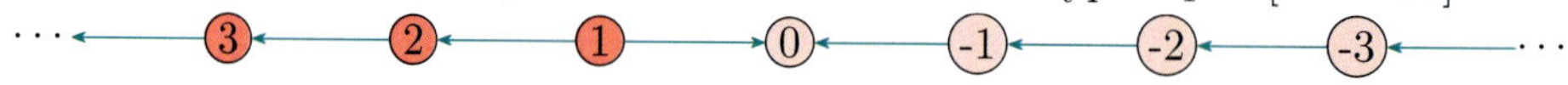

EXAMPLE 2.2 (A seed from $\mathbb{C}[SL_3]$). We consider the following seed arising from the cluster structure on $\mathbb{C}[SL_3]$. $I = \{k \in \mathbb{Z} | k \in [-1, 6]\}$, $I_{\mathrm{f}} = \{-1, 0, 4, 6\}$, all $\mathsf{d}_i = 1$, and $(b_{ij})_{-1 \leq i,j \leq 6} = \begin{pmatrix} 0 & -\frac{1}{2} & 1 & 0 & 0 & 0 & 0 & 0 \\ \frac{1}{2} & 0 & -1 & 0 & 1 & 0 & 0 & 0 \\ -1 & 1 & 0 & -1 & 0 & 0 & 0 & 0 \\ 0 & 0 & 1 & 0 & -1 & 0 & 1 & 0 \\ 0 & -1 & 0 & 1 & 0 & -1 & 0 & 0 \\ 0 & 0 & 0 & 0 & 1 & 0 & -1 & \frac{1}{2} \\ 0 & 0 & 0 & -1 & 0 & 1 & 0 & -1 \\ 0 & 0 & 0 & 0 & 0 & -\frac{1}{2} & 1 & 0 \end{pmatrix}$. The associated ice quiver is depicted in Figure 2.1.

EXAMPLE 2.3 (An infinite seed of type A_1). Now consider an infinite seed appearing in [**GHL24**], such that $I = I_{\mathrm{uf}} = \mathbb{Z}$, all $\mathsf{d}_i = 1$, and $(b_{ij})_{i,j \in \mathbb{Z}}$ is given by

$$(b_{ij})_{i,j \in \mathbb{Z}} = \begin{pmatrix} \cdots & \cdots & & & & & & & \\ \cdots & 0 & 1 & & & & & & \\ & -1 & 0 & 1 & & & & & \\ & & -1 & 0 & 1 & & & & \\ & & & -1 & 0 & -1 & & & \\ & & & & 1 & 0 & 1 & & \\ & & & & & -1 & 0 & 1 & \\ & & & & & & -1 & 0 & \cdots \\ & & & & & & & \cdots & \cdots \end{pmatrix} \tag{2.1}$$

is determined by the quiver in Figure 2.2 (the central entry in (2.1) is the $b_{00} = 0$). Note that almost all arrows point to the left except the one between the nodes 0 and 1.

If σ is a permutation on I such that $\sigma I_{\mathrm{uf}} = I_{\mathrm{uf}}$, we define the seed $\mathbf{s}' = \sigma \mathbf{s}$ such that $I' = I$, $I'_{\mathrm{uf}} = I_{\mathrm{uf}}$, $\mathsf{d}'_{\sigma i} = \mathsf{d}_i$, and $b'_{\sigma i, \sigma j} = b_{i,j}$. In the quantum case, we endow $\sigma \mathbf{s}$ with the compatible Poisson structure $\sigma \lambda$ whose associated matrix is $\sigma \Lambda$.

Torus algebras. Following [**GHKK18**], define $M^\circ := \oplus_{i \in I} \mathbb{Z} f_i \simeq \mathbb{Z}^I$ and $N := \oplus_{i \in I} \mathbb{Z} e_i \simeq \mathbb{Z}^I$, where f_i and e_i are the standard basis vectors. Define the linear map $p^* : N \to M^\circ$ such that

$$p^*(e_j) = \sum b_{ij} f_i.$$

Therefore, we have $p^* n = \widetilde{B} n$ for $n \in N$ as coordinate (column) vectors. We will denote $N_{\mathrm{uf}} := \oplus_{k \in I_{\mathrm{uf}}} \mathbb{Z} e_k$, $N^\oplus := \oplus_{k \in I_{\mathrm{uf}}} \mathbb{N} e_k$, and $M^\oplus := p^* N^\oplus$. We could view $M^\circ \otimes \mathbb{R}$ and $N \otimes \mathbb{R}$ as dual spaces by introduce the pairing $\langle f_i, e_j \rangle = \delta_{ij} \frac{\mathsf{d}_i}{\mathsf{d}}$ for any chosen constant $\mathsf{d} \neq 0$. When the set $\{\mathsf{d}_i, i \in I\}$ is finite, one could choose d to be the least common multiplier of $\{\mathsf{d}_i, i \in I\}$ as in [**Qin24d**, Section 2.1] (our d_i were denoted by $d_i^\vee$ and $\frac{\mathsf{d}}{\mathsf{d}_i}$ were denoted by d_i in loc. cit., so that the conventions were compatible with [**GHKK18**]).

Unless otherwise specified, we will make the following assumption.

ASSUMPTION 2.4 (Injectivity assumption). The map p^* is injective on N_{uf}.

Let $\Bbbk$ denote a commutative ring, which we choose to be $\mathbb{Z}$ for the classical case and $\mathbb{Z}[v^{\pm}]$ for the quantum case unless otherwise specified. Define the torus algebra $\boldsymbol{LP}$ to be the Laurent polynomial ring $\Bbbk[x_i^{\pm}]_{i\in I}$, whose commutative multiplication is denoted by $\cdot$. We naturally identify $\boldsymbol{LP}$ with the group ring $\Bbbk[M^\circ] := \oplus_{m\in M^\circ}\Bbbk x^m$ such that $\prod_{i\in I} x_i^{m_i} = x^m$ for any $m = (m_i)_{i\in I} = \sum m_i f_i \in M^\circ$. Similarly, we identify $\Bbbk[y_i^{\pm}]_{i\in I}$ with $\Bbbk[N] := \oplus \Bbbk y^n$ such that $\prod_i y_i^{n_i} = y^n$ for $n = (n_i)_{i\in I} = \sum n_i e_i \in N$. The linear map p^* induces a homomorphism $p^* : \Bbbk[N] \to \Bbbk[M]$ such that

$$p^*(y^n) := x^{p^*(n)} = x^{\widetilde{B}n}.$$

In particular, $p^*(y_k) = x^{\mathrm{col}_k \widetilde{B}}$, $\forall k \in I_{\mathrm{uf}}$. When there is no confusion, we might denote the Laurent monomial $x^{p^*(n)}$ by y^n for simplicity.

DEFINITION 2.5. The cluster monomials of $\mathbf{s}$ are x^m, $m \in \mathbb{N}^I$. Its localized cluster monomials are x^m, $m \in \mathbb{N}^{I_{\mathrm{uf}}} \oplus \mathbb{Z}^{I_{\mathrm{f}}}$, and its frozen factors are $x^p, p \in \mathbb{Z}^{I_{\mathrm{f}}}$.

Let $\widehat{\Bbbk[N^\oplus]}$ denote the completion of $\Bbbk[N^\oplus]$ with respect to the maximal ideal generated by y_k, $k \in I_{\mathrm{uf}}$, i.e., it is the ring of formal series $\Bbbk[\![y_k]\!]_{k\in I_{\mathrm{uf}}}$. Since p^* is injective, all $p^*(y_k)$ for $k \in I_{\mathrm{uf}}$ generate a maximal ideal for $\Bbbk[M^\oplus]$. Let $\widehat{\Bbbk[M^\oplus]}$ denote the correspond formal completion. We have $p^* : \widehat{\Bbbk[N^\oplus]} \simeq \widehat{\Bbbk[M^\oplus]}$. Finally, we define the following formal completion:

$$\widehat{\boldsymbol{LP}} := \boldsymbol{LP} \otimes_{\Bbbk[M^\oplus]} \widehat{\Bbbk[M^\oplus]}.$$

We will also denote $\overline{\boldsymbol{LP}} := \Bbbk[x_j]_{j\in I_{\mathrm{f}}}[x_k^{\pm}]_{k\in I_{\mathrm{uf}}}$, called the partially compactified torus algebra.

Quantization. A compatible Poisson structure (or a compatible Poisson form) for $\mathbf{s}$ is a $\mathbb{Z}$-valued skew-symmetric bilinear form λ on M°, such that there exist $\delta_k > 0$ for $k \in I_{\mathrm{uf}}$ and

$$\lambda(f_i, p^* e_k) = -\delta_{ik}\delta_k,\ \forall i \in I, k \in I_{\mathrm{uf}}.$$

If such a λ exists, we define the skew-symmetric matrix $\Lambda = (\Lambda_{ij})_{i,j\in I}$, called the quantization matrix, such that $\Lambda_{ij} = \lambda(f_i, f_j)$. The pair $(\Lambda, \widetilde{B})$ will be called a compatible pair as in [**BZ05**]. Note that p^* must be injective if λ exists. By [**GSV03, GSV05**], when $|I| < \infty$ and p^* is injective, a compatible λ must exist.

A quantum seed is a seed $\mathbf{s}$ endowed with a compatible Poisson structure λ. When we work with a quantum seed $\mathbf{s}$ at the quantum level $\Bbbk = \mathbb{Z}[v^{\pm}]$, we will endow the torus algebra $\boldsymbol{LP}$ with the v-twisted product $*$ such that

$$x^m * x^n := v^{\lambda(m,n)} x^{m+n},\ \forall m, n \in M^\circ.$$

And we will use $*$ as the multiplication for the algebra $\boldsymbol{LP}$ unless otherwise specified.

We let $\mathcal{F}$ denote the skew field of fractions of $\boldsymbol{LP}$; see [**BZ05**]. We also introduce the bar involution $\overline{(\ )}$ on $\boldsymbol{LP}$, which is the $\mathbb{Z}$-linear map such that $\overline{v^{\alpha}x^{m}} = v^{-\alpha}x^{m}$. It is an anti-involution.

EXAMPLE 2.6. It is a good exercise to find a compatible Poisson form λ for the infinite seed in Example 2.3. For convenience of the reader, let us provide an explicit solution: the quantization matrix associated with λ is $\Lambda = (\Lambda_{ij})_{i,j\in\mathbb{Z}}$:

$$\Lambda = \begin{pmatrix} & & & & & \cdots & & & & & \\ & 0 & 1 & 0 & 1 & 0 & -1 & 0 & -1 & 0 & \\ & -1 & 0 & 1 & 0 & 1 & 0 & -1 & 0 & -1 & \\ & 0 & -1 & 0 & 1 & 0 & -1 & 0 & -1 & 0 & \\ & -1 & 0 & -1 & 0 & 1 & 0 & -1 & 0 & -1 & \\ \cdots & 0 & -1 & 0 & -1 & 0 & -1 & 0 & -1 & 0 & \cdots \\ & 1 & 0 & 1 & 0 & 1 & 0 & 1 & 0 & 1 & \\ & 0 & 1 & 0 & 1 & 0 & -1 & 0 & 1 & 0 & \\ & 1 & 0 & 1 & 0 & 1 & 0 & -1 & 0 & 1 & \\ & 0 & 1 & 0 & 1 & 0 & -1 & 0 & -1 & 0 & \\ & & & & & \cdots & & & & & \end{pmatrix}. \tag{2.2}$$

The central entry in (2.2) is $\Lambda_{00} = 0$. The nonzero entries of Λ are given by

$$\begin{aligned} \Lambda_{i,i+2d+1} &= \begin{cases} 1 & i < i+2d+1 \leq 0 \\ -1 & \text{else} \end{cases} & \forall i < 0, d \in \mathbb{Z} \\ \Lambda_{0,2d+1} &= -1 & \forall d \in \mathbb{Z} \\ \Lambda_{i,i-2d-1} &= \begin{cases} -1 & 0 < i-2d-1 < i \\ 1 & \text{else} \end{cases} & \forall i > 0, d \in \mathbb{Z} \end{aligned}$$

This matrix Λ can be computed inductively as in Example 7.5.

Note that the matrix $\widetilde{B} = B$ in (2.1) is given by

$$\begin{aligned} \mathrm{col}_i \widetilde{B} &= e_{i-1} - e_{i+1} & \forall i \neq 0, 1 \\ \mathrm{col}_0 \widetilde{B} &= e_{-1} + e_1 \\ \mathrm{col}_1 \widetilde{B} &= -e_0 - e_2 \end{aligned}$$

It is straightforward to verify that $\Lambda\widetilde{B}$ is an $\mathbb{Z}\times\mathbb{Z}$ diagonal matrix whose diagonal entries are -2, i.e., all $\delta_k = 2$. Therefore, λ is a compatible Poisson pair. We will see in Example 7.5 that how to determine all solutions.

We refer the reader to [**Pag25**] for another approach to λ with representation theoretic interpretations.

By changing the sign in $\mathbf{s}$, we obtain its opposite seed $\mathbf{s}^{\mathrm{op}} := (I, I_{\mathrm{uf}}, (\mathsf{d}_i)_{i\in I}, -\widetilde{B}, (x_i)_{i\in I})$. When $\mathbf{s}$ is endowed with a compatible Poisson form λ, we endow $\mathbf{s}^{\mathrm{op}}$ with $-\lambda$. There is a canonical $\Bbbk$-linear anti-isomorphism ([**Qin20a**, (3.1)])

$$\iota : \boldsymbol{LP}(\mathbf{s}) \simeq \boldsymbol{LP}(\mathbf{s}^{\mathrm{op}}),\ \iota(x^m) := x^m. \tag{2.3}$$

REMARK 2.7. At the classical level, seeds for the same cluster algebras in different literature sometimes differ by a sign, because ι is an isomorphism and the sign difference is negligible. However, ι is not an isomorphism at the quantum level. We will pay attention to distinguish $\mathbf{s}$ and $\mathbf{s}^{\mathrm{op}}$, though it usually does not bring new information in practice.

Mutations. Denote $[\ ]_+ := \max(\ ,0)_+$. Let $\mathbf{s}$ be a given seed. For any unfrozen vertex $k \in I_{\mathrm{uf}}$, we have an operation μ_k called mutation, which produces a new seed $\mathbf{s}' = (I, I_{\mathrm{uf}}, (d_i)_{i\in I}, \widetilde{B}', (x'_i)_{i\in I})$ such that x'_i are indeterminates, and $\widetilde{B}' = (b'_{ij})_{i\in I, j\in I_{\mathrm{uf}}}$. Here, we define a new matrix $(b'_{ij})_{i,j\in I} := \mu_k(b_{ij})_{i,j\in I}$ by

$$b'_{ij} = \begin{cases} -b_{ij} & \text{if } i = k \text{ or } j = k \\ b_{ij} + [b_{ik}]_+[b_{kj}]_+ - [-b_{ik}]_+[-b_{kj}]_+ & \text{else} \end{cases}.$$

We further define maps associated with mutations: $\psi^M : (M^\circ)' \simeq M^\circ$ and $\psi^N : N' \simeq N$ such that

$$\psi^M(f'_i) = \begin{cases} -f_k + \sum_{j\in I}[-b_{jk}]_+ f_j & \text{if } i = k \\ f_i & \text{else} \end{cases}$$

$$\psi^N(e'_i) = \begin{cases} -e_k & \text{if } i = k \\ e_i + [b_{ki}]_+ e_k & \text{else} \end{cases}$$

These constructions are associated with positive sign mutations; see [**KQW23**, Section 2.7] for a general definition.

When $\mathbf{s}$ is a quantum seed, its Poisson structure λ induces a compatible Poisson structure on $\mathbf{s}'$ via ψ^M, still denoted λ:

$$\lambda(m, m') := \lambda(\psi^M m, \psi^M m'),\ \forall m, m' \in (M^\circ)'. \tag{2.4}$$

We further introduce a $\Bbbk$-algebra isomorphism $\mu_k^* : \mathcal{F}' \simeq \mathcal{F}$ between the associated skew fields of fractions, called the mutation map, such that

$$\mu_k^*(x'_i) = \begin{cases} x^{-f_k + \sum_{j\in I}[-b_{jk}]_+ f_j} + x^{-f_k + \sum_{i\in I}[b_{ik}]_+ f_i} & \text{if } i = k \\ x_i & \text{else} \end{cases}. \tag{2.5}$$

We will identify $\mathcal{F}'$ and $\mathcal{F}$ via μ_k^* and omit the symbol μ_k^*. Note that the mutation is an involution on seeds and on skew fields, see [**FZ02, DZ05**].

In general, for any finite sequence of unfrozen vertices $\underline{k} = (k_1, \ldots, k_r)$, we have the mutation sequence $\boldsymbol{\mu} := \boldsymbol{\mu}_{\underline{k}} := \mu_{k_r} \cdots \mu_{k_1}$ (read from right to left). For $\mathbf{s}' := \boldsymbol{\mu}\mathbf{s}$, the mutation map $\boldsymbol{\mu}^* : \mathcal{F}' \simeq \mathcal{F}$ is defined as the composition $\mu_{k_1}^* \cdots \mu_{k_r}^*$.

Cluster algebras. Let $\mathbf{s}$ be any given (classical or quantum) initial seed. Let $\Delta^+ := \Delta^+_{\mathbf{s}}$ denote the set of seed $\boldsymbol{\mu}\mathbf{s}$, where $\boldsymbol{\mu}$ is any finite mutation sequence. Identify $\mathcal{F}$ and $\mathcal{F}(\boldsymbol{\mu}\mathbf{s})$ via $\boldsymbol{\mu}^*$ as before. Note that $x_j(\boldsymbol{\mu}\mathbf{s}) = x_j(\mathbf{s})$ for $j \in I_{\mathrm{f}}$, which we can simply denote by x_j.

We define the following versions of cluster algebras associated to $\mathbf{s}$:

- The partially compactified ordinary cluster algebra is
$$\overline{\boldsymbol{A}}(\mathbf{s}) := \Bbbk[x_i(\mathbf{s}')]_{i\in I, \mathbf{s}'\in\Delta^+},$$
where $x_i(\mathbf{s}')$ denote the cluster variables of the seed $\mathbf{s}'$.
- The partially compactified upper cluster algebra is
$$\overline{\boldsymbol{U}}(\mathbf{s}) := \cap_{\mathbf{s}'\in\Delta^+}\overline{\boldsymbol{LP}}(\mathbf{s}').$$

- The localized ordinary cluster algebra is

$$\boldsymbol{A}(\mathbf{s}) := \overline{\boldsymbol{A}}(\mathbf{s})[x_j^{-1}]_{j\in I_{\mathrm{f}}} = \Bbbk[x_j^{\pm}]_{j\in I_{\mathrm{f}}}[x_k(\mathbf{s}')]_{k\in I_{\mathrm{uf}},\mathbf{s}'\in\Delta^+}.$$

- The localized upper cluster algebra is

$$\boldsymbol{U}(\mathbf{s}) := \overline{\boldsymbol{U}}(\mathbf{s})[x_j^{-1}]_{j\in I_{\mathrm{f}}} = \cap_{\mathbf{s}'\in\Delta^+}\boldsymbol{LP}(\mathbf{s}').$$

The rank of a cluster algebra is defined to be that of $\mathbf{s}$.

We will drop the symbol $\mathbf{s}$ when we do not want to emphasize our choice of the initial seed $\mathbf{s}$. By the Laurent phenomenon [**FZ02**][**BZ05**], we have $\overline{\boldsymbol{A}} \subset \overline{\boldsymbol{U}}$ and $\boldsymbol{A} \subset \boldsymbol{U}$.

We define the frozen torus algebra to be $\mathsf{R} = \Bbbk[x_j^{\pm}]_{j\in I_{\mathrm{f}}}$. Then $\boldsymbol{A}$ and $\boldsymbol{U}$ are R-algebras.

2.2. Tropical points. For the case $\mathbf{s}' = \mu_k\mathbf{s}$, we introduce the tropical mutation $\phi_{\mathbf{s}',\mathbf{s}} : M^\circ \to (M^\circ)'$, which is a piecewise linear map such that $m = (m_i)_{i\in I}$ is sent to $m' = (m'_i)_{i\in I}$ given by

$$m'_i := \begin{cases} -m_k & \text{if } i = k \\ m_i + [b_{ik}]_+[m_k]_+ - [-b_{ik}]_+[-m_k]_+ & \text{else} \end{cases}.$$

It can viewed as the change of coordinates associated to the change of basis ψ^M; see [**Qin24d**, Lemma A.1.2].

Now, consider any mutation sequence $\boldsymbol{\mu} = \boldsymbol{\mu}_{\underline{k}}$, where $\underline{k} = (k_1, \ldots, k_r)$. We will denote $\mathbf{s}_0 = \mathbf{s}$ and $\mathbf{s}_t := \mu_{k_t}\cdots\mu_{k_2}\mu_{k_1}\mathbf{s}$ for $t \in [1, r]$. For $\mathbf{s}' = \boldsymbol{\mu}\mathbf{s}$, define the tropical mutation $\phi_{\mathbf{s}',\mathbf{s}} : M^\circ \to (M^\circ)'$ as composition $\phi_{\mathbf{s}_r,\mathbf{s}_{r-1}}\cdots\phi_{\mathbf{s}_1,\mathbf{s}_0}$. Note that $\phi_{\mathbf{s}',\mathbf{s}}$ can be realized as the tropicalization of $\boldsymbol{\mu}^*$; see [**GHK15**]. In particular, the tropical mutation $\phi_{\mathbf{s}',\mathbf{s}}$ only depends on the mutation map $\boldsymbol{\mu}^*$, not on the specific sequence $\underline{k}$.

As in [**Qin24d**], we define the set of tropical point as

$$M^{\mathrm{trop}} := (\sqcup_{\mathbf{s}'\in\Delta^+} M^\circ(\mathbf{s}'))/\sim,$$

where $\sim$ is the equivalence relations that $m \in M^\circ(\mathbf{s})$ and $m' \in M^\circ(\mathbf{s}')$ are equivalent if and only if $m' = \phi_{\mathbf{s}',\mathbf{s}}(m)$. The equivalence class of $m' \in M^\circ(\mathbf{s}')$ is called a tropical point, denoted $[m']$. Conversely, m' is called the representative of $[m'] \in M^{\mathrm{trop}}$ in $(M^\circ)'$. We refer the reader to [**GHKK18**] for the geometric definition of tropical points.

2.3. Degrees and pointedness. Let $\mathbf{s}$ be any given seed. For the construction below, we need $\mathbf{s}$ to satisfy the Injectivity Assumption.

DEFINITION 2.8. An element $z \in \widehat{\boldsymbol{LP}}$ is said to be m-pointed for some $m \in M^\circ$ if it takes the form

$$z = x^m \cdot (1 + \sum_{0\neq n\in N^\oplus} c_n p^*(y^n)).$$

If $z' = v^\alpha z$ for some $\alpha \in \mathbb{Z}$, we say z' has degree m and its (degree) normalization is z with respect to $\mathbf{s}$, denoted $\deg^{\mathbf{s}}(z') = m$ and $[z']^{\mathbf{s}} := z$. We call $F_z := 1 + \sum c_n y^n \in \Bbbk[\![N^\oplus]\!]$ the F-function of z. F_z is called the F-polynomial if $F_z \in \Bbbk[N^\oplus]$.

Similarly, a formal sum z of Laurent monomials is called h-copointed for some $h \in M^\circ$ if it takes the form

$$z = x^h \cdot (1 + \sum_{0 \neq n \in N^\oplus} c_n p^*(y^{-n})).$$

If $z' = v^\alpha z$ for some $\alpha \in \mathbb{Z}$, we say z' has codegree m, denoted $\mathrm{codeg}^{\mathbf{s}}(z') = m$.

If $z \in \boldsymbol{LP}$ has degree m and codegree h, we say it has bidegree (m, h). We further define its support dimension to be $\mathrm{suppDim} z \in N^\oplus$ such that $h = m + p^*(\mathrm{suppDim} z)$. If it is further m-pointed and h-copointed, we say it is bipointed at bidegree (m, h).

The notions of degree, codegree, and normalization depends on the seed $\mathbf{s}$. When the context is clear, we will drop the symbol $\mathbf{s}$.

We introduce the following partial order on M°, such that m becomes the highest (or leading) degree of an m-pointed element, and h becomes the lowest (or trailing) degree of an h-copointed element.

Definition 2.9 (Dominance order [**Qin17**] [**CILFS15**, Proof of Proposition 4.3]). For any $m, m' \in M^\circ(\mathbf{s})$, we say m dominates m', denoted $m' \prec_{\mathbf{s}} m$, if $m' = m + p^*(n)$ for some $0 \neq n \in N^\oplus$.

We will denote $\boldsymbol{LP}_{\preceq m} := \oplus_{m' \preceq m} \Bbbk x^{m'}$ and $\widehat{\boldsymbol{LP}}_{\preceq m} := \boldsymbol{LP}_{\preceq m} \otimes_{\Bbbk[M^\oplus]} \widehat{\Bbbk[M^\oplus]}$.

Assume $\mathbf{s}' = \boldsymbol{\mu}\mathbf{s}$ for some mutation sequence $\boldsymbol{\mu}$. We say $z \in \boldsymbol{LP}(\mathbf{s}) \cap \boldsymbol{LP}(\mathbf{s}')$ is compatibly pointed at $\mathbf{s}, \mathbf{s}'$, if z is m-pointed in $\boldsymbol{LP}(\mathbf{s})$, m'-pointed in $\boldsymbol{LP}(\mathbf{s}')$, such that $[m] = [m']$. For any tropical point $[m] \in M^{\mathrm{trop}}$, we say $z \in \boldsymbol{U} = \cap_{\mathbf{s}' \in \Delta^+} \boldsymbol{LP}(\mathbf{s}')$ is $[m]$-pointed if, for any $\mathbf{s}'$, it is m'-pointed in $\boldsymbol{LP}(\mathbf{s}')$ such that $[m'] = [m]$.

Definition 2.10. A subset Z of $\widehat{\boldsymbol{LP}}(\mathbf{s})$ is said to be Θ-pointed for some $\Theta \subset M^\circ(\mathbf{s})$ if Z takes the form $\{Z_m | m \in \Theta\}$ such that Z_m are m-pointed.

A subset Z of $\boldsymbol{U}$ is said to be Θ-pointed for some $\Theta \subset M^{\mathrm{trop}}$ if Z takes the form $\{Z_{[m]} | [m] \in \Theta\}$ such that $Z_{[m]}$ are $[m]$-pointed.

By tropical properties, we refer to the behavior of the degrees of pointed elements under mutations.

2.4. Injective reachability. We will often need the following assumption.

Assumption 2.11 (Injective reachability). The seed $\mathbf{s}$ is injective reachable, i.e., there exists a sequence of mutation Σ and a permutation σ on I_{uf} such that $\mathrm{pr}_{I_{\mathrm{uf}}} \deg^{\mathbf{s}} x_{\sigma k}(\Sigma \mathbf{s}) = -f_k$, $\forall k \in I_{\mathrm{uf}}$.

The mutation sequence Σ is called a green to red sequence in [**Kel11**]; see [**Qin24d**, Proposition 2.3.3]. The assumption is known to be satisfied by most known cluster algebras arising from representation theory or (higher) Teichmüller theory. The exceptions are those arising from once-puncture closed surfaces; see [**FST08**, **FT18**][**Yur20**].

The seed $\Sigma\mathbf{s}$ will be denoted by $\mathbf{s}[1]$. Recursively, we could define $\mathbf{s}[d]$ for $d \in \mathbb{Z}$, such that $\mathbf{s}[0] = \mathbf{s}$ and $\mathbf{s}[d+1] = (\sigma^d \Sigma)\mathbf{s}[d]$, where $\sigma \boldsymbol{\mu}_{(k_1, \ldots, k_r)} := \boldsymbol{\mu}_{(\sigma k_1, \ldots, \sigma k_r)}$. It is known that if one seed is injective reachable, then so are all seeds of the cluster algebra; see [**Qin17**][**Mul16a**].

The combination of injective reachability and tropical properties produce some useful results.

LEMMA 2.12 ([**Qin24d**, Lemma 3.4.12]). *Assume that $z \in \boldsymbol{U}$ and z' is a localized cluster monomial of a seed s'. If z and z' are both m-pointed in $\boldsymbol{LP}(\mathrm{s})$, and they are compatibly pointed at $\mathrm{s}, \mathrm{s}', \mathrm{s}'[-1]$, then $z = z'$.*

THEOREM 2.13 ([**Qin24d**, Theorem 4.3.1]). *Let $\boldsymbol{S}$ denote a M°-pointed subset of $\boldsymbol{U}(\mathrm{s})$. If $\boldsymbol{S}$ is compatibly pointed at the seeds appearing along a mutation sequence from s to $\mathrm{s}[-1]$, then $\boldsymbol{S}$ is a basis of $\boldsymbol{U}(\mathrm{s})$.*

Note that the basis $\mathbf{S}$ in Theorem 2.13 must contain all localized cluster monomials by Lemma 2.12.

2.5. Valuations, partial compactification, and optimization. Take any seed s. We do not assume p^* is injective for the moment. For any $j \in I$, we introduce the valuation ν_j on $\mathcal{F}(\mathrm{s})$ such that for any $z = x_j^d * P * Q^{-1}$, where $P, Q \in \Bbbk[x_i]_{i \in I}$ are not divisible by x_j, we set $\nu_j(z) := d$. We say z is regular at $x_j = 0$ if $\nu_j(z) \geq 0$.

LEMMA 2.14 ([**Qin24a**, Lemma 2.12]). *For $\mathrm{s}' = \mu_k \mathrm{s}$, the valuation ν_j' defined on $\mathcal{F}(\mathrm{s}')$ satisfy $\nu_j'(z') = \nu_j(\mu_k^* z')$ for any $z' \in \mathcal{F}(\mathrm{s}')$.*

Therefore, we can simply denote the valuation by ν_j without specifying it is defined on which $\mathcal{F}(\mathrm{s})$.

It follows from definition that we have $\overline{\boldsymbol{U}} = \{z \in \boldsymbol{U} | \nu_j(z) \geq 0, \forall j \in I_{\mathrm{f}}\}$ for upper cluster algebras. However, we only know $\overline{\boldsymbol{A}} \subset \{z \in \boldsymbol{A} | \nu_j(z) \geq 0, \forall j \in I_{\mathrm{f}}\}$ for ordinary cluster algebras.

REMARK 2.15. By using valuation, $\overline{\boldsymbol{U}}$ is easy to understand once we know $\boldsymbol{U}$, while our inability to characterize the algebra $\overline{\boldsymbol{A}}$ is often an obstruction to prove desired properties for $\overline{\boldsymbol{A}}$. For example, while we could show some quantized coordinate rings are $\overline{\boldsymbol{U}}$, we still could not show that they are $\overline{\boldsymbol{A}}$ without additional knowledge [**Qin24c**][**QY25**].

Let us make the Injectivity Assumption from now on.

DEFINITION 2.16 ([**Qin24a**, Definition 2.16]). Choose any M^{trop}-pointed basis $\mathbf{S} = \{\mathbf{S}_{[m]} | [m] \in M^{\mathrm{trop}}\}$ for $\boldsymbol{U}$, where $\mathbf{S}_{[m]}$ are $[m]$-pointed. We define the corresponding set of dominant tropical points to be $\overline{M}^{\mathrm{trop}} := \{[m] \in M^{\mathrm{trop}} | \nu_j(\mathbf{S}_{[m]}) \geq 0, \forall j \in I_{\mathrm{f}}\}$. Let $\overline{M}$ denote the set of representatives of $\overline{M}^{\mathrm{trop}}$ in M°.

In definition 2.16, we have $\mathbf{S} \cap \overline{\boldsymbol{U}} = \{\mathbf{S}_{[m]} | [m] \in \overline{M}^{\mathrm{trop}}\}$. It is desirable to ask if $\mathbf{S} \cap \overline{\boldsymbol{U}}$ is a basis for $\overline{\boldsymbol{U}}$.

DEFINITION 2.17. We say $j \in I_{\mathrm{f}}$ is optimized in a seed s if $b_{jk} \geq 0$ for all $k \in I_{\mathrm{uf}}$. We say $j \in I_{\mathrm{f}}$ can be optimized if it is optimized in some seed in Δ^+.

When j is optimized in s, we have $\nu_j(z) = m_j$ for any m-pointed element z in $\boldsymbol{LP}(\mathrm{s})$. We further deduce the following pleasant result.

PROPOSITION 2.18 ([**Qin24a**, Proposition 2.15][**GHKK18**, Proposition 9.7]). *Assume all frozen vertices can be optimized. If $\boldsymbol{S}$ is a M^{trop}-pointed basis of $\boldsymbol{U}$, then $\boldsymbol{S} \cap \overline{\boldsymbol{U}}$ is a basis for $\overline{\boldsymbol{U}}$.*

REMARK 2.19. Note that, when all frozen vertices $j \in I_{\mathrm{f}}$ can be optimized, $\overline{M}^{\mathrm{trop}}$ is independent of the choice of the basis $\mathbf{S}$ and can be calculated explicitly.

More precisely, assume j is optimized in $\mathbf{s}_j$, the set of representatives of $\overline{M}^{\mathrm{trop}}$ in $M^\circ(\mathbf{s})$ is given by $\overline{M}(\mathbf{s}) = \{m \in M^\circ(\mathbf{s}) | (\phi_{\mathbf{s}_j,\mathbf{s}}(m))_j \geq 0, \forall j \in I_{\mathrm{f}}\}$; see [**Qin24a**, Lemma 2.17].

REMARK 2.20. Many cluster algebras possess a well-behaved M°-pointed basis ϑ called the theta basis [**GHKK18**]. By a recent work [**CMMM25**], the statement of Proposition 2.18 is true for the theta basis ϑ even without the assumption that the frozen vertices can be optimized.

It is desirable to know if results analogous to that of [**CMMM25**] hold for other good bases $\mathbf{S}$, for example, for the common triangular basis (Section 6) or the generic basis (see [**Pla13**][**Qin24d**] for definitions and results). Note that $\nu_j(z)$ could be interpreted as the value of a tropical function on the Laurent degrees of z [**GHKK18**]. So analogous results hold for other M°-pointed bases $\mathbf{S}$, if we can prove the following conjecture:

For any $m \in M^\circ$, the m-pointed basis elements ϑ_m and $\mathbf{S}_m$ have the same Newton polytope.

3. Seeds associated with signed words

We will construct seeds from signed words following [**BFZ05**][**SW21**]. They will be useful to describe cluster algebras arising from Lie theory.

3.1. Seeds construction. Let $C = (C_{ab})_{a,b\in J}$ denote a generalized Cartan matrix, where $J \subset \mathbb{N}_{>0}$ is a finite set of index. Denote its symmetrizers by D_a, $a \in J$, i.e., we have $D_a C_{ab} = D_b C_{ba}$, $\forall a, b \in J$.

For any $r, s \in \mathbb{Z}$. We use $[r, s]$ to denote the integer interval $\{k \in \mathbb{Z} | r \leq k \leq s\}$. A word $\underline{\eta} = (\eta_k)_{k\in[r,s]}$ is a sequence of elements in J. A signed word $\underline{\mathbf{i}} = (\mathbf{i}_k)_{k\in[r,s]}$ is a sequence of elements in $\pm J$, where we denote $\mathbf{i}_k = \varepsilon_k i_k$ for $i_k \in J$ and $\varepsilon_k = \pm 1$. We denote their length $l(\underline{\eta}) = l(\underline{\mathbf{i}}) = l([r, s]) := r - s + 1$. When $[r, s] = \emptyset$, $\underline{\mathbf{i}}_{[r,s]}$ is an empty sequence.

For any subinterval $[r', s'] \subset [r, s]$, we use $\mathbf{i}_{[r',s']}$ to denote the subsequence $(\mathbf{i}_{r'}, \mathbf{i}_{r'+1}, \ldots, \mathbf{i}_{s'})$. By abuse of notation, we will write

$$\underline{\mathbf{i}} = (\mathbf{i}_{r'}, \ldots, \mathbf{i}_{r'-1}, \underline{\mathbf{i}}_{[r',s']}, \mathbf{i}_{s'+1}, \ldots, \mathbf{i}_s).$$

The opposite word $\underline{\mathbf{i}}^{\mathrm{op}}$ is defined such that $(\underline{\mathbf{i}}^{\mathrm{op}})_k := \mathbf{i}_{s-k+r}$, $\forall k \in [r, s]$. For any $N \in \mathbb{N}$, we let $\mathbf{i}^N$ denote the sequence $(\underline{\mathbf{i}}, \underline{\mathbf{i}}, \ldots, \underline{\mathbf{i}})$ without specifying its index.

We associate with $\underline{\mathbf{i}}$ a classical seed $\ddot{\mathbf{s}} = \ddot{\mathbf{s}}(\underline{\mathbf{i}})$ as follows.

Denote $\dot{I} := [r, s]$. The set of vertex for $\ddot{\mathbf{s}}$ is $I := \ddot{I} := [r - |J|, r - 1] \sqcup \dot{I} = [r - |J|, s]$. Choose a Coxeter word $\underline{c} = (c_1, \ldots, c_{|J|})$, i.e. each element of J appears in $\underline{c}$ exactly once. We define the skew-symmetrizers to be $\mathsf{d}_k := \begin{cases} D_{c_{k-r+|J|+1}} & k \in [r - |J|, r - 1] \\ D_{i_k} & k \in \dot{I} \end{cases}$. Associated with the signed word $(i_k)_{k\in\ddot{I}} := (\underline{c}, \underline{\mathbf{i}})$, we define $[1] : \ddot{I} \to \ddot{I} \cup \{+\infty\}$ such that

$$k[1] = \min(\{j \in \ddot{I} | j > k, i_j = i_k\} \cup \{+\infty\}). \tag{3.1}$$

Denote $k[0] = k$ for $k \in \ddot{I}$. Denote $k[-1] := k'$ if $k = k'[1] \in \ddot{I}$. For $d \in \mathbb{Z}$ and $\varepsilon \in \{\pm 1\}$, if $k[d]$ and $(k[d])[\varepsilon]$ are both well defined, we define $k[d + \varepsilon] := (k[d])[\varepsilon]$.

Define $\dot{I}_{\mathrm{f}} := \{k \in \dot{I} | k[1] = +\infty\}$ and the set of frozen vertices $\ddot{I}_{\mathrm{f}} := [r - |J|, r - 1] \sqcup \dot{I}_{\mathrm{f}}$. Then the set of unfrozen vertex is $\dot{I}_{\mathrm{uf}} := \ddot{I}_{\mathrm{uf}} := \{k \in \dot{I} | k[1] \in \dot{I}\}$.

Following [**BFZ05**][**SW21**], define the exchange matrix $\ddot{B} = (b_{jk})_{j\in\ddot{I},k\in\ddot{I}_{\mathrm{uf}}}$ associated to $\underline{\mathbf{i}}$ by

$$b_{jk} = \begin{cases} \varepsilon_k & \text{if } k = j[1] \\ -\varepsilon_j & \text{if } j = k[1] \\ \varepsilon_k C_{i_j,i_k} & \text{if } \varepsilon_{j[1]} = \varepsilon_k,\ j < k < j[1] < k[1] \\ \varepsilon_k C_{i_j,i_k} & \text{if } \varepsilon_k = -\varepsilon_{k[1]},\ j < k < k[1] < j[1] \\ -\varepsilon_j C_{i_j,i_k} & \text{if } \varepsilon_{k[1]} = \varepsilon_j,\ k < j < k[1] < j[1] \\ -\varepsilon_j C_{i_j,i_k} & \text{if } \varepsilon_j = -\varepsilon_{j[1]},\ k < j < j[1] < k[1] \\ 0 & \text{otherwiese} \end{cases}$$

We refer the reader to [**Qin24a**, Section 6.1] for an intuitive construction of $\ddot{B}$ using triangulated trapezoid following [**SW21**], which also leads to the construction of half weight entries b_{ij} for $i, j \in \ddot{I}_{\mathrm{f}}$.

The seed $\dot{\mathbf{s}}$ is obtained from $\ddot{\mathbf{s}}$ by restricting to the subset of vertices $\dot{I}$. More precisely, $I(\dot{\mathbf{s}}) := \dot{I}$, $I(\dot{\mathbf{s}})_{\mathrm{uf}} := \dot{I}_{\mathrm{uf}}$, $\widetilde{B}(\dot{\mathbf{s}}) := (b_{jk})_{j\in\dot{I},k\in\dot{I}_{\mathrm{uf}}}$, and the skew-symmetrizers are $\mathsf{d}_k = D_{i_k}$ for $k \in \dot{I}$.

The frozen vertices of $\ddot{\mathbf{s}}$ could be optimized; see [**Qin24a**, Proposition 6.16]. By [**BZ05**], there is a compatible Poisson structure for $\ddot{\mathbf{s}}$. The same is true for $\dot{\mathbf{s}}$ by [**Qin24a**, Lemma 6.4].

These definitions of $[r, s]$, (signed) words, and the associated (quantum) seeds could be naturally extend to the case where $r = -\infty$ and/or $s = +\infty$.

3.2. Operations on signed words.

Signed words and group elements. Recall that the monoid Br^+ of positive braids associated to C is the monoid generated by the generators $\{\sigma_a | a \in J\}$ subject to the following relations:

$$\begin{aligned} \sigma_a\sigma_b &= \sigma_b\sigma_a \text{ if } C_{ab} = 0 \\ \sigma_a\sigma_b\sigma_a &= \sigma_b\sigma_a\sigma_a \text{ if } C_{ab}C_{ba} = 1 \\ (\sigma_a\sigma_b)^2 &= (\sigma_b\sigma_a)^2 \text{ if } C_{ab}C_{ba} = 2 \\ (\sigma_a\sigma_b)^3 &= (\sigma_b\sigma_a)^3 \text{ if } C_{ab}C_{ba} = 3 \end{aligned}$$

The generalized Braid group Br is constructed from Br^+ by including the formal inverses $\sigma_{\mathrm{a}}^{\pm}$, where $a \in J$. The associated Coxeter group W is the quotient group of Br by the relation $\sigma_{\mathrm{a}}^2 = e$ for $a \in J$, where e denotes the identity element. For any $\beta \in \mathrm{Br}$, we denote its image in W by $[\beta]$. In particular, we have the simple reflections $s_a = [\sigma_a]$.

For any word $\underline{\eta} = (\eta_k)_{k\in[r_1,r_2]}$, denote its length by $l(\underline{\eta}) := l([r_1, r_2]) = r_2 - r_1 + 1$. We let $\beta_{\underline{\eta}}$ denote the positive braid $\sigma_{\eta_{r_1}} \cdots \sigma_{\eta_{r_2}}$ and $w_{\underline{\eta}}$ denote the Coxeter group element $s_{\eta_{r_1}} \cdots s_{\eta_{r_2}}$, which equals $[\beta_{\underline{\eta}}]$. For any $w \in W$, define its length to be $l(w) := \min\{r | s_{\eta_1} \cdots s_{\eta_r} = w,\ \eta_i \in J,\ \forall i \in [1, r]\}$. $\underline{\eta}$ is called a reduced word if $l(\underline{\eta}) = l(w_{\underline{\eta}})$. When W is finite, we use w_0 to denote its longest element.

Signed words Operations. Let $\underline{\mathbf{i}} = (\mathbf{i}_k)_{k\in[r,s]}$ denote a signed word. Let $\mathbf{s}$ denote $\dot{\mathbf{s}}$ or $\ddot{\mathbf{s}}$. Following [**SW21**], we could construct a new signed word $\underline{\mathbf{i}}' = (\mathbf{i}'_k)_{k\in[r,s]}$ and thus new classical seeds $\mathbf{s}(\underline{\mathbf{i}}')$ by one of the following operations.

- (Left reflection) Define $\underline{\mathbf{i}}' = (-\mathbf{i}_r, \mathbf{i}_{[r+1,s]})$. In this case, we have $\dot{\mathbf{s}}(\underline{\mathbf{i}}') = \dot{\mathbf{s}}(\underline{\mathbf{i}})$.

- (Flips) When $k \in [r, s-1]$ and $\varepsilon_k = -\varepsilon_{k+1}$, a flip at k gives a new signed word $\underline{\mathbf{i}}' = (\underline{\mathbf{i}}_{[r,k-1]}, \mathbf{i}_{k+1}, \mathbf{i}_k, \underline{\mathbf{i}}_{[k+2,s]})$. Let $\sigma_{k,k+1}$ denote the transposition $(k, k+1)$ on $\mathbb{Z}$. In this case, we have $\mathbf{s}(\underline{\mathbf{i}}') = \sigma_{k,k+1}\mathbf{s}(\underline{\mathbf{i}})$ if $i_k \neq i_{k+1}$ and $\mathbf{s}(\underline{\mathbf{i}}') = \mu_k \mathbf{s}(\underline{\mathbf{i}})$ if $i_k = i_{k+1}$.
- (Braid moves) If $\underline{\mathbf{i}}_{[j,k]} = \varepsilon\underline{\eta}$ for some word $\underline{\eta}$ in J, $\varepsilon \in \{\pm 1\}$, and $\beta_{\underline{\eta}} = \beta_{\underline{\eta}'}$ in Br^+, we could construct $\underline{\mathbf{i}}' = (\underline{\mathbf{i}}_{[r,j-1]}, \varepsilon\underline{\eta}', \underline{\mathbf{i}}_{[k+1,s]})$. Then there exists a sequence of mutation $\boldsymbol{\mu}_{\underline{\mathbf{i}}',\underline{\mathbf{i}}}$ on $\{k' \in [j,k] | k'[1] \leq k\}$ and a permutation $\sigma_{\underline{\mathbf{i}}',\underline{\mathbf{i}}}$ on $[j,k]$ such that $\mathbf{s}(\underline{\mathbf{i}}') = \sigma_{\underline{\mathbf{i}}',\underline{\mathbf{i}}}\boldsymbol{\mu}_{\underline{\mathbf{i}}',\underline{\mathbf{i}}}\mathbf{s}(\underline{\mathbf{i}})$.

Let $\underline{\mathbf{i}}'$ be a signed word obtained form $\underline{\mathbf{i}}$ by a finite sequence of operations above. If we endow $\mathbf{s}(\underline{\mathbf{i}})$ with a compatible Poisson structure, then we could obtain an induced compatible Poisson structure on $\mathbf{s}(\underline{\mathbf{i}}')$ by using mutations step by step (see (2.4)). Moreover, this induced Poisson structure is independent of the choice of the operation sequence; see [**Qin24c**, Lemma 3.2].

3.3. Injective reachable sequences and fundamental variables. For any signed word $\underline{\mathbf{i}} = (\mathbf{i}_k)_{k\in[r,s]}$, we introduce the following notation for any $k \in [r, s]$:

$$o_+(k) := |\{k' | k' > k, k' \in [r,s], i_k = i_{k'}\}|$$
$$k^{\max} := k[o_+(k)]$$

Recall that denote $k^{\max}[1] = +\infty$ by (3.1). We will work with $\dot{I} = [r, s]$. Correspondingly, we introduce the following

$$o_-(k) := |\{k' | k' < k, k' \in [r,s], i_k = i_{k'}\}|$$
$$k^{\min} := k[-o_-(k)]$$
$$k^{\min}[-1] := -\infty.$$

We will denote $f_{-\infty} = 0$.

Let $\underline{\eta} = (\eta_k)_{k\in[r,s]}$ denote a length l word in J. We claim that the seed $\dot{\mathbf{s}} := \dot{\mathbf{s}}(\eta)$ is injective reachable ([**SW21**, Theorem 4.1]). To see this, denote $\boldsymbol{\mu}_{[j,k]} := \mu_k \cdots \mu_{j[1]}\mu_j$ for $j, k \in [1, l]$ (read from right to left) such that $\eta_j = \eta_k$. Define $\Sigma_k := \boldsymbol{\mu}_{[k^{\min}, k^{\min}[o_+(k)-1]]}$, and $\Sigma := \Sigma_s \cdots \Sigma_{r+1}\Sigma_r$. In particular, Σ_k is an empty sequence if $k = k^{\max}$. As in [**Qin24a**, Section 8.2], let σ denote the permutation on $[1, l]$ such that $\sigma(k) = k$ if $k = k^{\max}$ and $\sigma(k) = k^{\max}[-1 - o_-(k)]$ if $k \neq k^{\max}$.

LEMMA 3.1 ([**Qin24a**, Lemma 8.4]). *The cluster variables of the seeds appearing along the mutation sequence Σ starting from $\dot{\mathbf{s}}$ have the degrees $f_k - f_{j[-1]}$ for $[j, k] \subset [r, s]$ such that $\eta_j = \eta_k$.*

In particular, we have $x_{\sigma k}(\Sigma\dot{\mathbf{s}}) = -f_k + f_{k^{\max}}$, $\forall k \in I_{\mathrm{uf}}$. Consequently, $\dot{\mathbf{s}}$ satisfies the injective-reachability assumption.

DEFINITION 3.2. For any $[j, k] \subset [r, s]$ such that $\eta_j = \eta_k$, let $W_{[j,k]}$ denote the cluster variable with degree $f_k - f_{j[-1]}$ (see Lemma 3.1). $W_{[j,k]}$ is called an interval variable. $W_k := W_{[k,k]}$ is called a fundamental variable.

REMARK 3.3. In special cases, W_k correspond to fundamental modules of quantum affine algebras; hence the name (see Section A.2).

FIGURE 3.1. The ice quiver for $\dot{\mathbf{s}}(\underline{\mathbf{j}})$, $\underline{\mathbf{j}} = (1,2,3)^4$.

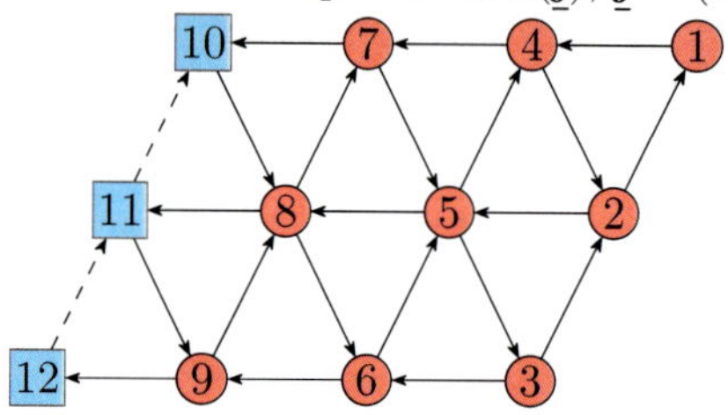

FIGURE 3.2. The ice quiver for $\dot{\mathbf{s}}(\underline{\mathbf{i}})$, $\underline{\mathbf{i}} = (1,2,3)^\infty$.

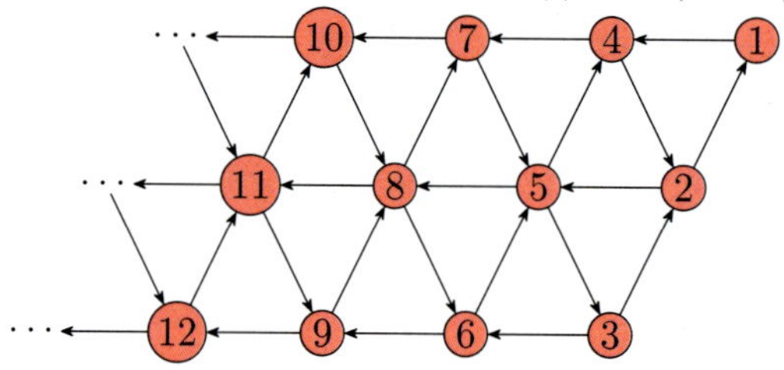

FIGURE 3.3. The ice quiver for $\ddot{\mathbf{s}}(\underline{\mathbf{i}})$, $\underline{\mathbf{i}} = ((1,2)^\infty, 1, -1, 2, -2, 1, -1, (2,1)^\infty)$

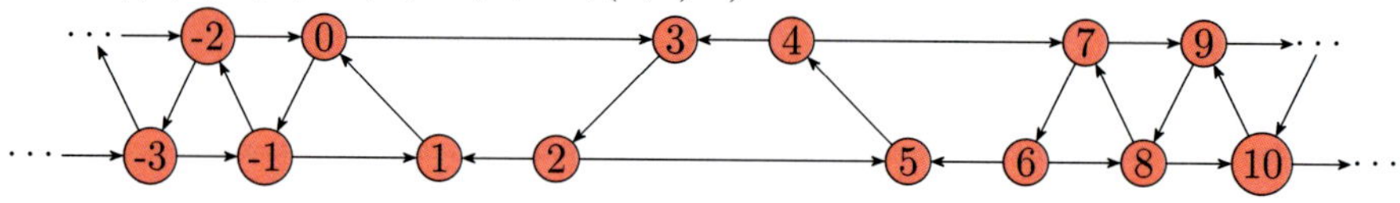

3.4. Examples.

EXAMPLE 3.4 (Seeds arising from $U_q(\hat{sl}_4)$-representations [**HL10**]). Choose the Cartan matrix $C = \begin{pmatrix} 2 & -1 & 0 \\ -1 & 2 & -1 \\ 0 & -1 & 2 \end{pmatrix}$ and the Coxeter word $\underline{c} = (1,2,3)$. Take the words $\underline{\mathbf{j}} = (\mathbf{j}_k)_{k\in[1,8]} = (\underline{c}^4)$, and $\underline{\mathbf{i}} = (\mathbf{i}_k)_{k\in[1,\infty)} = (\underline{c}^\infty)$. The ice quivers for $\dot{\mathbf{s}}(\underline{\mathbf{j}})$ and $\dot{\mathbf{s}}(\underline{\mathbf{i}})$ are depicted in Figure 3.1 and Figure 3.2, respectively.

EXAMPLE 3.5 (A seed for $\mathbb{C}[SL_3]$ [**BFZ05**]). Choose the Cartan matrix $C = \begin{pmatrix} 2 & -1 \\ -1 & 2 \end{pmatrix}$ and the signed word $\underline{\mathbf{i}} = (1,-1,2,-2,1,-1)$. The seed $\ddot{\mathbf{s}}(\underline{\mathbf{i}})$ is given Example 2.2 (Figure 2.1).

EXAMPLE 3.6 (An infinite seed of type A_2 [**GHL24**]). Choose the Cartan matrix $C = \begin{pmatrix} 2 & -1 \\ -1 & 2 \end{pmatrix}$, $\underline{c} = (1,2)$, and the signed word $\underline{\mathbf{i}} = (\underline{c}^\infty, 1, -1, 2, -2, 1, -1, (\underline{c}^{\mathrm{op}})^\infty) = (\ldots, 1, 2, 1, 2, 1, -1, 2, -2, 1, -1, 2, 1, 2, 1, \ldots)$. The ice quiver $\widetilde{Q}$ for the seed $\ddot{\mathbf{s}}(\underline{\mathbf{i}})$ is given Figure 3.3.

FIGURE 3.4. The ice quiver for a seed of $\mathbb{C}[SL_2]$.

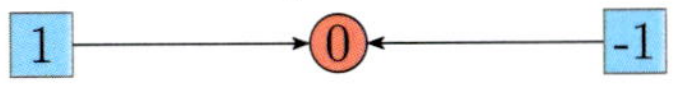

EXAMPLE 3.7 (A seed for $\mathbb{C}[SL_2]$). Choose the Cartan matrix $C = \begin{pmatrix}2\end{pmatrix}$ and the signed word $\underline{\mathbf{i}} = (\mathbf{i}_k)_{k\in[0,1]} = (1,-1)$. Then $\ddot{\mathbf{s}}(\underline{\mathbf{i}})$ is a seed for the cluster algebra $\mathbb{C}[SL_2]$; see Section A.1. The corresponding ice quiver is given Figure 3.4.

EXAMPLE 3.8 (An infinite seed of type A_1 [**GHL24**]). Choose the Cartan matrix $C = \begin{pmatrix}2\end{pmatrix}$ and the infinite signed word $\underline{\mathbf{i}} = (\ldots,1,1,-1,1,1,\ldots)$ where $\mathbf{i}_1 = -1$. The seed $\ddot{\mathbf{s}}(\underline{\mathbf{i}})$ is given in Example 2.3 (Figure 2.2).

4. Cluster operations

Following [**Qin24a**], we introduce some cluster operations that propagate structures from one cluster algebra to another cluster algebra whose seeds are closely related. These operations are crucial in the approach of [**Qin24a**] to Theorem 6.11 (results for cluster algebras from Lie theory).

This section is technical. Readers may skip it without loss of continuity in the subsequent sections.

4.1. Freezing. Let $\mathbf{s} = (I, I_{\mathrm{uf}}, (\mathsf{d}_i)_{i\in I}, \widetilde{B}, (x_i)_{i\in I})$ denote a seed. Let $(b_{ij})_{i,j\in I}$ be the associated matrix for its chosen ice quiver. Choose any $F \subset I_{\mathrm{uf}}$. By freezing the vertices in F, we obtain a new seed $\mathbf{s}' := \mathfrak{f}_F\mathbf{s} = (I, I_{\mathrm{uf}}\backslash F, (\mathsf{d}_i)_{i\in I}, \widetilde{B}, (x_i)_{i\in I})$. The matrix $(b_{ij})_{i,j\in I}$ is kept unchanged. When $\mathbf{s}$ is a quantum seed endowed with the compatible Poisson structure λ, we endow $\mathbf{s}'$ with the same compatible Poisson structure λ.

We have $M^\circ(\mathbf{s}') = M^\circ(\mathbf{s})$ but $N^\oplus(\mathbf{s}') \subset \mathrm{N}^\oplus(\mathbf{s})$. Note that the set of tropical points M^{trop} for $\boldsymbol{U}(\mathbf{s})$ and the set of tropical points $(M^{\mathrm{trop}})'$ for $\boldsymbol{U}(\mathbf{s}')$ are different, although they are both in bijection with $M^\circ(\mathbf{s}) = M^\circ(\mathbf{s}') \simeq \mathbb{Z}^I$.

Assume p^* is injective from now on. Note that $\preceq_{\mathbf{s}}$ implies $\preceq_{\mathbf{s}'}$, which are both denoted by $\preceq$. We define the freezing operator $\mathfrak{f}^{\mathbf{s}}_{F,m}$ to be a linear map from $\widehat{\boldsymbol{LP}}_{\preceq m}(\mathbf{s})$ to $\widehat{\boldsymbol{LP}}_{\preceq m}(\mathbf{s}')$ such that

$$\mathfrak{f}^{\mathbf{s}}_{F,m}\Big(\sum_{n\in N^\oplus} b_n x^{m+p^*(n)}\Big) := \sum_{0\neq n\in N^\oplus, \operatorname{supp} n\cap F=\emptyset} b_n x^{m+p^*(n)},\ \forall b_n \in \Bbbk.$$

Note that, if z is m-pointed, $\mathfrak{f}^{\mathbf{s}}_{F,m}(z)$ is m-pointed too.

When the context is clear, we might abbreviate the symbols $\mathbf{s}$, F, and/or m in $\mathfrak{f}^{\mathbf{s}}_{F,m}$. For a subset Z of $\widehat{\boldsymbol{LP}}(\mathbf{s})$ whose elements have unique $\prec$-maximal degrees, we denote $\mathfrak{f}Z := \{\mathfrak{f}z | z \in Z\}$. By freezing $F' \in I$, we mean freezing $F := F' \cap I_{\mathrm{uf}}$.

Note that $\mathfrak{f}$ is not a linear map from $\widehat{\boldsymbol{LP}}(\mathbf{s})$ to $\widehat{\boldsymbol{LP}}(\mathbf{s}')$. Nevertheless, Lemma 4.1 states that it still has some compatibility with algebra operations.

LEMMA 4.1 ([**Qin24a**, Lemma 3.6]). *(1) Assume that we have $m' \preceq m$ in M° such that $m' = m + p^*(n)$, $z \in \widehat{\boldsymbol{LP}}_{\preceq m'}(\mathbf{s}) \subset \widehat{\boldsymbol{LP}}_{\preceq m}(\mathbf{s})$, then we have*

$$\mathfrak{f}_m z = \begin{cases} \mathfrak{f}_{m'} z & \text{if } \operatorname{supp} n \cap F = \emptyset \\ 0 & \text{else} \end{cases}.$$

(2) For any $m_1, m_2 \in M^\circ$ and $z_1 \in \widehat{\boldsymbol{LP}}_{\preceq m_1}$, $z_2 \in \widehat{\boldsymbol{LP}}_{\preceq m_2}$, we have

$$\mathfrak{f}_{m_1}(z_1) * \mathfrak{f}_{m_2}(z_2) = \mathfrak{f}_{m_1+m_2}(z_1 * z_2).$$

PROPOSITION 4.2 ([**Qin24a**, Proposition 3.8]). *If $z \in \boldsymbol{LP}_{\preceq m}(\mathbf{s}) \cap \boldsymbol{U}(\mathbf{s})$, then $\mathfrak{f}_{F,m} z \in \boldsymbol{U}(\mathfrak{f}_F \mathbf{s})$.*

We have the following important result, which states that the freezing operator $\mathfrak{f}_F$ is compatible with mutations at $k \notin F$.

PROPOSITION 4.3 ([**Qin24a**, Lemma 3.9]). *Assume $k \notin F$ and take $\mathbf{s} = \mu_k \mathbf{s}'$. Then for any $z \in \boldsymbol{LP}(\mathbf{s}) \cap (\mu_k^*)^{-1} \boldsymbol{LP}(\mathbf{s}')$ such that z has a unique $\prec_{\mathbf{s}}$-maximal degree m in $\boldsymbol{LP}(\mathbf{s})$ and $\mu_k^* z$ has a unique $\prec_{\mathbf{s}'}$-maximal degree m' in $\boldsymbol{LP}(\mathbf{s}')$, then we have $\mathfrak{f}_{F,m'}^{\mathbf{s}'}(\mu_k^* z) = \mu_k^* \mathfrak{f}_{F,m}^{\mathbf{s}}(z)$.*

Denote $\mathbf{s}' = \mathfrak{f}_F \mathbf{s}$. Combining Proposition 4.2 and Proposition 4.3, we obtain the following result.

THEOREM 4.4 ([**Qin24a**, Theorem 3.10]). *If $z \in \boldsymbol{U}(\mathbf{s})$ is compatibly pointed at all seeds in $\Delta_{\mathbf{s}}^+$, then $\mathfrak{f}_F z \in \boldsymbol{U}(\mathbf{s}')$ is compatibly pointed at all seeds in $\Delta_{\mathbf{s}'}^+$.*

Theorem 4.4 and Theorem 2.13 implies the following.

THEOREM 4.5 ([**Qin24a**, Theorem 3.11]). *If $\boldsymbol{S}$ is a M^{trop}-pointed basis of $\boldsymbol{U}(\mathbf{s})$ and $\mathfrak{f}_F \mathbf{s}$ is injective reachable, then $\mathfrak{f}_F \boldsymbol{S}$ is a $(M^{\mathrm{trop}})'$-pointed basis of $\boldsymbol{U}(\mathbf{s}')$.*

We also have the following general result.

THEOREM 4.6 ([**Qin24a**, Theorem 3.12][**Qin24b**, Theorem 4.3.1]). *If z is a localized cluster monomial of $\boldsymbol{U}(\mathbf{s})$, then $\mathfrak{f}_F z$ is a localized cluster monomial of $\boldsymbol{U}(\mathfrak{f}_F \mathbf{s})$.*

In general, it is difficult to know if $z \in \boldsymbol{U}$ also belongs to $\boldsymbol{A}$. Though this happens if z has the following property.

DEFINITION 4.7 ([**Qin24a**, Definition 3.14]). Assume there is a finite decomposition $z = \sum b_i z_i$, such that $b_i \in \Bbbk$, z_i have distinct degrees in $\boldsymbol{LP}(\mathbf{s})$, and z_i are products of cluster variables and frozen factors, then we say z has a nice cluster decomposition in $\mathbf{s}$.

LEMMA 4.8 ([**Qin24a**, Lemma 3.15]). *For any $z \in \boldsymbol{LP}_{\preceq m}(\mathbf{s})$, if z has a nice cluster decomposition in $\mathbf{s}$, then $\mathfrak{f}_m z \in \boldsymbol{A}(\mathfrak{f}_F \mathbf{s})$.*

COROLLARY 4.9 ([**Qin24a**, Corollary 3.16]). *If $\boldsymbol{S}$ is a M^{trop}-pointed basis of $\boldsymbol{U}(\mathbf{s})$, such that its elements have nice cluster decomposition in $\mathbf{s}$, and if $\mathfrak{f}_F \mathbf{s}$ is injective reachable, then $\mathfrak{f}_F \boldsymbol{S}$ is a $(M^{\mathrm{trop}})'$-pointed basis of $\boldsymbol{U}(\mathfrak{f}_F \mathbf{s})$ and its elements have nice cluster decomposition. In particular, $\boldsymbol{A}(\mathfrak{f}_F \mathbf{s}) = \boldsymbol{U}(\mathfrak{f}_F \mathbf{s})$.*

EXAMPLE 4.10. Consider the seed $\mathbf{s} = \dot{\mathbf{s}}(\underline{\mathbf{i}})$ where $\underline{\mathbf{i}} = (\mathbf{i}_k)_{k \in [1,6]} = (1,2)^3$. The ice quiver for $\mathbf{s}$ is given in Figure 4.1. The element $z = x_2^{-1} \cdot x_6 \cdot (1 + y_2 + y_1 \cdot y_2 + y_2 \cdot y_4 + y_1 \cdot y_2 \cdot y_4 + y_1 \cdot y_2 \cdot y_3 \cdot y_4)$ is the $(-f_2 + f_6)$-pointed common triangular basis element. It can be identified with the (truncated) (q,t)-character of a Kirillov-Reishetikhin module (Section A.2).

Now freeze $\mathbf{s}$ at $F = \{4\}$. The ice quiver for $\mathfrak{f}_F \mathbf{s}$ is depicted in Figure 4.2. We obtain $\mathfrak{f}_F z = x_2^{-1} \cdot x_6 \cdot (1 + y_2 + y_1 \cdot y_2)$ be setting $y_4 = 0$.

FIGURE 4.1. The ice quiver for $\dot{\mathbf{s}}(\underline{\mathbf{i}})$, $\underline{\mathbf{i}} = (1,2)^3$

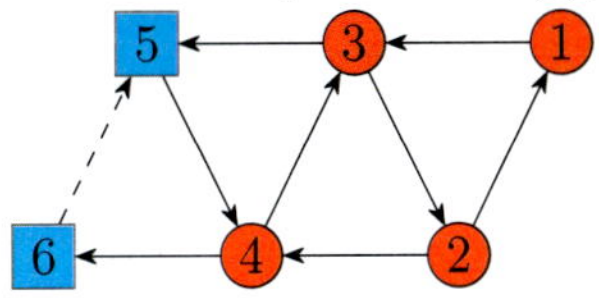

FIGURE 4.2. The ice quiver for $\mathfrak{f}_{\{4\}}\dot{\mathbf{s}}(\underline{\mathbf{i}})$, $\underline{\mathbf{i}} = (1,2)^3$

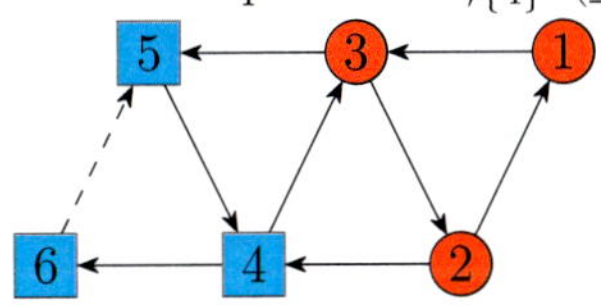

4.2. Similarity and base change. Recall that for a quantum seed $\mathbf{s}$, we have $\lambda(f_i, p^* e_k) = -\delta_{ik}\delta_k$ for $k \in I_{\mathrm{uf}}$, $\delta_k \in \mathbb{N}_{>0}$.

DEFINITION 4.11 ([**Qin14**][**Qin17**][**Qin24a**, Definition 4.1]). Let there be two seeds $\mathbf{s}$ and $\mathbf{s}'$, which are not necessarily related by a sequence of mutations, and σ a permutation of I'_{uf}. We say $\mathbf{s}$ and $\mathbf{s}'$ are similar (up to σ) if $b'_{\sigma i, \sigma j} = b_{ij}$ and $\mathsf{d}'_{\sigma i} = \mathsf{d}_i$, $\forall i, j \in I_{\mathrm{uf}}$. When $\mathbf{s}$ and $\mathbf{s}'$ are quantum seeds, we further require that there exists some scaling factor $\rho \in \mathbb{Q}_{>0}$ such that $\delta'_{\sigma k} = \rho\delta_k$, $\forall k \in I_{\mathrm{uf}}$.

EXAMPLE 4.12 (Principal coefficinet seed). For any seed $\mathbf{s}$, we can construct a new seed $\mathbf{s}^{\mathrm{prin}}$, called the seed of principal coefficients, as follows. Set $I_{\mathrm{uf}}(\mathbf{s}^{\mathrm{prin}}) := I_{\mathrm{uf}}$, $I_{\mathrm{f}}(\mathbf{s}^{\mathrm{prin}}) := \{k' | k \in I_{\mathrm{uf}}\}$, and $\mathsf{d}_{k'}(\mathbf{s}^{\mathrm{prin}}) = \mathsf{d}_k(\mathbf{s}^{\mathrm{prin}}) := \mathsf{d}_k(\mathbf{s})$, $\forall k \in I_{\mathrm{uf}}$. Define $b_{ij}(\mathbf{s}^{\mathrm{prin}}) = b_{ij}(\mathbf{s})$ for $i, j \in I_{\mathrm{uf}}$, and $b_{k'j} = \delta_{kj}$ for $j, k \in I_{\mathrm{uf}}$.

There is a map $\mathrm{var} : M^\circ(\mathbf{s}^{\mathrm{prin}}) \to M^\circ(\mathbf{s})$ such that

$$\mathrm{var}(f_i) = \begin{cases} f_i & \text{if } i \in I_{\mathrm{uf}} \\ \sum_{j \in I_{\mathrm{f}}} b_{jk} f_j & \text{if } i = k' \in I_{\mathrm{f}}(\mathbf{s}^{\mathrm{prin}}) \end{cases}. \tag{4.1}$$

It induces the $\Bbbk$-linear map var from $\boldsymbol{LP}(\mathbf{s}^{\mathrm{prin}})$ to $\boldsymbol{LP}(\mathbf{s})$ such that

$$\mathrm{var}(x^m) := x^{\mathrm{var}(m)}.$$

Note that $\mathrm{var}(x^{\widetilde{B}(\mathbf{s}^{\mathrm{prin}})e_k}) = x^{\widetilde{B}e_k}$, $\forall k \in I_{\mathrm{uf}}$.

If $\mathbf{s}$ is endowed with a compatible Poisson form λ, we endow $\mathbf{s}^{\mathrm{prin}}$ with the compatible Poisson form $\lambda(\mathbf{s}^{\mathrm{prin}})$ such that

$$\lambda(\mathbf{s}^{\mathrm{prin}})(m, m') := \lambda(\mathrm{var}(m), \mathrm{var}(m')).$$

The (classical or quantum) seed $\mathbf{s}$ and $\mathbf{s}'$ are similar with $\rho = 1$.

In practice, it usually suffices to consider $\rho = 1$. This happens when two similar $\mathbf{s}$ and $\mathbf{s}'$ related by a mutation sequence, for example, when $\mathbf{s}' = \mathbf{s}[1]$. Let us assume $\rho = 1$ from now on for simplicity. This assumption will greatly simplify treatments and statements, so that we do not need to extend $\Bbbk$ to a base ring $\Bbbk' = \Bbbk[v^{\frac{1}{r}}]$ for a large enough $r \in \mathbb{N}$ as in [**Qin24a**, Section 4].

Assume $\mathbf{s}$ and $\mathbf{s}'$ are similar (classical or quantum) seeds and $\rho = 1$. Replace $\mathbf{s}'$ by $\sigma^{-1}\mathbf{s}'$ if necessary, we can assume $\sigma = 1$ without loss of generality. We use $'$ to denote data associated with $\mathbf{s}'$ as usual.

DEFINITION 4.13 ([**Qin14**]). Let there be given any m-pointed element $z = x^m \cdot (1+\sum_{0\neq n\in N^\oplus} b_n \cdot x^{\widetilde{B}\cdot n})$ in $\widehat{\boldsymbol{LP}}$, where $b_n \in \Bbbk$. An m'-pointed element $z' \in \widehat{\boldsymbol{LP}}'$ is said to be similar to z if $\mathrm{pr}_{I_{\mathrm{uf}}} m' = \mathrm{pr}_{I_{\mathrm{uf}}} m$ and z' takes the form $z' = (x')^{m'} \cdot (1 + \sum_{0\neq n\in N^\oplus} b_n \cdot (x')^{\widetilde{B}'\cdot n})$.

The following correction technique is convenient for propagate algebra relations from $\widehat{\boldsymbol{LP}}$ to $\widehat{\boldsymbol{LP}}'$. For $m \in M^\circ$, let us define the following set

$$\widehat{V}_m := x^m \cdot (\Bbbk[p^*(N_{\mathrm{uf}})] \otimes_{\Bbbk[M^\oplus]} \widehat{\Bbbk[M^\oplus]}) \subset \widehat{\boldsymbol{LP}}$$
$$V_m := x^m \cdot \Bbbk[p^*(N_{\mathrm{uf}})] \subset \boldsymbol{LP}$$

For $m' \in (M^\circ)'$, $\widehat{V}'_{m'} \subset \widehat{\boldsymbol{LP}}'$ and $V'_{m'} \subset \boldsymbol{LP}'$ are defined similarly. Introduce the $\Bbbk$-linear isomorphism $\mathrm{var} := \mathrm{var}_{m',m} : \widehat{V}_m \to \widehat{V}'_{m'}$ such that

$$\mathrm{var}(x^{m+\widetilde{B}n}) := (x')^{m'+\widetilde{B}'n}$$

Then var sends a pointed element to a pointed element with the same F-function.

THEOREM 4.14 (Correction technique [**Qin14**][**Qin17**][**Qin24a**, Theorem 4.6]). *Let there be given m_s-pointed elements $z_s \in \widehat{\boldsymbol{LP}}$ and m'_s-pointed elements $z'_s \in \widehat{\boldsymbol{LP}}$ such that z_s and z'_s are similar, where $s \in [1, r]$ for some $r \in \mathbb{N}$. We allow repetition in theses degrees. Take $z \in \widehat{V}_m$ and $z' \in \widehat{V}'_{m'}$ such that $\mathrm{pr}_{I_{\mathrm{uf}}} m = \mathrm{pr}_{I'_{\mathrm{uf}}} m'$ and $z' = \mathrm{var}_{m',m} z$.*

(1) *If $z = [z_1 * \cdots * z_r]^{\mathbf{s}}$, then $z' = p' \cdot [z'_1 * \cdots * z'_r]^{\mathbf{s}'}$ for the frozen factor $p' \in \mathsf{R}'$, such that $\deg^{\mathbf{s}'} p' = m' - \sum_{s=1}^{r} m'_s$.*
(2) *If $z = \sum_{s=1}^{r} b_s z_s$ for some $b_s \in \Bbbk$, then $z = \sum_J b_s z_s$ for $J = \{s | b_s \neq 0,\ z_s \in \widehat{V}_m\}$. Moreover, $z' = \sum_{s\in J} \mathrm{var}_{m',m}(b_s z_s) = \sum_{s\in J} b_s p'_s z'_s$, for the frozen factors p'_s such that $\deg^{\mathbf{s}'} p'_s = \deg^{\mathbf{s}'} \mathrm{var}(z_s) - m'_s$.*

THEOREM 4.15 ([**FZ07**][**Tra11**][**GHKK18**]). *Let $\boldsymbol{\mu}$ be any mutation sequence. For any $i \in I$, the cluster variable $x_i(\boldsymbol{\mu}\mathbf{s}) \in \boldsymbol{LP}(\mathbf{s})$ and the cluster variable $x_i(\boldsymbol{\mu}\mathbf{s}') \in \boldsymbol{LP}(\mathbf{s}')$ are similar.*

Let A denote the ordinary cluster algebra $\boldsymbol{A}$ or the upper cluster algebra $\boldsymbol{U}$. The following statements are implied by [**Qin24a**, Lemma 4.7], Theorem 4.14, and Theorem 4.15.

PROPOSITION 4.16. *Let $z \in \boldsymbol{LP}$ and $z' \in \boldsymbol{LP}'$ be similar pointed elements.*
(1) If $z \in \mathsf{A}$, then if $z' \in \mathsf{A}'$.
(2) If z has a nice cluster decomposition in $\mathbf{s}$, then z' has a nice cluster decomposition in $\mathbf{s}'$.

Now, let $\mathcal{Z}$ be a subset of A such that it a $\Bbbk$-basis [5] and the following conditions are satisfied:

(C1) The elements of $\mathcal{Z}$ are pointed.
(C2) $\mathcal{Z}$ is closed under the commutative multiplication by frozen factors:

$$x_j^{\pm} \cdot \mathcal{Z} \subset \mathcal{Z}, \quad \forall j \in I_{\mathrm{f}}.$$

[5] [**Qin24a**, Section 4.2] does not assume $\mathcal{Z}$ to be a basis. The corresponding statements change slightly.

Let $\mathcal{Z}'$ denote the set of the elements in $\boldsymbol{LP}'$ that are similar to elements of $\mathcal{Z}$. Then $\mathcal{Z}'$ is also closed under the commutative multiplication by frozen factors.

PROPOSITION 4.17 ([**Qin24a**, Lemmas 4.9, 4.10]). *The set $\mathcal{Z}'$ is a $\Bbbk$-basis of A'.*

Finally, let us give a diagrammatic description of the base change. Assume that we have a $\Bbbk$-algebra homomorphism $\mathrm{var} : \boldsymbol{LP} \to \boldsymbol{LP}'$ such that $\mathrm{var}(x_k) = p_k \cdot x'_k$, $\mathrm{var}(x^{\mathrm{col}_k \widetilde{B}}) = (x')^{\mathrm{col}_k \widetilde{B}'}$, $\forall k \in I_{\mathrm{uf}}$, and var sends any frozen factor to a frozen factor. Such a map var is called a variation map following [**KQW23**] (or a quasi homomorphism in [**Fra16**] in the classical case).

Note that $\boldsymbol{LP}$ is a free R-bimodule, where $\mathsf{R} = \Bbbk[x_j^{\pm}]_{j \in I_{\mathrm{f}}}$. Similarly, $\boldsymbol{LP}'$ is a free R'-bimodule. Then we have the following left R'-module isomorphism

$$\varphi : \mathsf{R}' \otimes_{\mathsf{R}} \boldsymbol{LP} \simeq \boldsymbol{LP}'$$

such that $\varphi(p' \otimes x^m) := p' * \mathrm{var}(x^m)$, for any frozen factor $p' \in \mathsf{R}'$ and $m \in M^\circ$.

DEFINITION 4.18 ([**Qin24a**, Definition 4.13]). The map φ is called the base change from $\boldsymbol{LP}$ to $\boldsymbol{LP}'$ associated with var.

THEOREM 4.19 ([**Qin24a**, Theorems 4.14, 4.15]). *(1) φ restricts to a $\Bbbk$-algebra isomorphism $\mathsf{R}' \otimes_{\mathsf{R}} \boldsymbol{A} \simeq \boldsymbol{A}'$.*

(2) If $\boldsymbol{U}$ has a $\Bbbk$-basis satisfying condition (C1)(C2), or if $\boldsymbol{U}'$ has this property, then φ restricts to a $\Bbbk$-algebra isomorphism $\mathsf{R}' \otimes_{\mathsf{R}} \boldsymbol{U} \simeq \boldsymbol{U}'$.

The map φ is called a base change for the following reason. Denote $\mathsf{A} = \boldsymbol{A}$ or $\boldsymbol{U}$. When $\Bbbk = \mathbb{Z}$ and the condition in Theorem 4.19(2) is satisfied, we have a base change of scheme

$$\varphi^* : \operatorname{Spec} \mathsf{A}' \simeq \operatorname{Spec} \mathsf{R}' \times_{\operatorname{Spec} \mathsf{R}} \mathsf{A}.$$

In general, we do not know if there exists a variation map var from $\boldsymbol{LP}$ to $\boldsymbol{LP}'$. However, it is always possible to choose a variation map $\mathrm{var}_1 : \boldsymbol{LP}(\mathbf{s}^{\mathrm{prin}}, \lambda_1) \to \boldsymbol{LP}(\mathbf{s})$ and a variation map $\mathrm{var}_2 : \boldsymbol{LP}(\mathbf{s}^{\mathrm{prin}}, \lambda_2) \to \boldsymbol{LP}(\mathbf{s}')$. Here, $\mathbf{s}^{\mathrm{prin}}$ is the classical seed of principal coefficient similar to $\mathbf{s}$ (and thus similar to $\mathbf{s}'$). For $i = 1, 2$, $(\mathbf{s}^{\mathrm{prin}}, \lambda_i)$ is the quantum seed whose Poisson form is λ_i. var_1 (resp. var_2) is the homomorphism from $\boldsymbol{LP}(\mathbf{s}^{\mathrm{prin}}, \lambda_1)$ to $\boldsymbol{LP}(\mathbf{s})$ (resp. to $\boldsymbol{LP}(\mathbf{s}')$) as in Example 4.12.

In many situations, we can easily propagate structures and properties between the cluster algebras $\mathsf{A}(\mathbf{s}^{\mathrm{prin}}, \lambda_1)$ and $\mathsf{A}(\mathbf{s}^{\mathrm{prin}}, \lambda_2)$, because they only differ by a choice quantization. Furthermore, they are related to A and A' by the base change φ_i associated with var_i, $i = 1, 2$, respectively. And we have seen that base change also propagates interesting structures and properties. See the commutative diagram (4.2) for a summery.

$$\begin{array}{ccccccc} \mathsf{A} & \xleftarrow{\varphi_1} & \mathsf{A}(\mathbf{s}^{\mathrm{prin}}, \lambda_1) & \underset{\text{change}}{\overset{\text{quantization}}{\sim}} & \mathsf{A}(\mathbf{s}^{\mathrm{prin}}, \lambda_2) & \xrightarrow{\varphi_2} & \mathsf{A}' \\ \uparrow & & \uparrow & & \uparrow & & \uparrow \\ \mathsf{R} & \xleftarrow{\mathrm{var}_1} & \mathsf{R}(\mathbf{s}^{\mathrm{prin}}, \lambda_1) & \underset{\text{change}}{\overset{\text{quantization}}{\sim}} & \mathsf{R}(\mathbf{s}^{\mathrm{prin}}, \lambda_2) & \xrightarrow{\mathrm{var}_2} & \mathsf{R}' \end{array} \tag{4.2}$$

FIGURE 4.3. The quiver for a seed of $\Bbbk[N]$, $N \subset SL_4$ [**GLS13**]

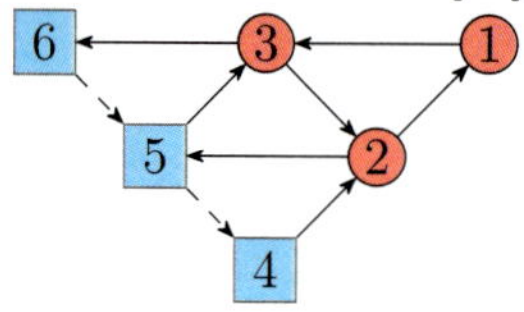

EXAMPLE 4.20. Choose $C = \begin{pmatrix} 2 & -1 & 0 & 0 \\ -1 & 2 & -1 & 0 \\ 0 & -1 & 2 & -1 \\ 0 & 0 & -1 & 2 \end{pmatrix}$ and $\underline{\mathbf{j}} = (\mathbf{j}_k)_{k \in [1,6]} = (1,2,1,3,2,1)$. Then $\mathbf{s}' = \dot{\mathbf{s}}(\underline{\mathbf{j}})$ is similar to $\mathfrak{f}_{\{4\}}\mathbf{s}$ in Example 4.10. Its ice quiver is depicted in Figure 4.3.

The cluster algebra $\overline{\boldsymbol{U}}(\mathfrak{f}_{\{4\}}\mathbf{s})$ possesses a common triangular basis element $\mathfrak{f}_{\{4\}}z = x_2^{-1} \cdot x_6 \cdot (1 + y_2 + y_1 \cdot y_2)$ (see Section 6). Take the similar element $z' = x_2^{-1} \cdot x_4 \cdot (1 + y_2 + y_1 \cdot y_2)$ in $\boldsymbol{LP}(\mathbf{s}')$. Then z' is the $(-f_2 + f_4)$-pointed common triangular basis element of $\overline{\boldsymbol{U}}(\mathbf{s}')$. Under the bijection $\kappa : \overline{\boldsymbol{U}}(\mathbf{s}') \simeq \Bbbk[N]$, z' is identified with a dual canonical basis element of $\Bbbk[N]$ up to a scalar multiple of v ([**Qin17**]), where N is the subgroup consisting of unipotent upper triangular matrices of SL_3 (Section A.1).

5. A review of Kazhdan-Lusztig algorithms

5.1. Kazhdan-Lusztig algorithm. Let E denote a free $\mathbb{Z}[v^{\pm}]$-module, endowed with a $\mathbb{Z}$-linear involution $\overline{(\ )}$ sending v to v^{-1}, called the bar involution. We make the following assumptions:

- E has a $\mathbb{Z}[v^{\pm}]$-basis $\mathbf{M} = \{\mathbf{M}_w | w \in W\}$, where W denotes an index set.
- W has an partial order $<$ which is bounded from below. It induces a natural order on $\mathbf{M}$.
- For any $\mathbf{M}_w$, we have $\overline{\mathbf{M}_w} - \mathbf{M}_w \in \sum_{w' < w} b_{w',w} \mathbf{M}_{w'}$, where $b_{w',w} \in \mathbb{Z}[v^{\pm}]$.

LEMMA 5.1 ([**Nak04**, Lemma 8.4][**Lus90**, 7.10]). *Choose $\mathbf{m} = v\mathbb{Z}[v]$ or $v^{-1}\mathbb{Z}[v^{-1}]$. The module E has a unique $\mathbb{Z}[v^{\pm}]$-basis $\boldsymbol{S} = \{\boldsymbol{S}_w | w \in W\}$, such that*

- $\overline{\boldsymbol{S}_w} = \boldsymbol{S}_w$.
- $\boldsymbol{S}_w = \boldsymbol{M}_w + \sum_{w' < w} c_{w',w} \boldsymbol{M}_{w'}$, *where* $c_{w',w} \in \mathbf{m}$.

PROOF. Since $<$ is bounded from below on W, the basis $\mathbf{S}$ could be recursively determined by $\mathbf{M}$.

More precisely, for any given w, assume that $\mathbf{S}_{w'}$ have been determined for all $w' < w$. We decompose $\overline{\mathbf{M}_w} - \mathbf{M}_w = \sum_{w' < w} d_{w',w} \mathbf{S}_{w'}$ where $d_{w',w} \in \mathbb{Z}[v^{\pm}]$. Note that $\overline{d_{w',w}} = -d_{w',w}$. Uniquely decompose $d_{w',w} = \overline{e_{w',w}} - e_{w',w}$ such that $e_{w',w} \in \mathbf{m}$. Then $\mathbf{S}_w := \mathbf{M}_w - \sum_{w' < w} e_{w',w} \mathbf{S}_{w'}$ is bar-invariant. In this way, we recursively construct the desired basis $\mathbf{S}$ satisfying all the conditions.

To show the uniqueness, assume that $\mathbf{S}' = \{\mathbf{S}'_w | w \in W\}$ is another such basis. If w is minimal in W, we have $\mathbf{S}'_w = \mathbf{M}_w = \mathbf{S}_w$. Inductively, assume we have shown $\mathbf{S}_{w'} = \mathbf{S}'_{w'}$ for all $w' < w$. Denote $\mathbf{M}_w = \mathbf{S}'_w + \sum e'_{w',w} \mathbf{S}'_{w'}$, where $e'_{w',w} \in \mathbf{m}$. Then we have $0 = \mathbf{M}_w - \mathbf{M}_w = (\mathbf{S}_w - \mathbf{S}'_w) + \sum_{w' < w} (e_{w',w} - e'_{w',w}) \mathbf{S}_{w'}$. The bar-invariance implies that $e_{w',w} - e'_{w',w} = 0$ and, consequently, $\mathbf{S}_w = \mathbf{S}'_w$. □

When E is the Hecke algebra of the Coxeter group W, we can choose $\mathbf{M}$ to be its standard basis $\{T_w | w \in W\}$, $<$ to be the Bruhat order, and $\mathbf{m} = v\mathbb{Z}[v]$. In this case, the resulting basis $\mathbf{S}$ is called its canonical basis, and the transition coefficients $c_{w',w}$ for $\mathbf{S}_w = \mathbf{M}_w + \sum_{w'<w} c_{w,w'}\mathbf{M}_{w'}$ are called the Kazhdan-Lusztig polynomials.

From now on, we call the recursive algorithm for constructing $\mathbf{S}$ in the proof of Lemma 5.1 the Kazhdan-Lusztig algorithm (the KL algorithm for short). The basis $\mathbf{S}$ will be called the Kazhdan-Lusztig-type basis (the KL basis for short).

REMARK 5.2. It may be helpful to think of the KL algorithm as sharing, in some sense, the spirit of correction as the classical Gram–Schmidt procedure:

- Let there be given an initial basis $\mathbf{M}$ which is sorted by some (partial) order $<$.
- Modifying each the basis elements by adding correction terms (basis elements inferior in $<$), we obtain a new basis $\mathbf{S}$ with desired properties.

REMARK 5.3. The above algorithm works for $\mathbb{Q}(v)$-modules as well. For example, let W be a finite Weyl gorup and $l(w_0)$ be the length of its longest element. When E is the quantized enveloping algebra $U_q(\mathfrak{g})$, we can choose $\mathbf{M}$ to be its PBW basis, endow $W = \mathbb{N}^{l(w_0)}$ with the lexicographical order, and take $\mathbf{m} = v^{-1}\mathbb{Z}[v^{-1}]$. Then the resulting basis $\mathbf{S}$ is the canonical basis.

5.2. Categorification. Consider the base change in the KL algorithm:

$$\mathbf{M}_w = \mathbf{S}_w + \sum_{w'<w} e_{w',w}\mathbf{S}_{w'} \tag{5.1}$$

In literature, it often happens that the coefficients $e_{w',w}$ take values in $\mathbf{m}_+ := v\mathbb{N}[v]$ or $v^{-1}\mathbb{N}[v^{-1}]$. Moreover, $\mathbf{S}$ often has positive structures constants: $\forall w, w' \in W$, we have

$$\mathbf{S}_w * \mathbf{S}_{w'} = \sum_{w''\in W} b_{w''}\mathbf{S}_{w''}, \quad b_{w''} \in \mathbb{N}[v^{\pm}].$$

These properties are far from obvious from the KL algorithm. They are often implied by categorification, which we will explain below.

Let $\mathbb{F}$ denote a field. Let $(\mathcal{T}, \otimes)$ denote a $\mathbb{F}$-linear category endowed with an exact bifunctort $\otimes$ called the tensor functor. We assume $\mathcal{T}$ is a tensor category in the sense of [**KKK18**, Section A.1], which we will also call a monoidal category. We endow its Grothendieck ring $K_0(\mathcal{M})$ with the multiplication $*$ given by the tensor product $\otimes$:

$$[X] * [Y] := [X \otimes Y]$$

where $X, Y \in \mathcal{T}$ and $[X], [Y]$ denote their isoclassess in $K_0(\mathcal{T})$. Under our definition of $\mathcal{T}$, $K_0(\mathcal{T})$ is associative, unital, and has a $\mathbb{Z}$-basis $\mathsf{S} = \{[S] | S \text{ is a simple object}\}$. Moreover, there is an object Q in $\mathcal{T}$ such that $\mathbf{q} := Q \otimes (\ )$ is an autoequivalence functor on $\mathcal{T}$. This makes $K_0(\mathcal{T})$ a $\mathbb{Z}[q^{\pm}]$-algebra where $q[X] := [\mathbf{q}X]$.

Denote $\Bbbk = \mathbb{Z}[v^{\pm}]$, $\mathbb{Q}[v^{\pm}]$, $\mathbb{Q}(v)$. Take the $\Bbbk$-module $K = \Bbbk \otimes_{\mathbb{Z}[q^{\pm}]} K_0(\mathcal{T})$ such that $v = q^{\frac{1}{2}}$. If $K_0(\mathcal{T})$ is non-commutative, it will often induces a multiplication for K. Otherwise, we could often endow K with a non-commutative multiplication $*$ (with desired properties), such that the classical limit $K|_{v\mapsto 1}$ is K_0. The algebra K is called the quantum Grothendieck ring of $\mathcal{T}$.

Let $\mathcal{T}^{\mathrm{op}}$ denote the opposite monoidal category of $\mathcal{T}$ in the sense of [**EGNO15**, Definition 2.1.5]. In particular, its tensor product $\otimes^{\mathrm{op}}$ is given by $X \otimes^{\mathrm{op}} Y := Y \otimes X$ for $X, Y \in \mathcal{T}$. Define its quantum Grothendieck ring to be the opposite algebra K^{op}.

Let R be an $\mathbb{k}$-algebra with a KL basis $\mathbf{S}$.

DEFINITION 5.4. We say a monoidal category $\mathcal{T}$ categorifies the pair $(R, \mathbf{S})$ if the following conditions are satisfied.

- There is a $\mathbb{k}$-algebra isomorphism $\kappa : R \simeq K$.
- For any simple object S of $\mathcal{T}$, we have $v^{\alpha}[S] \in \kappa\mathbf{S}$ for some $\alpha \in \mathbb{Z}$.

Assume $\mathcal{T}$ categorifies $(R, \mathbf{S})$. If $\mathbf{M}_w$ also corresponds to an object in $\mathcal{T}$, then coefficients $e_{w',w}$ are the Jordan–Hölder multiplicities of its composition factors. Thus we must have $e_{w',w} \in \mathbf{m}_+$.

In analogous to based modules in [**Lus93**], we might call the pair $(R, \mathbf{S})$ a based algebra, if it has some additional desired properties.

REMARK 5.5. We have extended the Grothendieck ring $K_0(\mathcal{T})$ to a $\mathbb{k}$-algebra K for applications in quantum cluster algebras; see [**Qin24a**, Section 5.1]. This technical treatment might not be necessary for other purposes.

Sometimes, the construction of a suitable $\mathbb{k}$-algebra K from $\mathcal{T}$ might be highly non-trivial, e.g., one might want to deform a commutative Grothendieck ring $K_0(\mathcal{T})$ to a non-commutative ring K when studying representations of quantum affine algebras [**Nak01**] [**VV03**] [**Her04**] (Section A.2).

6. Common triangular bases

Let A denote a quantum ordinary cluster algebra $\boldsymbol{A}$ or a quantum upper cluster algebra $\boldsymbol{U}$. In general, there is no known initial basis $\mathbf{M}$ for A on which we could apply the KL algorithm (Section 5). Following [**Qin17**], we hope to construct its KL basis in the following procedure:

(1) We choose any seed $\mathbf{s}$.
(2) We will construct a distinguished linearly independent set $\mathbf{I}^{\mathbf{s}}$ in $\boldsymbol{LP}(\mathbf{s})$ instead of an initial basis $\mathbf{M}$. For our construction in this step, **we need to assume that the set of vertices I is finite.**
(3) We sort the elements of $\mathbf{I}^{\mathbf{s}}$ by an unbounded partial order $\prec_{\mathbf{s}}$.
(4) We add correction terms to $\mathbf{I}^{\mathbf{s}}$ using the KL algorithm (Remark 5.2), which might not stop since $\prec_{\mathbf{s}}$ is unbounded. We will obtain a linearly independent set $\mathbf{L}^{\mathbf{s}}$ in the completion $\widehat{\boldsymbol{LP}}(\mathbf{s})$.

It might happen that $\mathbf{L}^{\mathbf{s}}$ is actually a basis for A. Then it will be called the triangular basis for A with respect to $\mathbf{s}$ if it satisfies some additional technical condition (Definition 6.3). If A has a basis $\mathbf{L}$ such that $\mathbf{L} = \mathbf{L}^{\mathbf{s}}$ for any seed $\mathbf{s}$, it will be call the common triangular basis (Definition 6.4).

The KL-algorithm implies that the triangular basis $\mathbf{L}^{\mathbf{s}}$ is unique if it exists. The uniqueness of $\mathbf{L}$ follows.

REMARK 6.1. In [**BZ14**], Berenstein and Zelevinsky constructed a KL basis for the quantum cluster algebra $\boldsymbol{A}(\mathbf{s})$ where $\mathbf{s}$ is any acyclic seed. By acyclic, we mean that there is an order $<$ on I_{uf} such that the exchange matrix B satisfies $b_{ij} = 0$ whenever $i \leq j$. Using this order, they constructed a basis $\mathbf{M}$ consisting of products of cluster variables in $\mathbf{s}$ and $\mu_k\mathbf{s}$, where $k \in I_{\mathrm{uf}}$. Sorting the elements of $\mathbf{M}$

lexicographically and applying the KL algorithm, they obtained a basis **S**, which we call the BZ-triangular basis $\mathbf{L}^{\mathrm{BZ}}$. They further showed that different acyclic seeds produce the same basis.

By [**Qin19, Qin20b**], $\mathbf{L}^{\mathrm{BZ}}$ coincides with the common triangular basis **L** for the acyclic quantum cluster algebra.

6.1. Triangular bases. Let us explain how to construct bases in details. Choose any seed $\mathbf{s}$, which is assumed to be injective reachable. Note that $|I|$ must be finite under this assumption.

Let $\mathbf{I}^{\mathbf{s}}$ denote the set of all normalized products of the form $[p * x(\mathbf{s})^{g_+} * x(\mathbf{s}[1])^{g_-}]^{\mathbf{s}}$, where p is a frozen factor, $g_\pm \in \mathbb{N}^{I_{\mathrm{uf}}}$, and $(g_+)_k(g_-)_k = 0\ \forall k \in I_{\mathrm{uf}}$ (i.e., the product are reduced). Then $\mathbf{I}^{\mathbf{s}}$ is a $M^\circ(\mathbf{s})$-pointed set. In particular, it is linearly independent. Let $\mathbf{I}^{\mathbf{s}}_m$ denote the m-pointed element in $\mathbf{I}^{\mathbf{s}}$ as before.

Recall that we have the dominance order $\prec_{\mathbf{s}}$ on $M(\mathbf{s})$. Choose $\mathbf{m} = v^{-1}\mathbb{Z}[v^{-1}]$. In analogous to Lemma 5.1, we have the following result.

LEMMA 6.2. *There is a unique subset* $\boldsymbol{L}^{\mathbf{s}} = \{\boldsymbol{L}^{\mathbf{s}}_m | m \in M(\mathbf{s})\}$ *of* $\widehat{\boldsymbol{LP}}(\mathbf{s})$ *such that*

- $\overline{\boldsymbol{L}^{\mathbf{s}}_m} = \boldsymbol{L}^{\mathbf{s}}_m$.
- $\boldsymbol{L}^{\mathbf{s}}_m = \boldsymbol{I}^{\mathbf{s}}_m + \sum_{m' \prec_{\mathbf{s}} m} c_{m',m} \boldsymbol{I}^{\mathbf{s}}_{m'}$ *in* $\widehat{\boldsymbol{LP}}(\mathbf{s})$, *where* $c_{m',m} \in \mathbf{m}$.

The decomposition in the second condition in Lemma 6.2 will be called a $(\prec_{\mathbf{s}}, \mathbf{m})$-decomposition. Note that it could be an infinite sum. Under an appropriate topology of $\widehat{\boldsymbol{LP}}(\mathbf{s})$, **I** becomes a topological basis of $\widehat{\boldsymbol{LP}}(\mathbf{s})$, and this infinite sum becomes convergent; see [**DM21**].

PROOF. The basis could be constructed via the KL algorithm, which should be viewed as an infinite correction process (Remark 5.2). See [**Qin20b**, Theorem 6.1.3, Lemma 6.1.4] or [**Lus90**, 7.10] for a detailed treatment. □

Let A denotes the quantum cluster algebra $\boldsymbol{A}$ or $\boldsymbol{U}$.

DEFINITION 6.3 ([**Qin17**, Definition 6.1.1]). Assume $\mathbf{s}$ is finite-rank and injective reachable. Let $\mathbf{L}^{\mathbf{s}} = \{\mathbf{L}^{\mathbf{s}}_m | m \in M^\circ(\mathbf{s})\}$ denote a $\Bbbk$-basis of A, such that $\mathbf{L}^{\mathbf{s}}_m$ are m-pointed. It is called the triangular basis $\mathbf{L}^{\mathbf{s}}$ of A with respect to $\mathbf{s}$, if it satisfies the following conditions.

- Bar-invariance: $\overline{\mathbf{L}^{\mathbf{s}}_m} = \mathbf{L}^{\mathbf{s}}_m$.
- $\mathbf{L}^{\mathbf{s}}$ contains all cluster monomials of $\mathbf{s}$ and $\mathbf{s}[1]$.
- Triangularity: $\forall m \in M^\circ(\mathbf{s})$ and $i \in I$, we have the following $(\prec_{\mathbf{s}}, \mathbf{m})$-triangular decomposition

$$[x_i(\mathbf{s}) * \mathbf{L}^{\mathbf{s}}_m]^{\mathbf{s}} = \mathbf{L}^{\mathbf{s}}_{m+f_i} + \sum_{m' \prec_{\mathbf{s}} m} b_{m'} \mathbf{L}^{\mathbf{s}}_{m'},\ b_{m'} \in \mathbf{m}.$$

Let $\mathbf{L}^{\mathbf{s}}$ denote the triangular basis with respect to $\mathbf{s}$. By definition, it must be the unique subset in Lemma 6.2.

6.2. Common triangular bases.

DEFINITION 6.4 (Common triangular basis [**Qin17**][**Qin23**]). If **L** is a $\Bbbk$-basis of A such that **L** is the triangular basis $\mathbf{L}^{\mathbf{s}}$ for any seed $\mathbf{s}$, it will be call the common triangular basis of A. If further, $\mathbf{L} \cap \overline{\mathsf{A}}$ is a basis of $\overline{\mathsf{A}}$, $\mathbf{L} \cap \overline{\mathsf{A}}$ is called the common triangular basis of $\overline{\mathsf{A}}$.

Assume A has the common triangular basis $\mathbf{L}$. Then it contains all cluster monomials by definition. [**Qin20b**, Proposition 6.4.3] implies the following.

PROPOSITION 6.5. *The common triangular basis* $\boldsymbol{L}$ *is* M^{trop}*-pointed.*

REMARK 6.6. By [**Qin24d**, Theorem 4.3.1], when $\mathbf{s}$ is injective reachable, any M^{trop}-pointed subset of $\boldsymbol{U}$ is a basis for $\boldsymbol{U}$. We deduce that $\mathsf{A} = \boldsymbol{U}$ in this case.

Therefore, for studying common triangular bases, it is sufficient to consider the upper cluster algebra $\mathsf{A} = \boldsymbol{U}$. In practice, we might also start with the ordinary cluster algebra $\mathsf{A} = \boldsymbol{A}$, since $\boldsymbol{A}$ naturally appears and is well understood in some situations. On the contrary, it might not worth the effort to study the triangular basis for a general subalgebra of $\boldsymbol{U}$.

THEOREM 6.7 ([**Qin17**][**Qin20b**]). *Let C be any generalized Cartan matrix.*

*(1) Take any Weyl group element $w \in W$. Endow the quantized coordinate ring $\Bbbk[N(w)]$ of a unipotent subgroup $N(w)$ with the canonical cluster structure ([**Kim12**] [**GLS13**] [**GY16, GY21**]). Then the dual canonical basis of $\Bbbk[N(w)]$ is its common triangular basis up to v-multiples.*

*(2) Assume C is of type ADE. Let $\mathcal{O}_l$ denote a level-l category consisting of representations of the quantum affine algebra $U_q(\hat{\mathfrak{g}})$ in the sense of [**HL10**] (Section A.2). Endow the quantized Grothendieck ring $K_t(\mathcal{O}_l)$ with the canonical cluster structure ([**HL10**][**Qin17**]). Then the set of isoclasses of simple objects is its common triangular basis.*

REMARK 6.8. Take any reduced word $\underline{\eta}$ such that $w_{\underline{\eta}} = w$. Then we have a canonical bijection $\kappa : \overline{\boldsymbol{U}}(\dot{\mathbf{s}}(\underline{\eta})) \simeq \Bbbk[N(w)]$ (see [**GLS13**][**GY16, GY21**] or [**Qin20b**, Section 8]). We also have the De Concini-Procesi isomorphism $\Bbbk[N(w)][(\kappa x_j)^{-1}]_{j\in I_{\mathrm{f}}} \simeq \Bbbk[N^w]$, where $\Bbbk[N^w]$ is the quantized coordinate ring of the unipotent cell [**KO21**]. Thus we obtain a cluster structure $\boldsymbol{U}(\dot{\mathbf{s}}(\underline{\eta})) \simeq \Bbbk[N^w]$. Up to v-multiples, the common triangular basis of cluster algebra $\boldsymbol{U}(\dot{\mathbf{s}}(\underline{\eta}))$ is obtained from the dual canonical basis after localization.

In practice, it is convenient to work with either the cluster algebra $\overline{\boldsymbol{U}}(\dot{\mathbf{s}}(\underline{\eta}))$, which has the standard basis (Section 6.6, dual PBW basis), or with $\boldsymbol{U}(\dot{\mathbf{s}}(\underline{\eta}))$, to which we could apply cluster operations (Section 4).

6.3. Existence of the common triangular bases. The common triangular basis is defined to have strong properties at the cost that its existence is difficult to prove. Note that $\mathbf{s}[1] = \Sigma\mathbf{s}$ and $\mathbf{s} = (\sigma^{-1}\Sigma)\mathbf{s}[-1]$ (or equivalently, $\mathbf{s}[-1] = (\sigma^{-1}\Sigma)^{-1}\mathbf{s}$). We have the following necessary and sufficient condition for its existence.

THEOREM 6.9 (A slightly different version of [**Qin20b**, Theorem 6.5.4]). *If* $\boldsymbol{L}$ *is the triangular basis* $\boldsymbol{L}^{\mathbf{s}}$ *for* A *with respect to the seed* $\mathbf{s}$*, and it contains the cluster monomials of the seeds appearing along the mutation sequences* Σ *and* $(\sigma^{-1}\Sigma)^{-1}$ *starting from* $\mathbf{s}$*. Then* $\boldsymbol{L}$ *is the common triangular basis for* A.

Cluster operations propagate the existence of (common) triangular basis. Assume $\mathbf{s}'$ is a seed similar to $\mathbf{s}$, and $\mathfrak{f}_F\mathbf{s}$ is obtained from $\mathbf{s}$ by freezing $F \subset I_{\mathrm{uf}}$. Let A denote $\boldsymbol{A}$ or $\boldsymbol{U}$. Assume $\mathbf{L}$ is the triangular basis for $\mathsf{A}(\mathbf{s})$ with respect to $\mathbf{s}$. Let $\mathbf{L}'$ denote the set of similar elements in $\boldsymbol{LP}(\mathbf{s}')$ and $\mathfrak{f}_F\mathbf{L}$ denote the set obtained from $\mathbf{L}$ by freezing.

THEOREM 6.10 ([**Qin24a**]). *The following statements are true.*

(1) $\boldsymbol{L}'$ *is the triangular basis for* $\mathsf{A}(\mathbf{s}')$ *with respect to* $\mathbf{s}'$*.*

(2) If $\boldsymbol{L}$ *is the common triangular basis, then* $\boldsymbol{L}'$ *is the common triangular basis too.*

(3) $\mathfrak{f}_F\boldsymbol{L}$ *is the triangular basis for* $\mathsf{A}(\mathfrak{f}_F\mathbf{s})$ *with respect to* $\mathfrak{f}_F\mathbf{s}$*.*

(4) If $\boldsymbol{L}$ *is the common triangular basis, then* $\mathfrak{f}_F\boldsymbol{L}$ *is the common triangular basis too.*

PROOF. The desired statements could be deduced from a combination of the correction technique (Theorem 4.14), similarity of the localized cluster monomials (Theorem 4.15), and freezing of cluster monomials (Theorem 4.6). □

6.4. Common triangular bases for known cluster algebras. There is an ever growing list of cluster algebras that arise from Lie theory. More precisely, we could construct (classical and quantum) upper cluster algebra $\boldsymbol{U}(\mathbf{s})$ whose initial seed $\mathbf{s}$ arises from the consideration of the following varieties from Lie theory.

(1) Algebra group G: Assume the generalized Cartan matrix is of finite type. Let G be the associated connected, simply connected, complex semisimple algebraic group. The coordinate ring $\mathbb{C}[G]$ and the quantized coordinate ring $\mathbb{Q}(v)[G]$ are upper cluster algebras $\overline{\boldsymbol{U}}(\mathbf{s})$ defined over $\mathbb{C}$ and $\mathbb{Q}(v)$, respectively; see [**QY25**] (the cluster structure on $\mathbb{C}[SL_n]$ were shown in [**FWZ20**]; see [**Oya25**] for another approach to $\mathbb{C}[G]$ when G is not of type F_4).
(2) Subvarieties $\mathscr{A}$ of G: Let $\mathscr{A}$ denote unipotent subgroups $N(w)$, unipotent cells N^w, or double Bruhat cell $G^{u,v}$, the (quantized) coordinate rings $\Bbbk[\mathscr{A}]$ are know to be cluster algebras $\overline{\boldsymbol{U}}(\mathbf{s})$ or $\boldsymbol{U}(\mathbf{s})$ (see [**GLS11**, **GLS13**][**GY16**, **GY21**]).
(3) Spaces of configurations of flags $\mathscr{A}$: Let $\mathscr{A}$ denote Grassmannians, open Positroid varieties, open Richardson varieties, double Bott-Samelson cells, or braid varieties. Its coordinate ring are know to be classical upper cluster algebras $\boldsymbol{U}(\mathbf{s})$; see [**Sco06**][**SSBW19**][**GL19**][**EL21**][**SW21**][**GLSBS22**] [**CGG$^+$25**].
(4) Simple Lie group G endowed with non-standard Lie-Poisson brackets [**GSV23**].

Note that Family (3) includes several (quantum) cluster algebras arising from monoidal categorifications, in particular, those from representations of quantum affine algebras [**HL10**][**Qin17**] (Section A.2) and in type $A^1_{(1)}$, those from Satake categories [**CW19**].

THEOREM 6.11 ([**Qin24a**]). *For almost all known quantum upper cluster algebras* $\boldsymbol{U}$ *arising from Lie theory (families (1)(2)(3)), we have the following results.*

(1) $\boldsymbol{U}$ *possesses the common triangular basis* $\boldsymbol{L}$*.*

(2) We have $\boldsymbol{A} = \boldsymbol{U}$*.*

(3) When the generalized Cartan matrix C *is symmetric,* $\boldsymbol{L}$ *has positive structure constants.*

Family (4) leads to exotic cluster structures on $\mathbb{C}[G]$, which are not examined in Theorem 6.11.

PROOF. We refer the reader to [**Qin24a**] for detailed arguments. We sketch its approach below.

The existence of common triangular bases is already known for those from quantum unipotent cells $\Bbbk[N^w]$ (Theorem 6.7(1), Remark 6.8). We wish to propagate structures and results from $\Bbbk[N^w]$ to other cluster algebra $\Bbbk[\mathscr{A}]$ where $\mathscr{A}$ arises from Lie theory. But $\mathscr{A}$ might be more complicated than N^w, for example, we cannot expect to recover G from N^w. To circumvent this difficulty, we introduce and use the following **extension and reduction technique**:

Extend C to a larger generalized Cartan matrix $\widetilde{C}$ by adding rows and columns. Then we might find a quantum unipotent cell $\Bbbk[\widetilde{N}^{\widetilde{w}}]$ associated with $\widetilde{C}$ such that it is complicated enough to recover $\Bbbk[\mathscr{A}]$ associated with C.

More precisely, we study $\Bbbk[\mathscr{A}]$ as below:

- Choose $\Bbbk[\widetilde{N}^{\widetilde{w}}]$ *associated with an extended generalized Cartan matrix* $\widetilde{C}$.
- Applying cluster operations (freezing and base changes in Section 4), we could propagate structures and results from $\Bbbk[\widetilde{N}^{\widetilde{w}}]$ to $\boldsymbol{U}(\dot{\mathbf{s}})$, where $\dot{\mathbf{s}} = \dot{\mathbf{s}}(\underline{\mathbf{i}})$ is a seed associated with a signed word.[6]
- Propagate structures and results from $\boldsymbol{U}(\dot{\mathbf{s}})$ to $\Bbbk[\mathscr{A}]$.

□

It is desirable to find a natural Lie theoretic interpretation of the approach in the proof of Theorem 6.11.

6.5. Quasi-categorifications. On the one hand, there are many cluster algebras, because one can construct a cluster algebra from any given skew-symmetrizable matrix. On the other hand, we only know a limited families of categories from representation theory that possess good properties. In the following, we relax the conditions on categorification, so that we could use categories to study more cluster algebras.

Let A denote a (classical or quantum) cluster algebra and $\mathcal{T}$ be a monoidal category. Let K be the (classical or quantum) Grothendieck ring of $\mathcal{T}$. Let $[V]$ denote the element in K corresponding to $V \in \mathcal{T}$.

DEFINITION 6.12 ([**Qin24a**, Definition 5.3]). We say $\mathcal{T}$ admits a cluster structure A if the following conditions are satisfied:

- There is a $\Bbbk$-algebra injective map $\kappa : \mathsf{A} \hookrightarrow K$.
- For any cluster monomial z, there exists a simple object S of $\mathcal{T}$ such that $\kappa z = v^{\alpha}[S]$ for some $\alpha \in \mathbb{Z}$.

We further say $\mathcal{T}$ categorifies A if κ is an isomorphism. This is a weaker version of the categorification in [**HL10**]. We say $\mathcal{T}$ categorifies A after localization if κ extends to an isomorphism $\kappa : \mathsf{A}[x_j^{-1}]_{j\in I_{\mathrm{f}}} \simeq K[\kappa(x_j)^{-1}]_{j\in I_{\mathrm{f}}}$.

DEFINITION 6.13 ([**Qin24a**, Definition 5.1]). Let $\mathbf{S}$ be a $\Bbbk$-basis of the cluster algebra A. The pair $(\mathsf{A}, \mathbf{S})$ is called a based cluster algebra if the following conditions hold.

- $\mathbf{S}$ contains all cluster monomials.
- (Frozen torus action) $\mathbf{S}$ is a closed under commutative multiplication by frozen variables: $\forall j \in I_{\mathrm{f}}$, $x_j \cdot \mathbf{S} \subset \mathbf{S}$.
- (Homogeneity) Any $b \in \mathbf{S}$ is contained in some $x^m \cdot \Bbbk[N_{\mathrm{uf}}]$.

[6] $\dot{\mathbf{s}}(\underline{\mathbf{i}})$ is a seed for the coordinate ring of a (half decorated) double Bott-Samelson variety [**SW21**].

- (Bar invariance) At the quantum level, the elements of $\mathbf{S}$ are further assumed to be bar-invariant.

DEFINITION 6.14 ([**Qin24a**, Definition 5.4]). Assume $\mathcal{T}$ is a monoidal category. We say $\mathcal{T}$ categorifies a based cluster algebra $(\mathsf{A}, \mathbf{S})$ (resp. after localization) if the following conditions are satisfied:

- $\mathcal{T}$ categorifies A (resp. after localization).
- For any simple object S of $\mathcal{T}$, we have $v^{\alpha}[S] \in \kappa\mathbf{S}$ for some $\alpha \in \mathbb{Z}$.

DEFINITION 6.15 ([**Qin24a**, Definition 5.5]). We say a monoidal category $\mathcal{T}$ quasi-categorifies a based cluster algebra $(\mathsf{A}, \mathbf{S})$, if there exists a based cluster algebra $(\mathsf{A}', \mathbf{S}')$ which is connected to $(\mathsf{A}, \mathbf{S})$ by a finite sequence of base changes and quantization changes, such that $\mathcal{T}$ categorifies $(\mathsf{A}', \mathbf{S}')$ after localization.

Now we have the following categorification result for the bases in Theorem 6.11.

THEOREM 6.16 ([**Qin24a**]). *For almost all known cluster algebras* A *arising from Lie theory (families (1)(2)(3)), when the generalized Cartan matrix is symmetric, the based cluster algebra* $(\mathsf{A}, \boldsymbol{L})$ *is quasi-categorified by some non semisimple monoidal category* $(\mathcal{T}, \otimes)$.

PROOF. We refer the reader to [**Qin24a**] for detailed treatments. The approach in loc. cit. was based on the observation that freezing corresponds to the restriction on a subcategory: if $\mathcal{T}$ categories $(\boldsymbol{U}, \mathbf{S})$, $(\mathfrak{f}_F \boldsymbol{U}, \mathfrak{f}_F \mathbf{S})$ is categorified by a monoidal subcategory $\mathcal{T}'$ of $\mathcal{T}$. □

6.6. Standard bases. In general, cluster algebras do not have PBW type basis (i.e, a basis consisting of all monomials of the chosen distinguished elements), because they are usually not polynomial rings at the classical level.

Nevertheless, let $\overline{\boldsymbol{U}}(\dot{\mathrm{s}})$ denote the (partially) compactified upper cluster algebra associated with a quantum seed $\dot{\mathrm{s}} = \dot{\mathrm{s}}(\underline{\eta})$ for some word $\underline{\eta} = (\eta_k)_{k \in [1,l]}$. Then it has the standard basis (PBW type basis), whose construction is given as follows.

For any $w \in \mathbb{N}^I$, define the corresponding standard monomial to be $\mathbf{M}(w) = [W_1^{w_1} * \cdots * W_l^{w_l}]^{\dot{\mathrm{s}}}$, where W_k are the fundamental variables (Section 3.3). Define $\mathbf{M} = \{\mathbf{M}(w) | w = (w_k)_{k \in [1,l]} \in \oplus_{k=1}^{l} \mathbb{N}\beta_k\}$, where β_k denotes the k-th standard unit vector in $\mathbb{Z}^l$. We will sort its elements by the lexicographical order $\mathbf{M}(w) < \mathbf{M}(w')$ if $w <_{\mathrm{lex}} w'$, or by the reverse lexicographical order $\mathbf{M}(w) < \mathbf{M}(w')$ if $w <_{\mathrm{rev}} w'$. Recall that we say $w <_{\mathrm{lex}} w'$ if $w_{[1,r]} = w'_{[1,r]}$ and $w_{r+1} < w'_{r+1}$ for some $r \in [0, l-1]$, and $w <_{\mathrm{rev}} w'$ if $w_{[r,l]} = w'_{[r,l]}$ and $w_{r-1} < w'_{r-1}$ for some $r \in [2, l+1]$.

Following [**Qin17**][**Qin20b**, Section 9.1], define the linear map

$$\theta^{-1} : \oplus_{k=1}^{l} \mathbb{Z}\beta_k \simeq M^{\circ}$$

such that

$$\theta^{-1}(\beta_k) = f_k - f_{k[-1]},\ \forall k \in [1, l]. \tag{6.1}$$

where we denote $f_{k^{\min}[-1]} = f_{-\infty} = 0$ as before. Let $\mathbf{L}_m$ denote the m-pointed common triangular basis element of $\overline{\boldsymbol{U}}(\dot{\mathrm{s}})$. We denote $\mathbf{L}(w) := \mathbf{L}_{\theta^{-1}(w)}$.

THEOREM 6.17 ([**Qin24a**, Theorem 8.10]). *For any* $\dot{\mathrm{s}} = \dot{\mathrm{s}}(\underline{\eta})$, *where* $\underline{\eta}$ *is a finite word in* J, *the following statements are true.*

(1) $\boldsymbol{M}$ *is a* $\Bbbk$*-basis of* $\overline{\boldsymbol{U}}(\dot{\mathrm{s}})$, *called the standard basis.*
(2) *We have* $\overline{\boldsymbol{A}}(\dot{\mathrm{s}}) = \overline{\boldsymbol{U}}(\dot{\mathrm{s}})$.

(3) *The standard basis satisfies the analog of the Levendorskii-Soibelman straightening law:*

$$W_k * W_j - v^{\lambda(\deg W_k, \deg W_j)} W_j * W_k = \sum_{w \in \mathbb{N}^{[j+1,k-1]}} b_w \boldsymbol{M}(w),\ b_w \in \Bbbk,\ \forall j \leq k.$$

(4) *The common triangular basis of $\overline{\boldsymbol{U}}(\dot{\mathbf{s}})$ is $\{\boldsymbol{L}(w)|w \in \oplus_{k=1}^{l} \mathbb{N}\beta_k\}$. Moreover, it is the Kazhdan-Lusztig-type basis associated with the standard basis $\boldsymbol{M}$, whose basis elements are sorted by the lexicographical order or by the reverse lexicographical order.*

THEOREM 6.18 ([**Qin24a**, Theorem 8.10]). *There is a non-semisimple monoidal category $\mathcal{T}(\beta_{\underline{\eta}})$ which categorifies $(\overline{\boldsymbol{U}}(\dot{\mathbf{s}}(\underline{\eta})), \boldsymbol{L})$.*

By [**Qin24a**], we could construct $\mathcal{T}(\beta_{\underline{\eta}})$ as a monoidal subcategory of $\mathcal{C}_{\mathbb{Z}}$, which is the monoidal category consisting of finite dimensional modules of quantum affine algebras $U_q(\hat{\mathfrak{g}})$ (Section A.2). In this construction, the monoidal category $\mathcal{T}(\beta_{\underline{\eta}})$ depends on the braid element $\beta_{\underline{\eta}}$, not on the choice of the word $\underline{\eta}$.

Assume C is of finite type. We could further compute the fundamental variables as below. First, W_{η_1} is an initial cluster variable $x_1(\dot{\mathbf{s}})$. Next, we can embed $\overline{\boldsymbol{U}}(\dot{\mathbf{s}}) = \overline{\boldsymbol{A}}(\dot{\mathbf{s}})$ as a subalgebra of an infinite rank cluster algebra $\overline{\boldsymbol{A}}_\infty$ ([**Qin24c**], Section 7). The cluster algebra $\overline{\boldsymbol{A}}_\infty$ is isomorphic to the virtual quantum Grothendieck ring K_∞ of a quantum affine algebra (also called the Bosonic extension of the quantized enveloping algebra); see [**JLO23b**]. Moreover, the braid group Br acts on K_∞ by [**JLO23a**][**KKOP24a**]. Then the remaining W_k could be computed by using the action of Br as follows.

THEOREM 6.19 ([**Qin24c**, Theorem 1.4]). *Assume C is of finite type, then the fundamental variables of $\overline{\boldsymbol{U}}(\dot{\mathbf{s}}(\underline{\eta}))$ could be computed by using the braid group action:*

$$W_k = \sigma_{\eta_1} \cdots \sigma_{\eta_{k-1}} W_{\eta_k},\ \forall k \in [2, l].$$

7. Extension to infinite ranks

7.1. Colimit of seeds.

DEFINITION 7.1 ([**Qin24c**]). We say $\mathbf{s}$ is a good subseed of $\mathbf{s}'$, if we have $I \subset I'$, $I_{\mathrm{uf}} \subset I'_{\mathrm{uf}}$, $\mathsf{d}_i = \mathsf{d}'_i\ \forall i \in I$, $\widetilde{B} = (\widetilde{B}')_{I \times I_{\mathrm{uf}}}$, and $b'_{jk} = 0\ \forall (j,k) \in (I' \backslash I) \times I_{\mathrm{uf}}$.

If $\mathbf{s}$ and $\mathbf{s}'$ are quantum seeds, we further require $\Lambda = (\Lambda')_{I \times I}$.

Now, let $\mathbf{s}$ be a good subseed of $\mathbf{s}'$. View M° as a sublattice of $M^{\circ\prime}$ naturally. We obtain natural inclusions $\boldsymbol{LP} \subset \boldsymbol{LP}'$ and $\mathcal{F} \subset \mathcal{F}'$. We further have the following results.

PROPOSITION 7.2 ([**Qin24a**, Section 3.1]). *We have $\overline{\boldsymbol{A}} \subset \overline{\boldsymbol{A}}'$ and $\overline{\boldsymbol{LP}}(\boldsymbol{\mu}\mathbf{s}) \subset \overline{\boldsymbol{LP}}(\boldsymbol{\mu}\mathbf{s}')$ for any mutation sequence $\boldsymbol{\mu}$ on I_{uf}. Moreover, if $|I'| < \infty$ and $\mathbf{s}'$ satisfies the injectivity assumption, we have $\overline{\boldsymbol{U}} \subset \overline{\boldsymbol{U}}'$.*

Let there be a chain of quantum or classical seeds $(\mathbf{s}_r)_{r \in \mathbb{N}}$ such that $\mathbf{s}_r$ is a good subseed of $\mathbf{s}_{r+1}$.

First assume $\mathbf{s}_r$ are classical seed. Then we could always construct their colimit seed $\mathbf{s}_\infty$ such that $I(\mathbf{s}_\infty) := I_\infty := \cup_r I(\mathbf{s}_r)$, $I_{\mathrm{uf}}(\mathbf{s}_\infty) := \cup_r I_{\mathrm{uf}}(\mathbf{s}_r)$, $\mathsf{d}_i = \mathsf{d}_i(\mathbf{s}_r)$ when $i \in I(\mathbf{s}_r)$, and $\widetilde{B} = (b_{ij})_{i \in I_\infty, j \in I_{\mathrm{uf}}(\mathbf{s}_\infty)}$ such that $b_{ij} = \widetilde{B}(\mathbf{s}_r)_{ij}$ when $(i,j) \in I(\mathbf{s}_r) \times I_{\mathrm{uf}}(\mathbf{s}_r)$.

Next assume $\mathbf{s}_r$ are quantum seed. If we can equip the classical seed $\mathbf{s}_\infty$ with a compatible Poisson structure $\lambda(\mathbf{s}_\infty)$, such that $\Lambda(\mathbf{s}_r) = \Lambda(\mathbf{s}_\infty)_{I(\mathbf{s}_r)\times I(\mathbf{s}_r)}$ for any r, we call the corresponding quantum seed the colimit $\mathbf{s}_\infty$.

7.2. Extension of quantizations. If $\mathbf{s}$ is a good subseed of $\mathbf{s}'$ as classical seeds, it is nontrivial to extend a compatible Poisson structure on $\mathbf{s}$ to that of $\mathbf{s}'$ such that $\mathbf{s}$ becomes a good subseed of $\mathbf{s}'$ as a quantum seed. Nevertheless, we could restrict on special cases that suffice for many applications.

DEFINITION 7.3. A matrix (b_{ij}) is said to be connected if, $\forall(i,j)$, there exists a sequence of letters $i_0 = i, i_1, \ldots, i_{s-1}, i_s = j$, such that $b_{i_r i_{r+1}} \neq 0$, $\forall r \in [0, s-1]$.

Assume that we have a partition $I' = I_1 \sqcup I_2 \sqcup I_3$ such that $I'_{\mathrm{uf}} = I = I_1 \sqcup I_2$, $I'_{\mathrm{f}} = I_3$, $I_{\mathrm{uf}} = I_1$, $I_{\mathrm{f}} = I_2$. Then we have $\widetilde{B}_{I_3,I_1} = 0$ since $\mathbf{s}$ is a good subseed of $\mathbf{s}'$.

LEMMA 7.4. *Assume that $\widetilde{B}_{I'_{\mathrm{uf}},I'_{\mathrm{uf}}}$ and $\widetilde{B}_{I_{\mathrm{uf}},I_{\mathrm{uf}}}$ are connected, $\widetilde{B}_{I_3,I_2}$ is of full rank. Then we can uniquely extends the compatible Poisson structure λ for $\mathbf{s}$ to a compatible Poisson structure λ' for $\mathbf{s}'$, such that $\Lambda = \Lambda'_{I\times I}$.*

EXAMPLE 7.5. We can solve the compatible Poisson form λ in Example 2.6 such that $\delta_k = 2$. Let λ be such a Poisson form and Λ its matrix. Then, for any $0 < r \in \mathbb{N}$, we have $\Lambda_{[-r,r]\times[-r,r]} B_{[-r,r]\times[-r+1,r-1]} = \begin{pmatrix} 0 \\ -2\mathrm{Id}_{2r-1} \\ 0 \end{pmatrix}$. In particular, when $r = 1$, we can obtain a one parameter family of solutions $\Lambda_{[-1,1]} = \begin{pmatrix} 0 & 1+c & 0 \\ -1-c & 0 & -1+c \\ 0 & 1-c & 0 \end{pmatrix}$, $\forall c \in \mathbb{Z}$. Using Lemma 7.4 and increasing r by 1 each time, we could inductively determine all $\Lambda_{[-r,r]}$ step by step for all $r > 0$. The resulting matrix Λ is given by (2.2) when we choose $c = 0$.

7.3. Colimit of algebras. Now assume $(\mathbf{s}_r)_{r\in\mathbb{N}}$ is a chain of quantum or classical seeds, such that $\mathbf{s}_r$ is a good subseed of $\mathbf{s}_{r+1}$, and it has the colimit seed $\mathbf{s}_\infty$. Note that $\boldsymbol{LP}(\mathbf{s}_\infty) = \cup_r \boldsymbol{LP}(\mathbf{s}_r)$. We define the colimit $\overline{\boldsymbol{A}}_\infty := \cup_r \overline{\boldsymbol{A}}(\mathbf{s}_r)$ and $\overline{\boldsymbol{U}}_\infty := \cup_r \overline{\boldsymbol{U}}(\mathbf{s}_r)$

LEMMA 7.6 ([**Qin24c**]). *(1) We have $\overline{\boldsymbol{A}}(\mathbf{s}_\infty) := \overline{\boldsymbol{A}}_\infty$.*

(2) Assume that $|I(\mathbf{s}_r)| < \infty$ and $\mathbf{s}_r$ satisfies the Injectivity Assumption, $\forall r$. Then we have $\overline{\boldsymbol{U}}(\mathbf{s}_\infty) = \overline{\boldsymbol{U}}_\infty$.

THEOREM 7.7 (Starfish Lemma). *Under the assumption in Lemma 7.6(2), we have*

$$\overline{\boldsymbol{U}}(\mathbf{s}_\infty) = \overline{\boldsymbol{LP}}(\mathbf{s}_\infty) \cap \left(\cap_{k\in I_{\mathrm{uf}}(\mathbf{s}_\infty)} \overline{\boldsymbol{LP}}(\mu_k \mathbf{s}_\infty)\right).$$

Theorem 7.7 was previously known as the starfish lemma when $|I| < \infty$ [**BFZ05, BZ05**].

Now, let A denotes $\overline{\boldsymbol{A}}$ or $\overline{\boldsymbol{U}}$. Assume $\mathsf{A}(\mathbf{s}_r)$ has the $\mathbb{k}$-basis $\mathbf{S}_r$, such that $\mathbf{S}_{r+1} \cap \mathsf{A}(\mathbf{s}_r) = \mathbf{S}_r$. Denote $\mathbf{S}_\infty = \cup_r \mathbf{S}_r$. Then it is a basis of $\mathsf{A}_\infty = \cup_r \mathsf{A}(\mathbf{s}_r)$. Now we could extend our constructions of common triangular bases to infinite ranks, despite that $\mathbf{s}_\infty$ is not injective reachable in general.

DEFINITION 7.8. Assume that we have $\mathsf{A}(\mathbf{s}_\infty) = \cup_r \mathsf{A}(\mathbf{s}_r)$. If $\mathbf{S}_r$ are the triangular basis of $\mathsf{A}(\mathbf{s}_r)$ with respect to $\mathbf{s}_r$ (resp. the common triangular basis), then

$\mathbf{S}_\infty$ is called the triangular basis of $\mathsf{A}(\mathbf{s}_\infty)$ with respect to $\mathbf{s}_\infty$ (resp. the common triangular basis).

EXAMPLE 7.9. For $r \in \mathbb{N}$, denote $I_r = [-1-r, 1+r]$ and $(I_r)_{\mathrm{f}} = \{-1-r, 1+r\}$. Denote the (classical or quantum) seed in Examples 2.3 2.6 by $\mathbf{s}_\infty$. Let $\mathbf{s}_r$ denote the seed obtained from $\mathbf{s}_\infty$ by restricting to the set of vertices I_r, where the frozen vertices are in $(I_r)_{\mathrm{f}}$. The associated ice quiver, denoted $\widetilde{Q}_r$, is the full subquiver on the vertices $[-1-r, 1+r]$ of the ice quiver of Figure 2.2, such that $\{-1-r, 1+r\}$ becomes frozen.

Note that we can identify $\mathbf{s}_r$ with the seed $\ddot{\mathbf{s}}(\underline{\mathbf{i}})$ where $\underline{\mathbf{i}} = (\mathbf{i}_k)_{k\in[-r,1+r]} = ((1)^r, 1, -1, (1)^r)$ where $\mathbf{i}_1 = -1$. $(\mathbf{s}_r)_{r\in\mathbb{N}}$ is a chain of seeds such that $\mathbf{s}_r$ is a good subseed of $\mathbf{s}_{r+1}$, and we have $\overline{\mathcal{U}}(\mathbf{s}_\infty) = \cup_{r\in\mathbb{N}}\overline{\mathcal{U}}(\mathbf{s}_r)$ by Lemma 7.6. The common triangular basis $\mathbf{L}_r$ of $\overline{\mathcal{U}}(\mathbf{s}_r)$ extends to the common triangular basis $\mathbf{L}_\infty$ of $\overline{\mathcal{U}}(\mathbf{s}_\infty)$.

Appendix A. Examples of Cluster algebras from Lie theory

A.1. Cluster structures on $\Bbbk[G^{u,w}]$ and $\Bbbk[G]$.

Root systems and generalized minors. Following [**BFZ05**, Section 2][**Hum72**], we review basics of root systems and generalized minors for convenience of the reader.

Let G be a simple, semisimple, complex algebra group of rank $|J|$. Let B, B_+ be two opposite Borel subgroups. Let N and N_- be their unipotent radicals. Take the maximal torus $H = B \cap B_-$.

Let $\mathfrak{g} = \mathrm{Lie}(G)$ be the Lie algebra of G. $\mathfrak{h} = \mathrm{Lie}(H)$ is the Cartan subalgebra. Recall that we have the exponential map $\exp : \mathfrak{h} \to G$. Let $\Pi = \{\alpha_a | a \in J\} \subset \mathfrak{h}^*$ be the set of simple roots. Denote the Cartan matrix by $C = (C_{ab})_{a,b\in J}$, where $C_{ab} = \frac{2(\alpha_a,\alpha_b)}{(\alpha_a,\alpha_a)}$. Let $\{\alpha_a^\vee | a \in J\} \subset \mathfrak{h}$ be the set of simple coroots, such that $C_{ab} = \langle \alpha_a^\vee, \alpha_b\rangle$.

The fundamental weights $\varpi_a \in \mathfrak{h}^*$ satisfy $\langle \alpha_a^\vee, \varpi_b\rangle = \delta_{ab}$, $\forall a, b \in J$. Equivalently, $s_a(\varpi_b) = \varpi_b - \delta_{ab}\alpha_a$.

The Weyl group is $\mathrm{Norm}_G(H)/H$. We will identify it with the corresponding Coxeter group W generated by s_a, $a \in J$ (see Section 3.2). The conjugation action of the Weyl group on H induces the action of W on $\mathfrak{h}^*$, such that the simple reflection is given by $s_a(\beta) = \beta - \langle \alpha_a^\vee, \beta\rangle\alpha_a$, $\forall \beta \in \mathfrak{h}^*$.

For any $a \in J$, let ϕ_a denote the SL_2-embedding which sends upper (resp. lower) triangular matrices into B (resp. B_-). We identify s_a with $\overline{s}_a H$ whose representative is chosen as

$$\overline{s}_a := \phi_a \begin{pmatrix} 0 & -1 \\ 1 & 0 \end{pmatrix} \in G.$$

For any $g = s_{a_1} \cdots s_{a_r} \in W$, let us denote $\overline{g} = \overline{s}_{a_1} \cdots \overline{s}_{a_r}$. We can also identify s_a with $H\overline{\overline{s}}_a$ where $\overline{\overline{s}}_a := (\overline{s}_a)^{-1}$, and denote $\overline{\overline{g}} := \overline{\overline{s}}_{a_1} \cdots \overline{\overline{s}}_{a_r}$. Note that $(\overline{g})^{-1} = \overline{\overline{g^{-1}}}$.

For any $g \in N_-HN$, denote the factorization $g = [g]_-[g]_0[g]_+$ such that $[g]_- \in N_-$, $[g]_0 \in H$, $[g]_+ \in N$.

Each weight $\gamma = \sum \gamma_a \varpi_a$, $\gamma_a \in \mathbb{Z}$, determines a character $(\ )^\gamma$ on H such that $(\exp(x))^\gamma := e^{\langle x,\gamma\rangle}$, $\forall x \in \mathfrak{h}$. $\forall u, w \in W$, we define the generalized minor $\Delta_{u\varpi_a,w\varpi_a}$ to be the regular function on G such that its value on $g \in \overline{u}N_-HN\overline{w}^{-1}$ is given by:

$$\Delta_{u\varpi_a,w\varpi_a}(g) = ([\overline{u}^{-1}g\overline{w}]_0)^{\varpi_a}.$$

The function $\Delta_{u\varpi_a,w\varpi_a}$ depends only on $(u\varpi_a, w\varpi_a)$, and not on the choice of (u,v).

When we work at the quantum level ($\Bbbk = \mathbb{Q}(v)$), we let $\Bbbk[G]$ denote the quantized coordinate ring of G; see [**Kas93**]. We also use $\Delta_{u\varpi_a,w\varpi_a}$ to denote the corresponding *generalized quantum minor*; see [**BZ05**].

We will also consider the double Bruhat cell $G^{u',w'} = Bu'B \cap B_-w'B_-$. Note that G^{w_0,w_0} is an open dense subset of G. We let $\Bbbk[G^{u',w'}]$ denote the (classical or quantized) coordinate ring of $G^{u,w}$; see [**BFZ05**][**BZ05**][**GY20**]. By abuse of notation, we let $\Delta_{u\varpi_a,w\varpi_a}$ denote the restriction of the generalized (classical or quantum) minor on $\Bbbk[G^{u,w}]$.

EXAMPLE A.1 (Type A_n). We briefly recall constructions for $G = SL_n(\mathbb{C})$ following [**Hum72**].

- Choose the Euclidean vector space $\mathbb{R}^{n+1}$ with standard basis $\{\epsilon_a | a \in [1, n+1]\}$ and the standard inner product $(\ ,\ )$.
- Choose the simple roots $\alpha_a = \epsilon_a - \epsilon_{a+1}$, $\forall a \in [1,n]$. They span the subspace $\mathfrak{h}^* \subset \mathbb{R}^{l+1}$. The Cartan matrix C is given by $C_{ab} = \frac{2(\alpha_a,\alpha_b)}{(\alpha_a,\alpha_a)}$.
- The simple reflection s_a is the reflection of $\mathbb{R}^{n+1}$ with respect to the hyperplane whose normal vector is α_a. In the standard basis, it is represented by the following matrix
$$\overline{s}_a = \begin{pmatrix} \ddots & & & & & \\ & 1 & & & & \\ & & 0 & -1 & & \\ & & 1 & 0 & & \\ & & & & 1 & \\ & & & & & \ddots \end{pmatrix}$$
such that $\overline{s}_a(\epsilon_a) = \epsilon_{a+1}$ and $\overline{s}_a(\epsilon_{a+1}) = -\epsilon_a$.
- Let $\{\epsilon_a^\vee | a \in [1, n+1]\}$ be the standard basis in the dual space $(\mathbb{R}^{n+1})^*$. We have $\alpha_a^\vee = \epsilon_a^\vee - \epsilon_{a+1}^\vee$.
- We can calculate ϖ_a explicitly. For example, when $n = 1$, we have $\varpi_1 = \frac{1}{2}\alpha_1 = \frac{1}{2}(\epsilon_1 - \epsilon_2)$; when $n = 2$, we have $\varpi_1 = \frac{1}{3}(2\alpha_1 + \alpha_2) = \frac{1}{3}(2\epsilon_1 - \epsilon_2 - \epsilon_3)$, $\varpi_2 = \frac{1}{3}(\alpha_1 + 2\alpha_2) = \frac{1}{3}(\epsilon_1 + \epsilon_2 - 2\epsilon_3)$

Choose the maximal torus H of $G = SL_{n+1}$ to be the subgroup of diagonal matrices, and N (reps. N_-) to be the subgroups of unipotent upper (resp. lower) triangular matrices. Identify $\epsilon_a^\vee$ with the size $(n+1)$ elementary matrix E_{aa}, whose (b,c)-entries are $(E_{aa})_{bc} = \delta_{ab}\delta_{ac}$, $\forall b, c \in [1, n+1]$. Then we have $\exp(x) \in H$ for $x = \sum x_i \alpha_i^\vee$, $x_i \in \mathbb{C}$. Each weight $\gamma = \sum \gamma_i \varpi_i$ determines a character $(\ )^\gamma$ on H such that $(\exp(x))^\gamma = e^{\langle x, \gamma \rangle} = e^{\sum x_i \gamma_i}$.

Note that W can be naturally identified with the permutation group S_{n+1} on $[1, n+1]$, such that s_i is identified with the transposition $(i, i+1)$. It is known that $\Delta_{u\varpi_i,w\varpi_i}(g)$ equals the determinant of the submatrix $g_{u[1,i]\times w[1,i]}$, See [**BFZ05**, Section 2.2]. So we can denote $\Delta_{u[1,i],w[1,i]} = \Delta_{u\varpi_i,w\varpi_i}$. We will further abbreviate $\Delta_{\{r_1,\ldots,r_s\},\{c_1,\ldots,c_s\}}$ by $\Delta_{r_1\ldots r_s,c_1\ldots c_s}$.

Cluster structures. Consider the group $W \times W$. Denote the simple reflections in the first copy by s_{-a} and those in the second copy by s_a. Let $\underline{\mathbf{i}} = (\mathbf{i}_1, \ldots, \mathbf{i}_l)$ be a signed word such that it is a reduced word for $(u,w) \in W \times W$, i.e., $s_{\mathbf{i}_1} \cdots s_{\mathbf{i}_l}$ is a minimal length factorization of (u,w) into simple reflections. Denote $(u_{\leq k}, w_{\leq k}) =$

$s_{\mathbf{i}_1}\cdots s_{\mathbf{i}_k}$, $\forall k \in [1,l]$. Let $(\mathbf{i}_{-|J|+1},\ldots,\mathbf{i}_{-1},\mathbf{i}_0)$ be any chosen Coxeter word (i.e., any permutation of J). We have the seed $\ddot{\mathbf{s}}(\underline{\mathbf{i}})$ (Section 3).

THEOREM A.2 ([**QY25**]). *Take $\Bbbk = \mathbb{C}$ or $\mathbb{Q}(v)$. Let $\underline{\mathbf{i}}$ be any reduced word for (w_0, w_0) and take $\mathbf{s} = \ddot{\mathbf{s}}(\underline{\mathbf{i}})^{\mathrm{op}}$. Then we have an isomorphism $\kappa : \boldsymbol{U}(\mathbf{s}) \simeq \Bbbk[G^{u,w}]$ such that $\kappa x_k = \Delta_{\gamma_k,\delta_k}$, $\forall k \in I$, where*

$$\text{(A.1)}\quad \gamma_k = \begin{cases} \varpi_{\mathbf{i}_k} & k \leq 0 \\ u_{\leq k}\varpi_{|\mathbf{i}_k|} & k \in [1,l] \end{cases} \quad \text{and} \quad \delta_k = \begin{cases} w^{-1}\varpi_{\mathbf{i}_k} & k \leq 0 \\ w^{-1}w_{\leq k}\varpi_{|\mathbf{i}_k|} & k \in [1,l] \end{cases}.$$

And the compatible Poisson structure $\lambda(\mathbf{s})$ has the skew-symmetric quantization matrix Λ given by

$$\Lambda_{kj} = (\gamma_k, \gamma_j) - (\delta_k, \delta_j), \ \forall k > j.$$

Moreover, take $\mathbf{s}' = \ddot{\mathbf{s}}(\underline{\mathbf{j}})^{\mathrm{op}}$ for another reduced signed word of (w_0, w_0), and let κ' denote the corresponding isomorphism $\kappa' : \boldsymbol{U}(\mathbf{s}') \simeq \Bbbk[G^{u,w}]$, then $(\kappa')^{-1}\kappa(x_k)$ are the cluster variables of the seed $\ddot{\mathbf{s}}(\underline{\mathbf{i}})^{\mathrm{op}}$ in $\overline{\boldsymbol{U}}(\mathbf{s}')$.

Finally, in the special case $(u, w) = (w_0, w_0)$, κ restrict a cluster structure on $\Bbbk[G]$:

$$\kappa : \overline{\boldsymbol{U}}(\mathbf{s}) \simeq \Bbbk[G].$$

By Theorem A.2, different choices of the the reduced words provide the same cluster structures on $\Bbbk[G^{u,w}]$, and the same is true for $\Bbbk[G]$. We will omit the symbol κ and identify $\Bbbk[G^{u,w}]$ and $\Bbbk[G]$ with the corresponding cluster algebras, respectively.

REMARK A.3. The statements about $\mathbb{C}[G^{u,w}]$ were first proved in [**BFZ05**] [**SW21**]. The statements for quantum $\Bbbk[G^{u,w}]$ was partially proved in [**BZ05**] [**GY20**][**GLS13**]. The cluster structure on $\mathbb{C}[SL_n]$ could be found in [**FWZ20**].

Based on an approach different from that of [**QY25**], [**Oya25**] proved that $\mathbb{C}[G] = \overline{\boldsymbol{A}}(\mathbf{s}) = \overline{\boldsymbol{U}}(\mathbf{s})$ when G is not of type F_4.

EXAMPLE A.4 ([**Qin24a**, Example 7.8]). Let us consider the case $G = SL_2$ and choose the reduced word $\underline{\mathbf{i}} = (1, -1)$ for $(w_0, w_0) \in W \times W$. The quiver for the seed $\ddot{\mathbf{s}}(\underline{\mathbf{i}})$ is given in Figure 3.4. Take the (classical or quantum) seed $\mathbf{s} = \ddot{\mathbf{s}}(\underline{\mathbf{i}})^{\mathrm{op}}$. Then we have an isomorphism $\kappa : \overline{\boldsymbol{U}}(\mathbf{s}) \simeq \Bbbk[SL_2]$, such that κ send x_k, $k \in [0, 2]$, to:

$$\Delta_{1,2}, \Delta_{1,1}, \Delta_{2,1}.$$

The frozen variables are $\Delta_{1,2}, \Delta_{2,1}$. The quantization matrix Λ is

$$\Lambda = \begin{pmatrix} 0 & -1 & 0 \\ 1 & 0 & 1 \\ 0 & -1 & 0 \end{pmatrix}.$$

The matrix $(b_{ij})_{i,j \in [-1,1]}$ is

$$\widetilde{B} = \begin{pmatrix} 0 & -1 & 0 \\ 1 & 0 & 1 \\ 0 & -1 & 0 \end{pmatrix}.$$

Let us denote $u = x_0(\mathbf{s})$, $x = x_1(\mathbf{s})$, $v = x_2(\mathbf{s})$, $y = x_1(\mu_1 \mathbf{s})$. The common triangular basis of $\overline{\boldsymbol{A}} = \overline{\boldsymbol{U}}$ is $\{[u^n x^m v^l] | n, m, l \in \mathbb{N}\} \sqcup \{[u^n y^m v^l] | n, m, l \in \mathbb{N}, m > 0\}$.

FIGURE A.1. The ice quiver for a seed of $\Bbbk[SL_3]$

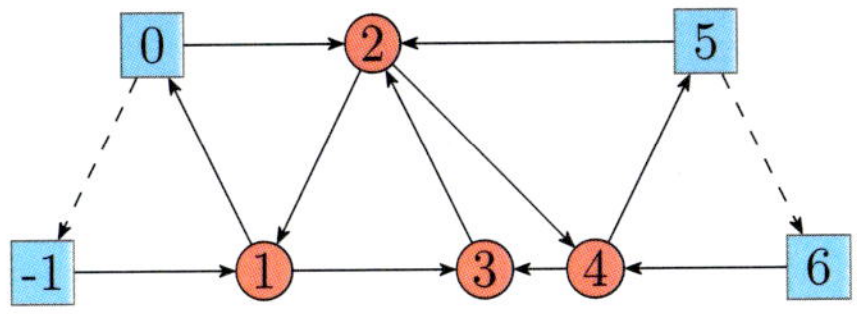

FIGURE A.2. The ice quiver for a seed of $\Bbbk[SL_3^{e,w_0}]$

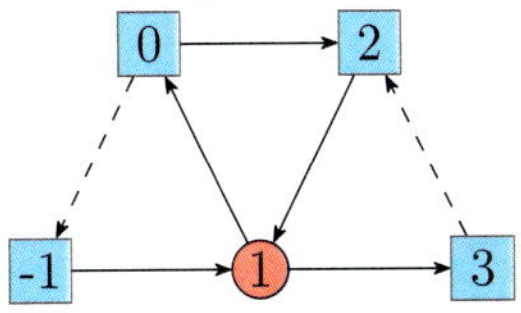

It coincides with the global crystal basis of the $\mathbb{Q}(v)[SL_2]$ in [**Kas93**, Section 9] up to v-multiples.

EXAMPLE A.5. Let us consider the case $G = SL_3$ and choose the reduced word $\underline{\mathbf{i}} = (1, -1, 2, -2, 1, -1)$ for $(w_0, w_0) \in W \times W$. The quiver for the seed $\ddot{\mathbf{s}}(\underline{\mathbf{i}})$ is given in Figure 2.1. Take the (classical or quantum) seed $\ddot{\mathbf{s}} = \ddot{\mathbf{s}}(\underline{\mathbf{i}})^{\mathrm{op}}$. We have $\kappa : \overline{\boldsymbol{U}}(\mathbf{s}) \simeq \Bbbk[SL_3]$, such that κ send x_k, $k \in [-1, 6]$, to:

$$\Delta_{1,3}, \Delta_{12,23}, \Delta_{1,2}, \Delta_{2,2}, \Delta_{12,12}, \Delta_{23,12}, \Delta_{2,1}, \Delta_{3,1}.$$

The frozen variables are $\Delta_{1,3}, \Delta_{12,23}, \Delta_{23,12}, \Delta_{3,1}$.

EXAMPLE A.6 ([**BFZ05**, Example 2.9]). Let us consider the case $G = SL_3$ and choose the reduced word $\underline{\mathbf{i}} = (1, 2, 1, -1, -2, -1)$ for $(w_0, w_0) \in W \times W$. The quiver for the seed $\ddot{\mathbf{s}}(\underline{\mathbf{i}})$ is given in Figure A.1. Take the (classical or quantum) seed $\mathbf{s} = \ddot{\mathbf{s}}(\underline{\mathbf{i}})^{\mathrm{op}}$. We have $\kappa : \overline{\boldsymbol{U}}(\mathbf{s}) \simeq \Bbbk[SL_3]$, such that κ send x_k, $k \in [-1, 6]$, to:

$$\Delta_{1,3}, \Delta_{12,23}, \Delta_{1,2}, \Delta_{12,12}, \Delta_{1,1}, \Delta_{2,1}, \Delta_{23,12}, \Delta_{3,1}.$$

The frozen variables are $\Delta_{1,3}, \Delta_{12,23}, \Delta_{23,12}, \Delta_{3,1}$.

EXAMPLE A.7. Let us consider the case $G = SL_3$ and choose the reduced word $\underline{\mathbf{i}} = (1, 2, 1)$ for $(e, w_0) \in W \times W$. The quiver for the seed $\ddot{\mathbf{s}}(\underline{\mathbf{i}})$ is given in Figure A.2. Take the (classical or quantum) seed $\mathbf{s} = \ddot{\mathbf{s}}(\underline{\mathbf{i}})^{\mathrm{op}}$. We have $\kappa : \boldsymbol{U}(\mathbf{s}) \simeq \Bbbk[SL_3^{e,w_0}]$, such that κ send x_k, $k \in [-1, 3]$, to:

$$\Delta_{1,3}, \Delta_{12,23}, \Delta_{1,2}, \Delta_{12,12}, \Delta_{1,1}.$$

We have $I_{\mathrm{f}} = \{-1, 0, 2, 3\}$, $I_{\mathrm{uf}} = \{1\}$, $\widetilde{B} = \begin{pmatrix} -1 \\ 1 \\ 0 \\ -1 \\ 1 \end{pmatrix}$.

Note that the cluster variables are

$$\begin{aligned}\Delta_{1,3} &= \Delta_{\varpi_1, w_0^{-1}\varpi_1} = \Delta_{\varpi_1,\varpi_1-\alpha_1-\alpha_2} = \Delta_{\frac{2}{3}\alpha_1+\frac{1}{3}\alpha_2, -\frac{1}{3}\alpha_1-\frac{2}{3}\alpha_2}\\ \Delta_{12,23} &= \Delta_{\varpi_2, w_0^{-1}\varpi_2} = \Delta_{\varpi_2,\varpi_2-\alpha_1-\alpha_2} = \Delta_{\frac{1}{3}\alpha_1+\frac{2}{3}\alpha_2, -\frac{2}{3}\alpha_1-\frac{1}{3}\alpha_2}\\ \Delta_{1,2} &= \Delta_{\varpi_1, s_1 s_2\varpi_1} = \Delta_{\varpi_1,\varpi_1-\alpha_1} = \Delta_{\frac{2}{3}\alpha_1+\frac{1}{3}\alpha_2, -\frac{1}{3}\alpha_1+\frac{1}{3}\alpha_2}\\ \Delta_{12,12} &= \Delta_{\varpi_2, s_1\varpi_2} = \Delta_{\frac{1}{3}\alpha_1+\frac{2}{3}\alpha_2, \frac{1}{3}\alpha_1+\frac{2}{3}\alpha_2}\\ \Delta_{1,1} &= \Delta_{\varpi_1,\varpi_1} = \Delta_{\frac{2}{3}\alpha_1+\frac{1}{3}\alpha_2, \frac{2}{3}\alpha_1+\frac{1}{3}\alpha_2}.\end{aligned}$$

The quantization matrix is

$$\Lambda = \begin{pmatrix} 0 & 0 & -1 & -1 & -1\\ 0 & 0 & 0 & -1 & -1\\ 1 & 0 & 0 & 0 & -1\\ 1 & 1 & 0 & 0 & 0\\ 1 & 1 & 1 & 0 & 0\end{pmatrix}.$$

A.2. Cluster algebras from representations of simply-laced quantum affine algebras. Let $\mathfrak{g}$ denote a finite dimensional complex semisimple Lie algebra and whose Cartan matrix C is symmetric ($\mathfrak{g}$ is called simply laced). Let $U_\epsilon(\hat{\mathfrak{g}})$ denote the quantum affine algebra, which is a Hopf algebra, where the parameter $\epsilon \in \mathbb{C}^\times \backslash \{0\}$ is not a root of unity. Let $\mathcal{O}$ denote the abelian category consisting of its (type 1) finite dimensional representations, which has the tensor product $\otimes$. The simple modules S in $\mathcal{O}$ are parameterized by their Drinfeld polynomials $P_S = \{P_a(z) | a \in J\}$, where z is a formal variable, such that $P_a(z)$ take the form $P_a(z) = \prod_i (1-\lambda_i z)$ for some $\lambda_i \in \mathbb{C}^\times$.

We can always choose a height function $\xi : J \to \mathbb{Z}$ such that $|\xi(a) - \xi(b)| = 1$ whenever $C_{ab} \neq 0$. Let $\mathcal{O}_{\mathbb{Z}}$ be the abelian subcategory of $\mathcal{O}$ such that the roots of the Drinfeld polynomials P_a of the simple objects in $\mathcal{O}_{\mathbb{Z}}$ belong to $\epsilon^{2\mathbb{Z}+\xi(a)}$. $\mathcal{O}_{\mathbb{Z}}$ is a tensor subcategory.

Denote $\hat{J} = \{(a,d) | a \in J, d \in \xi(a) + 2\mathbb{Z}\}$ and $W = \oplus_{(a,d)\in\hat{J}} \mathbb{N} e_{a,d}$. For $w = \sum w_{a,d} e_{a,d} \in W$, let $S(w)$ denote the simple module whose Drinfeld polynomials are $P_a(z) = \prod_{d\in\xi(a)+2\mathbb{Z}} (1-\varepsilon^d z)^{w_{a,d}}$, $a \in J$. In particular, the modules $S(e_{a,d})$ are called the fundamental modules, denoted $W_{a,d}$. For $k \in \mathbb{Z}_{>0}$, the modules $S(e_{a,d} + e_{a,d+2} + \cdots + e_{a,d+2k-2})$ are called the Kirillov-Reshetikhin modules, denoted $W^k_{a,d}$. We also introduce the standard module arising from quiver varieties

(A.2) $$M(w) := \overrightarrow{\otimes}_d (\otimes_a {W_{a,d}}^{\otimes w_{a,d}}),$$

where we take the increasing order in $\overrightarrow{\otimes}_d$ (different orders in $\otimes_a$ produce the same result); see [**Nak04**, Axiom 3][**VV02**, Corollary 7.16].

Although $V \otimes W \not\cong W \otimes V$ in $\mathcal{O}_{\mathbb{Z}}$ in general, the Grothendieck ring $K_0(\mathcal{O}_{\mathbb{Z}})$ is a commutative algebra. For simplicity, we will denote the isoclass of any module V in $K_0(\mathcal{O}_{\mathbb{Z}})$ by V. Recall that there is an embedding called the q-character $\chi_q : K_0(\mathcal{O}_{\mathbb{Z}}) \hookrightarrow \mathcal{Y} := \mathbb{Z}[Y^{\pm}_{a,d}]_{(a,d)\in\hat{J}}$. Define the Laurent monomial

$$A_{a,h} := Y_{a,h+1} Y_{a,h-1} \prod_{b\neq a} Y_{b,h}^{C_{ba}},$$

where $(b, h-1) \in \hat{J}$. Denote $Y^w = \prod Y_{a,d}^{w_{a,d}}$ and $A^{-v} = \prod A_{b,h}^{-v_{b,h}}$ for $v_{b,h} \in \mathbb{Z}$. We introduce the dominance order $\prec$ on W such that $w' \preceq w$ if $Y^{w'} = Y^w (A^{-1})^v$ for some $v = (v_{a,h}) \geq 0$. Then $\chi_q(S(w))$ is w-pointed, i.e., Y^w is its unique $\prec$-leading term with coefficient 1. Denote $\chi_q(V)$ by $[V]$. We will identify $K_0(\mathcal{O}_{\mathbb{Z}})$ with its image under χ_q.

We can introduce a quantum torus algebra $\mathcal{Y}_t := \mathbb{Z}[t^{\pm\frac{1}{2}}][Y^{\pm}_{a,d}]_{(a,d)\in\hat{J}}$ using the t-twisted multiplication $*$ in [**Qin17**, Section 7.3][**HL15**, (6)], viewed as a t-deformation of $\mathcal{Y}$. The bar-involution $\overline{(\ )}$ is the $\mathbb{Z}$-linear automorphism on $\mathcal{Y}_t$ such that $\overline{t^\alpha Y^w} = t^{-\alpha} Y^w$. Following [**Nak04**][**VV03**][**Her04**], we have a Z-linear injective map, called a t-analog of q-characters:

$$\chi_{q,t} : K_0(\mathcal{O}_{\mathbb{Z}}) \hookrightarrow \mathcal{Y}_t.$$

At the classical limit $t^{\frac{1}{2}} \mapsto 1$, $\chi_{q,t}$ becomes χ_q. We further have

$$\chi_{q,t} M(w) = t^\alpha \overleftarrow{\prod_d} (\prod_a \chi_{q,t}(W_{a,d})^{w_{a,d}}). \tag{A.3}$$

for appropriate $\alpha \in \frac{1}{2}\mathbb{Z}$ such $M(w)$ is w-pointed, where the product $\overleftarrow{\prod_d}$ is taking in the **decreasing** order.

REMARK A.8. The equation (A.3) was known in [**Nak04**][**VV02**] for a slightly different t-deformation and the increasing order $\overrightarrow{\prod_d}$. It holds for our t-deformation and the decreasing order $\overleftarrow{\prod_d}$; see [**HL15**, Sections 5.6][**Qin14**, Proposition 5.3.6].

Let $K_t(\mathcal{O}_{\mathbb{Z}})^{\mathrm{op}}$ denote the free $\mathbb{Z}[t^{\pm\frac{1}{2}}]$-module spanned by $\{\chi_{q,t}(S) | S \text{ is simple in } \mathcal{O}_{\mathbb{Z}}\}$. It is a subalgebra of $\mathcal{Y}_t$. We will call its opposite algebra $K_t(\mathcal{O}_{\mathbb{Z}})$ the quantum Grothendieck ring of $\mathcal{O}_{\mathbb{Z}}$, which is viewed as a t-deformation of $K_0(\mathcal{O}_{\mathbb{Z}})$. Correspondingly, $K_t(\mathcal{O}_{\mathbb{Z}}^{\mathrm{op}}) = K_t(\mathcal{O}_{\mathbb{Z}})^{\mathrm{op}}$ is called the quantum Grothendieck ring of $\mathcal{O}_{\mathbb{Z}}^{\mathrm{op}}$. Recall that we have the canonical anti-isomorphism $\iota : \mathcal{Y}_t \simeq \mathcal{Y}_t^{\mathrm{op}}$ between $\mathbb{Z}[t^{\pm\frac{1}{2}}]$-algebras, such that $\iota(Y^w) := Y^w$ (see (2.3)). It induces $\iota : K_t(\mathcal{O}_{\mathbb{Z}})^{\mathrm{op}} \simeq K_t(\mathcal{O}_{\mathbb{Z}})$. For any $V \in \mathcal{O}_{\mathbb{Z}}$, we denote $[V]_t := \iota\chi_{q,t}(V) \in K_t(\mathcal{O}_{\mathbb{Z}})$.

REMARK A.9. Our quantum Grothendieck ring $K_t(\mathcal{O}_{\mathbb{Z}})$ is the opposite algebra of the one used in [**HL15**][**Qin17**]. We make this choice because it leads to the same product order on the both sides below in $K_t(\mathcal{O}_{\mathbb{Z}})$:

$$[\overrightarrow{\otimes}_d (\otimes_a W_{a,d}{}^{\otimes w_{a,d}})]_t = t^\alpha \overrightarrow{\prod_d} (\prod_a [W_{a,d}]_t^{w_{a,d}}).$$

Correspondingly, if an algebra A was previously categorified by $\mathcal{O}_{\mathbb{Z}}$ in the convention of [**HL15**][**Qin17**], A^{op} is said to be categorified by $\mathcal{O}_{\mathbb{Z}}$ in our convention.

THEOREM A.10 ([**Nak04**]). *For any $w \in W$, the following statements are true in the quantum Grothendieck ring $K_t(\mathcal{O}_{\mathbb{Z}})$.*

(1) $\overline{[S(w)]_t} = [S(w)]_t$.
(2) $[M(w)]_t = [S(w)]_t + \sum_{w' < w} a_{w,w'}(t)[S(w')]_t$.
(3) $a_{w,w'}(t) \in t^{-1}\mathbb{N}[t^{-1}]$.
(4) $a_{w,w'}(1)$ *equals the multiplicity* $[M(w) : S(w')]$.

Cluster structure. Choose any height function ξ. Let $\underline{c} = (c_1, \ldots, c_{|J|})$ be a Coxeter word and $\underline{\eta} = (\eta_i)_{i\in[r,s]}$ be a $\underline{c}$-adapted word, i.e., $\underline{\eta}$ is a subsequence of $\underline{c}^N$ for some N.

For any interval $[j,k] \subset [r,s]$ such that $\eta_j = \eta_k =: a$, let $W_{[j,k]}(\xi)$ denote the Kirillov-Reshetikhin module $W^{k-j+1}_{a,-2k+\xi(a)} = S(e_{a,-2k+\xi(a)} + e_{a,-2k+2+\xi(a)} +$

$\cdots + e_{a,-2j+\xi(a)})$. Denote the fundamental module by $W_k(\xi) := W_{[k,k]}(\xi) = S(e_{a,-2k+\xi(a)})$. Let $\mathcal{O}_{\underline{\eta}} = \mathcal{O}_{\underline{\eta}}(\xi)$ be the smallest tensor subcategory of $\mathcal{O}_{\mathbb{Z}}$ containing the fundamental modules $W_k(\xi)$, $\forall k \in [r,s]$, and $K_t(\mathcal{O}_{\underline{\eta}})$ the corresponding subalgebra of $K_t(\mathcal{O}_{\mathbb{Z}})$ generated by $[W_k(\xi)]_t$, $k \in [r,s]$.

REMARK A.11. Hernandez and Leclerc called $\mathcal{O}_{(\underline{c})^{l+1}}$ a level-l subcategory, denoted $\mathcal{O}_l$ ([**HL10**]). When $\overrightarrow{w}$ is a reduced word of w_0 and it is $\underline{c}$-adapted, they also considered the subcategory $\mathcal{O}_{\overrightarrow{w}}$ and showed that $K_t(\mathcal{O}_{\overrightarrow{w}})^{\mathrm{op}} \otimes \mathbb{Q}(v)$ is isomorphic to the quantized coordinate ring $\mathbb{Q}(v)[N]$ [**HL15**].

When $S(e_{a,d})$ is the fundamental module $W_k(\xi)$ for $k \in [r,s]$, denote $\beta_k := e_{a,d}$. As in (6.1), denote the linear map $\theta^{-1} : \oplus_{k\in[r,s]}\mathbb{Z}\beta_k \simeq M^\circ$ such that

$$\theta^{-1}(\beta_k) = f_k - f_{k[-1]}, \forall k \in [r,s],$$

where $f_{k[-1]} = 0$ if $k = k^{\min}$ in $[r,s]$.

THEOREM A.12 ([**Qin17**]). *Take* $\mathbf{s} = \dot{\mathbf{s}}(\underline{\eta})^{\mathrm{op}}$ *and the canonical anti-isomorphism* $\iota : \overline{\boldsymbol{U}}(\mathbf{s}^{\mathrm{op}}) \simeq \overline{\boldsymbol{U}}(\mathbf{s})$. *Then we have the* $\mathbb{Z}$*-algebra isomorphism* $\kappa : \overline{\boldsymbol{U}}(\mathbf{s}) \simeq K_t(\mathcal{O}_{\underline{\eta}})$ *such that* $\kappa(v) = t^{\frac{1}{2}}$, $\kappa\iota(W_{[j,k]}) = [W_{[j,k]}(\xi)]_t$. *Moreover, the common triangular basis* $\boldsymbol{L}$ *of* $\overline{\boldsymbol{U}}(\mathbf{s})$ *is* $\{(\kappa)^{-1}[S(w)]_t | S(w) \text{ is simple in } \mathcal{O}_{\underline{\eta}}\}$, *such that* $(\kappa)^{-1}[S(w)]_t$ *is* $\theta^{-1}(w)$*-pointed.*

REMARK A.13. Note that we always have $\overline{\boldsymbol{U}}(\mathbf{s}) = \overline{\boldsymbol{A}}(\mathbf{s})$.

In the classical case ($\Bbbk = \mathbb{Z}$), [**HL10**] showed that $\mathcal{O}_{(\underline{c})^{l+1}}$ categorifies $\overline{\boldsymbol{A}}(\mathbf{s})$ for $\mathfrak{g}$ of type A_1 or $l = 1$. [**Qin17**] proved this categorification result (at the classical or quantum level) for all simply laced $\mathfrak{g}$. The categorification result is true for all semisimple $\mathfrak{g}$ by [**KKOP24b**].

Extending the range $[r,s]$ for an appropriate signed word $\underline{\mathbf{i}} = (\mathbf{i}_k)_{k\in[r,s]}$ to $(-\infty,+\infty)$ as in Section 7, one could easily extend the statements in Theorem 6.7(2) to $K_t(\mathcal{O}_{\mathbb{Z}})$; see [**Qin24c**]. One could still establish the cluster structure when $\mathfrak{g}$ is not simply laced (see [**HL16**]). In this case, $\kappa\mathbf{L}$ still equals the Kazhdan-Lusztig-type basis $\{[S(w)]_t | S(w) \text{ is simple}\}$ ([**Qin24a**]). But it is still conjectured that $\chi_{q,t}(S(w))|_{t\mapsto 1} = \chi_q(S(w))$ ([**FHOO22**] proved it in type B).

EXAMPLE A.14. Let us continue the seed $\mathbf{s} = \dot{\mathbf{s}}(\underline{\mathbf{i}})$ in Example 3.4, where $\underline{c} = (1,2,3)$, $\underline{\mathbf{i}} = (\mathbf{i}_k)_{k\in[1,\infty)} = (\underline{c})^\infty$. Choose $\xi(a) = -a$ for $a \in \{1,2,3\}$, such that $\xi(c_j) \geq \xi(c_k)$ when $j < k$. We have $\kappa\iota : \overline{\boldsymbol{U}}(\mathbf{s}) \subset K_t(\mathcal{O}_{\mathbb{Z}})$. The interval variables are $W_{[a+2r,a+2s]}$ for $0 \leq r \leq s$. Then we have

$$\kappa\iota W_{[a+2r,a+2s]} = [S(e_{a,-2s+\xi(a)} + e_{a,-2s+2+\xi(a)} + \cdots + e_{a,-2r+\xi(a)}]_t \in K_t(\mathcal{O}_{\mathbb{Z}}).$$

References

[BFZ05] Arkady Berenstein, Sergey Fomin, and Andrei Zelevinsky, *Cluster algebras. III. Upper bounds and double Bruhat cells*, Duke Math. J. **126** (2005), no. 1, 1–52, DOI 10.1215/S0012-7094-04-12611-9. MR2110627

[BZ05] Arkady Berenstein and Andrei Zelevinsky, *Quantum cluster algebras*, Adv. Math. **195** (2005), no. 2, 405–455, DOI 10.1016/j.aim.2004.08.003. MR2146350

[BZ14] Arkady Berenstein and Andrei Zelevinsky, *Triangular bases in quantum cluster algebras*, Int. Math. Res. Not. IMRN **6** (2014), 1651–1688, DOI 10.1093/imrn/rns268. MR3180605

[CG22] Roger Casals and Honghao Gao, *Infinitely many Lagrangian fillings*, Ann. of Math. (2) **195** (2022), no. 1, 207–249, DOI 10.4007/annals.2022.195.1.3. MR4358415

[CGG+25] Roger Casals, Eugene Gorsky, Mikhail Gorsky, Ian Le, Linhui Shen, and José Simental, *Cluster structures on braid varieties*, J. Amer. Math. Soc. **38** (2025), no. 2, 369–479, DOI 10.1090/jams/1048. MR4868947

[CGM+17] Man Wai Cheung, Mark Gross, Greg Muller, Gregg Musiker, Dylan Rupel, Salvatore Stella, and Harold Williams, *The greedy basis equals the theta basis: a rank two haiku*, J. Combin. Theory Ser. A **145** (2017), 150–171, DOI 10.1016/j.jcta.2016.08.004. MR3551649

[CILFS15] Giovanni Cerulli Irelli, Daniel Labardini-Fragoso, and Jan Schröer, *Caldero-Chapoton algebras*, Trans. Amer. Math. Soc. **367** (2015), no. 4, 2787–2822, DOI 10.1090/S0002-9947-2014-06175-8. MR3301882

[CMMM25] Man-Wai Cheung, Timothy Magee, Travis Mandel, and Greg Muller, *Valuative independence and cluster theta reciprocity*, preprint (2025).

[CS14] Leonid Chekhov and Michael Shapiro, *Teichmüller spaces of Riemann surfaces with orbifold points of arbitrary order and cluster variables*, Int. Math. Res. Not. IMRN **10** (2014), 2746–2772, DOI 10.1093/imrn/rnt016. MR3214284

[CW19] Sabin Cautis and Harold Williams, *Cluster theory of the coherent Satake category*, J. Amer. Math. Soc. **32** (2019), no. 3, 709–778, DOI 10.1090/jams/918. MR3981987

[DM21] Ben Davison and Travis Mandel, *Strong positivity for quantum theta bases of quantum cluster algebras*, Invent. Math. **226** (2021), no. 3, 725–843, DOI 10.1007/s00222-021-01061-1. MR4337972

[EGNO15] Pavel Etingof, Shlomo Gelaki, Dmitri Nikshych, and Victor Ostrik, *Tensor categories*, Mathematical Surveys and Monographs, vol. 205, American Mathematical Society, Providence, RI, 2015, DOI 10.1090/surv/205. MR3242743

[EL21] Balázs Elek and Jiang-Hua Lu, *Bott-Samelson varieties and Poisson Ore extensions*, Int. Math. Res. Not. IMRN **14** (2021), 10745–10797, DOI 10.1093/imrn/rnz127. MR4285734

[FG06] Vladimir Fock and Alexander Goncharov, *Moduli spaces of local systems and higher Teichmüller theory*, Publ. Math. Inst. Hautes Études Sci. **103** (2006), 1–211, DOI 10.1007/s10240-006-0039-4. MR2233852

[FHOO22] Ryo Fujita, David Hernandez, Se-jin Oh, and Hironori Oya, *Isomorphisms among quantum Grothendieck rings and propagation of positivity*, J. Reine Angew. Math. **785** (2022), 117–185, DOI 10.1515/crelle-2021-0088. MR4402493

[Fom10] Sergey Fomin, *Total positivity and cluster algebras*, Proceedings of the International Congress of Mathematicians. Volume II, Hindustan Book Agency, New Delhi, 2010, pp. 125–145. MR2827788

[Fra16] Chris Fraser, *Quasi-homomorphisms of cluster algebras*, Adv. in Appl. Math. **81** (2016), 40–77, DOI 10.1016/j.aam.2016.06.005. MR3551663

[FST08] Sergey Fomin, Michael Shapiro, and Dylan Thurston, *Cluster algebras and triangulated surfaces. I. Cluster complexes*, Acta Math. **201** (2008), no. 1, 83–146, DOI 10.1007/s11511-008-0030-7. MR2448067

[FT18] Sergey Fomin and Dylan Thurston, *Cluster algebras and triangulated surfaces Part II: Lambda lengths*, Mem. Amer. Math. Soc. **255** (2018), no. 1223, v+97, DOI 10.1090/memo/1223. MR3852257

[FWZ16] Sergey Fomin, Lauren Williams, and Andrei Zelevinsky, *Introduction to Cluster Algebras. chapters 1-3*, preprint (2016).

[FWZ17] Sergey Fomin, Lauren Williams, and Andrei Zelevinsky, *Introduction to Cluster Algebras. Chapters 4-5*, preprint (2017).

[FWZ20] Sergey Fomin, Lauren Williams, and Andrei Zelevinsky, *Introduction to Cluster Algebras. Chapter 6*, preprint (2020).

[FWZ21] Sergey Fomin, Lauren Williams, and Andrei Zelevinsky, *Introduction to Cluster Algebras. Chapter 7*, preprint (2021).

[FZ02] Sergey Fomin and Andrei Zelevinsky, *Cluster algebras. I. Foundations*, J. Amer. Math. Soc. **15** (2002), no. 2, 497–529, DOI 10.1090/S0894-0347-01-00385-X. MR1887642

[FZ03] Sergey Fomin and Andrei Zelevinsky, *Cluster algebras. II. Finite type classification*, Invent. Math. **154** (2003), no. 1, 63–121, DOI 10.1007/s00222-003-0302-y. MR2004457

[FZ07] Sergey Fomin and Andrei Zelevinsky, *Cluster algebras. IV. Coefficients*, Compos. Math. **143** (2007), no. 1, 112–164, DOI 10.1112/S0010437X06002521. MR2295199

[GHK15] Mark Gross, Paul Hacking, and Sean Keel, *Birational geometry of cluster algebras*, Algebr. Geom. **2** (2015), no. 2, 137–175, DOI 10.14231/AG-2015-007. MR3350154

[GHKK18] Mark Gross, Paul Hacking, Sean Keel, and Maxim Kontsevich, *Canonical bases for cluster algebras*, J. Amer. Math. Soc. **31** (2018), no. 2, 497–608, DOI 10.1090/jams/890. MR3758151

[GHL24] Christof Geiss, David Hernandez, and Bernard Leclerc, *Representations of shifted quantum affine algebras and cluster algebras I: The simply laced case*, Proc. Lond. Math. Soc. (3) **129** (2024), no. 3, Paper No. e12630, 75, DOI 10.1112/plms.12630. MR4793282

[GL19] Pavel Galashin and Thomas Lam, *Positroid varieties and cluster algebras* (English, with English and French summaries), Ann. Sci. Éc. Norm. Supér. (4) **56** (2023), no. 3, 859–884, DOI 10.24033/asens.2545. MR4650160

[GLS11] Christof Geiß, Bernard Leclerc, and Jan Schröer, *Kac-Moody groups and cluster algebras*, Adv. Math. **228** (2011), no. 1, 329–433, DOI 10.1016/j.aim.2011.05.011. MR2822235

[GLS13] C. Geiß, B. Leclerc, and J. Schröer, *Cluster structures on quantum coordinate rings*, Selecta Math. (N.S.) **19** (2013), no. 2, 337–397, DOI 10.1007/s00029-012-0099-x. MR3090232

[GLSBS22] Pavel Galashin, Thomas Lam, Melissa Sherman-Bennett, and David Speyer, *Braid variety cluster structures, I: 3D plabic graphs*, preprint (2022).

[Gra15] Sira Gratz, *Cluster algebras of infinite rank as colimits*, Math. Z. **281** (2015), no. 3-4, 1137–1169, DOI 10.1007/s00209-015-1524-6. MR3421657

[GSV03] Michael Gekhtman, Michael Shapiro, and Alek Vainshtein, *Cluster algebras and Poisson geometry* (English, with English and Russian summaries), Mosc. Math. J. **3** (2003), no. 3, 899–934, 1199, DOI 10.17323/1609-4514-2003-3-3-899-934. {Dedicated to Vladimir Igorevich Arnold on the occasion of his 65th birthday}. MR2078567

[GSV05] Michael Gekhtman, Michael Shapiro, and Alek Vainshtein, *Cluster algebras and Weil-Petersson forms*, Duke Math. J. **127** (2005), no. 2, 291–311, DOI 10.1215/S0012-7094-04-12723-X. MR2130414

[GSV21] Misha Gekhtman, Michael Shapiro, and Alek Vainshtein, *Generalized cluster structures related to the Drinfeld double of GL_n*, J. Lond. Math. Soc. (2) **105** (2022), no. 3, 1601–1633, DOI 10.1112/jlms.12542. MR4407106

[GSV23] Misha Gekhtman, Michael Shapiro, and Alek Vainshtein, *A unified approach to exotic cluster structures on simple Lie groups*, preprint (2023).

[GY16] K. R. Goodearl and M. T. Yakimov, *Quantum cluster algebra structures on quantum nilpotent algebras*, Mem. Amer. Math. Soc. **247** (2017), no. 1169, vii+119, DOI 10.1090/memo/1169. MR3633289

[GY20] K. R. Goodearl and M. T. Yakimov, *The Berenstein-Zelevinsky quantum cluster algebra conjecture*, J. Eur. Math. Soc. (JEMS) **22** (2020), no. 8, 2453–2509, DOI 10.4171/JEMS/969. MR4118613

[GY21] K. R. Goodearl and M. T. Yakimov, *Integral quantum cluster structures*, Duke Math. J. **170** (2021), no. 6, 1137–1200, DOI 10.1215/00127094-2020-0061. MR4255051

[Her04] David Hernandez, *Algebraic approach to q,t-characters*, Adv. Math. **187** (2004), no. 1, 1–52, DOI 10.1016/j.aim.2003.07.016. MR2074171

[HL10] David Hernandez and Bernard Leclerc, *Cluster algebras and quantum affine algebras*, Duke Math. J. **154** (2010), no. 2, 265–341, DOI 10.1215/00127094-2010-040. MR2682185

[HL15] David Hernandez and Bernard Leclerc, *Quantum Grothendieck rings and derived Hall algebras*, J. Reine Angew. Math. **701** (2015), 77–126, DOI 10.1515/crelle-2013-0020. MR3331727

[HL16] David Hernandez and Bernard Leclerc, *A cluster algebra approach to q-characters of Kirillov-Reshetikhin modules*, J. Eur. Math. Soc. (JEMS) **18** (2016), no. 5, 1113–1159, DOI 10.4171/JEMS/609. MR3500832

[Hum72] James E. Humphreys, *Introduction to Lie algebras and representation theory*, Graduate Texts in Mathematics, Vol. 9, Springer-Verlag, New York-Berlin, 1972. MR323842

[IIK+10] Rei Inoue, Osamu Iyama, Atsuo Kuniba, Tomoki Nakanishi, and Junji Suzuki, *Periodicities of T-systems and Y-systems*, Nagoya Math. J. **197** (2010), 59–174, DOI 10.1215/00277630-2009-003. MR2649278

[JLO23a] Il-Seung Jang, Kyu-Hwan Lee, and Se-jin Oh, *Braid group action on quantum virtual Grothendieck ring through constructing presentations*, preprint (2023).

[JLO23b] Il-Seung Jang, Kyu-Hwan Lee, and Se-jin Oh, *Quantization of virtual Grothendieck rings and their structure including quantum cluster algebras*, Comm. Math. Phys. **405** (2024), no. 7, Paper No. 173, 83, DOI 10.1007/s00220-024-05037-z. MR4773249

[Kas91] M. Kashiwara, *On crystal bases of the Q-analogue of universal enveloping algebras*, Duke Math. J. **63** (1991), no. 2, 465–516, DOI 10.1215/S0012-7094-91-06321-0. MR1115118

[Kas93] Masaki Kashiwara, *Global crystal bases of quantum groups*, Duke Math. J. **69** (1993), no. 2, 455–485, DOI 10.1215/S0012-7094-93-06920-7. MR1203234

[Kel08] Bernhard Keller, *Cluster algebras, quiver representations and triangulated categories*, Triangulated categories, London Math. Soc. Lecture Note Ser., vol. 375, Cambridge Univ. Press, Cambridge, 2010, pp. 76–160. MR2681708

[Kel11] Bernhard Keller, *On cluster theory and quantum dilogarithm identities*, Representations of Algebras and Related Topics, Editors A. Skowronski and K. Yamagata, EMS Series of Congress Reports, European Mathematical Society, 2011, pp. 85–11.

[Kel13] Bernhard Keller, *The periodicity conjecture for pairs of Dynkin diagrams*, Ann. of Math. (2) **177** (2013), no. 1, 111–170, DOI 10.4007/annals.2013.177.1.3. MR2999039

[Kim12] Yoshiyuki Kimura, *Quantum unipotent subgroup and dual canonical basis*, Kyoto J. Math. **52** (2012), no. 2, 277–331, DOI 10.1215/21562261-1550976. MR2914878

[KKK18] Seok-Jin Kang, Masaki Kashiwara, and Myungho Kim, *Symmetric quiver Hecke algebras and R-matrices of quantum affine algebras*, Invent. Math. **211** (2018), no. 2, 591–685, DOI 10.1007/s00222-017-0754-0. MR3748315

[KKKO18] Seok-Jin Kang, Masaki Kashiwara, Myungho Kim, and Se-jin Oh, *Monoidal categorification of cluster algebras*, J. Amer. Math. Soc. **31** (2018), no. 2, 349–426, DOI 10.1090/jams/895. MR3758148

[KKOP24a] Masaki Kashiwara, Myungho Kim, Se-jin Oh, and Euiyong Park, *Braid symmetries on bosonic extensions*, preprint (2024).

[KKOP24b] Masaki Kashiwara, Myungho Kim, Se-jin Oh, and Euiyong Park, *Monoidal categorification and quantum affine algebras II*, Invent. Math. **236** (2024), no. 2, 837–924, DOI 10.1007/s00222-024-01249-1. MR4720249

[KL09] Mikhail Khovanov and Aaron D. Lauda, *A diagrammatic approach to categorification of quantum groups. I*, Represent. Theory **13** (2009), 309–347, DOI 10.1090/S1088-4165-09-00346-X. MR2525917

[KL11] Mikhail Khovanov and Aaron D. Lauda, *A diagrammatic approach to categorification of quantum groups II*, Trans. Amer. Math. Soc. **363** (2011), no. 5, 2685–2700, DOI 10.1090/S0002-9947-2010-05210-9. MR2763732

[KNS11] Atsuo Kuniba, Tomoki Nakanishi, and Junji Suzuki, *T-systems and Y-systems in integrable systems*, J. Phys. A **44** (2011), no. 10, 103001, 146, DOI 10.1088/1751-8113/44/10/103001. MR2773889

[KO21] Yoshiyuki Kimura and Hironori Oya, *Twist automorphisms on quantum unipotent cells and dual canonical bases*, Int. Math. Res. Not. IMRN **9** (2021), 6772–6847, DOI 10.1093/imrn/rnz040. MR4251290

[KQW23] Yoshiyuki Kimura, Fan Qin, and Qiaoling Wei, *Twist automorphisms and Poisson structures*, SIGMA Symmetry Integrability Geom. Methods Appl. **19** (2023), Paper No. 105, 39, DOI 10.3842/SIGMA.2023.105. MR4681480

[Lec10] Bernard Leclerc, *Cluster algebras and representation theory*, Proceedings of the International Congress of Mathematicians. Volume IV, Hindustan Book Agency, New Delhi, 2010, pp. 2471–2488. MR2827980

[LLRZ14] Kyungyong Lee, Li Li, Dylan Rupel, and Andrei Zelevinsky, *Greedy bases in rank 2 quantum cluster algebras*, Proc. Natl. Acad. Sci. USA **111** (2014), no. 27, 9712–9716, DOI 10.1073/pnas.1313078111. MR3263303

[LLZ14] Kyungyong Lee, Li Li, and Andrei Zelevinsky, *Greedy elements in rank 2 cluster algebras*, Selecta Math. (N.S.) **20** (2014), no. 1, 57–82, DOI 10.1007/s00029-012-0115-1. MR3147413

[Lus90] G. Lusztig, *Canonical bases arising from quantized enveloping algebras*, J. Amer. Math. Soc. **3** (1990), no. 2, 447–498, DOI 10.2307/1990961. MR1035415

[Lus91] G. Lusztig, *Quivers, perverse sheaves, and quantized enveloping algebras*, J. Amer. Math. Soc. **4** (1991), no. 2, 365–421, DOI 10.2307/2939279. MR1088333

[Lus93] George Lusztig, *Introduction to quantum groups*, Progress in Mathematics, vol. 110, Birkhäuser Boston, Inc., Boston, MA, 1993. MR1227098

[Lus94] G. Lusztig, *Total positivity in reductive groups*, Lie theory and geometry, Progr. Math., vol. 123, Birkhäuser Boston, Boston, MA, 1994, pp. 531–568, DOI 10.1007/978-1-4612-0261-5_20. MR1327548

[Man17] Travis Mandel, *Theta bases are atomic*, Compos. Math. **153** (2017), no. 6, 1217–1219, DOI 10.1112/S0010437X17007060. MR3705255

[McN21] Peter J McNamara, *Cluster monomials are dual canonical*, preprint (2021).

[MQ23] Travis Mandel and Fan Qin, *Bracelets bases are theta bases*, preprint (2023), 131 pages.

[MSW13] Gregg Musiker, Ralf Schiffler, and Lauren Williams, *Bases for cluster algebras from surfaces*, Compos. Math. **149** (2013), no. 2, 217–263, DOI 10.1112/S0010437X12000450. MR3020308

[Mul16a] Greg Muller, *The existence of a maximal green sequence is not invariant under quiver mutation*, Electron. J. Combin. **23** (2016), no. 2, Paper 2.47, 23, DOI 10.37236/5412. MR3512669

[Mul16b] Greg Muller, *Skein and cluster algebras of marked surfaces*, Quantum Topol. **7** (2016), no. 3, 435–503, DOI 10.4171/QT/79. MR3551171

[Nak01] Hiraku Nakajima, *Quiver varieties and finite-dimensional representations of quantum affine algebras*, J. Amer. Math. Soc. **14** (2001), no. 1, 145–238, DOI 10.1090/S0894-0347-00-00353-2. MR1808477

[Nak04] Hiraku Nakajima, *Quiver varieties and t-analogs of q-characters of quantum affine algebras*, Ann. of Math. (2) **160** (2004), no. 3, 1057–1097, DOI 10.4007/annals.2004.160.1057. MR2144973

[Oya25] Hironori Oya, *A note on a cluster structure of the coordinate ring of a simple algebraic group*, Proc. Amer. Math. Soc. **154** (2026), no. 5, 1867–1879, DOI 10.1090/proc/17551. MR5060395

[Pag25] Francesca Paganelli, *Quantum cluster algebras and representations of shifted quantum affine algebras*, preprint (2025).

[Pla13] Pierre-Guy Plamondon, *Generic bases for cluster algebras from the cluster category*, Int. Math. Res. Not. IMRN **10** (2013), 2368–2420, DOI 10.1093/imrn/rns102. MR3061943

[Qin14] Fan Qin, *t-analog of q-characters, bases of quantum cluster algebras, and a correction technique*, Int. Math. Res. Not. IMRN **22** (2014), 6175–6232, DOI 10.1093/imrn/rnt115. MR3283002

[Qin17] Fan Qin, *Triangular bases in quantum cluster algebras and monoidal categorification conjectures*, Duke Math. J. **166** (2017), no. 12, 2337–2442, DOI 10.1215/00127094-2017-0006. MR3694569

[Qin19] Fan Qin, *Compare triangular bases of acyclic quantum cluster algebras*, Trans. Amer. Math. Soc. **372** (2019), no. 1, 485–501, DOI 10.1090/tran/7610. MR3968777

[Qin20a] Fan Qin, *An analog of Leclerc's conjecture for bases of quantum cluster algebras*, SIGMA Symmetry Integrability Geom. Methods Appl. **16** (2020), Paper No. 122, 22, DOI 10.3842/SIGMA.2020.122. MR4178877

[Qin20b] Fan Qin, *Dual canonical bases and quantum cluster algebras*, preprint (2020).

[Qin23] Fan Qin, *Cluster algebras and their bases*, Representations of algebras and related structures, EMS Ser. Congr. Rep., EMS Press, Berlin, [2023] ©2023, pp. 335–369. MR4693645

[Qin24a] Fan Qin, *Analogs of dual canonical bases for cluster algebras from Lie theory*, preprint (2024).

[Qin24b] Fan Qin, *Applications of the freezing operators on cluster algebras*, J. Algebra Appl. **25** (2026), no. 3, Paper No. 2550357, 30, DOI 10.1142/S0219498825503578. MR5002858

[Qin24c] Fan Qin, *Based cluster algebras of infinite ranks*, preprint (2024).

[Qin24d] Fan Qin, *Bases for upper cluster algebras and tropical points*, J. Eur. Math. Soc. (JEMS) **26** (2024), no. 4, 1255–1312, DOI 10.4171/jems/1308. MR4721032

[QY25] Fan Qin and Milen Yakimov, *Partially compactified quantum cluster structures on simple algebraic groups and the full Berenstein–Zelevinsky conjecture*, preprint (2025).

[Rei10] Idun Reiten, *Cluster categories*, Proceedings of the International Congress of Mathematicians. Volume I, Hindustan Book Agency, New Delhi, 2010, pp. 558–594. MR2827905

[Rou08] Raphaël Rouquier, *2-Kac-Moody algebras*, preprint (2008).

[Sco06] Joshua S. Scott, *Grassmannians and cluster algebras*, Proc. London Math. Soc. (3) **92** (2006), no. 2, 345–380, DOI 10.1112/S0024611505015571. MR2205721

[SSBW19] K. Serhiyenko, M. Sherman-Bennett, and L. Williams, *Cluster structures in Schubert varieties in the Grassmannian*, Proc. Lond. Math. Soc. (3) **119** (2019), no. 6, 1694–1744, DOI 10.1112/plms.12281. MR4295518

[SW21] Linhui Shen and Daping Weng, *Cluster structures on double Bott-Samelson cells*, Forum Math. Sigma **9** (2021), Paper No. e66, 89, DOI 10.1017/fms.2021.59. MR4321011

[Thu14] Dylan Paul Thurston, *Positive basis for surface skein algebras*, Proc. Natl. Acad. Sci. USA **111** (2014), no. 27, 9725–9732, DOI 10.1073/pnas.1313070111. MR3263305

[Tra11] Thao Tran, *F-polynomials in quantum cluster algebras*, Algebr. Represent. Theory **14** (2011), no. 6, 1025–1061, DOI 10.1007/s10468-010-9226-6. MR2844755

[VV02] M. Varagnolo and E. Vasserot, *Standard modules of quantum affine algebras*, Duke Math. J. **111** (2002), no. 3, 509–533, DOI 10.1215/S0012-7094-02-11135-1. MR1885830

[VV03] M. Varagnolo and E. Vasserot, *Perverse sheaves and quantum Grothendieck rings*, Studies in memory of Issai Schur (Chevaleret/Rehovot, 2000), Progr. Math., vol. 210, Birkhäuser Boston, Boston, MA, 2003, pp. 345–365. MR1985732

[Yur20] Toshiya Yurikusa, *Density of g-vector cones from triangulated surfaces*, Int. Math. Res. Not. IMRN **21** (2020), 8081–8119, DOI 10.1093/imrn/rnaa008. MR4176846

Beijing Normal University, China
Email address: qin.fan.math@gmail.com

Contemporary Mathematics
Volume **845**, 2026
https://doi.org/10.1090/conm/845/16959

Hall algebras and edge contractions with loops

Adhish Rele

ABSTRACT. We extend the study of Hall algebras and edge contractions by generalizing Yiqiang Li's work to contraction along vertices with multiple edges. Using the edge contractions, we establish new embeddings among Hall algebras in this broader setting. Our results demonstrate that these embeddings preserve key algebraic structures, including Hopf algebra operations.

Introduction

In 1900, while serving as a Privatdozent at Technische Hochschule Berlin Charlottenburg, Steinitz presented a talk at the Deutsche Mathematiker-Vereinigung meeting in Aachen [**Ste01**], where he introduced an algebra over the integers based on isomorphism classes of finite abelian groups, now known as the Hall algebra. He also proposed several conjectures that were later proven by Phillip Hall in the 1950s [**Hal59**] in the context of finite abelian p-groups. It has since emerged as a fundamental structure in representation theory, particularly in the study of quantum groups and geometric representation theory. The connection between Hall algebras and quantum groups was established by Ringel [**Rin90**], who extended the concept to more general abelian categories, such as the category of representations of quivers. He demonstrated that the composition algebra of the category of representations of a quiver, specifically of type ADE, can be viewed as a realization of the positive part of the quantum group associated with the underlying graph. Lusztig further expanded this perspective in [**Lus91**], providing a geometric approach to Hall algebras via perverse sheaves and canonical bases. These structures play a significant role in understanding quantum groups, categorification, and cluster algebras, leading to applications including, but not limited to, symplectic geometry, algebraic geometry and mathematical physics.

Edge contraction is a fundamental operation in graph theory where an edge is removed, and its incident vertices are merged into a single vertex. This operation has been extensively studied in various mathematical contexts, including topology, combinatorics, and representation theory. In the setting of Hall algebras and quantum groups, edge contraction provides insights into how algebraic structures associated with graphs behave under simplifications of the graph. Recent studies, including Li's solo work [**Li23**] and his collaboration with Ren [**LR24**], have explored how Hall algebras and cohomological Hall algebras behave under

2020 *Mathematics Subject Classification.* Primary 17B37; Secondary 16G20.

edge contractions, demonstrating induced embeddings that inform the structure of quantum groups, among other algebraic structures.

This paper extends Li's work on embeddings among Hall algebras via edge contractions by relaxing the assumption that there is exactly one edge connecting the vertices involved in the contraction. We generalize this setting to cases where multiple edges exist between these vertices, leading to the formation of loops, a new phenomenon, after contraction. This generalization requires revisiting fundamental structures, including the definition of the Cartan datum, quiver representation spaces, and Hall algebras. A central question explored in this work is whether the embedding phenomenon established in Li's work persists under this broader framework. We prove the following key result:

THEOREM A (Theorem 2.3.1). *For a quiver with multiple edges between two vertices along which the contraction takes place, the edge contraction operation induces an embedding of the Hall algebra of the contracted quiver into the associated Hall algebra prior to the edge contraction.*

It turns out that the embedding is a result of a composition of two injective algebra homomorphisms, namely $j_!$ and $\mu^\star$, described in Section 2.3. The proof that these maps are algebra homomorphisms relies on the four lemmas in Section 2.2.

THEOREM B (Theorem 2.5.2). *Hall algebra of a contracted quiver is a split subquotient of the Hall algebra prior to the edge contraction.*

The proof of this theorem follows directly from [**Li23**], Proposition 2.4.1 on page 17.

Looking ahead, our work opens up several exciting directions for further research. One promising avenue is to investigate the subquotient structure predicted by Li within the broader setting of Hall algebras that include loops. We aim to determine whether this phenomenon is a universal feature that can be observed across various algebraic systems. Additionally, exploring the interplay between edge contraction and modern categorification frameworks—such as Khovanov-Lauda-Rouquier algebras [**Mak18**] and quiver Hecke algebras—could provide deeper insights into both the combinatorial and geometric aspects of representation theory. By clarifying these connections, we hope to pave the way for new applications in mathematical physics and a more unified understanding of the underlying algebraic structures.

This paper starts by defining the generalized Cartan datum, introducing key modifications necessary to accommodate quivers with multiple edges and loops. It also provides an overview of root data and Weyl groups associated with such quivers. Following this, we develop the representation spaces corresponding to quivers with loops, laying the groundwork for our study of Hall algebras in this setting. We then review Lusztig's induction and restriction diagrams, which play a crucial role in constructing and understanding embeddings among Hall algebras. Using these tools, we define Hall algebras associated with quivers containing loops and establish our main result: the embedding of Hall algebras under edge contractions in this generalized framework. Finally, we demonstrate the compatibility of this embedding with the PBW basis and analyze its interaction with split short exact sequences.

Acknowledgments. I was first introduced to Hall algebras by Yiqiang Li when he shared his paper on Quantum Groups and Edge Contractions with me during my time as his student. Shortly thereafter, he became my research advisor and encouraged me to generalize his work to graphs with multiple edges between pairs of vertices. This work, which forms a part of my thesis, is the result of his unwavering encouragement, patience, and insightful guidance, which have been instrumental in bringing this project to fruition.

1. Foundations

In this section, we define the generalized Cartan data, study quivers and Weyl groups under edge contractions.

1.1. Generalized Cartan data. Define the set of natural numbers as $\mathbb{N} = \{0, 1, 2, \dots\}$. The generalized Cartan datum is defined to be a quadruple $(I, \cdot, \phi_1, \phi_2)$, where I is a finite set, $\phi_1 : I \longrightarrow \mathbb{N}\backslash\{0\}$, $\phi_2 : I \longrightarrow \mathbb{N}$, along with a symmetric bilinear form "$\cdot$" that maps to values in $\mathbb{Z}$ and is defined on the free abelian group $\mathbb{Z}[I]$, meeting the following conditions:

1. For every $i \in I$, we have,

$$i \cdot i = 2(\phi_1(i) - \phi_1(i)\phi_2(i)) \in 2\mathbb{Z}; \tag{1.1.a}$$

2. For distinct elements $i, j \in I$,

$$\frac{i \cdot j}{\phi_1(i)} \in -\mathbb{N}. \tag{1.1.b}$$

For a directed graph, the maps $\phi_1(i)$ and $\phi_2(i)$ count the number of vertices in the $a-$orbit $[i]$ and the number of loops on i respectively. See Section 1.2 for further details. The generalized Cartan datum restricts to Lusztig's Cartan datum when ϕ_2 is identically equal to 0.

Example 1.1.1. Here is an example of a generalized Cartan datum for $I = \{a, b, c\}$

$$\begin{aligned} &\phi_1(a) = 2 \qquad &\phi_2(a) &= 2 \qquad &a \cdot b = -6 \\ &\phi_1(b) = 3 \qquad &\phi_2(b) &= 1 \qquad &a \cdot c = 0 \\ &\phi_1(c) = 1 \qquad &\phi_2(c) &= 3 \qquad &b \cdot c = -3 \end{aligned} \tag{1.1.c}$$

Figure 1 in Section 1.2 is an example of a graph whose Cartan datum would be the one in the above example.

Now, suppose there exists a particular pair $(i_+, i_-) \in I$ that satisfies the following:

$$\begin{aligned} &\phi_1(i_+) = \phi_1(i_-), \\ &\phi_2(i_+) = \phi_2(i_-) = 0, \\ &i_+ \cdot i_- \neq 0. \end{aligned} \tag{1.1.d}$$

Let us fix such a pair. Define

$$i_0 = i_+ + i_- \in \mathbb{Z}[I]$$

Next, construct the set $\hat{I} = I \cup \{i_0\} - \{i_+, i_-\}$. Note that the free abelian group $\mathbb{Z}[\hat{I}]$ forms a subgroup of $\mathbb{Z}[I]$ and the bilinear form $\cdot$ on $\mathbb{Z}[I]$ naturally restricts to a symmetric bilinear form on $\mathbb{Z}[\hat{I}]$, still denoted by $\cdot$.

With these definitions, we obtain the following lemma:

LEMMA 1.1.2. *The quadruple* $(\hat{I}, \cdot, \widehat{\phi_1}, \widehat{\phi_2})$ *is a generalized Cartan datum where* $\widehat{\phi_1} : \hat{I} \longrightarrow \mathbb{N}\backslash\{0\}$ *is defined as* $\widehat{\phi_1}(i) = \begin{cases} \phi_1(i) & \text{if } i \neq i_0 \\ \phi_1(i_+)\ (= \phi_1(i_-)) & \text{if } i = i_0 \end{cases}$

and $\widehat{\phi_2} : \hat{I} \longrightarrow \mathbb{N}$ is defined as $\widehat{\phi_2}(i) = \begin{cases} \phi_2(i) & \text{if } i \neq i_0 \\ \dfrac{-i_+ \cdot i_-}{\phi_1(i_+)} - 1 & \text{if } i = i_0 \end{cases}.$

PROOF. We only need to show the conditions hold when i_0 is involved. First to show that $i_0 \cdot i_0 = 2(\widehat{\phi_1}(i_0) - \widehat{\phi_1}(i_0)\widehat{\phi_2}(i_0))$. The left hand side of the equation is evaluated as follows:

$$\begin{aligned} i_0 \cdot i_0 &= (i_+ + i_-) \cdot (i_+ + i_-) \\ &= i_+ \cdot i_+ + 2i_+ \cdot i_- + i_- \cdot i_- \\ &= 2\phi_1(i_+) + 2i_+ \cdot i_- + 2\phi_1(i_-). \end{aligned}$$

And the right hand side is

$$\begin{aligned} 2[\widehat{\phi_1}(i_0) - \widehat{\phi_1}(i_0)\widehat{\phi_2}(i_0)] &= 2\left[\phi_1(i_+) - \phi_1(i_+)\left(\frac{-i_+ \cdot i_-}{\phi_1(i_+)} - 1\right)\right] \\ &= 2\phi_1(i_+) + 2i_+ \cdot i_- + 2\phi_1(i_+) \\ &= 2\phi_1(i_+) + 2i_+ \cdot i_- + 2\phi_1(i_-). \end{aligned}$$

Hence, $i_0 \cdot i_0 = 2(\widehat{\phi_1}(i_0) - \widehat{\phi_1}(i_0)\widehat{\phi_2}(i_0))$. Also, we have,

$$\begin{aligned} \frac{i_0 \cdot j}{\phi_1(i_0)} &= \frac{(i_+ + i_-) \cdot j}{\phi_1(i_0)} = \frac{i_+ \cdot j}{\phi_1(i_+)} + \frac{i_- \cdot j}{\phi_1(i_-)} \in -\mathbb{N} \quad \forall j \in \hat{I} - \{i_0\}, \\ \frac{j \cdot i_0}{\phi_1(j)} &= \frac{j \cdot (i_+ + i_-)}{\phi_1(j)} = \frac{j \cdot i_+}{\phi_1(j)} + \frac{j \cdot i_-}{\phi_1(j)} \in -\mathbb{N} \quad \forall j \in \hat{I} - \{i_0\}. \end{aligned}$$

This proves the lemma. □

The generalized Cartan datum $(\hat{I}, \cdot, \widehat{\phi_1}, \widehat{\phi_2})$ is called the edge contraction of $(I, \cdot, \phi_1, \phi_2)$ along the pair $\{i_+, i_-\}$.

1.2. Graph Structures. Consider a finite oriented graph that may include loops, represented by the quadruple $(\mathbf{I}, \Omega,\ ' : \Omega \to \mathbf{I},\ '' : \Omega \to \mathbf{I})$, where $\mathbf{I}$ and Ω are two finite sets with $\mathbf{I} \neq \emptyset$. We define an admissible automorphism a for this oriented graph as a pair of bijections, $(a : \mathbf{I} \to \mathbf{I}, a : \Omega \to \Omega)$, satisfying:

- $a(h') = a(h)', a(h'') = a(h)'',\ \forall h \in \Omega$,
- $\{a(h)', a(h)''\} \not\subseteq [\mathbf{i}]$ for all $h \in \Omega$ with $h' \neq h''$ and for any $\mathbf{i} \in \mathbf{I}$ where $[\mathbf{i}]$ is the a-orbit of $\mathbf{i}$.

Henceforth, $[\mathbf{i}]$ is replaced by $\mathbf{i}$ where the context is clear.

With the oriented graph $(\mathbf{I}, \Omega)$ and its admissible automorphism a, we can derive a generalized Cartan datum $(\mathbf{I}/a, \cdot, \phi_1, \phi_2)$ where $\mathbf{I}/a$ represents the set of a-orbits in $\mathbf{I}$, and where:

(1) $\phi_1 : \mathbf{I}/a \to \mathbb{N}\backslash\{0\}$ defined as $\phi_1(\mathbf{i}) = \#[\mathbf{i}]$.
(2) $\phi_2 : \mathbf{I}/a \to \mathbb{N}$ is defined as $\phi_2(\mathbf{i}) = \#\{h \in \Omega \mid h' = h'' = j$ for some fixed $j \in [\mathbf{i}]\}$.
(3) $[\mathbf{i}] \cdot [\mathbf{i}] = 2\#[\mathbf{i}](1 - \#\{h \in \Omega \mid h' = h'' = j$ for some fixed $j \in [\mathbf{i}]\}) \in 2\mathbb{Z}$ for all $[\mathbf{i}] \in \mathbf{I}/a$. This means, $[\mathbf{i}] \cdot [\mathbf{i}] = 2\phi_1(\mathbf{i})(1 - \phi_2(\mathbf{i}))$.
(4) $[\mathbf{i}] \cdot [\mathbf{j}] = -\#\{h \in \Omega \mid h', h'' \in [\mathbf{i}] \cup [\mathbf{j}], h' \neq h''\}$ for all $[\mathbf{i}] \neq [\mathbf{j}] \in \mathbf{I}/a$.

It is evident that $\dfrac{[\mathbf{i}]\cdot[\mathbf{j}]}{\phi_1(\mathbf{i})} \in -\mathbb{N} \ \ \forall[\mathbf{i}] \neq [\mathbf{j}] \in \mathbf{I}/a$.

Example of an oriented graph based on the Cartan datum in Example 1.1.1 with loops respecting the action of an admissible automorphism a is as follows:

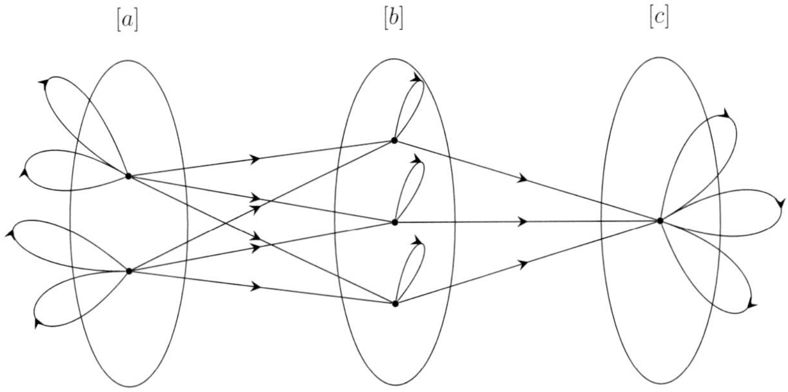

FIGURE 1. An example of a graph with 6 vertices and 3 a-orbits drawn as ellipses. Here the edges and loops are drawn in a way that respects the action of an admissible automorphism.

Assume that there exist two orbits $[\mathbf{i}_+], [\mathbf{i}_-] \in \mathbf{I}/a$ such that

(1.2.a)

1. $\#[\mathbf{i}_+] = \#[\mathbf{i}_-]$ i.e. $\phi_1(\mathbf{i}_+) = \phi_1(\mathbf{i}_-)$.
2. If $h \in \Omega$ such that $h', h'' \in [\mathbf{i}_+] \cup [\mathbf{i}_-], h' \neq h''$ then there does not exist $l \in \Omega$ such that $l', l'' \in [\mathbf{i}_+] \cup [\mathbf{i}_-], l' \in \{h', h''\}$ but $l'' \notin \{h', h''\}$.
3. $\phi_2(\mathbf{i}_+) = \phi_2(\mathbf{i}_-) = 0$.

Fix the pair $[\mathbf{i}_+], [\mathbf{i}_-] \in \mathbf{I}/a$. Choose an edge $e \in \Omega$ such that $e' \in [\mathbf{i}_+]$ and $e'' \in [\mathbf{i}_-]$. If there exists no such edge, rename $[\mathbf{i}_+]$ to $[\mathbf{i}_-]$ and vice versa. Define $h_k = a^k(e) \ \ \forall\, 1 \leq k \leq \phi_1([\mathbf{i}])$. Now, for any edge $h \in \Omega$, let $\overline{h}$ be a corresponding formal symbol that will represent the edge opposite to h, i.e., $h' = \overline{h}''$ and $h'' = \overline{h}'$. We construct a new oriented graph as follows:

(1.2.b)

$$\widehat{\mathbf{I}} = \mathbf{I} - [\mathbf{i}_-],$$

$$\widehat{\Omega} = \Omega \ \cup \ \{l_1 h_k \mid h_k'' = l_1'\}_{\{1\leq k\leq \phi_1([\mathbf{i}_+])\}} \ \cup \ \{\overline{h_k}\, l_2 \mid h_k'' = l_2''\}_{\{1\leq k\leq \phi_1([\mathbf{i}_+])\}} - \{h \mid h' \text{ or } h'' \in [\mathbf{i}_-]\}.$$

The new maps $' : \widehat{\Omega} \longrightarrow \widehat{I}$ and $'' : \widehat{\Omega} \longrightarrow \widehat{I}$ on $(\widehat{I}, \widehat{\Omega})$ are induced by the original maps $'$ and $''$ on the graph (I, Ω) and by defining

$$(l_1 h_k)' = h_k', \qquad (l_1 h_k)'' = l_1'' \qquad \text{whenever } h_k'' = l_1',$$

$$(\overline{h_k}\, l_2)' = l_2', \qquad (\overline{h_k}\, l_2)'' = h_k' \qquad \text{whenever } h_k'' = l_1''.$$

We further define an automorphism $\widehat{a}$ on the new graph $(\widehat{I}, \widehat{\Omega})$ by defining $\widehat{a} : \widehat{I} \longrightarrow \widehat{I}$ as the restriction of $a : I \longrightarrow I$ and $a : \widehat{\Omega} \longrightarrow \widehat{\Omega}$ as follows:

$$\widehat{a}(h) = a(h) \qquad \text{if } h \in \Omega \cap \widehat{\Omega},$$

$$\widehat{a}(l_1 h_k) = a(l_1)a(h_k) \qquad \text{if } l_1 h_k \in \widehat{\Omega},\ h_k'' = l_1',$$

$$\widehat{a}(\overline{h_k l_2}) = \overline{a(h_k)a(l_2)} \qquad \text{if } \overline{h_k l_2} \in \widehat{\Omega},\ h_k'' = l_2''.$$

LEMMA 1.2.1. *The automorphism $\widehat{a}$ is admissible.*

PROOF. It is clear that for any edge $h \in \Omega \cap \widehat{\Omega}$, $\widehat{a}(h') = \widehat{a}(h)'$, $\widehat{a}(h'') = \widehat{a}(h)''$ and $\{\widehat{a}(h)', \widehat{a}(h)''\} \nsubseteq [\mathbf{i}]$ for any $[\mathbf{i}] \in \widehat{\mathbf{I}} \backslash \widehat{a}$. Consider the edge $l_1 h_k \in \widehat{\Omega}$, $h_k'' = l_1'$. To show $\widehat{a}(l_1 h_k)' = \widehat{a}((l_1 h_k)')$. Notice that $(\widehat{a}(l_1 h_k))' = (a(l_1)a(h_k))' = a(h_k)' = a(h_k')$ and $\widehat{a}((l_1 h_k)') = \widehat{a}(h_k') = a(h_k')$. Similarly, one can show that $\widehat{a}(l_1 h_k)'' = \widehat{a}((l_1 h_k)'')$.

Finally we show that $\widehat{a}(l_1 h_k)'$ and $\widehat{a}(l_1 h_k)''$ are in different $\widehat{a}$–orbits. Suppose $l_1'' \neq h_k'$. By assumption (2) in (1.2.a), $l_1'' \notin [\mathbf{i}_+]$. From above, $\widehat{a}(l_1 h_k)' = a(h_k')$ and $\widehat{a}(l_1 h_k)'' = a(l_1'')$. Since h_k' and l_1'' are in different a–orbits, we can conclude that $\widehat{a}(l_1 h_k)'$ and $\widehat{a}(l_1 h_k)''$ are in different $\widehat{a}$–orbits. If $l_1'' = h_k'$ then there is nothing to prove since that will imply $(l_1 h_k)' = (l_1 h_k)''$, i.e. $l_1 h_k$ is a loop and by definition of admissible automorphism, loops under the action of a land up in the same orbit. Similarly, one can prove the above for the edge $\overline{h_k l_2}$. Hence, the automorphism $\widehat{a}$ is admissible. □

PROPOSITION 1.2.2. *Let $(I, \cdot, \phi_1, \phi_2)$ be a generalized Cartan datum. There exists a finite graph $(\mathbf{I}, H,', '')$ and an admissible automorphism a of this graph such that $(I, \cdot, \phi_1, \phi_2)$ is obtained from them by the construction in the second paragraph in Section 1.2.*

PROOF. For each $i \in I$, we consider the set D_i such that $|D_i| = \phi_1(i)$ and a cyclic permutation $a : D_i \to D_i$. Let $\mathbf{I} = \sqcup_{i \in I} D_i$ and let $a : \mathbf{I} \to \mathbf{I}$ be the permutation whose restriction to each D_i is the permutation $a : D_i \to D_i$ considered above. The proof in the case when there are no loops can be found in [**Lus10**] (Chap. 14). There are certain additions to be made in the case of graphs with loops and they are presented below.

Choose $i \in I$ such that $\phi_2(i) \neq 0$. Let $K = \{(\alpha, \alpha) \mid \alpha \in D_i\}$. Consider $a \times a : K \longrightarrow K$. Let ρ_1 be the a-orbit. This implies $|\rho_1| = |D_i| = \phi_1(i)$. Let $H_i = \sqcup_{\phi_2(i)} \rho_1$. Hence, $|H_i| = \phi_2(i)\phi_1(i)$. We have,

$$\begin{aligned}
\mathbf{I}/a &= \{D_i \mid i \in I\}, \\
D_i \cdot D_i &= 2(|D_i| - |H_i|) \\
&= 2(\phi_1(i) - \phi_1(i)\phi_2(i)) \\
&= i \cdot i, \\
D_i \cdot D_j &= -|\{h \in H \mid h', h'' \in \{D_i, D_j\}\}| \\
&= -|H_{ij}| \\
&= -\frac{-i \cdot j}{\operatorname{lcm}(|D_i|, |D_j|)} \operatorname{lcm}(|D_i|, |D_j|) \\
&= i \cdot j \quad (i \neq j).
\end{aligned}$$

where $H_{ij} = \bigsqcup_{\frac{-i\cdot j}{\mathrm{lcm}(\phi_1(i),\phi_1(j))}} \rho$ where ρ is an orbit of the permutation $a : D_i \times D_j \longrightarrow D_i \times D_j$ and $H = \sqcup H_{ij}$ where the union is over all unordered distinct pairs $i, j \in I$.

Let $\widehat{\phi_1}$ and $\widehat{\phi_2}$ be the maps obtained from the constructed graph. To show that $\widehat{\phi_1} = \phi_1$ and $\widehat{\phi_2} = \phi_2$. We can see that,

$$\widehat{\phi_1}(i) = \#[i] = |D_i| = \phi_1(i)$$

$$\widehat{\phi_2}(i) = \frac{|H_i|}{\widehat{\phi_1}(i)} = \frac{\phi_1(i)\phi_2(i)}{\phi_1(i)} = \phi_2(i).$$

This proves the proposition. □

DEFINITION 1.2.3. A Cartan datum $(I, \cdot, \phi_1, \phi_2)$ is said to be isomorphic to another Cartan datum $(J, \circ, \widehat{\phi_1}, \widehat{\phi_2})$ if there exists a bijection from I to J preserving the bilinear forms, $\phi_1 = \widehat{\phi_1}$ and $\phi_2 = \widehat{\phi_2}$.

LEMMA 1.2.4. *Consider a graph* $(\mathbf{I}, \Omega, \ ' : \Omega \to \mathbf{I}, \ '' : \Omega \to \mathbf{I})$ *with the corresponding generalized Cartan datum* $(\mathbf{I}/a, \cdot, \phi_1, \phi_2)$. *Then the edge contraction of the above Cartan datum* $(\tilde{\mathbf{I}}/\tilde{a}, \cdot, \widetilde{\phi_1}, \widetilde{\phi_2})$ *is isomorphic to the Cartan datum* $(\hat{\mathbf{I}}/\hat{a}, \circ, \widehat{\phi_1}, \widehat{\phi_2})$ *of the graph obtained from an edge contraction of the above graph.*

PROOF. First we need to check that $[i_+] \circ [j] = [i_0] \cdot [j]$ for every $[j] \neq [i_+], [j] \neq [i_0]$. Now by definition,

$$\begin{aligned}\circ[j] &= -\#\{h \in \widehat{\Omega} \mid h', h'' \in [i_+] \cup [j], h' \neq h''\} \\ &= -\#\{h \in \Omega \mid h', h'' \in [i_+] \cup [j], h' \neq h''\} - \\ &\quad \#\{h \in \Omega \mid h', h'' \in [i_-] \cup [j], h' \neq h''\} \\ &= [i_+] \cdot [j] + [i_-] \cdot [j] \\ &= [i_0] \cdot [j]\end{aligned}$$

Next, $\widehat{\phi_1}([i_+]) = \#[i_+] = \phi_1([i_+]) = \widetilde{\phi_1}([i_0])$. Also,

$$\widehat{\phi_2}([i_+]) = \#\{h \in \widehat{\Omega} \mid h' = h'' = j \text{ for some fixed } j \in [i_+]\} = \frac{-[i_+]\cdot[i_-]}{\phi_1(i_+)} - 1 = \widetilde{\phi_2}(i_0)$$

The lemma is proved. □

1.3. Root datum. We define root datum associated with a generalized Cartan datum.

DEFINITION 1.3.1. Root datum associated with $(I, \cdot, \phi_1, \phi_2)$ consists of:

1) Finitely generated free abelian groups Y and X
2) A perfect bilinear pairing $\langle \cdot, \cdot \rangle : Y \times X \longrightarrow \mathbb{Z}$
3) An embedding $I \hookrightarrow X$ $(i \mapsto i')$ and an embedding $I \hookrightarrow Y$ $(i \mapsto i)$
4) $\langle i, j' \rangle = \frac{i \cdot j}{\phi_1(i)}$ $\forall i, j \in I$

The elements of Y are called roots and the elements of X are called co-roots. Let $(\widehat{I}, \cdot, \widehat{\phi_1}, \widehat{\phi_2})$ be the edge contraction of $(I, \cdot, \phi_1, \phi_2)$ along the pair (i_+, i_-). Embedding $I \hookrightarrow X$ defines a group homomorphism $\mathbb{Z}[I] \longrightarrow X$. Define an embedding $\widehat{I} \hookrightarrow X$ to be the restriction of the homomorphism $\mathbb{Z}[I] \longrightarrow X$ to $\widehat{I}$, i.e. $i \mapsto i'$ if $i \in \widehat{I} \backslash i_0$ and $i_0 \mapsto (i_+)' + (i_-)'$. Embedding $I \hookrightarrow Y$ defines a group homomorphism

$\mathbb{Z}[I] \longrightarrow Y$. Define an embedding $\widehat{I} \hookrightarrow Y$ to be the restriction of the homomorphism $\mathbb{Z}[I] \longrightarrow Y$ to $\widehat{I}$, i.e. $i \mapsto i$ if $i \in \widehat{I}\backslash i_0$ and $i_0 \mapsto i_+ + i_-$. It is easy to check that the collection $(Y, X, \langle\cdot,\cdot\rangle, \widehat{I} \hookrightarrow Y, \widehat{I} \hookrightarrow X)$ is a root datum associated to the Cartan datum $(\widehat{I}, \cdot)$.

A root datum as above is called $X-$regular (resp. $Y-$regular) if the image of the embedding $I \subset X$ (resp. $I \subset Y$) is linearly independent in X (resp. Y). A $Y-$regular root datum is called simply connected root datum and a $X-$regular root datum is called adjoint root datum.

1.4. Weyl Groups. Let $(Y, X, \langle\cdot,\cdot\rangle, I \hookrightarrow Y, I \hookrightarrow X)$ be a simply connected root datum associated to $(I, \cdot)$. We define a reflection $s_i : Y \to Y$ by $s_i(y) = y - \langle y, i'\rangle i$ for each $i \in I$. The Weyl group is a group generated by these reflections and is denoted by W_I. We call W_I to be the Weyl group associated to the Cartan datum $(I, \cdot, \phi_1, \phi_2)$. Let $(\widehat{I}, \cdot, \widehat{\phi_1}, \widehat{\phi_2})$ be the edge contraction of $(I, \cdot, \phi_1, \phi_2)$ along the pair $\{i_+, i_-\}$. Let $W_{\widehat{I}}$ be the Weyl group associated to $(\widehat{I}, \cdot, \widehat{\phi_1}, \widehat{\phi_2})$.

LEMMA 1.4.1. *The function* $\psi : W_{\widehat{I}} \longrightarrow \mathrm{Aut}(Y)$ *defined as* $s_i \mapsto s_i$ *for all* $i \in \widehat{I} - i_0$ *and* $s_{i_0} \mapsto s_{i_+} s_{i_- + \widehat{\phi_2}(i_0) i_+} s_{i_+}$ *defines a group embedding where* $s_{i_- + \widehat{\phi_2}(i_0) i_+}(y) = y - \langle y, i'_- + \widehat{\phi_2}(i_0) i'_+\rangle (i_- + \widehat{\phi_2}(i_0) i_+)$.

PROOF. We only need to show that $s_{i_0} = s_{i_+} s_{i_- + \widehat{\phi_2}(i_0) i_+} s_{i_+}$. We divide the proof in 5 steps:

STEP 1.

$$\begin{aligned} s_{i_-}(i_+) &= i_+ - \langle i_+, i'_-\rangle i_- \\ &= i_+ - \frac{i_+ \cdot i_-}{\phi_1(i_+)} i_- \\ &= i_+ + (\widehat{\phi_2}(i_0) + 1) i_- \\ &= i_0 + \widehat{\phi_2}(i_0) i_-. \end{aligned}$$

STEP 2. Similarly, we can show, $s_{i_+}(i_-) = i_0 + \widehat{\phi_2}(i_0) i_+$.

STEP 3.

$$\begin{aligned} s_{i_+}(i_0) &= i_0 - \langle i_0, i'_+\rangle i_+ \\ &= i_0 - \frac{i_0 \cdot i_+}{\phi_1(i_0)} i_+ \\ &= i_0 - (1 - \widehat{\phi_2}(i_0)) i_+ \\ &= (i_0 - i_+) + \widehat{\phi_2}(i_0) i_+. \end{aligned}$$

The third equality holds because

$$\frac{i_0 \cdot i_+}{\phi_1(i_0)} = \frac{(i_+ + i_-) \cdot i_+}{\phi_1(i_0)} = \frac{i_+ + i_+}{\phi_1(i_0)} + \frac{i_+ \cdot i_-}{\phi_1(i_+)} = 2 - \widehat{\phi_2}(i_0) - 1 = 1 - \widehat{\phi_2}(i_0).$$

STEP 4.

$$\begin{aligned} s_{i_- + \widehat{\phi_2}(i_0) i_+}(i_+) &= i_+ - \langle i_+, i'_- + \widehat{\phi_2}(i_0) i'_+\rangle (i_- + \widehat{\phi_2}(i_0) i_+) \\ &= i_+ - (\langle i_+, i'_-\rangle + \widehat{\phi_2}(i_0)\langle i_+, i'_+\rangle)(i_- + \widehat{\phi_2}(i_0) i_+) \\ &= i_+ + (1 - \widehat{\phi_2}(i_0))(i_- + \widehat{\phi_2}(i_0) i_+) \\ &= i_0 + \widehat{\phi_2}(i_0)(i_+ - i_-) - \widehat{\phi_2}^{\,2}(i_0) i_+. \end{aligned}$$

STEP 5. Applying s_{i_+} to Step 4, we get,

$$s_{i_+}s_{i_-+\widehat{\phi_2}(i_0)i_+}(i_+) = (i_0-i_+)+\widehat{\phi_2}(i_0)i_+ + \widehat{\phi_2}(i_0)(-i_+-i_0-\widehat{\phi_2}(i_0)i_+ + \widehat{\phi_2}^2(i_0)i_+$$
$$= i_0 - i_+ - \widehat{\phi_2}(i_0)i_0.$$

The first equality is because of Step 3. Combining all the above steps we get,

$$\begin{aligned}
&s_{i_+}s_{i_-+\widehat{\phi_2}(i_0)i_+}s_{i_+}(y)\\
&= s_{i_+}s_{i_-+\widehat{\phi_2}(i_0)i_+}(y-\langle y, i'_+\rangle i_+)\\
&= s_{i_+}[y-\langle y, i'_- + \widehat{\phi_2}(i_0)i'_+\rangle(i_- + \widehat{\phi_2}(i_0)i_+) - \langle y, i'_+\rangle s_{i_-+\widehat{\phi_2}(i_0)i_+}(i_+)]\\
&= y - \langle y, i'_+\rangle i_+ - \langle y, i'_- + \widehat{\phi_2}(i_0)i'_+\rangle i_0 - \langle y, i'_+\rangle(i_0 - i_+ - \widehat{\phi_2}(i_0)i_0)\\
&= y - \langle y, i'_- + i'_+\rangle i_0\\
&= y + \langle y, i_0\rangle i_0\\
&= s_{i_0}(y).
\end{aligned}$$

Third equality is from using Step 5. This proves the lemma. □

EXAMPLE 1.4.2. Note that the above map is not an embedding into the Weyl group W_I as demonstrated by the following example.

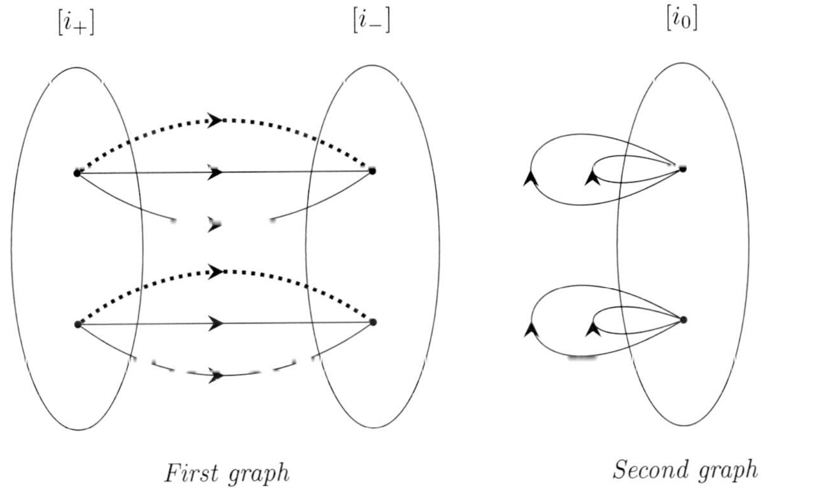

FIGURE 2. The first graph is before the edge contraction and the second graph is after the edge contraction. The dotted edges denote the edges along which the contraction takes place. The solid edges in the first graph form loops after the contraction.

Consider the graph with $I = \{i_+, i_-\}$ as shown in Figure 2 with the first ellipse denoting $[i_+]$ and the second $[i_-]$. After applying an admissible automorphism a we get a Cartan datum $(I/a, \cdot)$ abbv. as $(I, \cdot)$. Let $Y = \mathbb{Z}[I]$ and $X = \mathrm{Hom}(Y, \mathbb{Z})$ and $\langle \cdot, \cdot \rangle : Y \times X \longrightarrow \mathbb{Z}$ be defined as $\langle y, x'\rangle = x(y)$, $I \hookrightarrow Y$, $I \hookrightarrow X$ as $i(j') = \frac{i\cdot j}{\phi_1(i)} \in \mathbb{Z}$.

To show that $s_{i_-+2i_+} \notin W_I$. Now by definition, W_I is a group generated by s_{i_+}, s_{i_-}. This is because the group generated by reflection corresponding to roots is same as the group generated by reflections corresponding to simple roots. Therefore, we show that $s_{i_-+2i_+}$ cannot be written as any composition of the reflections s_{i_+} and s_{i_-}. But any composition of reflections reduces to alternating compositions of the reflections $\{s_{i_+}, s_{i_-}\}$, since $s^2 = 1$. First we see that,

$$\begin{aligned} s_{i_-+2i_+}(i_+) &= i_+ - \frac{i_+(i_- + 2i_+)}{\phi_1(i_+)}(i_- + 2i_+) \\ &= i_+ - \frac{-6+8}{2}(i_- + 2i_+) \\ &= i_+ - (i_- + 2i_+) \\ &= -i_+ - i_-. \end{aligned}$$

Case (1): Start with the reflection s_{i_-}.

$$s_{i_-}(i_+) = i_+ + 3i_- \neq -i_+ - i_-$$

As mentioned before, it only makes sense to now apply s_{i_+} (since $s_{i_-}s_{i_-} = 1$).

$$s_{i_+}s_{i_-}(i_+) = 8i_+ + 3i_- \neq -i_+ - i_-.$$

We will use induction to show that if $s_{i_-}(i_+) = ai_+ + bi_-$ with $b > a > 0$, then no matter the number of reflections applied, we never end up with $-i_+ - i_-$. So assuming, $s_{i_-}(i_+) = ai_+ + bi_-$ we get,

$$s_{i_+}[s_{i_-}(i_+)] = (3b - a)i_+ + bi_-$$

and it is clear that coefficients of both i_+ and i_- are positive and hence it cannot equal $-i_+ - i_-$. Next, to this we apply the reflection s_{i_-} to get,

$$s_{i_-}[s_{i_+}s_{i_-}(i_+)] = (3b - a)i_+ + (8b - 3a)i_-$$

Here, if we let $\alpha = 3b - a$ and $\beta = 8b - 3a$, we get back to the first step where $\beta > \alpha > 0$ and hence cannot equal $-i_+ - i_-$. We basically start all over again where we apply s_{i_+} to $\alpha i_+ + \beta i_-$ with $\beta > \alpha > 0$ and repeat the above steps to get the required result.

Case (2): Start with reflection s_{i_+}. Notice that,

$$\begin{aligned} s_{i_+}(i_+) &= -i_+ \neq -i_+ - i_-, \\ s_{i_-}s_{i_+}(i_+) &= -(i_+ + 3i_-) \neq -i_+ - i_-, \\ s_{i_+}s_{i_-}s_{i_+}(i_+) &= -(8i_+ + 3i_-) \neq -i_+ - i_-. \end{aligned}$$

We get the same coefficients with the sign reversed. We cannot use the same technique as above the relied on getting positive coefficients. Here we try to look at the coefficients of i_+ instead. We showed that $s_{i_-+2i_+}(i_+) = -i_+ - i_-$. Therefore, the coefficient of $i_+ = -1$. From above we have,

$$s_{i_-}s_{i_+}(i_+) = -(i_+ + 3i_-).$$

This is an expression of the form $-(ai_+ + bi_-)$ with $b > a > 0$ and to this we will apply s_{i_+} to get,

$$s_{i_+}(-(ai_+ + bi_-)) = -[(3b - a)i_+ + bi_-]$$

Here since $b > a > 0$, this implies $3b > 2a \implies 3b - a > a$. In the previous iterations we had, $a = 1$ and $3b - a$ must be greater than 1. Next we apply s_{i_-} to the above expression to get,

$$s_{i_-}s_{i_+}(-(ai_+ + bi_-)) = -[(3b - a)i_+ + (8b - 3a)i_-]$$

Here the coefficient of i_+ is the same as the previous iteration and hence must be greater than 1. If we let $\alpha = 3b - a$ and $\beta = 8b - 3a$, we end up where we started with $\beta > \alpha > 0$ and the next step is the application of the reflection s_{i_+}.

In summary, the coefficient of i_+ increases or remains the same at each application of the reflections and hence it cannot be equal to -1 once it's absolute value greater than 1. Hence, $s_{i_- + 2i_+} \notin W_I$. □

2. Embedding among Hall Algebras associated to Quivers with Loops

This section starts with an introduction to representation spaces of quivers with loops. Using Lusztig's induction and restriction diagram, we give a formulation of Hall algebra associated to quivers with loops and prove the embedding among Hall algebras induced by an edge contraction. We end the section by constructing a split short exact sequence of Hall algebras.

2.1. Quiver Representation Spaces. Let p be a prime number and $q = p^e$ for some $e \in \mathbb{N}\backslash\{0\}$.

Consider a finite oriented graph $(\mathbf{I}, \Omega, ', '')$ such that $|\mathbf{I}| = n$ and let a be the admissible automorphism as in section (1.2). Let $\mathcal{V}_a$ be a set of $\mathbf{I}$-graded K-Vector spaces $\mathbf{V} = \bigoplus_{\mathbf{i}\in\mathbf{I}} \mathbf{V}_\mathbf{i}$ where K is the algebraic closure of a finite field $\mathbb{F}_p$ and $a : \mathbf{V} \longrightarrow \mathbf{V}$ is a linear map such that $a(\mathbf{V})_\mathbf{i} = \mathbf{V}_{a(\mathbf{i})}$ and $a^j|_{\mathbf{V}_\mathbf{i}} = 1|_{\mathbf{V}_\mathbf{i}}$ $\forall$ $\mathbf{i} \in \mathbf{I}$ and $\forall$ $j \geq 1$ such that $a^j(\mathbf{i}) = \mathbf{i}$. Now to this oriented graph we define a representation space

$$\mathbf{E}_{\mathbf{V},\Omega} = \oplus_{h\in\Omega} \mathrm{Hom}(V_{h'}, V_{h''}).$$

The algebraic group $\mathbf{G}_\mathbf{V} = \prod_{\mathbf{i}\in\mathbf{I}} \mathrm{GL}(\mathbf{V}_\mathbf{i})$ acts on the $\mathbf{E}_{\mathbf{V},\Omega}$ by conjugation, i.e., for any $g \in \mathbf{G}_\mathbf{V}$ and $x = (x_h)_{h\in\Omega} \in \mathbf{E}_{\mathbf{V},\Omega}$ $(g.x)_h = g_{h''} x_h g_{h'}^{-1}$.

Let $\mathrm{Fr} : K \longrightarrow K$ be the Frobenius map defined by raising the elements to a power of q. This induces natural Frobenius maps on $\mathbf{E}_{\mathbf{V},\Omega}$ and $\mathbf{G}_\mathbf{V}$, denoted by the same notation, by raising the entries of the matrix representation of the above spaces to the power of q. This is the same Frobenius map as defined in [**Lus98**].

Let $I = \mathbf{I}/a$ be the a-orbits of $\mathbf{I}$ from section (1.2). Let $\nu = \sum_{i\in I} \nu_i \in \mathbb{N}[I]$ Then $\mathcal{V}_{I,\nu} = \{\mathbf{V} \in \mathcal{V}_a \mid \dim \mathbf{V}_\mathbf{i} = \nu_i \ \forall \mathbf{i} \in i, i \in I = \mathbf{I}/a\}$.

The admissible automorphism a induces an automorphism on $\mathbf{E}_{\mathbf{V},\Omega}$ and $\mathbf{G}_\mathbf{V}$ by permuting the edges and the grading on $\mathbf{V}$ respectively, i.e. $a : \mathbf{E}_{\mathbf{V},\Omega} \longrightarrow \mathbf{E}_{\mathbf{V},\Omega}$ is defined as $a(x_h)_{h\in\Omega} = (x_{a(h)})_{h\in\Omega}$ and $a : \mathbf{G}_\mathbf{V} \longrightarrow \mathbf{G}_\mathbf{V}$ given by:

$$a(g_1, g_2, \ldots, g_n)(v_1, v_2, \ldots, v_n) = (a(g_1)a(v_1), a(g_2)a(v_2), \ldots, a(g_n)a(v_n))$$

where $a(g_i) = g_{a(i)}$ for $1 \leq i \leq n$.

The maps Fr and a are commuting automorphisms on $\mathbf{E}_{\mathbf{V},\Omega}$ and $\mathbf{G}_\mathbf{V}$. Let $\mathbf{F} = \mathrm{Fr} \circ a$ and call it the twisted Frobenius as seen in [**Li23**]. Let $\mathbf{E}^\mathbf{F}_{\mathbf{V},\Omega}$ and $\mathbf{G}^\mathbf{F}_\mathbf{V}$ be the fixed points under the action of $\mathbf{F}$. Hence $\mathbf{G}^\mathbf{F}_\mathbf{V}$ acts on $\mathbf{E}^\mathbf{F}_{\mathbf{V},\Omega}$ induced by the action of $\mathbf{G}_\mathbf{V}$ on $\mathbf{E}_{\mathbf{V},\Omega}$.

We restrict the further discussion to a graph satisfying the conditions in (1.2.a). Let $\widehat{I} = I - \{i_+, i_-\} \cup \{i_0\}$. Fix $e \in \Omega$ such that $e' \in i_+$ and $e'' \in i_-$. Let $\mathbf{E}^\heartsuit_{\mathbf{V},\Omega} = \{x \in \mathbf{E}_{\mathbf{V},\Omega} \mid x_h \text{ is an isomorphism } \forall h = a^k(e) \text{ for some } 1 \leq k \leq \phi_1(i_+)\}$ where $\mathbf{V} \in \mathcal{V}_{I,\nu}$ such that $\dim V_{i_+} = \dim V_{i_-}$.

Clearly, the space $\mathbf{E}^{\heartsuit}_{\mathbf{V},\boldsymbol{\Omega}}$ is a $\mathbf{G_V}$-invariant subvariety of $\mathbf{E}_{\mathbf{V},\boldsymbol{\Omega}}$. Moreover, we can naturally define a free action of the subgroup $\mathbf{G_V}^{[\mathbf{i}_-]} = \prod_{\mathbf{i}\in[\mathbf{i}_-]} \mathrm{GL}(\mathbf{V}_i)$ of $\mathbf{G_V}$ on $\mathbf{E}^{\heartsuit}_{\mathbf{V},\boldsymbol{\Omega}}$.

Consider the quotient map $\mathbf{q_V} : \mathbf{E}^{\heartsuit}_{\mathbf{V},\boldsymbol{\Omega}} \longleftrightarrow \mathbf{G_V}^{[\mathbf{i}_-]}\backslash\mathbf{E}^{\heartsuit}_{\mathbf{V},\boldsymbol{\Omega}}$. Let $\widehat{\mathbf{V}}=\oplus_{\mathbf{i}\in\mathbf{I}-[\mathbf{i}_-]}\mathbf{V}_\mathbf{i}$. Define a morphism of varieties: $\mu_\nu : \mathbf{E}^{\heartsuit}_{\mathbf{V},\boldsymbol{\Omega}} \longrightarrow \mathbf{E}_{\widehat{\mathbf{V}},\widehat{\boldsymbol{\Omega}}}$, $x \mapsto \widehat{x}$ by

$$
\begin{aligned}
\widehat{x}_h &= x_h && \text{if } h \in \widehat{\Omega}\cap\Omega,\\
\widehat{x}_{l_1 h_k} &= x_{l_1}x_{h_k} && \text{if } h_k'' = l_1',\\
\widehat{x}_{\overline{h_k} l_2} &= x_{h_k}^{-1}x_{l_2} && \text{if } h_k'' = l_2''.
\end{aligned}
\tag{2.1.a}
$$

There is also a natural projection $\widehat{\ }: \mathbf{G_V} \longrightarrow \mathbf{G}_{\widehat{\mathbf{V}}}$ by forgetting the $[\mathbf{i}_-]$ components. It is straightforward to check the compatibility relation: $\mu_\nu(g.x) = \widehat{g}.\mu_\nu(x)\ \ \forall g \in \mathbf{G_V},\ x \in \mathbf{E}^{\heartsuit}_{\mathbf{V},\boldsymbol{\Omega}}$. This relation implies that the map μ_ν factors through the quotient map $\mathbf{q_V}$:

$$
\begin{array}{ccc}
\mathbf{E}^{\heartsuit}_{\mathbf{V},\boldsymbol{\Omega}} & \xrightarrow{\ \mathbf{q_V}\ } & \mathbf{G_V}^{[\mathbf{i}_-]}\backslash\mathbf{E}^{\heartsuit}_{\mathbf{V},\boldsymbol{\Omega}} \\
& \searrow^{\mu_\nu} & \downarrow{\scriptstyle \widetilde{\mu}_\nu} \\
& & \mathbf{E}_{\widehat{\mathbf{V}},\widehat{\boldsymbol{\Omega}}}
\end{array}
\tag{2.1.b}
$$

Following [**Li23**] (Lemma 2.1.1), we can prove that $\widetilde{\mu}_\nu$ is an isomorphism and hence μ_ν gets identified with the quotient map $\mathbf{q_V}$.

If $\mathbf{V} \in \mathcal{V}_{I,\nu}$, then the twisted Frobenius map $\mathbf{F}$ restricts to a twisted Frobenius $\mathbf{F}$ on $\mathbf{E}^{\heartsuit}_{\mathbf{V},\boldsymbol{\Omega}}$. Similarly, we can define a twisted Frobenius $\widehat{\mathbf{F}}$ on $\mathbf{E}_{\widehat{\mathbf{V}},\widehat{\boldsymbol{\Omega}}}$ and $\mathbf{G}_{\widehat{\mathbf{V}}}$. The twisted Frobenius is further compatible with μ_ν, and hence μ_ν induces a map, again call it μ_ν, from $\mathbf{E}^{\heartsuit,\mathbf{F}}_{\mathbf{V},\boldsymbol{\Omega}}$ to $\mathbf{E}^{\widehat{\mathbf{F}}}_{\widehat{\mathbf{V}},\widehat{\boldsymbol{\Omega}}}$. Due to the compatibility of the projection map $\mathbf{G}^{\mathbf{F}}_{\mathbf{V}} \longrightarrow \mathbf{G}^{\widehat{\mathbf{F}}}_{\widehat{\mathbf{V}}}$, we have the following commutative diagram:

$$
\begin{array}{ccc}
\mathbf{E}^{\heartsuit,\mathbf{F}}_{\mathbf{V},\boldsymbol{\Omega}} & \xrightarrow{\ \mathbf{q_V}\ } & \mathbf{G_V}^{[\mathbf{i}_-],\mathbf{F}}\backslash\mathbf{E}^{\heartsuit,\mathbf{F}}_{\mathbf{V},\boldsymbol{\Omega}} \\
& \searrow^{\mu_\nu} & \downarrow{\scriptstyle \cong\ \widetilde{\mu}_\nu} \\
& & \mathbf{E}^{\widehat{\mathbf{F}}}_{\widehat{\mathbf{V}},\widehat{\boldsymbol{\Omega}}}
\end{array}
\tag{2.1.c}
$$

Like [**Li23**], we will also call μ_ν, a contraction map of $\mathbf{E}^{\mathbf{F}}_{\mathbf{V},\boldsymbol{\Omega}}$ along $\{[\mathbf{i}_+],[\mathbf{i}_-]\}$.

Having defined μ_ν, we can also define the inclusion map, denoted by j_ν, from $\mathbf{E}^{\heartsuit}_{\mathbf{V},\boldsymbol{\Omega}}$ to $\mathbf{E}_{\mathbf{V},\boldsymbol{\Omega}}$. As it is compatible with the twisted Frobenius, we get a map from $\mathbf{E}^{\heartsuit,\mathbf{F}}_{\mathbf{V},\boldsymbol{\Omega}}$ to $\mathbf{E}^{\mathbf{F}}_{\mathbf{V},\boldsymbol{\Omega}}$ again denoted by the same notation.

2.2. Lusztig's Induction and Restriction Diagram. Let $\mathbf{V}$ be an $\mathbf{I}$ graded vector space over K. We can decompose $\mathbf{V}$ as $\mathbf{V} = \mathbf{T}\oplus\mathbf{W}$. Let $\mathbf{S_W} = \{x \in \mathbf{E}_{\mathbf{V},\boldsymbol{\Omega}} \mid x_h(\mathbf{W}_{h'}) \subseteq \mathbf{W}_{h''}\}$, i.e., W is x-invariant. Hence we can define $x^{\mathbf{W}} \in \mathbf{E}_{\mathbf{W},\boldsymbol{\Omega}}$ as the restriction of $x \in \mathbf{S_W}$ to $\mathbf{W}$. We can further define $x^{\mathbf{T}} \in \mathbf{E}_{\mathbf{W},\boldsymbol{\Omega}}$ by passage through the quotient $\mathbf{V}/\mathbf{W}$. Following [**Lus10**], we have the following restriction diagram,

$$\mathbf{E}_{\mathbf{V},\Omega} \xleftarrow{\iota} \mathbf{S}_{\mathbf{W}} \xrightarrow{\kappa} \mathbf{E}_{\mathbf{T},\Omega} \times \mathbf{E}_{\mathbf{W},\Omega} \tag{2.2.a}$$

where ι is the inclusion map and κ is defined by $x \mapsto (x^{\mathbf{T}}, x^{\mathbf{W}})$.

Now to define the induction diagram, we need a stabilizer of $\mathbf{W}$ in $\mathbf{G}_{\mathbf{V}}$. Call it Q. Q is a parabolic subgroup of $\mathbf{G}_{\mathbf{V}}$. We denote by U the unipotent radical of Q. Now any element of Q can be written as a block matrix:

$$\left[\begin{array}{c|c} \text{Morphism from } T \text{ to } T & 0 \\ \hline \text{Morphism from } T \text{ to } W & \text{Morphism from } W \text{ to } W \end{array}\right]$$

This induces a surjective map from Q to $\mathbf{G}_{\mathbf{T}} \times \mathbf{G}_{\mathbf{W}}$ with kernel U. Hence, $Q/U \cong \mathbf{G}_{\mathbf{T}} \times \mathbf{G}_{\mathbf{W}}$.

Now, we can define the action of Q on the set $\mathbf{G}_{\mathbf{V}} \times \mathbf{S}_{\mathbf{W}}$ by $b.(g,x) = (gb^{-1}, b.x)$ where the action on the second component is the restriction of the action of $\mathbf{G}_{\mathbf{V}}$ on $\mathbf{E}_{\mathbf{V},\Omega}$. U also acts on $\mathbf{G}_{\mathbf{V}} \times \mathbf{S}_{\mathbf{W}}$ via restriction. Let $\mathbf{E}'$ and $\mathbf{E}''$ be the orbits of the action of U and Q on $\mathbf{G}_{\mathbf{V}} \times \mathbf{S}_{\mathbf{W}}$ respectively. Let p_1 be the map from $\mathbf{E}'$ to $\mathbf{E}_{\mathbf{T},\Omega} \times \mathbf{E}_{\mathbf{W},\Omega}$ defined as $[g,x]_U \mapsto (x^{\mathbf{T}}, x^{\mathbf{W}})$. p_2 is the map that takes the U-orbit to the corresponding Q-orbit. Note that p_2 is a $Q/U(\cong \mathbf{G}_{\mathbf{T}} \times \mathbf{G}_{\mathbf{W}})$ principal bundle. Further, p_3 maps an element of the Q-orbit, $[g,x]_Q$, to $g.x$. Combining these maps gives rise to the induction diagram shown below:

$$\mathbf{E}_{\mathbf{T},\Omega} \times \mathbf{E}_{\mathbf{W},\Omega} \xleftarrow{p_1} \mathbf{E}' \xrightarrow{p_2} \mathbf{E}'' \xrightarrow{p_3} \mathbf{E}_{\mathbf{V},\Omega} \tag{2.2.b}$$

Note the the restriction and induction diagrams are compatible with the twisted Frobenius $\mathbf{F}$ provided $\mathbf{V}, \mathbf{T}, \mathbf{W}$ are in $\mathcal{V}_a$. So we have the following twisted induction and restriction diagrams with the maps being restriction of the corresponding maps in the restriction and induction diagram:

$$\mathbf{E}^{\mathbf{F}}_{\mathbf{V},\Omega} \xleftarrow{\iota} \mathbf{S}^{\mathbf{F}}_{\mathbf{W}} \xrightarrow{\kappa} \mathbf{E}^{\mathbf{F}}_{\mathbf{T},\Omega} \times \mathbf{E}^{\mathbf{F}}_{\mathbf{W},\Omega} \tag{2.2.c}$$

$$\mathbf{E}^{\mathbf{F}}_{\mathbf{T},\Omega} \times \mathbf{E}^{\mathbf{F}}_{\mathbf{W},\Omega} \xleftarrow{p_1} \mathbf{E}'^{\mathbf{F}} \xrightarrow{p_2} \mathbf{E}''^{\mathbf{F}} \xrightarrow{p_3} \mathbf{E}^{\mathbf{F}}_{\mathbf{V},\Omega} \tag{2.2.d}$$

Suppose $\nu = \tau + \omega$ where $\nu = \dim \mathbf{V}$, $\tau = \dim \mathbf{T}$ and $\omega = \dim \mathbf{W}$, each compatible with the grading. Let $(\mathbf{V}, \mathbf{T}, \mathbf{W})$ be the vector spaces in $\mathcal{V}_{I,\nu} \times \mathcal{V}_{I,\tau} \times \mathcal{V}_{I,\omega}$. Fix this triple. We can extend the restriction diagram by introducing the $\heartsuit$ and edge contracted (or $\widehat{\ }$) versions of ι and κ, denoted by $\iota^{\heartsuit}$, $\kappa^{\heartsuit}$, $\widehat{\iota}$ and $\widehat{\kappa}$ respectively. Here $\iota^{\heartsuit}$ and $\kappa^{\heartsuit}$ are just the restrictions of ι and κ respectively. Following, [**Li23**], we have the extended restriction diagram:

$$\begin{array}{ccccc}
\mathbf{E}^{\mathbf{F}}_{\mathbf{V},\Omega} & \xleftarrow{\quad\iota\quad} & \mathbf{S}^{\mathbf{F}}_{\mathbf{W}} & \xrightarrow{\quad\kappa\quad} & \mathbf{E}^{\mathbf{F}}_{\mathbf{T},\Omega} \times \mathbf{E}^{\mathbf{F}}_{\mathbf{W},\Omega} \\
\uparrow{\scriptstyle j_\nu} & & \uparrow{\scriptstyle j'_\nu} & & \uparrow{\scriptstyle j_\tau \times j_\omega} \\
\mathbf{E}^{\heartsuit,\mathbf{F}}_{\mathbf{V},\Omega} & \xleftarrow{\quad\iota^{\heartsuit}\quad} & \mathbf{S}^{\heartsuit,\mathbf{F}}_{\mathbf{W}} & \xrightarrow{\quad\kappa^{\heartsuit}\quad} & \mathbf{E}^{\heartsuit,\mathbf{F}}_{\mathbf{T},\Omega} \times \mathbf{E}^{\heartsuit,\mathbf{F}}_{\mathbf{W},\Omega} \\
\downarrow{\scriptstyle \mu_\nu} & & \downarrow{\scriptstyle \mu'_\nu} & & \downarrow{\scriptstyle \mu_\tau \times \mu_\omega} \\
\mathbf{E}^{\widehat{\mathbf{F}}}_{\widehat{\mathbf{V}},\widehat{\Omega}} & \xleftarrow{\quad\widehat{\iota}\quad} & \mathbf{S}^{\widehat{\mathbf{F}}}_{\widehat{\mathbf{W}}} & \xrightarrow{\quad\widehat{\kappa}\quad} & \mathbf{E}^{\widehat{\mathbf{F}}}_{\widehat{\mathbf{T}},\widehat{\Omega}} \times \mathbf{E}^{\widehat{\mathbf{F}}}_{\widehat{\mathbf{W}},\widehat{\Omega}}
\end{array} \tag{2.2.e}$$

The vertical maps j'_ν and μ'_ν are the restriction maps of j_ν and μ_ν respectively.

We make the following observations about the extended restriction diagram.

LEMMA 2.2.1. *All the squares in the diagram* (2.2.e) *are commutative. Further, the top two squares are Cartesian.*

PROOF. It is clear that the top two squares and the bottom left square are commutative, since we are dealing with restrictions and inclusions. For the bottom right square, since restriction to $\mathbf{W}$ and contraction map are commutative, we have the commutativity.

Let us prove that the top left square is Cartesian. Let C' be the fibered product of $\mathbf{S}^{\mathbf{F}}_{\mathbf{W}}$ and $\mathbf{E}^{\heartsuit,\mathbf{F}}_{\mathbf{V},\boldsymbol{\Omega}}$. Consider the following diagram where π_1 and π_2 are the projections onto the first and second components respectively.

(2.2.f)

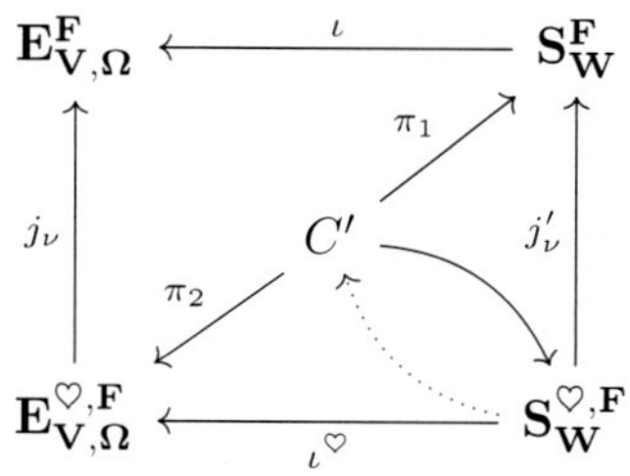

Since C' is the fibered product, we have $\iota\pi_1(x,\overline{x}) = j_\nu\pi_2(x,\overline{x})$ where $x \in \mathbf{S}^{\mathbf{F}}_{\mathbf{W}}$ and $\overline{x} \in \mathbf{E}^{\heartsuit,\mathbf{F}}_{\mathbf{V},\boldsymbol{\Omega}}$. This implies $x = \overline{x}$ and hence $x \in \mathbf{S}^{\mathbf{F}}_{\mathbf{W}} \cap \mathbf{E}^{\heartsuit,\mathbf{F}}_{\mathbf{V},\boldsymbol{\Omega}} = \mathbf{S}^{\heartsuit,\mathbf{F}}_{\mathbf{W}}$. This makes the top left Cartesian.

Next we prove that the top right square is Cartesian. Again, we let $C'' = \mathbf{S}^{\mathbf{F}}_{\mathbf{W}} \times (\mathbf{E}^{\heartsuit,\mathbf{F}}_{\mathbf{T},\boldsymbol{\Omega}} \times \mathbf{E}^{\heartsuit,\mathbf{F}}_{\mathbf{W},\boldsymbol{\Omega}})$, a fibered product. Similarly to the above, consider the following diagram:

(2.2.g)

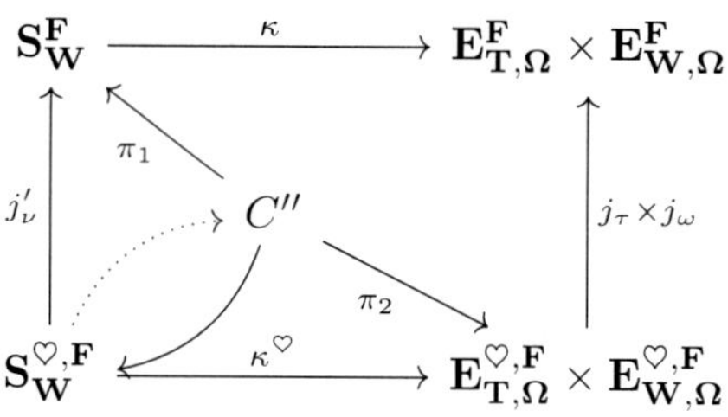

Let $x \in \mathbf{S}^{\mathbf{F}}_{\mathbf{W}}$ and $(y^{\mathbf{T}}, y^{\mathbf{W}}) \in \mathbf{E}^{\heartsuit,\mathbf{F}}_{\mathbf{T},\boldsymbol{\Omega}} \times \mathbf{E}^{\heartsuit,\mathbf{F}}_{\mathbf{W},\boldsymbol{\Omega}}$. Then we have, $\kappa(x) = j_\tau \times j_\omega(y^{\mathbf{T}}, y^{\mathbf{W}})$. Hence, $(x^{\mathbf{T}}, x^{\mathbf{W}}) = (y^{\mathbf{T}}, y^{\mathbf{W}})$, i.e., $x^{\mathbf{T}} \in \mathbf{E}^{\heartsuit,\mathbf{F}}_{\mathbf{T},\boldsymbol{\Omega}}$, $x^{\mathbf{W}} \in \mathbf{E}^{\heartsuit,\mathbf{F}}_{\mathbf{W},\boldsymbol{\Omega}}$. Now using that fact that x_h is an isomorphism iff $x_h^{\mathbf{T}}$ and $x_h^{\mathbf{W}}$ are isomorphisms, we get $x \in \mathbf{S}^{\heartsuit,\mathbf{F}}_{\mathbf{W}}$. The lemma is proved. □

LEMMA 2.2.2.

(2.2.h)

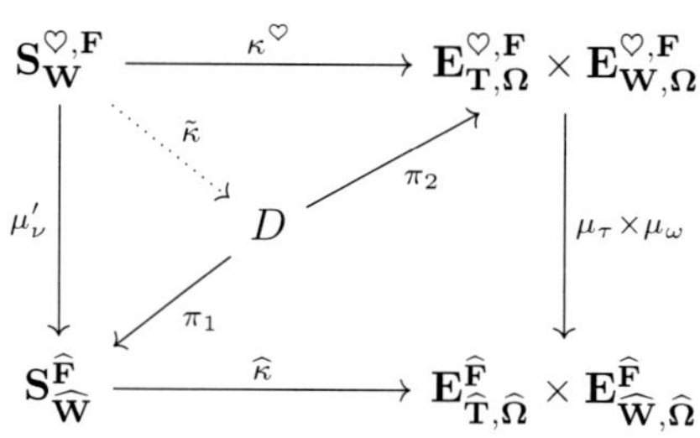

Consider the bottom right square of the extended restriction diagram. Let D be the fibered product of $\mathbf{S}^{\widehat{\mathbf{F}}}_{\widehat{\mathbf{W}}}$ and $\mathbf{E}^{\heartsuit,\mathbf{F}}_{\mathbf{T},\mathbf{\Omega}} \times \mathbf{E}^{\heartsuit,\mathbf{F}}_{\mathbf{W},\mathbf{\Omega}}$. Then the unique map $\widetilde{\kappa} : \mathbf{S}^{\heartsuit,\mathbf{F}}_{\mathbf{W}} \longrightarrow D$ is a vector bundle whose fiber is isomorphic to $(\mathbb{F}_{q^{\phi_1(i_+)}})^{\tau_{i_+}\cdot\omega_{i_-}}$.

PROOF. The proof is similar to the proof in [**Li23**] (Lemma 2.2.2), however, the proof is provided for the reader's convenience and completeness. Let $(\widehat{x}, x^T, x^W) \in D\ (= \mathbf{S}^{\widehat{\mathbf{F}}}_{\widehat{\mathbf{W}}} \times (\mathbf{E}^{\heartsuit,\mathbf{F}}_{\mathbf{T},\mathbf{\Omega}} \times \mathbf{E}^{\heartsuit,\mathbf{F}}_{\mathbf{W},\mathbf{\Omega}}))$. By definition of the fibered product, we have $\widehat{\kappa} \circ \pi_1(\widehat{x}, x^T, x^W) = \mu_\tau \times \mu_\omega \circ \pi_2(\widehat{x}, x^T, x^W) \implies \widehat{\kappa}(\widehat{x}) = \mu_\tau \times \mu_\omega(x^T, x^W)$. In order to determine the fiber of such a triple, we need to determine the $(x_h)_{h\in\Omega}$-components of a $x \in \mathbf{S}^{\heartsuit,\mathbf{F}}_{\mathbf{W}}$ such that $\widetilde{\kappa}(x)$ is equal to that triple. From (2.1.a), $\widehat{x}_h = x_h$ if $h \in \widehat{\Omega} \cap \Omega$, results in the determination of all components of the form x_h where $h \in \widehat{\Omega} \cap \Omega$. We are left to determine the components that are of the form x_{h_k}, x_{l_1} and x_{l_2} where $h_k'' = l_1'$ and $h_k'' = l_2''$. These exactly correspond to those edges which are incident with $[i_-]$. Since, $\mathbf{V}_h = \mathbf{T}_h \oplus \mathbf{W}_h$ for every edge h, we can see that,

$$x_{h_k} = \begin{bmatrix} x^T_{h_k} & 0 \\ x^{T\to W}_{h_k} & x^W_{h_k} \end{bmatrix} \quad x_{l_1} = \begin{bmatrix} x^T_{l_1} & 0 \\ x^{T\to W}_{l_1} & x^W_{l_1} \end{bmatrix} \quad x_{l_2} = \begin{bmatrix} x^T_{l_2} & 0 \\ x^{T\to W}_{l_2} & x^W_{h_k} \end{bmatrix}$$

We notice that the only thing here left to be determined is $x^{T\to W}_{h_k}, x^{T\to W}_{l_1}$ and $x^{T\to W}_{l_2}$. Using the second and the third condition in (2.1.a) yields,

$$\widehat{x}_{l_1 h_k} = x_{l_1} x_{h_k} \implies \begin{bmatrix} \widehat{x}^T_{l_1 h_k} & 0 \\ \widehat{x}^{T\to W}_{l_1 h_k} & \widehat{x}^W_{l_1 h_k} \end{bmatrix} = \begin{bmatrix} x^T_{l_1} & 0 \\ x^{T\to W}_{l_1} & x^W_{l_1} \end{bmatrix} \begin{bmatrix} x^T_{h_k} & 0 \\ x^{T\to W}_{h_k} & x^W_{h_k} \end{bmatrix}$$

$$\widehat{x}_{\overline{h_k} l_2} = x^{-1}_{h_k} x_{l_2} \implies \widehat{x}_{l_1 h_k} = x_{l_1} x_{h_k}$$

$$\implies \begin{bmatrix} \widehat{x}^T_{h_k l_2} & 0 \\ \widehat{x}^{T\to W}_{h_k l_2} & \widehat{x}^W_{h_k l_2} \end{bmatrix} = \begin{bmatrix} x^T_{h_k} & 0 \\ x^{T\to W}_{h_k} & x^W_{h_k} \end{bmatrix}^{-1} \begin{bmatrix} x^T_{l_2} & 0 \\ x^{T\to W}_{l_2} & x^W_{l_2} \end{bmatrix}$$

These lead to the following equations:

$$\widehat{x}^{T\to W}_{l_1 h_k} = x^{T\to W}_{l_1} x^T_{h_k} + x^W_{l_1} x^{T\to W}_{h_k} \quad \text{if } h_k'' = l_1'$$

$$\widehat{x}^{T\to W}_{h_k l_2} = x^T_{l_2}(-(x^W_{h_k})^{-1} x^{T\to W}_{h_k} (x^T_{h_k})^{-1}) + x^{T\to W}_{l_2} (x^W_{h_k})^{-1} \quad \text{if } h_k'' = l_2''$$

Finally,

$$x^{T\to W}_{l_1} = (\widehat{x}^{T\to W}_{l_1 h_k} - x^W_{l_1} x^{T\to W}_{h_k})(x^T_{h_k})^{-1} \quad \text{if } h_k'' = l_1'$$

$$x^{T\to W}_{l_2} = [\widehat{x}^{T\to W}_{h_k l_2} + x^T_{l_2}((x^W_{h_k})^{-1} x^{T\to W}_{h_k} (x^T_{h_k})^{-1})](x^W_{h_k}) \quad \text{if } h_k'' = l_2''$$

This shows that $x^{T\to W}_{l_1}$ and $x^{T\to W}_{l_2}$ are dependent on $x^{T\to W}_{h_k}$. Hence, the fiber is completely determined by the element $x^{T\to W}_{h_k}$. Notice that $x^{T\to W}_{h_k} : \mathbf{T}_{h_k'} \longrightarrow \mathbf{W}_{h_k''}$ implying it is a $\omega_{i_-} \times \tau_{i_+}$ matrix, whose entries are in $(\mathbb{F}_{q^{\phi_1(i_+)}})$. Therefore, the fiber is isomorphic to $(\mathbb{F}_{q^{\phi_1(i_+)}})^{\tau_{i_+}\cdot\omega_{i_-}}$. $\square$

Similar to the extension of the restriction diagram, we can extend the induction diagram as shown in [**Li23**] and note down the observations in the lemma below the diagram.

(2.2.i)
$$\begin{array}{ccccccc}
\mathbf{E}^{\mathbf{F}}_{\mathbf{T},\boldsymbol{\Omega}} \times \mathbf{E}^{\mathbf{F}}_{\mathbf{W},\boldsymbol{\Omega}} & \xleftarrow{p_1} & \mathbf{E}'^{\mathbf{F}} & \xrightarrow{p_2} & \mathbf{E}''^{\mathbf{F}} & \xrightarrow{p_3} & \mathbf{E}^{\mathbf{F}}_{\mathbf{V},\boldsymbol{\Omega}} \\
\uparrow{\scriptstyle j_\tau \times j_\omega} & & \uparrow{\scriptstyle j'_\nu} & & \uparrow{\scriptstyle j''_\nu} & & \uparrow{\scriptstyle j_\nu} \\
\mathbf{E}^{\heartsuit,\mathbf{F}}_{\mathbf{T},\boldsymbol{\Omega}} \times \mathbf{E}^{\heartsuit,\mathbf{F}}_{\mathbf{W},\boldsymbol{\Omega}} & \xleftarrow{p_1^\heartsuit} & \mathbf{E}'^{\heartsuit,\mathbf{F}} & \xrightarrow{p_2^\heartsuit} & \mathbf{E}''^{\heartsuit\mathbf{F}} & \xrightarrow{p_3^\heartsuit} & \mathbf{E}^{\heartsuit,\mathbf{F}}_{\mathbf{V},\boldsymbol{\Omega}} \\
\downarrow{\scriptstyle \mu_\tau \times \mu_\omega} & & \downarrow{\scriptstyle \mu'_\nu} & & \downarrow{\scriptstyle \mu''_\nu} & & \downarrow{\scriptstyle \mu_\nu} \\
\mathbf{E}^{\widehat{\mathbf{F}}}_{\widehat{\mathbf{T}},\widehat{\boldsymbol{\Omega}}} \times \mathbf{E}^{\widehat{\mathbf{F}}}_{\widehat{\mathbf{W}},\widehat{\boldsymbol{\Omega}}} & \xleftarrow{\widehat{p_1}} & \widehat{\mathbf{E}}'^{\widehat{\mathbf{F}}} & \xrightarrow{\widehat{p_2}} & \widehat{\mathbf{E}}''^{\widehat{\mathbf{F}}} & \xrightarrow{\widehat{p_3}} & \mathbf{E}^{\widehat{\mathbf{F}}}_{\widehat{\mathbf{V}},\widehat{\boldsymbol{\Omega}}}
\end{array}$$

Here, the first row is (2.2.d), the second row is just the restriction of the maps on the first row and the third row represents the edge contracted versions of the first row. Also, the maps j'_ν and j''_ν are inclusion maps and μ'_ν and μ''_ν are the maps defined by applying μ_ν and the natural projection $(\,\widehat{}\,)$ to its components.

LEMMA 2.2.3. *All the squares in the extended induction diagram* (2.2.i)*) are commutative. Further, all the top squares and the bottom right square is Cartesian.*

PROOF. Commutativity follows from the definition. The proof of top left square being Cartesian follows closely to the proof of the top right square of the extended restriction diagram (2.2.e) being Cartesian.

To prove that the top middle square is Cartesian, we let $C = \mathbf{E}'^{\mathbf{F}} \times \mathbf{E}''^{\heartsuit\mathbf{F}}$, a fibered product. Consider the following diagram:

(2.2.j)

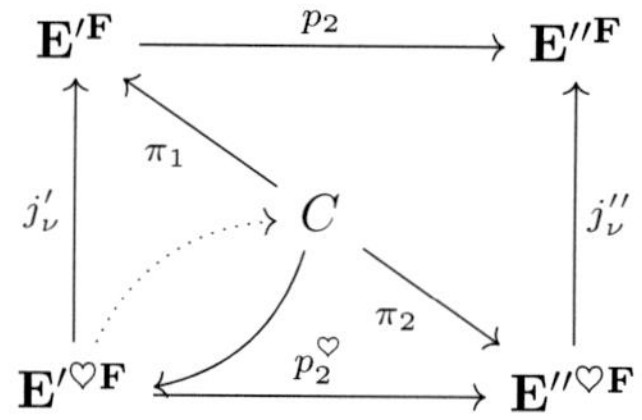

By the definition of a fibered product, we have $p_2([g,x]_U) = j''_\nu([\overline{g},\overline{x}]_Q)$ where $[g,x]_U \in \mathbf{E}'^{\mathbf{F}}$ and $[\overline{g},\overline{x}]_Q \in \mathbf{E}''^{\heartsuit\mathbf{F}}$. This implies, $[g,x]_Q = [\overline{g},\overline{x}]_Q$. Hence, there exists a $r \in Q$ such that $(g,x) = r.(\overline{g},\overline{x}) = (\overline{g}r^{-1}, r.\overline{x})$. Therefore, $x = r.\overline{x}$, implying $x \in \mathbf{S}^{\heartsuit,\mathbf{F}}_{\mathbf{W}}$ and we finally have $[g,x]_U \in \mathbf{E}'^{\heartsuit,\mathbf{F}}$. In a similar fashion, the top right square can be proved Cartesian.

Next, we proceed to prove that the bottom right square is Cartesian. To prove this, we will use the following:

$$\mathbf{E}''^{\mathbf{F}} \cong \{(x, \mathbf{U}) \mid x \in \mathbf{E}^{\mathbf{F}}_{\mathbf{V},\boldsymbol{\Omega}}, U \text{ is } x-\text{invariant}, \mathbf{V}/\mathbf{U} \cong \mathbf{T}, \mathbf{U} \cong \mathbf{W}\}.$$

On a side note, we can also show that,

$$\mathbf{E}'^{\mathbf{F}} \cong \{(x, \mathbf{U}, g', g'') \mid (x, \mathbf{U}) \in \mathbf{E}''^{\mathbf{F}},\ g' : \mathbf{V}/\mathbf{U} \cong \mathbf{T}, g'' : \mathbf{U} \cong \mathbf{W}\}.$$

Hence, we have

$$\widehat{\mathbf{E}}''^{\widehat{\mathbf{F}}} \cong \{(\widehat{x}, \widehat{\mathbf{U}}) \mid \widehat{x} \in \mathbf{E}^{\widehat{\mathbf{F}}}_{\widehat{\mathbf{V}},\widehat{\boldsymbol{\Omega}}}, \widehat{\mathbf{V}}/\widehat{\mathbf{U}} \cong \widehat{\mathbf{T}}, \widehat{\mathbf{U}} \cong \widehat{\mathbf{W}}\} \quad \text{and}$$
$$\mathbf{E}''^{\heartsuit,\mathbf{F}} \cong \{(x, \mathbf{U}) \mid x \in \mathbf{E}^{\heartsuit,\mathbf{F}}_{\mathbf{V},\boldsymbol{\Omega}}, U \text{ is } x-\text{invariant}, \mathbf{V}/\mathbf{U} \cong \mathbf{T}, \mathbf{U} \cong \mathbf{W}\}.$$

Let $C' = \mathbf{E}^{\heartsuit,\mathbf{F}}_{\mathbf{V},\mathbf{\Omega}} \times \widehat{\mathbf{E}}''^{\widehat{\mathbf{F}}}$.

(2.2.k)

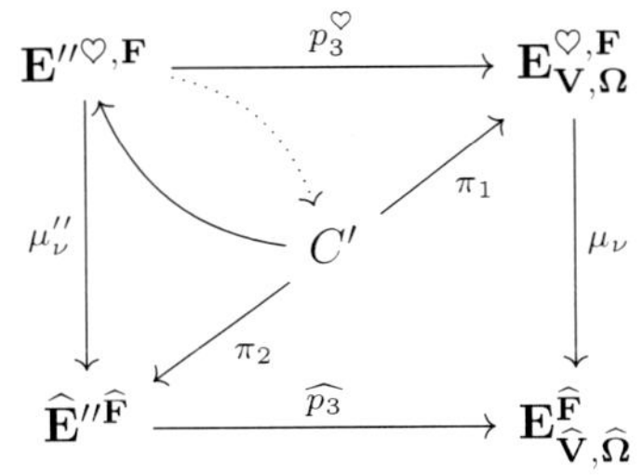

We need to construct a unique element $(x, U) \in \mathbf{E}''^{\heartsuit,\mathbf{F}}$ from C'. We get x from $\mathbf{E}^{\heartsuit,\mathbf{F}}_{\mathbf{V},\mathbf{\Omega}}$. Now to get U, define $\mathbf{U_i} = \widehat{\mathbf{U}}_\mathbf{i}$, $\mathbf{i} \neq \mathbf{i}_-$ and $\mathbf{U}_{\mathbf{i}_-} = x_{h_k}(\mathbf{U}_{h'_k})$ where h_k is the edge along which the contraction takes place. Let l_1 be the edge which starts at $\mathbf{i}_-$ as described in (1.2.b). To show $x_{l_1}(\mathbf{U}_{\mathbf{i}_-}) \subseteq \mathbf{U}_{l''_1}$. This follows as: $x_{l_1}(\mathbf{U}_{\mathbf{i}_-}) = \widehat{x}_{l_1 h_k} x^{-1}_{h_k}(\mathbf{U}_{\mathbf{i}_-}) = \widehat{x}_{l_1 h_k}(\mathbf{U}_{h'_k}) = \widehat{x}_{l_1 h_k}(\widehat{\mathbf{U}}_{h'_k}) \subseteq \widehat{\mathbf{U}}_{l''_1} = \mathbf{U}_{l''_1}$. Therefore, we get a map from $C' \to \mathbf{E}''^{\heartsuit,\mathbf{F}}$ defined by $(x, (\widehat{x}, \widehat{\mathbf{U}})) \mapsto (x, \mathbf{U})$ and the result follows. □

LEMMA 2.2.4. *Let C be the fibered product of $\mathbf{E}''^{\heartsuit\mathbf{F}}$ and $\widehat{\mathbf{E}}'^{\widehat{\mathbf{F}}}$. Then there exists maps $p' : \mathbf{E}'^{\heartsuit\mathbf{F}} \to C$, $\widehat{p} : C \to \mathbf{E}''^{\heartsuit,\mathbf{F}}$ and $\widehat{\mu} : C \to \widehat{\mathbf{E}}'^{\widehat{\mathbf{F}}}$ such that the middle square in the extended induction diagram factors through C and makes the following diagram commute:*

(2.2.l)

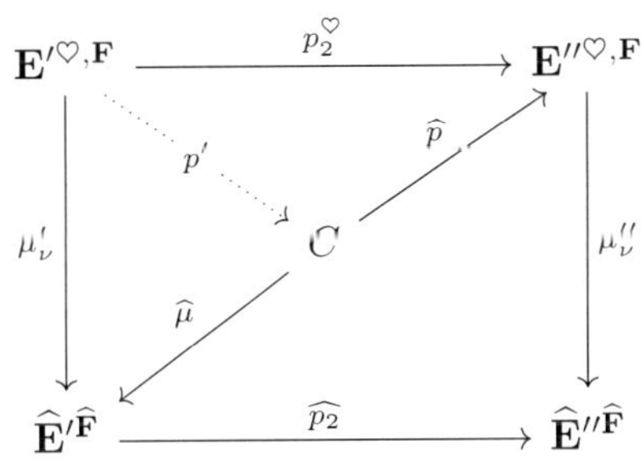

Moreover, the map p' defines a $\mathbf{G}^{[\mathbf{i}_-],\mathbf{F}}_{\mathbf{T}} \times \mathbf{G}^{[\mathbf{i}_-],\mathbf{F}}_{\mathbf{W}}$-bundle.

PROOF. The first part of the proof follows from the universal property of fibered products.

When defining the induction diagram, we noted that p_2 is a $Q^{\mathbf{F}}/U^{\mathbf{F}}(\cong \mathbf{G}^{\mathbf{F}}_{\mathbf{T}} \times \mathbf{G}^{\mathbf{F}}_{\mathbf{W}})$-bundle. It follows that $p_2^{\heartsuit}$ is a $Q^{\mathbf{F}}/U^{\mathbf{F}}(\cong \mathbf{G}^{\mathbf{F}}_{\mathbf{T}} \times \mathbf{G}^{\mathbf{F}}_{\mathbf{W}})$-bundle. This implies that $\mathbf{E}'^{\heartsuit,\mathbf{F}}$ has a natural fiber structure compatible with the group actions of $\mathbf{G}^{\mathbf{F}}_{\mathbf{T}}$ and $\mathbf{G}^{\mathbf{F}}_{\mathbf{W}}$. Note that the map, $\widehat{p} : C \to \mathbf{E}''^{\heartsuit,\mathbf{F}}$ defined as $\widehat{p}([g, x]_Q, [h, y]_{\widehat{U}}) = [g, x]_Q$ has fibers in $\widehat{\mathbf{E}}'^{\widehat{\mathbf{F}}}$, implying it is a $\widehat{Q}^{\widehat{\mathbf{F}}}/\widehat{U}^{\widehat{\mathbf{F}}}(\cong \mathbf{G}^{\widehat{\mathbf{F}}}_{\widehat{\mathbf{T}}} \times \mathbf{G}^{\widehat{\mathbf{F}}}_{\widehat{\mathbf{W}}})$-bundle. Therefore, by the properties of fiber bundles, p' inherits the same bundle structure as a $\mathbf{G}^{[\mathbf{i}_-],\mathbf{F}}_{\mathbf{T}} \times \mathbf{G}^{[\mathbf{i}_-],\mathbf{F}}_{\mathbf{W}}$-bundle and the lemma is proved. □

2.3. Embedding among Hall Algebras. Fix a square root $q^{1/2}$ of q in the field $\mathbb{C}$ of complex numbers and let $\mathbb{Q}(q^{1/2})$ be the subfield of $\mathbb{C}$ generated by $\mathbb{Q}$ and $q^{1/2}$.

Let $\phi : X \to Y$ be a map between two finite sets. If $f : Y \to \mathbb{Q}(q^{1/2})$ is a function, we define the pullback of f along ϕ by $\phi^*(f) = f\phi$. If $f' : X \to \mathbb{Q}(q^{1/2})$ is a function, we define a push-forward function $\phi_!(f') : Y \to \mathbb{Q}(q^{1/2})$ by $\phi_!(f')(y) = \sum_{x\in\phi^{-1}(y)} f'(x)$ for all $y \in Y$.

Fix $\mathbf{V} \in \mathcal{V}_{I,\nu}$ where $\nu \in \mathbb{N}[I]$. To each $\nu \in \mathbb{N}[I]$ we define a vector space associated with it as follows:

$$\text{(2.3.a)} \qquad \mathbf{H}_{\nu,\Omega} = \{f : \mathbf{E}^{\mathbf{F}}_{\mathbf{V},\mathbf{\Omega}} \to \mathbb{Q}(q^{1/2}) \mid f(g.x) = f(x)\ \forall g \in \mathbf{G}^{\mathbf{F}}_{\mathbf{V}}, x \in \mathbf{E}^{\mathbf{F}}_{\mathbf{V},\mathbf{\Omega}}\}.$$

i.e., f is $\mathbf{G}^{\mathbf{F}}_{\mathbf{V}}$-invariant. Let $\nu, \tau, \omega \in \mathbb{N}[I]$ such that $\nu = \tau + \omega$. Further, we let $(\mathbf{V}, \mathbf{W}, \mathbf{T}) \in \mathcal{V}_{I,\nu} \times \mathcal{V}_{I,\tau} \times \mathcal{V}_{I,\omega}$. Now, we define the Hall algebra $\mathbf{H}_\Omega$ by taking $\mathbf{H}_{\nu,\Omega}$ and summing over all the dimension vectors ν, i.e.,

$$\text{(2.3.b)} \qquad \mathbf{H}_\Omega = \bigoplus_{\nu\in\mathbb{N}[I]} \mathbf{H}_{\nu,\Omega}.$$

and defining multiplication as follows:

$$\text{(2.3.c)} \qquad \begin{aligned} &\mathrm{Ind}^{\nu}_{\tau,\omega} : \mathbf{H}_{\tau,\Omega} \otimes \mathbf{H}_{\omega,\Omega} \longrightarrow \mathbf{H}_{\nu,\Omega}; \\ &\mathrm{Ind}^{\nu}_{\tau,\omega} = (q^{1/2})^{-m_\Omega(\tau,\omega)} \frac{1}{\#\mathbf{G}^{\mathbf{F}}_{\mathbf{T}} \times \mathbf{G}^{\mathbf{F}}_{\mathbf{W}}} (p_3)_!(p_2)_!(p_1)^* \end{aligned}$$

where $m_\Omega(\tau, \omega) = \sum_{\mathbf{i}\in\mathbf{I}} \tau_{\mathbf{i}}\omega_{\mathbf{i}} + \sum_{h\in\Omega} \tau_{h'}\omega_{h''}$ and the maps p_1, p_2 and p_3 are from the induction diagram.

In [**Lus98**], Lusztig proved that the above-mentioned Hall algebra is a unital associative algebra over $\mathbb{Q}(q^{1/2})$. Let $f_1 \in \mathbf{H}_{\tau,\Omega}$ and $f_2 \in \mathbf{H}_{\omega,\Omega}$. We then write,

$$\text{(2.3.d)} \qquad \begin{aligned} &f_1 \star f_2 = \frac{1}{\#\mathbf{G}^{\mathbf{F}}_{\mathbf{T}} \times \mathbf{G}^{\mathbf{F}}_{\mathbf{W}}} (p_3)_!(p_2)_!(p_1)^*(f_1 \otimes f_2) \quad \text{and} \\ &f_1 \circ f_2 = (q^{1/2})^{-m_\Omega(\tau,\omega)} f_1 \star f_2 = \mathrm{Ind}^{\nu}_{\tau,\omega}(f_1 \otimes f_2). \end{aligned}$$

Similarly, the restriction diagram helps to define the map dual to $\mathrm{Ind}^{\nu}_{\tau,\omega}$ as follows:

$$\text{(2.3.e)} \qquad \begin{aligned} &\mathrm{Res}^{\nu}_{\tau,\omega} : \mathbf{H}_{\nu,\Omega} \longrightarrow \mathbf{H}_{\tau,\Omega} \otimes \mathbf{H}_{\omega,\Omega} \\ &\mathrm{Res}^{\nu}_{\tau,\omega} = (q^{1/2})^{-m^*_\Omega(\tau,\omega)} \kappa_! \iota^* \end{aligned}$$

where $m^*_\Omega(\tau, \omega) = -\sum_{\mathbf{i}\in\mathbf{I}} \tau_{\mathbf{i}}\omega_{\mathbf{i}} + \sum_{h\in\Omega} \tau_{h'}\omega_{h''}$ and the maps κ and ι are from the restriction diagram.

Let $r_\Omega = \bigoplus_{\nu=\tau+\omega} \mathrm{Res}^{\nu}_{\tau,\omega} : \mathbf{H}_\Omega \to \mathbf{H}_\Omega \otimes \mathbf{H}_\Omega$. In [**Lus98**], we see that $(\mathbf{H}_\Omega, r_\Omega)$ forms an associative algebra by defining multiplication in $\mathbf{H}_\Omega \otimes \mathbf{H}_\Omega$ as $(f_1 \otimes f_2)(g_1 \otimes g_2) = (q^{1/2})^{\nu_2\cdot\mu_1} f_1 \circ g_1 \otimes f_2 \circ g_2$ for all $f_1, f_2, g_1, g_2 \in \mathbf{H}_\Omega$ such that $(f_1, f_2) \in \mathbf{H}_{\nu_1} \times \mathbf{H}_{\nu_2}$ and $(g_1, g_2) \in \mathbf{H}_{\mu_1} \times \mathbf{H}_{\mu_2}$.

Having defined $\mathbf{H}_{\nu,\Omega}$, we define the following:

$$\mathbf{H}^{\heartsuit}_{\nu,\Omega} = \{f : \mathbf{E}^{\heartsuit,\mathbf{F}}_{\mathbf{V},\mathbf{\Omega}} \to \mathbb{Q}(q^{1/2}) \mid f(g.x) = f(x)\ \forall g \in \mathbf{G}^{\mathbf{F}}_{\mathbf{V}}, x \in \mathbf{E}^{\heartsuit,\mathbf{F}}_{\mathbf{V},\mathbf{\Omega}}\},$$

$$\mathbf{H}_{\nu,\widehat{\Omega}} = \{f : \mathbf{E}^{\widehat{\mathbf{F}}}_{\widehat{\mathbf{V}},\widehat{\mathbf{\Omega}}} \to \mathbb{Q}(q^{1/2}) \mid f(g.x) = f(x)\ \forall g \in \mathbf{G}^{\widehat{\mathbf{F}}}_{\widehat{\mathbf{V}}}, x \in \mathbf{E}^{\widehat{\mathbf{F}}}_{\widehat{\mathbf{V}},\widehat{\mathbf{\Omega}}}\},$$

where $\nu \in \mathbb{N}[\widehat{I}]$. Just like in [**Li23**], this gives 2 new Hall algebras,

$$\mathbf{H}_{\Omega}^{\heartsuit} = \bigoplus_{\nu \in \mathbb{N}[\widehat{I}]} \mathbf{H}_{\nu,\Omega}^{\heartsuit} \quad \text{and} \quad \mathbf{H}_{\widehat{\Omega}} = \bigoplus_{\nu \in \mathbb{N}[\widehat{I}]} \mathbf{H}_{\nu,\widehat{\Omega}},$$

where multiplication is defined using the second and the third row in the extended induction diagram (2.2.i) respectively.

The inclusion map $j_\nu : \mathbf{E}_{\mathbf{V},\mathbf{\Omega}}^{\heartsuit,\mathbf{F}} \to \mathbf{E}_{\mathbf{V},\mathbf{\Omega}}^{\mathbf{F}}$, induces the map, $j_{\nu!} : \mathbf{H}_{\nu,\Omega}^{\heartsuit} \to \mathbf{H}_{\nu,\Omega}$ by extension by zero, i.e., $j_{\nu!}(f)(x) = \begin{cases} f(x) & \text{if } x \in \mathbf{E}_{\mathbf{V},\mathbf{\Omega}}^{\heartsuit,\mathbf{F}} \\ 0 & \text{if } x \in \mathbf{E}_{\mathbf{V},\mathbf{\Omega}}^{\mathbf{F}} \backslash \mathbf{E}_{\mathbf{V},\mathbf{\Omega}}^{\heartsuit,\mathbf{F}} \end{cases}$.

Taking sum over $\nu \in \mathbb{N}[\widehat{I}]$, we get $j_! : \mathbf{H}_{\Omega}^{\heartsuit} \to \mathbf{H}_{\Omega}$, and it can be shown that it is an algebra homomorphism over $\mathbb{Q}(q^{1/2})$, by using the fact that the top square in the extended induction diagram is Cartesian.

We may also define the pullback of the contraction map $\mu_\nu : \mathbf{E}_{\mathbf{V},\mathbf{\Omega}}^{\heartsuit,\mathbf{F}} \to \mathbf{E}_{\widehat{\mathbf{V}},\widehat{\mathbf{\Omega}}}^{\widehat{\mathbf{F}}}$ by $\mu_\nu^* : \mathbf{H}_{\nu,\widehat{\Omega}} \to \mathbf{H}_{\nu,\Omega}^{\heartsuit}$, $f \mapsto \mu_\nu \circ f$. Note that this map is an isomorphism of vector spaces. We twist this map and define

$$\mu_\nu^\star = (q^{-1/2})^{\nu_{i_-}^2 - \phi_1(i_-)} \mu_\nu^*.$$

Again, taking sum over $\nu \in \mathbb{N}[\widehat{I}]$, we get

$$\mu^\star : \mathbf{H}_{\widehat{\Omega}} \to \mathbf{H}_{\Omega}^{\heartsuit}$$

We will later prove that this is an algebra homomorphism, but this is in fact an isomorphism.

The next theorem gives the main result of this paper. The theorem confers a embedding among the Hall algebras $\mathbf{H}_{\widehat{\Omega}}$ and $\mathbf{H}_{\Omega}$. The proof of the theorem is similar to the proof of theorem 2.3.1 in [**Li23**] with some changes accounting for additional edges between the orbits, about which edges are contracted, and formation of loops after the edge contraction. Nonetheless, the entire proof is written here for convenience.

THEOREM 2.3.1. *The composition of maps defined by $\psi = j_! \circ \mu^\star : \mathbf{H}_{\widehat{\Omega}} \to \mathbf{H}_{\Omega}$ is an embedding of Hall algebras.*

PROOF. Above we have proved that $j_! : \mathbf{H}_{\Omega}^{\heartsuit} \to \mathbf{H}_{\Omega}$, is an algebra homomorphism. It remains to show that $\mu^\star$ is also a homomorphism. To do that, we need to show the multiplicative compatibility of $\mu^\star$, i.e., $\mu_\nu^\star \widehat{\mathrm{Ind}}_{\tau,\omega}^{\nu} = \mathrm{Ind}_{\tau,\omega}^{\nu,\heartsuit}(\mu_\tau^\star \otimes \mu_\omega^\star)$. From the extended induction diagram (2.2.i) we have,

$$\begin{aligned} \mu_\nu^*(\widehat{p}_3)_!(\widehat{p}_2)_!\widehat{p}_1^* &= (p_3^{\heartsuit})_! \mu_\nu''^* (\widehat{p}_2)_! \widehat{p}_1^* \\ &= (p_3^{\heartsuit})_! \widehat{p}_! \widehat{\mu}^* \widehat{p}_1^* \\ &= \frac{1}{\#\mathbf{G}_{\mathbf{T}}^{\mathbf{i}_-,\mathbf{F}} \times \mathbf{G}_{\mathbf{W}}^{\mathbf{i}_-,\mathbf{F}}} (p_3^{\heartsuit})_!(p_2^{\heartsuit})_! \mu_\nu'^* \widehat{p}_1^* \\ &= \frac{1}{\#\mathbf{G}_{\mathbf{T}}^{\mathbf{i}_-,\mathbf{F}} \times \mathbf{G}_{\mathbf{W}}^{\mathbf{i}_-,\mathbf{F}}} (p_3^{\heartsuit})_!(p_2^{\heartsuit})_!(p_1^{\heartsuit})^*(\mu_\tau^* \otimes \mu_\omega^*), \end{aligned} \tag{2.3.f}$$

where the third equality is due to Lemma (2.2.4),

$$(p_2^{\heartsuit})_! \mu_\nu'^* = \widehat{p}_! p'_! p'^* \widehat{\mu}^* = \#(\mathbf{G}_{\mathbf{T}}^{\mathbf{i}_-,\mathbf{F}} \times \mathbf{G}_{\mathbf{W}}^{\mathbf{i}_-,\mathbf{F}}) \widehat{p}_! \widehat{\mu}^*.$$

By (2.3.f), we have,

$$\begin{aligned}\mu_\nu^\star\widehat{\mathrm{Ind}}^{\nu}_{\tau,\omega} &= (q^{1/2})^{-\nu_{i_-}^2\phi_1(i_-)-m_{\widehat{\Omega}}(\tau,\omega)}\frac{1}{\#\mathbf{G}^{\widehat{\mathbf{F}}}_{\widehat{\mathbf{T}}}\times\mathbf{G}^{\widehat{\mathbf{F}}}_{\widehat{\mathbf{W}}}}\mu_\nu^*(\widehat{p}_3)_!(\widehat{p}_2)_!(\widehat{p}_1)^*\\ &= (q^{1/2})^{-\nu_{i_-}^2\phi_1(i_-)-m_{\widehat{\Omega}}(\tau,\omega)}\frac{1}{\#\mathbf{G}^{\mathbf{F}}_{\mathbf{T}}\times\mathbf{G}^{\mathbf{F}}_{\mathbf{W}}}(p_3^\heartsuit)_!(p_2^\heartsuit)_!(p_1^\heartsuit)^*(\mu_\tau^*\otimes\mu_\omega^*)\\ &= (q^{1/2})^{-N}\,\mathrm{Ind}^{\nu,\heartsuit}_{\tau,\omega}(\mu_\tau^\star\otimes\mu_\omega^\star),\end{aligned}$$

where

$$\begin{aligned}N &= (\nu_{i_-}^2-\tau_{i_-}^2-\omega_{i_-}^2)\phi_1(i_-)+m_{\widehat{\Omega}}(\tau,\omega)-m_\Omega(\tau,\omega)\\ &= (\nu_{i_-}^2-\tau_{i_-}^2-\omega_{i_-}^2)\phi_1(i_-)-2\tau_{i_-}\omega_{i_-}(\phi_1(i_-))\\ &= 0.\end{aligned}$$

The theorem is proved. □

The calculation $m_{\widehat{\Omega}}(\tau,\omega)-m_\Omega(\tau,\omega) = -2\tau_{i_-}\omega_{i_-}(\phi_1(i_-))$ can be illustrated using Figure 3 as follows:

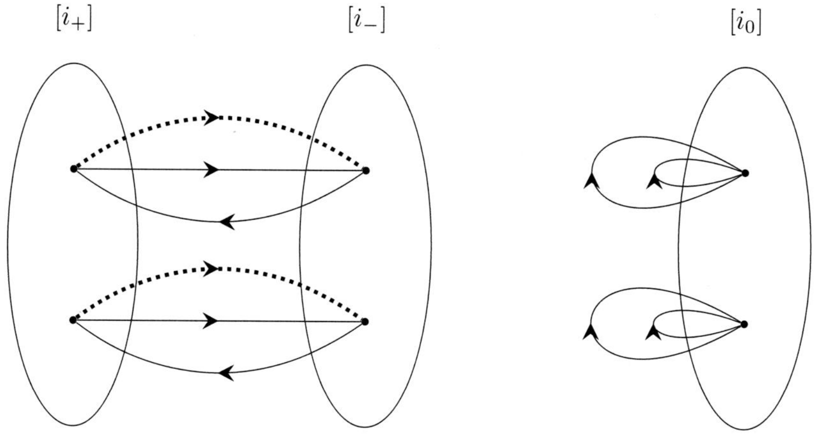

FIGURE 3. The first graph is before the edge contraction and the second graph is after the edge contraction. The dotted edges denote the edges along which the contraction takes place. The solid edges in the first graph form loops after the dotted edges contract.

The formula for m_Ω and $m_{\widehat{\Omega}}$ assigns weights to vertices and edges of the graph. Because of the vertices on the second ellipse before the edge contraction, we get $\phi_1(i_-)\tau_{i_-}\omega_{i_-}$ as additional weights on the vertices compared to the second graph after the edge contraction. The weights on the solid edges cancel out and we are left with the weights on the dotted edges which is precisely, $\phi_1(i_-)\tau_{i_+}\omega_{i_-}$. Since, $\tau_{i_+}=\tau_{i_-}$, we get $m_{\widehat{\Omega}}(\tau,\omega)-m_\Omega(\tau,\omega) = -2\tau_{i_-}\omega_{i_-}(\phi_1(i_-))$.

2.4. Compatibility of $j_!\mu_\nu^*$ with the PBW basis. Let $\mathcal{O}$ be a $\mathbf{G}^{\mathbf{F}}_{\mathbf{V}}$- orbit in $\mathbf{E}^{\mathbf{F}}_{\mathbf{V},\boldsymbol{\Omega}}$. Let $P_{\mathcal{O}}:\mathbf{G}^{\mathbf{F}}_{\mathbf{V}}\backslash\mathbf{E}^{\mathbf{F}}_{\mathbf{V},\boldsymbol{\Omega}}\to\{0,1\}$ be defined as $P_{\mathcal{O}}(\mathcal{O})=1$ and 0 otherwise, i.e., it is the characteristic function of the orbit $\mathcal{O}$. Since, $\mathbf{E}^{\mathbf{F}}_{\mathbf{V},\boldsymbol{\Omega}}$ is finite, the number of orbits is also finite. Let $\epsilon = \{P_{\mathcal{O}} \mid \mathcal{O} \in \mathbf{G}^{\mathbf{F}}_{\mathbf{V}} \backslash \mathbf{E}^{\mathbf{F}}_{\mathbf{V},\boldsymbol{\Omega}}\}$.

Now, $P_{\mathcal{O}}$ induces a map $\zeta_{\mathcal{O}} : \mathbf{E}^{\mathbf{F}}_{\mathbf{V},\boldsymbol{\Omega}} \to \mathbb{Q}(q^{1/2})$ by defining $\zeta_{\mathcal{O}}(x) = P_{\mathcal{O}}([x])$ where $[x]$ denotes the $\mathbf{G}^{\mathbf{F}}_{\mathbf{V}}$-orbit of x. The set $\{\zeta_{\mathcal{O}}\}_{\mathcal{O} \in \mathbf{G}^{\mathbf{F}}_{\mathbf{V}} \backslash \mathbf{E}^{\mathbf{F}}_{\mathbf{V},\boldsymbol{\Omega}}}$ forms the PBW basis of $\mathbf{H}_{\Omega}$.

Remark: Let X and Y be finite sets with a group action by G. A map on sets, say $\phi : X \to Y$ induces a well defined map on $\mu : G\backslash X \to G\backslash Y$ by defining $\mu([x]) = [\phi(x)]$ if ϕ preserves the group action, i.e., $\phi(g.x) = g.\phi(x)$.

Since $\mu_\nu : \mathbf{E}^{\heartsuit,\mathbf{F}}_{\mathbf{V},\boldsymbol{\Omega}} \to \mathbf{E}^{\widehat{\mathbf{F}}}_{\widehat{\mathbf{V}},\widehat{\boldsymbol{\Omega}}}$ defined by $\mu_\nu(x) = \widehat{x}$ is compatible with the group action $\mathbf{G}^{\mathbf{F}}_{\mathbf{V}}$, by the above remark, we have a bijection $\mu_\nu : \mathbf{G}^{\mathbf{F}}_{\mathbf{V}} \backslash \mathbf{E}^{\heartsuit,\mathbf{F}}_{\mathbf{V},\boldsymbol{\Omega}} \to \mathbf{G}^{\widehat{\mathbf{F}}}_{\widehat{\mathbf{V}}} \backslash \mathbf{E}^{\widehat{\mathbf{F}}}_{\widehat{\mathbf{V}},\widehat{\boldsymbol{\Omega}}}$ which is defined as $\mu_\nu([x]) = [\widehat{x}]$. Next we show that the map $j_! \mu^*_\nu$ preserves the PBW basis.

PROPOSITION 2.4.1. *If $\mu_\nu(\mathcal{O}) = \mathcal{O}'$ then, $j_! \mu^*_\nu(P_{\mathcal{O}'}) = P_{\mathcal{O}}$.*

PROOF. Let $x \in \mathbf{E}^{\mathbf{F}}_{\mathbf{V},\boldsymbol{\Omega}}$ such that $[x] = \mathcal{O}$ and $x \in \mathbf{E}^{\heartsuit,\mathbf{F}}_{\mathbf{V},\boldsymbol{\Omega}}$. This implies $\mu_\nu([x]) = \mu_\nu(\mathcal{O}) = \mathcal{O}'$. The computation is worked out as follows:

$$\begin{aligned}
j_! \mu^*_\nu(P_{\mathcal{O}'})([x]) &= j_!(\mu^*_\nu P_{\mathcal{O}'})([x]) \\
&= \sum_{x^\heartsuit \in j_\nu^{-1}([x])} \mu^*_\nu(P_{\mathcal{O}'})([x^\heartsuit]) \\
&= P_{\mathcal{O}'} \mu_\nu([x]) \\
&= P_{\mathcal{O}'}(\mathcal{O}') \\
&= 1 \\
&= P_{\mathcal{O}}(\mathcal{O}) \\
&= P_{\mathcal{O}}([x]).
\end{aligned}$$

If $x \in \mathbf{E}^{\mathbf{F}}_{\mathbf{V},\boldsymbol{\Omega}}$ such that either $x \notin \mathbf{E}^{\heartsuit,\mathbf{F}}_{\mathbf{V},\boldsymbol{\Omega}}$ or $[x] \neq \mathcal{O}$ then $j_! \mu^*_\nu(P_{\mathcal{O}'})([x]) = 0 = P_{\mathcal{O}}([x])$.

Therefore we have, $j_! \mu^*_\nu(P_{\mathcal{O}'}) = P_{\mathcal{O}}$. □

2.5. Split short exact sequence of Hall Algebras. We will show that $\mathbf{H}_{\widehat{\Omega}}$ is a split subquotient of $\mathbf{H}_{\Omega}$, which was proved in [**Li23**].

Recall the representation space with isomorphisms on the edges along which the contaction takes place, i.e., $\mathbf{E}^{\heartsuit,\mathbf{F}}_{\mathbf{V},\boldsymbol{\Omega}}$. Let $\mathbf{E}^{\mathbf{c},\mathbf{F}}_{\mathbf{V},\boldsymbol{\Omega}}$ be its compliment in $\mathbf{E}^{\mathbf{F}}_{\mathbf{V},\boldsymbol{\Omega}}$. This space is $\mathbf{G}^{\mathbf{F}}_{\mathbf{V}}$-stable. Just like we constructed a Hall algebra associated to the representation spaces $\mathbf{E}^{\mathbf{F}}_{\mathbf{V},\boldsymbol{\Omega}}$ and $\mathbf{E}^{\heartsuit,\mathbf{F}}_{\mathbf{V},\boldsymbol{\Omega}}$, we do the same for $\mathbf{E}^{\mathbf{c},\mathbf{F}}_{\mathbf{V},\boldsymbol{\Omega}}$ as well. Let $\mathbf{H}^c_{\nu,\Omega} = \{f : \mathbf{E}^{\mathbf{c},\mathbf{F}}_{\mathbf{V},\boldsymbol{\Omega}} \to \mathbb{Q}(q^{1/2}) \mid f(g.x) = f(x)\ \forall g \in \mathbf{G}^{\mathbf{F}}_{\mathbf{V}}, x \in \mathbf{E}^{\mathbf{c},\mathbf{F}}_{\mathbf{V},\boldsymbol{\Omega}}\}$. We define,

$$\mathbf{H}^c_{\Omega} = \oplus_{\nu \in \mathbb{N}[\widehat{I}]} \mathbf{H}^c_{\nu,\Omega}.$$

We further define a new space

$$\mathbf{H}^{\widehat{I}}_{\Omega} = \oplus_{\nu \in \mathbb{N}[\widehat{I}]} \mathbf{H}_{\nu,\Omega}.$$

By definition, we have

$$\mathbf{H}^{\widehat{I}}_{\Omega} = \mathbf{H}^{\heartsuit}_{\Omega} \oplus \mathbf{H}^c_{\Omega}.$$

LEMMA 2.5.1. *$\mathbf{H}^c_{\Omega}$ is a two-sided ideal of $\mathbf{H}^{\widehat{I}}_{\Omega}$.*

PROOF. We need to show that if $f \in \mathbf{H}^c_\Omega$ and $g \in \mathbf{H}^{\widehat{I}}_\Omega$ then $f \circ g \in \mathbf{H}^c_{\widehat{\Omega}}$. Without loss of generality, assume that $f \in \mathbf{H}^c_{\omega,\Omega}$ and $g \in \mathbf{H}_{\tau,\Omega}$ where $\omega, \tau \in \widehat{I}$. So it is sufficient to prove that $f \circ g \in \mathbf{H}^c_{\tau+\omega,\Omega}$, i.e., to show that $f \circ g$ is a map on $\mathbf{E}^{\mathbf{c},\mathbf{F}}_{\mathbf{V},\mathbf{\Omega}}$ where the dimension vector of $V = \nu = \tau + \omega$. Consider $(x', x'') \in \mathbf{E}^{\mathbf{F}}_{\mathbf{W},\mathbf{\Omega}} \times \mathbf{E}^{\mathbf{F}}_{\mathbf{T},\mathbf{\Omega}}$ where $\dim(\mathbf{W}) = \omega$ and $\dim(\mathbf{T}) = \tau$. Suppose that one of the components in the pair (x', x'') is either in $\mathbf{E}^{\mathbf{c},\mathbf{F}}_{\mathbf{T},\mathbf{\Omega}}$ or $\mathbf{E}^{\mathbf{c},\mathbf{F}}_{\mathbf{W},\mathbf{\Omega}}$. Suppose $x \in \mathbf{E}^{\mathbf{F}}_{\mathbf{V},\mathbf{\Omega}}$, such that $x = p_3 p_2 p_1^{-1}(x', x'')$. Since $V = W \oplus T$, we have $x = \left[\begin{array}{c|c} x' & x^{\mathbf{T}\to\mathbf{W}} \\ \hline 0 & x'' \end{array}\right]$. Therefore, x is invertible iff x' and x'' is invertible. This shows that $x \in \mathbf{E}^{\mathbf{c},\mathbf{F}}_{\mathbf{V},\mathbf{\Omega}}$. Hence, $\mathbf{H}^c_\Omega$ is a two-sided ideal of $\mathbf{H}^{\widehat{I}}_\Omega$. □

Note that the map $j^*_\nu : \mathbf{H}_{\nu,\Omega} \to \mathbf{H}^{\heartsuit}_{\nu,\Omega}$ is a restriction map and is also surjective by extension by zero. Taking direct sum over $\nu \in \widehat{I}$, we define a new map,

$$j^* = \oplus_{\nu \in \widehat{I}}\, j^*_\nu : \mathbf{H}^{\widehat{I}}_\Omega \to \mathbf{H}^{\heartsuit}_\Omega.$$

THEOREM 2.5.2. *We have the following split short exact sequence of Hall algebras*

$$0 \longrightarrow \mathbf{H}^c_\Omega \longrightarrow \mathbf{H}^{\widehat{I}}_\Omega \underset{j_!}{\overset{j^*}{\rightleftarrows}} \mathbf{H}^{\heartsuit}_\Omega \longrightarrow 0$$

Furthermore, $\mathbf{H}_{\widehat{\Omega}}$ *is a split subquotient of* $\mathbf{H}_\Omega$, *and* $\mathbf{H}_{\widehat{\Omega}} \cong \mathbf{H}^{\widehat{I}}_\Omega / \mathbf{H}^c_\Omega$.

PROOF. The proof is identical to the proof in [**Li23**] page 17 proposition 2.4.1. □

References

[Hal59] P. Hall, *The algebra of partitions*, Proceedings of the 4th Canadian Mathematical Congress, Banff, 1957, University of Toronto Press, Toronto, 1959, pp. 147–159.

[Li23] Y. Li, *Quantum groups and edge contractions*, Preprint, arXiv:2308.16306 [math.QA], 2023.

[Lus91] G. Lusztig, *Quivers, perverse sheaves, and quantized enveloping algebras*, J. Amer. Math. Soc. **4** (1991), no. 2, 365–421, DOI 10.2307/2939279. MR1088333

[Lus98] G. Lusztig, *Canonical bases and hall algebras*, Representation Theories and Algebraic Geometry (A. Broer, A. Daigneault, and G. Sabidussi, eds.), Nato ASI Series C: Mathematical and Physical Sciences, vol. 514, Springer, Dordrecht, 1998, pp. 365–399.

[LR24] Y. Li and J. Ren *Critical cohomological Hall algebra and edge contraction*, Preprint, arXiv:2401.04839, 2024.

[Lus10] G. Lusztig, *Introduction to quantum groups*, Birkhäuser Boston, Boston, MA, 2010.

[Mak18] Ruslan Maksimau, *Categorical representations and KLR algebras*, Algebra Number Theory **12** (2018), no. 8, 1887–1921, DOI 10.2140/ant.2018.12.1887. MR3892967

[Rin90] Claus Michael Ringel, *Hall algebras and quantum groups*, Invent. Math. **101** (1990), no. 3, 583–591, DOI 10.1007/BF01231516. MR1062796

[Ste01] E. Steinitz, *Zur theorie der abel'schen gruppen*, Jahresbericht der Deutschen Mathematiker-Vereinigung **9** (1901), 80–85.

DEPARTMENT OF MATHEMATICS, UNIVERSITY AT BUFFALO, THE STATE UNIVERSITY OF NEW YORK, BUFFALO, NEW YORK 14260

Email address: adhishpr@buffalo.edu

Contemporary Mathematics
Volume **845**, 2026
https://doi.org/10.1090/conm/845/16960

Generalized Weyl-type algebras and Witt-type Lie algebras on expolynomials and their modules

Supriya Sharma, R. S. Raja Durai, and Ki-Bong Nam

Abstract. The functions $y = e^x$ and $y = x^x$ are interesting ones since some real solutions of the equations are transcendental numbers which can be used for a cryptography. So we need the expolynomial ring $\mathbb{F}(m^*, k, n)$ which includes the polynomial ring and the differential equation $y' - ix_t^{it-1}y = 0, 1 \leq t \leq m$, has the solution in $\mathbb{F}(m^*, k, n)$. We define the Weyl-type algebra $A(\mathbb{F}(m^*, k, n))$ and the Witt-type Lie algebra $W(\mathbb{F}(m^*, k, n))$ on $\mathbb{F}(m^*, k, n)$, and show that the algebras are simple. We also show that $\mathbb{F}(m^*, k, n)$ is a $A(\mathbb{F}(m^*, k, n))$-module, and by using an automorphism θ of $A(\mathbb{F}(m^*, k, n))$, we define twisted Weyl-type module $\mathbb{F}(m^*, k, n)_\theta$. We show that there are uncountably many pairwise non-isomorphic Weyl-type algebras and uncountably many pairwise non-isomorphic twisted Weyl-type modules of $\mathbb{F}(m^*, k, n)$ in this paper. We also show that there are uncountably many pairwise non-isomorphic Witt-type simple Lie algebras. We show that the real solution of the equation $x^{x^x} = 10$ is an irrational number in this work.

1. Introduction

Weyl-type algebras, Witt-type Lie algebras, and related non-associative algebras play an important role in connecting algebraic structures with differential operators and transcendental functions [**2**, **4**]. For a given $\mathbb{F}$-algebra A, if A is commutative and associative, then we define the Weyl-type algebra $A(A)$, the Witt-type Lie algebra $W(A)$, and the non-associative algebra $N(A)$ [**11**, **14**]. In this paper, we consider the $\mathbb{F}$-algebra $\mathbb{F}[m^*, k, n]$ and prove that the algebras $A(\mathbb{F}[m^*, k, n])$, $W(\mathbb{F}[m^*, k, n])$, and $N(\mathbb{F}[m^*, k, n])$ are simple. These results show that A, W, and N can be regarded as functors from the category of all associative–commutative algebras to the categories of Weyl-type algebras, Witt-type Lie algebras, and non-associative algebras, respectively. We show that there are uncountably many non-isomorphic Weyl-type algebras and Witt-type algebras. Beyond their intrinsic algebraic interest, these algebras connect to fundamental questions in number theory and transcendental analysis [**3**]. It is well known that the real solution of $x^x = 100$ is a transcendental number. In this work, we also prove that the real solution of $x^{ax^x} = b$ is an irrational number where a is a non-zero rational number and b is positive algebraic number.

Key words and phrases. expolynomial ring, Weyl-type algebra, automorphism, isomorphism, twisted Weyl-type algebra-module, Witt-type Lie algebra, Lie module, transcendental differentiable function.

2. Preliminaries

Let $\mathbb{Z}$ and $\mathbb{N}$ be the sets of all integers and all non-negative integers respectively. Let $\mathbb{F}$ be a field of characteristics zero that contains the field $\mathbb{Q}$ of all rational numbers, and $\mathbb{F}^*$ and $\mathbb{Q}^*$ be the sets of all the non-zero elements of $\mathbb{F}$ and $\mathbb{Q}$ respectively. Let $\mathbb{F}[x_1, \ldots, x_n]$ be the polynomial ring over $\mathbb{F}$. Throughout the paper, A denotes an associative commutative $\mathbb{F}$-algebra. Since there are uncountably many associative commutative $\mathbb{F}$-algebras, for some results of the paper, let us define a specific $\mathbb{F}$-algebra as follows;

$$\mathbb{F}(m^*, k, n) := \mathbb{F}[e^{x_1^{i_1}}, \ldots, e^{x_m^{i_m}}, x_1^{\pm 1}, \ldots, x_k^{\pm 1}, x_{k+1}, \ldots, x_{k+n}] \tag{2.1}$$

where $i_1, \ldots, i_m$ are given integers, m, k, n are given non-negative integers and $m \leq k+n$. From now on, let us assume that $m > 0$ without any further comments. Since the Lie algebras on $\mathbb{F}(m^*, k, n)$ are studied in the paper [**14**], and for $i_1 = \cdots = i_m = 1$, the Weyl-type algebra on the associative algebra $\mathbb{F}(m, k, n)$ is studied in the paper [**15**]. Throughout the paper, the $\mathbb{F}$-subalgebras $\mathbb{F}(m^*, 0, 0)$, $\mathbb{F}(0, k, n)$, $\mathbb{F}(0, k, 0)$, $\mathbb{F}(0, 0, n)$, and other subalgebras of $\mathbb{F}(m, k, n)$ can be defined without any ambiguity. An $\mathbb{F}$-subalgebra A of the Laurent formal power series ring $\mathbb{F}[[x_1^{\pm 1}, \ldots, x_n^{\pm 1}]]$ is stable with respect to the derivatives of the ring if for any f of A, $\frac{\partial^{i+\cdots+j}}{\partial x_1^i \cdots \partial x_n^j}(f) \in A$ holds for any $i, \ldots, j \in \mathbb{N}$ (see [**20**]). Let $\mathcal{T}$ be the set of all the transcendental numbers. Two transcendental numbers η_1 and η_2 are equivalent if there is a rational number r such that $\eta_1 = r\eta_2$ holds. This gives the equivalence class $\overline{\eta} = \{\eta_1 | \eta = r\eta_1, \eta_1 \in \mathcal{T} \text{ for } r \in \mathbb{Q}\}$ of η of $\mathcal{T}$. So we have the set of equivalent classes of $\mathcal{T}$:

$$\mathcal{E} = \{\overline{\eta} | \eta \in \mathcal{T}\}.$$

It is obvious that there are uncountably many non equivalent transcendental numbers.

3. Weyl-type Simple Algebras

For the future reference of a Weyl algebra or its generalization, let us give a general definition of the Weyl algebra on a commutative $\mathbb{F}$-algebra A. An $\mathbb{F}$-linear map $\partial : A \to A$ is a derivation if for all $a, b \in A$, $\partial(ab) = \partial(a)b + a\partial(b)$ holds. Obviously, we have that for any $\alpha \in \mathbb{F}$, $\partial(\alpha) = 0$. It is easy to check that for any $a \in A$ and ∂, $a\partial$ is a derivation of A as well. Throughtout the paper, D is the set of all derivations of A. For any $a \in A$, we define a map $\widehat{a}$ from A to itself as follows:

$$\widehat{a}(b) = ab$$

for any $b \in A$. We have the following obvious relation which generalizes the Weyl relation $\partial_i \circ \widehat{x_j} - \widehat{x_j} \circ \partial_i = \widehat{\delta_{i,j}}$ on $\mathbb{F}[x_1, \ldots, x_n]$, where $\partial_i := \frac{\partial}{\partial x_i}$ is the derivation on $\mathbb{F}[x_1, \ldots, x_n]$ with respect to x_i, $\widehat{x_j}$ is the muliplication map which is defined as $\widehat{a}$, $\delta_{i,j}$ is the Kronecker delta, and $\circ$ is the composition of maps (see [**2**], [**4**], [**5**], [**11**], [**15**]). While some references use $\widehat{\partial}$ instead of ∂, throughout the paper, we use ∂ for brevity. Since by definition of D, we have the following basic lemma for the Weyl algebra or its generalization.

LEMMA 3.1. *For any $\partial \in D$ and $a \in A$, the relation*

$$\partial \circ \widehat{a} - \widehat{a} \circ \partial = \widehat{\partial(a)} \tag{3.1}$$

holds. □

For an associative commutative algebra A with the identity element 1, if there is an element a_1 of A such that $\partial(a_1) = 1$, then the Weyl relation $\partial\widehat{a}_1 - \widehat{a}_1\partial = \widehat{1}$ holds where $\widehat{1} = 1$ is the identity map on A. From now on, we will omit the symbol $\circ$ for the sake of brevity. The Weyl relation

$$\partial_i\widehat{x}_j - \widehat{x}_j\partial_i = \widehat{\delta}_{i,j} \tag{3.2}$$

of $A_n = A(\mathbb{F}[x_1, \ldots, x_n])$ is a special case of the relation of Lemma 3.1 and for any $a, b \in A$, $\widehat{ab} = \widehat{ba}$ holds. By lemma 3.1, we define Weyl-type algebra $A(A)$ on the $\mathbb{F}$-algebra A as follows:

$$A(A) = \mathbb{F}[\widehat{a}, \partial]/\mathcal{R}$$

for $a \in A$ and $\partial \in D$ where $\mathcal{R}$ is the relation of lemma 3.1, $A(A)$ is obviously a vector space over $\mathbb{F}$, and $\circ$ is the multiplication of algebra. $A(A)$ is a non-commutative algebra which has no zero divisors. If we take $A = \mathbb{F}(m^*, k, n)$ in (2.1), then we have the relation $\mathcal{R}$ by Lemma 3.1 (see [**2**] and [**4**]). So we define the Weyl-type algebra $A(\mathbb{F}(m^*, k, n)) =< \mathbb{F}[\widehat{e^{x_1^{i_1}}}, \ldots, \widehat{e^{x_m^{i_m}}}, \widehat{x_1^{\pm 1}}, \ldots, \widehat{x_k^{\pm 1}}, \widehat{x_{k+1}}, \ldots, \widehat{x_{k+n}}, \partial_1, \ldots, \partial_{k+n}]/\mathcal{R}$, $\circ, + >$ as $A(A)$. From now on, let us omit the hats of $\widehat{e^{x_u^{i_u}}}$, $1 \le u \le m$, $\widehat{x_v^{\pm 1}}$, $1 \le v \le k$, and, $\widehat{x_w}$, $k+1 \le w \le k+n$, because it is just a multiplication map and throughout the paper, the deletion of the hat does not give any ambiguity. The algebra $A(\mathbb{F}(m^*, k, n))$ is $\mathbb{Z}^m$-graded as follows:

$$A(\mathbb{F}(m^*, k, n)) = \bigoplus_{(a_1, \ldots, a_m) \in \mathbb{Z}^m} A_{a_1, \ldots, a_m}, \tag{3.3}$$

where the $(a_1, \ldots, a_m)$-homogeneous component $A_{(a_1, \ldots, a_m)}$ is a vector subspace of $A(\mathbb{F}(m^*, k, n))$ which is spanned by

$$\begin{aligned} B = \{&e^{a_1 x_1^{i_1}} \cdots e^{a_m x_m^{i_m}} x_1^{j_1} \cdots x_k^{j_k} x_{k+1}^{j_{k+1}} \cdots x_{k+n}^{j_{k+n}} \partial_1^{s_1} \cdots \partial_{k+n}^{s_{k+n}}, \\ &j_1, \ldots, j_k \in \mathbb{Z}, j_{k+1}, \ldots, j_{k+n}, s_1, \ldots, s_{k+n} \in \mathbb{N}\}. \end{aligned} \tag{3.4}$$

Throughout the paper, we call

$$M = e^{a_1 x_1^{i_1}} \cdots e^{a_m x_m^{i_m}} x_1^{j_1} \cdots x_k^{j_k} x_{k+1}^{j_{k+1}} \cdots x_{k+n}^{j_{k+n}} \partial_1^{s_1} \cdots \partial_{k+n}^{s_{k+n}},$$

the regular monomial/expolynomial and write it as exactly the order of the variables in (3.4). For

$$M_1 = e^{a_1 x_1^{i_1}} \cdots e^{a_m x_m^{i_m}} x_1^{j_1} \cdots x_k^{j_k} x_{k+1}^{j_{k+1}} \cdots x_{k+n}^{j_{k+n}} \partial_1^{s_1} \cdots \partial_{k+n}^{s_{k+n}},$$

and

$$M_2 = e^{a'_1 x_1^{i_1}} \cdots e^{a'_m x_m^{i_m}} x_1^{j'_1} \cdots x_k^{j'_k} x_{k+1}^{j'_{k+1}} \cdots x_{k+n}^{j'_{k+n}} \partial_1^{s'_1} \cdots \partial_{k+n}^{s'_{k+n}}.$$

We define lexicographic order of them as follows: $M_1 > M_2$ if and only if $a_1 > a'_1$, or $a_1 = a'_1$ and $a_2 > a'_2$, and so on so forth. In this way we can define the lexicographic order of the elements of the algebra $A(\mathbb{F}(m^*, k, n))$. Since every element of the algebra $A(\mathbb{F}(m^*, k, n))$ is the linear combination of the elements of B, we write it using the lexicographic order. The element $e^{a_1 x_1^{i_1}} \cdots e^{a_m x_m^{i_m}} x_1^{j_1} \cdots x_k^{j_k}$ is an invertible element of the algebra. For the future reference, let us show that the algebra $A(\mathbb{F}(m^*, k, n))$ is simple. Note that the algebra $A(\mathbb{F}(m^*, k, n))$ does not have a zero divisor. Since the Weyl-type algebra $A(A)$ is a non-commutative algebra, $A(A)$ is simple if and only if it has only two two-sided ideals $\{0\}$ and itself. So if A is an $\mathbb{F}$-algebra, then we define $W(A)$ as the Weyl-type Lie algebra as well (see Section 3). From now on, we define subalgebras $A(\mathbb{F}(m^*, 0, n))$, $m \le n$, and $A(\mathbb{F}(m^*, k, 0))$,

$m \leq k$, of $A(\mathbb{F}(m^*, k, n))$ freely. The Weyl-type algebra $A(\mathbb{F}(m, k, n))$ is studied in the paper [**15**].

LEMMA 3.2. *For any element $f(x)$ of a stable subalgebra A of $\mathbb{F}[[x^{\pm 1}]]$,*

$$\partial^i x^a(f(x)) = (\binom{a}{0} x^a \partial^i + i \binom{a}{1} x^{a-1}\partial^{i-1} + i(i-1)\binom{a}{2} x^{a-2}\partial^{i-2} + \cdots$$

$$+ i(i-1)\cdots(i-b+1)\binom{a}{b} x^{a-b}\partial^{i-b} + \cdots$$

$$+ i(i-1)\cdots(i-a+1)\binom{a}{a}\partial^{i-a})(f(x)), \tag{3.5}$$

holds, where $\binom{a}{b} = \frac{a!}{(a-b)!b!}$ *for $a, b, i \in \mathbb{N}$. Furthermore, for $g \in A$, the followings*

(3.6)
$$\partial^i g = g\partial^i + \binom{i}{1} g'\partial^{i-1} + \cdots + \binom{i}{r} g^{(r)}\partial^{i-r} + \cdots + \binom{i}{i-1} g^{(i-1)}\partial +$$
$$\binom{i}{i} g^{(i)},$$

(3.7)
$$\partial x^k = x^k\partial + kx^{k-1},$$

(3.8)
$$\partial^i x = x\partial^i + i\partial^{i-1},$$

(3.9)
$$\partial x^{-s} = x^{-s}\partial - sx^{-s-1},$$

(3.10)
$$\partial^i x^{-k} = x^{-k}\partial^i - \binom{i}{1} kx^{-k-1}\partial^{i-1} - \binom{i}{2} k(-k-1)x^{-k-2}\partial^{i-2} - \cdots$$
$$- \binom{i}{t} k(-k-1)\cdots(-k-t+1)x^{-k-t}\partial^{i-t} - \cdots - \binom{i}{i} k(-k-1)\cdots$$
$$(-k-i+1)x^{-k-i}$$

hold where we denote $g(x)$ as g which is really an operator on A and $k, i, s \in \mathbb{N}$.

PROOF. The formula (3.5) can be proven by induction on a and i of $\partial^i x^a$ and the evaluations of both sides of (3.5) on $f(x)$, and the rest of the lemma can be proven by induction on i of $\partial^i g$ and k of ∂x^k. □

The Weyl realtion $\partial x - x\partial = 1$ is a special case of (2.8).

PROPOSITION 3.3. *The subalgebra $A(\mathbb{F}(0, k, n))$ of $A(\mathbb{F}(m^*, k, n))$ is simple and $A(\mathbb{F}(0, k, n))$ is the $(0, \ldots, 0)$-homogeneous component of the algebra $A(\mathbb{F}(m^*, k, n))$.*

PROOF. Let I be a non-zero ideal of $A(\mathbb{F}(0, k, n))$ and l be a non-zero element of I. It suffices to prove that I contains a non-zero element of $A_n = A(\mathbb{F}(0, 0, n))$ which is simple, i.e., I contains 1. Since l can be written as the linear combination of the monomials of $B \cap A(\mathbb{F}(0, k, n)$ of (3.4). If we take an appropriate non-zero element l_1 of $A(\mathbb{F}(0, 0, n))$, then since $A(\mathbb{F}(0, k, n))$ has no zero divisor, we have the non-zero element $l_1 l \in A(\mathbb{F}(0, 0, n))$. This implies that $1 \in I$, i.e., $I = A(\mathbb{F}(0, k, n))$.

So the algebra $A(\mathbb{F}(0,k,n))$ is simple. The proof of the rest of the proposition is straightforward by definition of the gradation of the algebra. □

As the simplicity of the Weyl-type algebra $A(\mathbb{F}[e^{\pm x_1},\dots,e^{\pm x_m}])$ on $\mathbb{F}[e^{\pm x_1},\dots, e^{\pm x_m}]$ is not found in the literatures, it may be worth to provide a proof here. Note that all the derivations of $\mathbb{F}[e^{\pm x_1},\dots,e^{\pm x_m}]$ are inner (see **[4]**.)

PROPOSITION 3.4. *The Weyl-type algebra $A(\mathbb{F}[e^{\pm x_1},\dots,e^{\pm x_m}])$ is simple.*

PROOF. The algebra $A(\mathbb{F}[e^{\pm x_1},\dots,e^{\pm x_m}])$ does not have the Weyl relation (3.2), but it has a similar commuting relation $\partial_i e^{x_j} - e^{x_j}\partial_i = \delta_{i,j}e^{x_j}$ for $1 \le i,j \le m$. The algebra $A(\mathbb{F}[e^{\pm x_1},\dots,e^{\pm x_m}])$ has a similar $\mathbb{Z}^m$-gradation as (3.3) and the $(0,\dots,0)$-homogeneous component is spanned by $\{\partial_1^{s_1}\cdots\partial_m^{s_m} | s_1,\dots,s_m \in \mathbb{N}\}$. Let I be a non-zero ideal of the algebra $A(\mathbb{F}[e^{\pm x_1},\dots,e^{\pm x_m}])$ and l be a non-zero element of I. We prove the proposition by induction on the number $\#(l)$ of different components which has non-zero terms of l. Let $\#(l) = 1$ and l is in the $(0,\dots,0)$-homogeneous component. We have that $l = c\partial_1^{s_1}\cdots\partial_m^{s_m} + l_1$ where l_1 is the sum of the remaining terms of l, $\partial_1^{s_1}\cdots\partial_m^{s_m}$ is the maximal term, and c is a non-zero scalar. Without loss of generality, we can assume that s_1 is not zero. By substituting g by e^{x_1} in (3.7), we have the non-zero element

$$le^{x_1} - e^{x_1}l = c(\partial_1^{s_1}e^{x_1} - e^{x_1}\partial_1^{s_1})\cdots\partial_m^{s_m} + l_2 \tag{3.11}$$

of I such that the degree of ∂_1 of $le^{x_1} - e^{x_1}l$ is less than the degree of ∂_1 of l where l_2 is the sum of the remaining terms of $le^{x_1} - e^{x_1}l$. Repeating the operation (3.11), we have a non-zero element l_3 of I such that l_3 can be a non-zero scalar, i.e., $1 \in I$. So we have proven the proposition in this case. Let us now assume that l is a linear combination of elements of the $(a_1,\dots,a_m)$-homogeneous component where at least one of $a_1,\dots,a_m$ is not zero. Then the element $e^{-a_1x_1}\cdots e^{-a_mx_m}l$ is a non-zero element of I and it is in the $(0,\dots,0)$-homogeneous component. So we have proven the proposition already. Let us assume that we have proven the proposition for $\#(l) = p$ where $p > 1$ is a given positive integer. Let us assume that $\#(l) = p+1$. If l does not have a term of the $(0,\dots,0)$-homogeneous component and it has a non-zero term of the $(b_1,\dots,b_m)$-homogeneous component. Then $e^{-b_1x_1}\cdots e^{-b_mx_m}l$ is a non-zero element and it has a term of the $(0,\dots,0)$-homogeneous component. So we can assume that the element l has a term of the $(0,\dots,0)$-homogeneous component. By repeating the operation (3.11) and by (3.7), we have a non-zero element l_4 of I such that l_4 has no term of the $(0,\dots,0)$-homogeneous component. This implies that $\#(l) \le p$. So by induction, the ideal I is the algebra $A(\mathbb{F}[e^{\pm x_1},\dots,e^{\pm x_m}])$, i.e., the algebra is simple, and the proof is complete. □

THEOREM 3.5. *The Weyl-type algebra $A(\mathbb{F}(m^*,k,n))$ is simple where $m \le k+n$.*

PROOF. Let I be a non-zero ideal of $A(\mathbb{F}(m^*,k,n))$ and l be a non-zero element of I. The element l is a linear combination of the elements of homogeneous components of the algebra with respect to the lexicographic order. Let us prove the theorem by induction on the number $\#(l)$ as in Proposition 3.4. Let us assume that $\#(l) = 1$. For this case, we are able to prove the theorem as the proof of $\#(l) = 1$ of Proposition 3.4. Let us assume that we have proven the theorem for $\#(l) = p$ where $p > 1$ is a given positive integer. Let us prove that l has non-zero terms of $(p+1)$-homogeneous components. By taking an appropriate element l_1, we have that $l_1 l$ has a non-zero term of the $(0,\dots,0)$-homogeneous component. Let $cx_1^{r_1}x_2^{r_2}\cdots x_{k+n}^{r_{k+n}}\partial_1^{q_1}\cdots x_{k+n}^{q_{k+n}}$ be the maximal term of the element $l_1 l$ in the $(0,\dots,0)$-homogeneous component where $r_1,\dots,r_{k+n}\geq 0$ by taking appropriate l_1 and $c\in\mathbb{F}^*$. We have the following non-zero element

$$\begin{aligned} l_2 &= \partial_1 l_1 l - l_1 l \partial_1 \\ &= l_2' + c(\partial_1 x_1^{r_1} - x_1^{r_1}\partial_1)x_2^{r_2}\cdots x_{k+n}^{r_{k+n}}\partial_1^{q_1}\cdots x_{k+n}^{q_{k+n}} + l_3 \\ &= l_2' + c_1 r_1 x_1^{r_1-1}x_2^{r_2}\cdots x_{k+n}^{r_{k+n}}\partial_1^{q_1}\cdots x_{k+n}^{q_{k+n}} + l_4 \in I \end{aligned} \tag{3.12}$$

where l_2' is the linear combinations of elements which are not in $(0,\dots,0)$-homogeneous components, and l_3 and l_4 are the sums of remaining terms of $\partial_1 l_1 l - l_1 l \partial_1$ which are in the $(0,\dots,0)$-homogeneous components with appropriate non-zero coefficients c_1 and others. Assume that the theorem holds for $\#(l)=p$ and that $\#(l_2)=p+1$. By repeating the similar operations of (3.12), we have the following non-zero element l_5 of I:

$$l_5 = l_6 + c_1\partial_1^{s_1}\cdots\partial_{k+n}^{s_{k+n}} + l_7,$$

where l_6 is the sum of elements which are not in the $(0,\dots,0)$-homogeneous component and $\partial_1^{s_1}\cdots\partial_{k+n}^{s_{k+n}}+l_7$ is the sum of elements of the $(0,\dots,0)$-homogeneous component such that $\partial_1^{s_1}\cdots\partial_{k+n}^{s_{k+n}}$ is the maximal term of the $(0,\dots,0)$-homogeneous component with appropriate non-zero coefficients. If $\#(l_5)\leq p$, then there is nothing to prove by induction. Let us assume that $\#(l_5)=p+1$. Without loss of generality, we can assume that l_5 has a non-zero term with x_1. Then $l_8=\partial_1 l_5 - l_5\partial_1$ is a non-zero element of I such that $\#(l_8)\leq p$. Thus by induction, we have proven the theorem. □

COROLLARY 3.6. *The Weyl-type algebra $A(\mathbb{F}(m^*,k,0))$ and the Weyl algebra $A(\mathbb{F}(0,0,m))=A_m$ are simple where $m\leq k$.*

PROOF. The proof of the corollary is straightforward from the proof of Theorem 1. □

It is well known that $A_n\otimes_{\mathbb{F}} A_1$ is isomorphic to A_{n+1} for any postive integer n.

PROPOSITION 3.7. *The algebras $A(\mathbb{F}[x_1^{\pm1},\dots,x_n^{\pm1}])\otimes A(\mathbb{F}[x_{n+1}^{\pm1}])$ and $A(\mathbb{F}[x_1^{\pm1},\dots,x_{n+1}^{\pm1}])$ are isomorphic.*

PROOF. For $\epsilon\in\{1,-1\}$, we define the maps

$$\theta : A(\mathbb{F}[x_1^{\pm1},\dots,x_n^{\pm1}]) \to A(\mathbb{F}[x_1^{\pm1},\dots,x_{n+1}^{\pm1}]),$$
$$\phi : A(\mathbb{F}[x_{n+1}^{\pm1}]) \to A(\mathbb{F}[x_1^{\pm1},\dots,x_{n+1}^{\pm1}]),$$

by

$$\theta(x_i^\epsilon)=x_i^\epsilon,\ \theta(\partial_i)=\partial_i, 1\leq i\leq n,$$
$$\phi(x_{n+1}^\epsilon)=x_{n+1}^\epsilon,\ \phi(\partial_{n+1})=\partial_{n+1}.$$

Then we have the map $\theta\otimes\phi : A(\mathbb{F}[x_1^{\pm1},\dots,x_n^{\pm1}])\otimes A(\mathbb{F}[x_{n+1}^{\pm1}])$. If we define a map ψ from $A(\mathbb{F}[x_1^{\pm1},\dots,x_{n+1}^{\pm1}])$ to $A(\mathbb{F}[x_1^{\pm1},\dots,x_n^{\pm1}])\otimes A(\mathbb{F}[x_{n+1}^{\pm1}])$ by

$$\psi(x_i^\epsilon)=x_i^\epsilon\otimes 1,\ 1\leq i\leq n,$$

$$\psi(\partial_i)=\partial_i\otimes 1,\ \psi(x_{n+1}^\epsilon)=1\otimes x_{n+1}^\epsilon \text{ and } \psi(\partial_{n+1})=1\otimes\partial_{n+1}.$$

Then the maps $\theta\otimes\psi$ and ψ are inverses of each other. So they are isomorphic. □

THEOREM 3.8. *If $m,k < n$, then the algebras $A(\mathbb{F}((m-1)^*,k,n-1)) \otimes_{\mathbb{F}} A(\mathbb{F}(1^*,0,1))$ and $A(\mathbb{F}(m^*,k,n))$ are isomorphic.*

PROOF. The proof of the theorem is similar to that of Proposition 3.7, and we omit it. □

By Theorem 3.5, the algebra $A(\mathbb{F}(m^*,k,n))$ is simple and the proof of Theorem 3.5 is necessary for the proof of Theorems 4.2, 5.1, and 5.2 in this paper. Actually, by Theorem 3.8, it suffices to prove that the algebra $A(\mathbb{F}(1^*,1,0))$, $A(\mathbb{F}(1^*,0,1))$, or $A(\mathbb{F}(1^*,1,1))$ is simple for the simplicity of $A(\mathbb{F}(m^*,k,n))$. For the given set $S = \{e^{x^4}x^2, e^{x^4}x^5, e^{x^4}x^{10}\}$, we can not use the Wronskian method for the linear independence of S. By the proof of Theorem 3.5, we are able to prove the linear independence of S easily.

4. Simple Witt-type Lie Algebras

For the future reference of the Witt-type Lie algebra or its generalization (see [**9**]), we need the following obvious lemma.

LEMMA 4.1. *For the elements a_1 and a_2 of the associative commutative algebra A, and for $a_1\partial_1, a_2\partial_2 \in D$,*

$$[a_1\partial_1, a_2\partial_2] = a_1\partial_1 \circ a_2\partial_2 - a_2\partial_2 \circ a_1\partial_1 = a_1\partial_1(a_2)\partial_2 - a_2\partial_2(a_1)\partial_1 \tag{4.1}$$

holds. □

We define the Witt-type Lie algebra $W(A)$ on an associative commutative algebra A is the set

$$\{\sum_i a_i\partial_i | a_i \in A, \partial_i \in D\}$$

with the Lie bracket $[,]$ of (4.1) such that the Jacobi identity and other properties of a Lie algebra hold with respect to the Lie bracket

$$[a\partial_1, b\partial_2] = a\partial_1 \circ b\partial_2 - b\partial_2 \circ a\partial_1 = a\partial_1(b)\partial_2 - b\partial_2(a)\partial_1 \text{ for } a\partial_1, b\partial_2 \in W(A).$$

This simple calculation gives the Witt-type Lie algebra $W(A)$ on the $\mathbb{F}$-algebra A. From now on, we omit the composition $\circ$ of maps generally because of the clear definition of it. The Witt type Lie algebra $W(\mathbb{F}(m^*,k,n))$ is spanned by

$$B_1 = \{e^{a_1x_1^{i_1}} \cdots e^{a_mx_m^{i_m}} x_1^{j_1} \cdots x_k^{j_k} x_{k+1}^{j_{k+1}} \cdots x_{k+n}^{j_{k+n}} \partial_u,$$
$$a_1, \dots, a_m, j_1, \dots, j_k \in \mathbb{Z}, j_{k+1}, \dots, j_{k+n} \in \mathbb{N}, 1 \le u \le k+n\}.$$

The well-known simple Witt algebra $W(n^+)$ on $\mathbb{F}[x_1,\dots,x_n]$ is spanned by the standard basis $\{x_1^{i_1}\dots x_n^{i_n}\partial_u | i_1,\dots,i_n \in \mathbb{N}, 1 \le u \le n\}$ with the Lie bracket (4.1) which gives the Witt-type Lie algebra. We denote the Witt algebra $W(n^+)$ on $\mathbb{F}[x_1,\dots,x_n]$ as $W(\mathbb{F}[x_1,\dots,x_n])$ because we will define many Witt-type Lie algebras on $\mathbb{F}$-algebras. Let S be a non-empty subset of a Lie algebra L. For a non-zero element l of L is ad-diagonal with respect to the set S if for all x of S, $[l,s] = c_s s$ holds where c_s is a scalar depending on the element s. For a given non-empty subset S of a Lie algebra L, S is a subalgebra of L if S is a vector subspace of L and for any $l_1, l_2 \in S$, $[l_1,l_2] \in S$ holds. For a given non-zero element l of a Lie algebra L, the centralizer $Cl_L(l)$ of l is a subalgebra of L. Since the Witt algebra $W^+(1)$ is spanned by $\{x^i\partial | i \in \mathbb{N}\}$, $W^+(1)$ is generated by $\{\partial, x^k\partial | k > 2, k \in \mathbb{N}\}$.

THEOREM 4.2. *The Witt-type Lie algebra $W(\mathbb{F}(m^*,k,n))$ is simple.*

PROOF. Since the Lie algebra has a similar $\mathbb{Z}^m$-gradation as (3.3), so the proof of the theorem is standard (see [**14**], [**13**]). □

COROLLARY 4.3. *The Witt-type algebras* $W(\mathbb{F}(0,m,0))$ *and* $W(\mathbb{F}(m,0,0))$ *are simple.*

PROOF. By Theorem 4.2, the proof of the corollary is obvious and well-known. □

PROPOSITION 4.4. *The Witt-type Lie algebras* $W(\mathbb{F}(0,m,0))$ *and* $W(\mathbb{F}(m,0,0))$ *are isomorphic.*

PROOF. If we define $\mathbb{F}$-linear map θ from $W(\mathbb{F}(0,m,0))$ to $W(\mathbb{F}(m,0,0))$ as follows:

$$\theta(x_1^{j_1}\cdots x_u^{j_u}x_{u+1}^{j_{u+1}}\cdots x_m^{j_m}\partial_u) = e^{j_1x_1}\cdots e_{u-1}^{j_{u-1}x_{u-1}}e_u^{(j_u-1)x_u}e_{u+1}^{j_{u+1}x_{u+1}}\cdots e^{j_mx_m}\partial_u,$$

then θ can be linearly extended to the Lie algebra isomorphism of the algebras for

$$x_1^{j_1}\cdots x_u^{j_u}x_{u+1}^{j_{u+1}}\cdots x_m^{j_m}\partial_u \in W(\mathbb{F}(0,m,0)),$$

and

$$e^{j_1x_1}\cdots e_{u-1}^{j_{u-1}x_{u-1}}e_u^{(j_u-1)x_u}e_{u+1}^{j_{u+1}x_{u+1}}\cdots e^{j_mx_m}\partial_u \in W(\mathbb{F}(m,0,0)).$$

□

COROLLARY 4.5. *The Lie algebra* $W(\mathbb{F}(m,0,0))$ *has a Lie subalgebra* L *which is isomorphic to* $W(\mathbb{F}(0,0,m))$.

PROOF. Let L be the Lie subalgebra of $W(\mathbb{F}(m,0,0))$ spanned by

$$\{e^{j_1x_1}\cdots e_{u-1}^{j_{u-1}x_{u-1}}e_u^{(j_u-1)x_u}e_{u+1}^{j_{u+1}x_{u+1}}\cdots e^{j_mx_m}\partial_u | j_1,\ldots,j_u,\ldots,j_m \in \mathbb{N}\}$$

which is closed under the Lie bracket. Then by Proposition 4.4, L is isomorphic to the Lie algebra $W(\mathbb{F}(0,0,m))$. □

Note that $W(\mathbb{F}(m^*,k,n))$ has a subalgebra $\mathfrak{S}$ which is spanned by $\{\partial_1, x_1\partial_1, x_1^2\partial\}$. $\mathfrak{S}$ is isomorphic to the Lie algebra $\mathfrak{sl}_2(\mathbb{F})$.

5. Localization

For a given non-empty subset S of $\mathbb{F}(m^*,k,n)$, we will define the localization $\mathbb{F}(m^*,k,n)_S$ of $\mathbb{F}(m^*,k,n)$ as follows: $\mathbb{F}(m^*,k,n)_S = \{\frac{f}{g} | f \in \mathbb{F}(m^*,k,n), g = g_1^{k_1}\cdots g_r^{k_r}$, where $g_1,\ldots,g_r \in S, k_1,\ldots,k_r \in \mathbb{N}\}$. So we define the Weyl-type algebra $A(\mathbb{F}(m^*,k,n)_S)$ and the Witt-type Lie algebra $W(\mathbb{F}(m^*,k,n)_S)$ on $\mathbb{F}(m^*,k,n)$ as $A(\mathbb{F}(m^*,k,n))$ and $W(\mathbb{F}(m^*,k,n))$ respectively. The algebra $A(\mathbb{F}(m^*,k,n)_S)$ has no zero divisor. For any non-zero element l of $A(\mathbb{F}(m^*,k,n)_S)$, if l is not in the subalgebra $A(\mathbb{F}(m^*,k,n))$ of $A(\mathbb{F}(m^*,k,n)_S)$, then we call l a local element.

THEOREM 5.1. *The Weyl-type algebra* $A(\mathbb{F}(m^*,k,n)_S)$ *is simple.*

PROOF. Let I be a non-zero ideal of $A(\mathbb{F}(m^*,k,n)_S)$ and l a non-zero element of I. Using the similar Weyl relation (3.1) on $A(\mathbb{F}(m^*,k,n)_S)$, we can write l as the sum of local terms and non-local terms of l. By taking an appropriate element l_1 of $A(\mathbb{F}(m^*,k,n)_S)$, we have a non-zero element $l_1l \in A(\mathbb{F}(m^*,k,n))$. By Theorem 3.5, I contains 1, i.e., $I = A(\mathbb{F}(m^*,k,n)_S)$. So the algebra $A(\mathbb{F}(m^*,k,n)_S)$ is simple. □

THEOREM 5.2. *The Witt-type Lie algebra* $W(\mathbb{F}(m^*,k,n)_S)$ *is simple.*

PROOF. The proof of the theorem is straightforward by Theorems 4.2 and 5.1, and we omit it. □

6. Automorphism Groups & Isomorphisms of Weyl-type Algebras and Witt-type Lie Algebras

It is easy to prove that the $\mathbb{F}$-algebras $\mathbb{F}[x^{\pm 1}]$, $\mathbb{F}[x^{\frac{\pm 1}{2}}]$, and $\mathbb{F}[e^{\pm x}]$ are isomorphic. By (3.9) and Proposition 6.1 of this section, the Weyl-type algebras $A(x^{\pm 1})$, $A(x^{\frac{\pm 1}{2}})$, and $A(e^{\pm x})$ are not isomorphic, and the Witt-type Lie algebras $W(x^{\frac{\pm 1}{2}})$ and $W(e^{\pm x})$ are not isomorphic as well. For the Weyl algebra $A(0,0,n) = A_n$, and for a given positive integer k, if we define $\mathbb{F}$-map $\theta_{\cdot,k}$ from A_n to itself as $\theta_{\cdot,k}(x_u) = x_u$ and $\theta_{\cdot,k}(\partial_u) = \partial_u + x_u^k$, $1 \leq u \leq n$, then $\theta_{\cdot,k}$ can be extended to an automorphism of A_n. Similarly, if we define $\mathbb{F}$-map $\theta_{l,\cdot}$ from A_n to itself as $\theta_{l,\cdot}(x_u) = x_u + \partial_u^l$ and $\theta_{l,\cdot}(\partial_u) = \partial_u$, $1 \leq u \leq n$, then $\theta_{l,\cdot}$ can be extended to an automorphism of A_n. It is well know that the automorphism group $Aut(A_1)$ is generated by the automorphisms $\theta_{\cdot,k}$ and $\theta_{l,\cdot}$ of A_1 (see [**4**], [**16**], and [**11**]). The automorphism τ of $Aut(A_n)$ is called the Fourier transformation such that $\tau(x_u) = -\partial_u$ and $\tau(\partial_u) = x_u$, $1 \leq u \leq n$. The automorphism group $Aut(A_n)$ of A_n is found in the paper [**1**]. For the Weyl-type algebra $A(\mathbb{F}[e^{\pm x^i}, x^{\pm 1}])$, if we define the $\mathbb{F}$-linear maps $\theta_{+,a,k}$ and $\theta_{-,b,l}$ from $A(\mathbb{F}[e^{\pm x^i}, x^{\pm 1}])$ to itself by:

$$\theta_{+,a,k}(x) = x, \theta_{+,a,k}(e^{x^i}) = e^{x^i}, \text{ and } \theta_{+,a,k}(\partial) = \partial + ax^k,$$

$$\theta_{-,b,l}(x) = x, \theta_{-,b,l}(e^{x^i}) = e^{-x^i}, \text{ and } \theta_{-,b,l}(\partial) = \partial + bx^l,$$

then they can be linearly extended to automorphisms of $A(\mathbb{F}[e^{\pm x^i}, x^{\pm 1}])$ where a and b are non-zero scalars for any given $k, l \in \mathbb{N}$. The automorphism group $Aut(A(\mathbb{F}[e^{\pm x^i}, x^{\pm 1}]))$ is generated by $\theta_{+,a,k}$ and $\theta_{-,b,l}$. Note that there is no Fourier transformation of $A(\mathbb{F}[e^{\pm x^i}, x^{\pm 1}])$ (see [**4**], [**9**]).

NOTE 1. The Weyl-type algebra $A(\mathbb{F}[e^{\pm x}])$ on $\mathbb{F}[e^{\pm x}]$ is generated by the elements e^x, e^{-x}, and ∂. If we define $\mathbb{F}$-linear maps θ and ϕ from $A(\mathbb{F}[e^{\pm x}])$ to itself as follows,

$$\theta(e^x) = e^x, \text{ and } \theta(\partial) = \partial + \sum c_i f_i,$$
$$\phi(e^x) = e^{-x}, \text{ and } \phi(\partial) = -\partial + \sum c_j g_j, \tag{6.1}$$

then θ and ϕ are linearly extended to the automorphisms of $A(\mathbb{F}[e^{\pm x}])$ where $c_i, c_j \in \mathbb{F}$, $f_i, g_j \in \mathbb{F}[e^{\pm x}]$. Note that f_i, g_j are operators on $\mathbb{F}[e^{\pm x}]$ because of clear definitions of them, and θ and ϕ have inverses. □

PROPOSITION 6.1. *The automorphism group $Aut(A(\mathbb{F}[e^{\pm x}]))$ of the simple algebrs $A(\mathbb{F}[e^{\pm x}])$ is generated by θ and ϕ which are defined in Note 1.*

PROOF. To find the automorphism group $Aut(A(\mathbb{F}[e^{\pm x}]))$, it is enough to find the automorphisms which are defined on the generators $\{e^x, e^{-x}, \partial\}$ of the algebra. Let θ be an automorphism of $Aut(A(\mathbb{F}[e^{\pm x}]))$. Since e^x is invertible and θ is onto, we have two cases, Case I: $\theta(e^x) = ce^x$ and Case II: $\theta(e^x) = c_1 e^{-x}$ hold where $c, c_1 \in \mathbb{F}^*$.

CASE I. Let us assume that $\theta(e^x) = ce^x$ holds. By $\theta(\partial)\theta(e^x) - \theta(e^x)\theta(\partial) = \theta(e^x)$, we have that

$$\theta(\partial)e^x - e^x\theta(\partial) = e^x. \tag{6.2}$$

So $\theta(\partial)$ has a term with the factor ∂, we put $\theta(\partial)$ as follows:

$$\theta(\partial) = c_{a,i}e^{ax}\partial^i + c_{a,i-1}e^{ax}\partial^{i-1} + \cdots + c_{a-1,j}e^{(a-1)x}\partial^j + \cdots \tag{6.3}$$

where $e^{ax}\partial^i$ has the maximal order of $\theta(\partial)$ with respect to the lexicographic order on ∂ as the first order and e^x as the second order, etc., and $\theta(\partial)$ is written as (6.3) with appropriate non-zero scalars. So by (3.9), (6.2), and (6.3), we have that $\theta(\partial) = \partial + c_{0,b}e^{bx} + \cdots$ with appropriate coefficients and the sum of the remaining terms of $\theta(\partial)$. This implies that $c = 1$, i.e., $\theta(e^x) = e^x$. The map θ such that $\theta(e^x) = e^x$ and $\theta(\partial) = \partial + c_{0,b}e^{bx} + \cdots$ induces an automorphism of $A(\mathbb{F}[e^{\pm x}])$ and this is the map θ in Note 1.

CASE II. Let us assume that $\theta(e^x) = c_1e^{-x}$ holds. By $\theta(\partial)\theta(e^x) - \theta(e^x)\theta(\partial) = \theta(e^x)$, we have that

$$\theta(\partial)e^{-x} - e^{-x}\theta(\partial) = e^{-x}. \tag{6.4}$$

So we have that $\theta(\partial) = -\partial + d_{0,b}e^{bx} + \cdots$ as Case I with appropriate coefficients. The map θ such that $\theta(e^x) = e^{-x}$ and $\theta(\partial) = -\partial + c_{0,b}e^{bx} + \cdots$ induces an automorphism of $A(\mathbb{F}[e^{\pm x}])$ and this is the map ϕ in Note 1. So we have proven the proposition. □

PROPOSITION 6.2. *For non-equivalent transcendental numbers a and b, the simple algebras $A(\mathbb{F}[(x^2 + ax + 1)^{\pm 1}, x^{\pm 1}])$ and $A(\mathbb{F}[(x^2 + bx + 1)^{\pm 1}, x^{\pm 1}])$ are not isomorphic.*

PROOF. Suppose that there is an isomorphism θ from $A(\mathbb{F}[(x^2+ax+1)^{\pm 1}, x^{\pm 1}])$ to $A(\mathbb{F}[(x^2 + bx + 1)^{\pm 1}, x^{\pm 1}])$. Since x is invertible, we have that $\theta(x) = cx^m(x^2 + bx + 1)^n$ for $m, n \in \mathbb{Z}$ and $c \in \mathbb{F}^*$. This implies that by

$$c\theta(\partial)x^m(x^2 + bx + 1)^n - cx^m(x^2 + bx + 1)^n\theta(\partial) = 1, \tag{6.5}$$

$\theta(\partial)$ has a term with the factor ∂, i.e.,

$$\theta(\partial) = g_t(x)\partial^t + \cdots + g_1(x)\partial + g_0(x)$$

where $g_t(x), \ldots, g_0(x) \in A(\mathbb{F}[(x^2 + bx + 1)^{\pm 1}, x^{\pm 1}])$ with appropriate indices. This implies, by (3.9) and (6.5), that $t = 1$, i.e., $\theta(\partial) = g_1(x)\partial + g_0(x)$. This implies, by (6.5), that $m = 1$, $n = 0$, $\theta(x) = cx$ and $g_1(x) = c^{-1}$ i.e., $\theta(x) = cx$ and $\theta(\partial) = c^{-1}\partial + g_0(x)$. By $\theta(x) = cx$, we have that $\theta(x^2 + ax + 1) = c^2x^2 + acx + 1$. Since $x^2 + ax + 1$ is invertible, $\theta(x^2 + ax + 1) = c_1(x^2 + bx + 1)^k$ holds for $k \in \mathbb{Z}$ and $c_1 \in \mathbb{F}^*$. By the Weyl relation on $x^2 + ax + 1$ and ∂, we have that $k = 1$. So we have that $c = \pm 1$. This implies that $a = \pm b$. By the assumption, we have a contradiction. So the algebras are not isomorphic. □

COROLLARY 6.3. *The automorphism group $Aut(A(\mathbb{F}[(x^2 + ax + 1)^{\pm 1}, x])$ of the algebra $A(\mathbb{F}[(x^2 + ax + 1)^{\pm 1}, x])$ is generated by the map θ_f from $A(\mathbb{F}[(x^2 + ax + 1)^{\pm 1}, x]$ to itself such that $\theta_f(x) = x$, $\theta_f(\partial) = \partial + f$, and θ_f can be linearly extended to the automorphism of the algebra for $f \in \mathbb{F}[(x^2 + ax + 1)^{\pm 1}, x]$.*

PROOF. In the proof of Proposition 6.2, if we take a instead of b of $x^2 + bx + 1$, then the proof of the corollary is straightforward by that of Proposition 6.2. □

THEOREM 6.4. *There are uncountably many pairwise non-isomorphic Weyl-type simple algebras.*

PROOF. Since there are uncountably many non-equivalent transcendental numbers, the proof of the theorem are straightforward by Propositions 3.3 and 6.2. □

In Section 2, we found countably many Weyl-type simple algebras, but Theorem 6.4 shows that there are uncountably many pairwise non-isomorphic Weyl-type simple algebras. By the proof of Proposition 6.2, it can be readily shown that there is no non-zero homomorphism from $A(\mathbb{F}[(x^2+ax+1)^{\pm 1}, x])$ to $A(\mathbb{F}[x^{\pm 1}])$ and vice versa (see [**21**]). It is easy to prove that the Lie algebra $W(\mathbb{F}[(x^2+ax+1)^{\pm 1}, x^{\pm 1}])$ is simple (see [**12**]).

PROPOSITION 6.5. *For any different scalars a and b, the simple Lie algebras $W(\mathbb{F}[(x^2+ax+1)^{\pm 1}, x^{\pm 1}])$ and $W(\mathbb{F}[(x^2+bx+1)^{\pm 1}, x^{\pm 1}])$ are not isomorphic.*

PROOF. For any non-zero element of the Lie algebras $W(\mathbb{F}[(x^2+ax+1)^{\pm 1}, x^{\pm 1}])$ and $W(\mathbb{F}[(x^2+bx+1)^{\pm 1}, x^{\pm 1}])$, since the centralizer of the element is one dimensional, the algebras are self-centralized (see [**9**] and [**14**]). Let us assume that there is a Lie algebra isomorphism θ from $W(\mathbb{F}[(x^2+ax+1)^{\pm 1}, x^{\pm 1}])$ to $W(\mathbb{F}[(x^2+bx+1)^{\pm 1}, x^{\pm 1}])$. Since θ is injective, by

$$\theta([x\partial, x^n\partial]) = (n-1)\theta(x^n\partial), \tag{6.6}$$

we have that $\theta(x\partial) \neq c\partial$ for any non-zero scalar c and by taking $n = 0$. If $\theta(x\partial) \neq c_1 x\partial + c_0\partial$, then by (6.6), the maximal terms of $\theta(x^n\partial)$, $n \in \mathbb{N}$, are the same with non-zero scalars where $c_1 \in \mathbb{F}^*$ and $c_0 \in \mathbb{F}$. If we take a positive integer m which is not one, then we have that

$$\theta([x\partial, x^m\partial]) = (m-1)\theta(x^m\partial). \tag{6.7}$$

Since $\theta(x\partial) \neq c_1 x\partial + c_0\partial$ holds and the algebras are self-centralized, the maximal term of the left side of (6.7) and the maximal term of the right side of (6.7) are different. This gives that the equality of (6.7) does not hold, i.e., we have a contradiction. So by (6.7), we have that $\theta(x\partial) = x\partial + c_0\partial$. This implies that $\theta(\partial) = d\partial$ holds for a non-zero scalar d. By $\theta([d\partial, \theta(x^2\partial)] = 2x\partial + 2c_0\partial$, we have that $\theta(x^2\partial) = x^2\partial + \frac{2c_0}{d}x\partial + c_0'\partial$ for $c_0' \in \mathbb{F}$. By $[x\partial + c_0\partial, x^2\partial + \frac{2c_0}{d}x\partial + c_0'\partial] = x^2\partial + \frac{2c_0}{d}x\partial + c_0'\partial$, we have that $\theta(x^2\partial) = kd^{-1}(x+c_0)^2\partial$. So by induction on $k \in \mathbb{N}$, we are able to prove that

$$\theta(x^k\partial) = d^{1-k}(x+c_0)^k\partial.$$

By $[\theta(x^{-1}\partial), \theta(x^2\partial)] = 3\theta(\partial)$, we have that $[\theta(x^{-1}\partial), d^{-1}(x+c_0)^2\partial] = 3d\partial$. This implies that $c_0 = 0$ and $\theta(x^{-1}\partial) = dx^{-1}\partial$. By induction on $k \in \mathbb{Z}$ of $x^k\partial$, we are able to prove that $\theta(x^k\partial) = d^{1-k}x^k\partial$ holds. There are two cases, Case I: $d = 1$ and Case II: $d \neq 1$.

CASE I. Let us assume that $d = 1$, i.e., $\theta(x^k\partial) = x^k\partial$ for any $k \in \mathbb{Z}$. This implies that $\theta((x^2+ax+1)\partial) = x^2\partial + ax\partial + \partial$. Since $\theta(x^2+ax+1)^r\partial) = (x^2+ax+1)^r\partial$ holds for $r \in \mathbb{N}$, we have that $\theta((x^2+ax+1)^{-1}\partial) = (x^2+ax+1)^{-1}\partial$. Since $(x^2+ax+1)^{-1}\partial$ is not an element of $W(\mathbb{F}[(x^2+bx+1)^{\pm 1}, x^{\pm 1}])$, we have a contradiction. So there is no isomorphism from $W(\mathbb{F}[(x^2+ax+1)^{\pm 1}, x^{\pm 1}])$ to $W(\mathbb{F}[(x^2+bx+1)^{\pm 1}, x^{\pm 1}])$ for this case.

CASE II. Let us assume that $d \neq 1$. By $\theta(x^k\partial) = d^{1-k}x^k\partial$, $k \in \mathbb{Z}$, we have that $\theta((x^2+ax+1)\partial) = (d^{-1}x^2+ax+d)\partial$. This implies that $\theta((x^2+ax+1)^2\partial) = d(d^{-2}x^2+ad^{-1}x+1)^2\partial$ holds. By $\theta([(x^2+ax+1)^{-1}\partial, (x^2+ax+1)^2\partial]) = 3\theta(2x+a)$, we have that

$$[\theta((x^2+ax+1)^{-1}\partial), d(d^{-2}x^2+ad^{-1}x+1)^2\partial)] = 3\theta(2x+a). \tag{6.8}$$

There is no element $\theta((x^2+ax+1)^{-1}\partial)$ of $W(\mathbb{F}[(x^2+bx+1)^{\pm 1}, x^{\pm 1}])$ which holds the equality (6.8). So there is no isomorphism from $W(\mathbb{F}[(x^2+ax+1)^{\pm 1}, x^{\pm 1}])$ to $W(\mathbb{F}[(x^2+bx+1)^{\pm 1}, x^{\pm 1}])$ for this case. Thus, by Cases I and II, we have proven the proposition. □

THEOREM 6.6. *There are uncountably many pairwise non-isomorphic Witt-type simple Lie algebras.*

PROOF. If we take two different transcendental numbers a and b in the statement of Proposition 6.5, then the simple Lie algebras $W(\mathbb{F}[(x^2+ax+1)^{\pm 1}, x^{\pm 1}])$ and $W(\mathbb{F}[(x^2+bx+1)^{\pm 1}, x^{\pm 1}])$ are not isomorphic. So by Proposition 3.3, there are uncountably many pairwise non-isomorphic Witt-type simple Lie algebras. □

For given positive algebraic numbers $\alpha_1, \ldots, \alpha_n$ of $\mathbb{F}$, if we define the $\mathbb{F}$-algebra $\mathbb{F}[e^{\pm\alpha_1 x_1}, \ldots, e^{\pm\alpha_n x_n}, x_1, \ldots, x_n]$ as $\mathbb{F}(m^*, k, n)$, then we can define the Weyl-type algebra $A(\mathbb{F}[e^{\pm\alpha_1 x_1}, \ldots, e^{\pm\alpha_n x_n}, x_1, \ldots, x_n])$ as $A(\mathbb{F}(m^*, k, n))$ and the Lie algebra $W(\mathbb{F}[e^{\pm\alpha_1 x_1}, \ldots, e^{\pm\alpha_n x_n}, x_1, \ldots, x_n])$ as $W(\mathbb{F}(m^*, k, n))$.

LEMMA 6.7. *If α is a non-zero algebraic number of $\mathbb{F}$, then the Weyl-type algebras $A(\mathbb{F}[e^{\pm x_1}, \ldots, e^{\pm x_n}])$ and $A(\mathbb{F}[e^{\pm\alpha x_1}, \ldots, e^{\pm\alpha x_n}])$ are isomorphic.*

PROOF. For elements $e^{a_u x_u}$ and ∂_v of $A(\mathbb{F}[e^{\pm x_1}, \ldots, e^{\pm x_n}])$, if we define the $\mathbb{F}$-linear map θ from algebra $A(\mathbb{F}[e^{\pm x_1}, \ldots, e^{\pm x_n}])$ to the algebra $A(\mathbb{F}[e^{\pm\alpha x_1}, \ldots, e^{\pm\alpha x_n}])$ by

$$\theta(e^{a_u x_u}) = \alpha^{a_u} e^{a_u \alpha x_u}, \quad \theta(\partial_v) = \alpha^{-1}\partial_v, \tag{6.9}$$

then θ can be linearly extended to the algebra isomorphism from $A(\mathbb{F}[e^{\pm x_1}, \ldots, e^{\pm x_n}])$ to $A(\mathbb{F}[e^{\pm\alpha x_1}, \ldots, e^{\pm\alpha x_n}])$ where $1 \leq u, v \leq n$ and $a_u \in \mathbb{Z}$. □

PROPOSITION 6.8. *The Weyl-type algebra $A(\mathbb{F}[e^{\pm\alpha x_1}, \ldots, e^{\pm\alpha x_n}])$ is simple where α is a given non-zero algebraic number of $\mathbb{F}$.*

PROOF. The proof of the proposition is straightforward by Theorem 4.2 and Lemma 6.7. □

LEMMA 6.9. *The Witt-type Lie algebras $W(\mathbb{F}[e^{\pm x_1}, \ldots, e^{\pm x_n}])$ and $W(\mathbb{F}[e^{\pm\alpha x_1}, \ldots, e^{\pm\alpha x_n}])$ are isomorphic where α is a given non-zero algebraic number of $\mathbb{F}$.*

PROOF. Define the $\mathbb{F}$-linear map θ from the Lie algebra $W(\mathbb{F}[e^{\pm x_1}, \ldots, e^{\pm x_n}])$ to the Lie algebra $W(\mathbb{F}[e^{\pm\alpha x_1}, \ldots, e^{\pm\alpha x_n}])$ by defining its action on any basis element $e^{a_1 x_1} \cdots e^{a_n x_n}\partial_u$ of $W(\mathbb{F}[e^{\pm x_1}, \ldots, e^{\pm x_n}])$ as

$$\theta(e^{a_1 x_1} \cdots e^{a_n x_n}\partial_u) = \alpha^{-1} e^{a_1 \alpha x_1} \cdots e^{a_n \alpha x_n}\partial_u. \tag{6.10}$$

Then θ can be linearly extended to the Lie algebra isomorphism between the algebras of the lemma where $1 \leq u \leq n$. □

PROPOSITION 6.10. *The Witt-type Lie algebra $W(\mathbb{F}[e^{\pm\alpha_1 x_1}, \ldots, e^{\pm\alpha_n x_n}, x_1, \ldots, x_n])$ is simple where $\alpha_1, \ldots, \alpha_n$ are given non-zero algebraic numbers.*

PROOF. The proof of the proposition is straightforward by the proof of Theorem 4.2. □

Proposition 6.8 is really a corollary of Proposition 6.10. So, there occurs a natural question whether or not the algebras $W(\mathbb{F}[e^{\pm\alpha_1 x_1},\dots,e^{\pm\alpha_n x_n},x_1,\dots,x_m])$ and $W(\mathbb{F}[e^{\pm x_1},\dots,e^{\pm x_n},x_1,\dots,x_m])$ are isomorphic where α_1 and α_2 are different integers not equal to 1 and $m\geq n$. For given non-zero algebraic numbers $\alpha_1,\dots,\alpha_n$ of $\mathbb{F}$, if we define the Weyl-type algebra $A(\mathbb{F}[e^{\pm\alpha_1 x_1},\dots,e^{\pm\alpha_n x_n},x_1^{\pm1},\dots,x_n^{\pm1}])$ as the Weyl-type algebra $A(\mathbb{F}[e^{\pm\alpha_1 x_1},\dots,e^{\pm\alpha_n x_n},x_1,\dots,x_n])$ and the Witt-type Lie algebra $W(\mathbb{F}[e^{\pm\alpha_1 x_1},\dots,e^{\pm\alpha_n x_n},x_1^{\pm1},\dots,x_n^{\pm1}])$ as the Witt-type Lie algebra $W(\mathbb{F}[e^{\pm\alpha_1 x_1},\dots,e^{\pm\alpha_n x_n},x_1,\dots,x_n])$, then by Propositions 6.8 and 6.10, the algebra $A(\mathbb{F}[e^{\pm\alpha_1 x_1},\dots,e^{\pm\alpha_n x_n},x_1^{\pm1},\dots,x_n^{\pm1}])$ and the Lie algebra $W(\mathbb{F}[e^{\pm\alpha_1 x_1},\dots,e^{\pm\alpha_n x_n},x_1^{\pm1},\dots,x_n^{\pm1}])$ are simple.

7. Lie algebras over a Field of Characteristic p

Throughout this section and Section 8, p is a prime number such that $p\neq 2$. Let $\mathbb{E}_p$ be a field of characteristic p containing the prime field $\mathbb{Z}_p$. For given $i_1,\dots,i_m\in\mathbb{Z}_p$ and $n\geq m$, we define the associative-commutative $\mathbb{E}_p$-algebra

$$\mathbb{E}_p(m^*,n):=\mathbb{E}_p[e^{x_1^{i_1}},\dots,e^{x_m^{i_m}},x_1,\dots,x_n] \tag{7.1}$$

as $\mathbb{F}(m^*,k,n)$ with the standard basis $\{e^{a_1x_1^{i_1}}\cdots e^{a_mx_m^{i_m}}x_1^{j_1}\cdots x_n^{j_n}|a_1,\dots,a_m,$ $j_1,\dots,$
$j_n\in\mathbb{Z}_p\}$ as (3.4). So we define the Lie algebra $W(\mathbb{E}_p(m^*,n))$ as $W(\mathbb{F}(m^*,k,n))$, and $W(\mathbb{E}_p(m^*,n))$ is $\mathbb{Z}_p^m$-graded as follows:

$$W(\mathbb{E}_p(m^*,n))=\bigoplus_{(a_1,\dots,a_m)\in\mathbb{Z}_p^m}W_{(a_1,\dots,a_m)} \tag{7.2}$$

where the component $W_{(a_1,\dots,a_m)}$ is a vector subspace of $W(\mathbb{Z}_p(m^*,n))$ which is spanned by the basis $B_p:=\{e^{a_1x_1^{i_1}}\cdots e^{a_mx_m^{i_m}}x_1^{j_1}\cdots x_n^{j_n}\partial_u|j_1,\dots,j_n\in\mathbb{Z}_p,1\leq u\leq n\}$ as (3.4). It is well-known that the Lie algebras $W(\mathbb{Z}_p(0,n))$ and $W(\mathbb{E}_p(0,n))$ are simple (see [**22**]). For a given basis B of a Lie algebra, an element l of L is an ad-diagonal element or a semisimple element of B (see [**19**]) if for any b of B, $[l,b]=c_bb$ holds where c_b is non-zero scalar and $b\neq l$. For any automorphism θ of L, the ad-diagonal element l is auto-invariant, i.e., $\theta([l,b])=[\theta(l),\theta(b)]=c_b\theta(b)$ holds for any $b\in B$. For the standard basis $B=\{\begin{pmatrix}1&0\\0&-1\end{pmatrix},\begin{pmatrix}0&1\\0&0\end{pmatrix},\begin{pmatrix}0&0\\1&0\end{pmatrix}\}$ of $\mathfrak{sl}_2(\mathbb{E}_p)$, $\begin{pmatrix}1&0\\0&-1\end{pmatrix}$ is an ad-diagonal element. Because of the recently fast computation, the automorphism group of a finite dimensional Lie algebra can be found well. Since the element $x=\begin{pmatrix}0&1\\0&0\end{pmatrix}$ of $\mathfrak{sl}_2(\mathbb{E}_p)$ is nilpotent, there is an automorphism $exp(x)$ of $\mathfrak{sl}_2(\mathbb{E}_p)$ (see [**21**]). Since $\mathfrak{sl}_2(\mathbb{E}_p)$ is the Lie algebra over a field $\mathbb{E}_p$ of characteristic p such that $p\neq 2$, the automorphism $exp(x)$ is just the automorphism θ of $\mathfrak{sl}_2(\mathbb{F})$ such that $\theta\begin{pmatrix}1&0\\0&-1\end{pmatrix}=-\begin{pmatrix}1&0\\0&-1\end{pmatrix}$ and it can be extended to the algebra automorphism of $\mathfrak{sl}_2(\mathbb{F})$ as in the proof of Proposition 6.1 (see [**6**]). In the next theorem we prove the simplicity of the algebra $W(\mathbb{E}_p(m^*,n))$ for its further results on the algebra. The dimension $dim(\mathfrak{sl}_{l+1}(\mathbb{E}_p))$ of the Lie algebra $\mathfrak{sl}_{2l}(\mathbb{E}_p)$ is $(l+1)^2-1$. The dimension $dim(\mathfrak{sp}_{2l}(\mathbb{E}_p))$ of the Lie algebra $\mathfrak{sp}_{2l}(\mathbb{E}_p)$ is $2l^2+l$. The dimension $dim(\mathfrak{o}_{2l+1}(\mathbb{E}_p))$ of the Lie algebra $\mathfrak{o}_{2l+1}(\mathbb{E}_p)$ is $2l^2+l$. The dimension $dim(\mathfrak{o}_{2l}(\mathbb{E}_p))$ of the Lie algebra $\mathfrak{o}_{2l}(\mathbb{E}_p)$ is $2l^2-l$ (see [**6**] and [**7**]).

THEOREM 7.1. *The Lie algebra $W(\mathbb{E}_p(m^*,n))$ is simple and the dimension of the Lie algebra $W(\mathbb{E}_p(m^*,n))$ is $p^{m+n}n$.*

PROOF. Since the algebra is $\mathbb{Z}_p^m$-graded, the proof of the theorem is standard as that of Theorem 4.2. □

COROLLARY 7.2. *The Lie algebra $W(\mathbb{Z}_p(m^*,n))$ is simple and the dimension of the algebra $W(\mathbb{Z}_p(m^*,n))$ is $p^{m+n}n$.*

PROOF. The proof of the corollary is straightforward by that of Theorem 7.1, and we omit. □

PROPOSITION 7.3. *If $n_1 \neq n_2$, then Lie algebras $W(\mathbb{Z}_p(m^*,n_1))$ and $W(\mathbb{Z}_p(m^*, n_2))$ are not isomorphic.*

PROOF. Since the algebras of the proposition have different dimensions they are not isomorphic as Lie algebras. □

COROLLARY 7.4. *The Lie algebras $W(\mathbb{Z}_p(1^*,1))$ and $W(\mathbb{Z}_p(0,2))$ are not isomorphic.*

PROOF. Since the dimension of the Lie algebra $W(\mathbb{Z}_p(1^*,1))$ is p^2 and the dimension of the Lie algebra $W(\mathbb{Z}_p(0,2))$ is $2p^2$, they are not isomorphic. □

COROLLARY 7.5. *The Lie algebra $W(\mathbb{Z}_p(1^*,1))$ is not isomorphic to any of the Lie algebras $\mathfrak{sl}_{2l}(\mathbb{E}_p)$, $\mathfrak{sp}_{2l}(\mathbb{E}_p)$, $\mathfrak{o}_{2l+1}(\mathbb{E}_p)$, and $\mathfrak{o}_{2l}(\mathbb{E}_p)$.*

PROOF. Since the dimensions of the Lie algebras $W(\mathbb{Z}_p(1^*,1))$, $\mathfrak{sl}_{2l}(\mathbb{E}_p)$, $\mathfrak{sp}_{2l}(\mathbb{E}_p)$, $\mathfrak{o}_{2l+1}(\mathbb{E}_p)$, and $\mathfrak{o}_{2l}(\mathbb{E}_p)$ are different, no two of them are isomorphic. □

PROPOSITION 7.6. *The Lie algebra $W(\mathbb{Z}_p(1^*,1))$ does not have an ad-diagonal element with respect to the standard basis B of the algebra.*

PROOF. Since $W(\mathbb{Z}_p(1^*,1))$ is $\mathbb{Z}_p$-graded, the ad-diagonal element is in the 0-homogeneous component. Obviously $x\partial$ is the ad-diagonal element with respect to the standard basis of the 0-homogeneous component. By the gradation of the algebra, it cannot be an ad-diagonal element with respect to B. □

COROLLARY 7.7. *The Lie algebra $W(\mathbb{Z}_p(m^*,n))$ has a Lie subalgebra $\mathfrak{S}$ which is spanned by $\partial_u, x_u\partial_u, x_u^2\partial_u$ for a given $u \in \{1,\dots,n\}$, and it is isomorphic to $\mathfrak{sl}_2(\mathbb{Z}_p)$.*

PROOF. Define a $\mathbb{Z}_p$-linear map θ from $\mathfrak{S}$ to $\mathfrak{sl}_2(\mathbb{Z}_p)$ by

$$\theta(\partial_u) = \begin{pmatrix} 0 & 0 \\ 1 & 0 \end{pmatrix}, \theta(x_u\partial_u) = \frac{1}{2}\begin{pmatrix} 1 & 0 \\ 0 & -1 \end{pmatrix}, \text{ and } \theta(x_u^2\partial_u) = \begin{pmatrix} 0 & 1 \\ 0 & 0 \end{pmatrix}.$$

Then θ can be linearly extended to the Lie algebra isomorphism from $\mathfrak{S}$ to $\mathfrak{sl}_2(\mathbb{Z}_p)$. □

8. Non-Associative Algebras, Representations, and Lie Modules

Let us define the non-associative algebra $N(m^*, k, n)$ on $\mathbb{F}(m^*, k, n)$ with the standard basis

$$\mathcal{B} = \{e^{a_1 x_1^{i_1}} \cdots e^{a_m x_m^{i_m}} x_1^{j_1} \cdots x_k^{j_k} x_{k+1}^{j_{k+1}} \cdots x_{k+n}^{j_{k+n}} \partial_u,$$
$$a_1, \ldots, a_m, j_1, \ldots, j_k \in \mathbb{Z}, j_{k+1}, \ldots, j_{k+n} \in \mathbb{N}, 1 \leq u \leq k+n\},$$

with the obvious vector space operations and for $e^{a_1 x_1^{i_1}} \cdots e^{a_m x_m^{i_m}} x_1^{j_1} \cdots x_k^{j_k} x_{k+1}^{j_{k+1}} \cdots x_{k+n}^{j_{k+n}} \partial_u$, $e^{b_1 x_1^{i_1}} \cdots e^{b_m x_m^{i_m}} x_1^{r_1} \cdots x_k^{r_k} x_{k+1}^{r_{k+1}} \cdots x_{k+n}^{r_{k+n}} \partial_v \in \mathcal{B}$, the multiplication $\cdot$ is defined as follows:

$$\begin{aligned} &e^{a_1 x_1^{i_1}} \cdots e^{a_m x_m^{i_m}} x_1^{j_1} \cdots x_k^{j_k} x_{k+1}^{j_{k+1}} \cdots x_{k+n}^{j_{k+n}} \partial_u \cdot e^{b_1 x_1^{i_1}} \cdots e^{b_m x_m^{i_m}} x_1^{r_1} \cdots \\ &\quad x_k^{r_k} x_{k+1}^{r_{k+1}} \cdots x_{k+n}^{r_{k+n}} \partial_v \\ &= e^{a_1 x_1^{i_1}} \cdots e^{a_m x_m^{i_m}} x_1^{j_1} \cdots x_k^{j_k} x_{k+1}^{j_{k+1}} \cdots x_{k+n}^{j_{k+n}} \partial_u (e^{b_1 x_1^{i_1}} \cdots e^{b_m x_m^{i_m}} x_1^{r_1} \cdots \\ &\quad x_k^{r_k} x_{k+1}^{r_{k+1}} \cdots x_{k+n}^{r_{k+n}}) \partial_v \\ &= b_u i_u e^{(a_1+b_1) x_1^{i_1}} \cdots e^{(a_m+b_m) x_m^{i_m}} x_1^{j_1+r_1} \cdots x_k^{j_k+r_k} x_{k+1}^{j_{k+1}+r_{k+1}} \cdots x_{k+n}^{j_{k+n}+r_{k+n}} \\ &\quad x_u^{i_u-1} \partial_v + r_u e^{(a_1+b_1) x_1^{i_1}} \cdots e^{(a_m+b_m) x_m^{i_m}} x_1^{j_1+r_1} \cdots x_k^{j_k+r_k} x_{k+1}^{j_{k+1}+r_{k+1}} \cdots \\ &\quad x_{k+n}^{j_{k+n}+r_{k+n}} x_u^{-1} \partial_v \end{aligned} \tag{8.1}$$

for $u \leq m$, where $i_1, \ldots, i_m$ are fixed non-zero integers. In case that $u > m$ the multiplication (8.1) can also be defined similarly. It is easy to prove that $\cdot$ can be linearly extended to the algebra $N(m^*, k, n)$. From now on we omit the multiplication $\cdot$. For any $f\partial_u$ and $g\partial_v$ of $N(m^*, k, n)$, if we define the Lie bracket as $[f\partial_u, g\partial_v] = f(\partial_u(g))\partial_v - g(\partial_v(f))\partial_u$, then by (4.1), we define the Lie algebra $N(m^*, k, n)^-$ which has the same set as $N(m^*, k, n)$. The Lie algebra $N(m^*, k, n)^-$ is the Witt-type Lie algebra $W(\mathbb{F}(m^*, k, n))$ and it is called the anti-symmetrized algebra with respect to the non-associative algebra $N(m^*, k, n)^-$ (see [**10**]). The Witt-type Lie algebra $W(\mathbb{F}(m^*, k, n))$ has a Lie subalgebra $\mathcal{S}$ which is spanned by $\{x_u\partial_u - x_v\partial_v, x_w\partial_z | 1 \leq u < v \leq m, 1 \leq w \neq z \leq m\}$. Note that the Lie algebra $\mathfrak{sl}_m(\mathbb{F})$ is spanned by $\{\epsilon_{uu} - \epsilon_{vv}, \epsilon_{wz} | 1 \leq u < v \leq m, 1 \leq w \neq z \leq m\}$ where ϵ_{uv} is the elementary matrix whose uv-entry is one and the other entries are zeros for $1 \leq u, v \leq m$. For a positive integer d, we define a vector subspace V_d of $\mathbb{F}(0, 0, m)$ which is spanned by $\{x_1^{i_1} \cdots x_m^{i_m} | i_1 + \cdots + i_m = d, i_1, \ldots, i_m \in \mathbb{N}\}$. For a given Lie algebra L, a representation of L is a Lie algebra homomorphism $\theta : L \to Hom_{\mathbb{F}}(V, V)$ where $Hom_{\mathbb{F}}(V, V)$ is the set of all linear maps from a finite dimensional vector space V over $\mathbb{F}$ to itself. A Lie module for a Lie algebra L over $\mathbb{F}$ (i.e., L-module) is a finite dimensional vector space V over $\mathbb{F}$ with a map

$$L \times V \to V, \quad (l, v) \to l \cdot v$$

with the conditions

$$\begin{aligned} &(c_1 l_1 + c_2 l_2) \cdot v = c_1(l_1 \cdot v) + c_2(l_2 \cdot v), \\ &l_1 \cdot (c_1 v_1 + c_2 v_2) = c_1(l_1 \cdot v_1) + c_2(l_1 \cdot v_2), \\ &[l_1, l_2] \cdot v_1 = l_1 \cdot (l_2 \cdot v_1) - l_2 \cdot (l_1 \cdot v_1), \end{aligned}$$

for all $l_1, l_2 \in L$, $v_1, v_2 \in V$, and $c_1, c_2 \in \mathbb{F}$. From now on we omit $\cdot$ as usual. For $d \geq 1$, V_d is a $\mathcal{S}$-module. For given two basis elements $x_1^{i_1} \cdots x_m^{i_m}$ and $x_1^{j_1} \cdots x_m^{j_m}$

of V_d, we define the lexicographic order of them with respect to the powers of the elements orderly where $i_1, \dots, i_n, j_1, \dots, j_m \in \mathbb{N}$, i.e., $i_1 > j_1$, or if $i_1 = j_1$, then $i_2 > j_2$, or etc.. For any element l of V_d, l can be written using the lexicographic order.

THEOREM 8.1. *The non-associative algebra $N(m^*, k, n)$ is simple.*

PROOF. The proof of the theorem is standard and we omit it (see [**10**]). □

LEMMA 8.2. *The matrix algebra $M_m(\mathbb{F})$ is isomorphic to the subalgebra $\mathcal{G}$ of $N(m^*, k, n)$ which is spanned by $\{x_u\partial_v | 1 \le u, v \le m\}$.*

PROOF. Notice that $\mathcal{G}$ is a subring of $Hom_{\mathbb{F}}(V_d, V_d)$. Define an $\mathbb{F}$-linear map from $M_m(\mathbb{F})$ to $\mathcal{G}$ by

$$\theta(\epsilon_{uv}) = x_u\partial_v.$$

Then θ can be linearly extended to the algebra isomorphism from $M_m(\mathbb{F})$ to $\mathcal{G}$. □

PROPOSITION 8.3. *The Lie algebras $\mathcal{G}$ has a subalgebra $\mathcal{S}$ which is isomorphic to $\mathfrak{sl}_m(\mathbb{F})$.*

PROOF. The proof of the proposition is straightforward by Lemma 8.2. □

THEOREM 8.4. *For $d \in \mathbb{N} - \{0\}$, the vector subspace V_d of $\mathbb{F}(0, 0, m)$ is a simple $\mathcal{S}$-module where the subalgebra $\mathcal{S}$ of $N(m^*, k, n)^-$ is isomorphic to $\mathfrak{sl}_m(\mathbb{F})$ where $m \ge 2$.*

PROOF. The theorem for the case that $m = 2$, is well-known, and we assume that $m \ge 3$. We prove the theorem by induction on d. If $d = 1$, then $x_u\partial_v$ annhilates x_w for $w \ne v$. It is easy to prove that each monomial x_z generates V_1 where $1 \le u \ne v, w, z \le m$, i.e., V_1 is a simple $\mathcal{S}$-module. Assume that V_k is a simple S-module. Let us show that V_{k+1} is a simple $\mathcal{S}$-module. Let us show that the submodule $< x_u^{k+1} >$ of V_{k+1} generated by x_u^{k+1} is V_{k+1}. By taking $x_v\partial_u$, $u \ne v$, we have that $(x_v\partial_u)^t x_u^{k+1} = x_v^t x_u^{k+1-t}$ for $t \le k+1$. Similarly, we have an element $x_1^{i_1} \cdots x_m^{i_m}$ of $< x_u^{k+1} >$ with $i_1 + \cdots + i_m = k+1$, i.e., $V_{k+1} =< x_u^{k+1} >$. This shows that the submodule of V_{k+1} which is generated by one basis element of V_{k+1} is V_{k+1} itself. Now let l be a non-zero element of a non-zero submodule I of V_{k+1}. If $l = cx_1^{i_1} \cdots x_m^{i_m}$, where $c \in \mathbb{F}^*$, then there is nothing to prove. We prove the theorem by induction on the number $\#(l)$ of non-zero basis terms of l. The theorem for the case $\#(l) = 1$ is already proven. Let us assume that the theorem holds for the case $\#(l) = t - 1$ and that $\#(l) = t \ge 2$. It can be written as

$$l = c_1 x_1^{i_{11}} \cdots x_m^{i_{1m}} + \cdots + c_t x_1^{i_{t1}} \cdots x_m^{i_{tm}}, \tag{8.2}$$

with appropriate non-zero coefficients and $t > 1$. Since the terms of l are different, we can assume that $i_{11} \ne i_{1s}$ where $s \ne 1$ by swapping terms of l in (8.2). This implies that

$$(x_1\partial_1 - x_s\partial_s)(c_1(i_{11} - i_{1s})x_1^{i_{11}} \cdots x_m^{i_{1m}} + \cdots$$

holds. If $\#((x_1\partial_1 - x_s\partial_s)(c_1(i_{11} - i_{1s})x_1^{i_{11}} \cdots x_m^{i_{1m}} + \cdots) < t$, then there is nothing to prove. Otherwise, we have that $(x_1\partial_1 - x_s\partial_s)l - (i_{11} - i_{1s})l \ne 0$ and $\#((x_1\partial_1 - x_s\partial_s)l - (i_{11} - i_{1s})) < t$ by taking an appropriate s. Since $\#((x_1\partial_1 - x_s\partial_s)l - (i_{11} - i_{1s})l) < t$, by induction, we have that $I = V_{k+1}$. So V_d is a simple $\mathcal{S}$-module. □

The $\mathbb{F}$-algebra $\mathbb{F}(m,k,n)$ is $\mathbb{Z}^m$-graded and its $(a_1,\dots,a_m)$-homogeneous component $V_{(a_1,\dots,a_m)}$ is a vector subspace of $\mathbb{F}(m,k,n)$ whose standard basis is

$$\{e^{a_1x_1}\cdots e^{a_mx_m}x_1^{i_1}\cdots x_{k+n}^{i_{k+n}}|i_1,\dots,i_k\in\mathbb{Z},i_{k+1},\dots,i_{k+n}\in\mathbb{N}\}.$$

For each basis element $e^{a_1x_1}\cdots e^{a_mx_m}x_1^{i_1}\cdots x_{k+n}^{i_{k+n}}$ of $V_{(a_1,\dots,a_m)}$, we define the degree $deg(e^{a_1x_1}\cdots e^{a_mx_m}\,x_1^{i_1}\cdots x_{k+n}^{i_{k+n}})$ of $e^{a_1x_1}\cdots e^{a_mx_m}x_1^{i_1}\cdots x_{k+n}^{i_{k+n}}$ as $\sum_{v=1}^{k+n}|i_v|$. We call $i_1,\dots,i_{k+n}$ the powers of $e^{a_1x_1}\cdots e^{a_mx_m}x_1^{i_1}\cdots x_{k+n}^{i_{k+n}}$.

THEOREM 8.5. *If at least one of $a_1,\dots,a_m$ is not zero, then the $(a_1,\dots,a_m)$-homogeneous component $V_{(a_1,\dots,a_m)}$ of $\mathbb{F}(m,k,n)$ is a simple $W(0,k,n)$-module.*

PROOF. Let I be a non-zero submodule of $V_{(a_1,\dots,a_m)}$ and l be a non-zero element of I. Then there exists an element l_1 of $W(0,k,n)$, such that the l_1l has non-negative powers. So we may assume that the powers of l are non-negative integers. For ∂_u, $1\le u\le m$, we have that $deg(\partial_u(l))\le deg(l)$, i.e., ∂_u is a degree stabilizer of each basis element of $V_{(a_1,\dots,a_m)}$ of $\mathbb{F}(m,k,n)$. So the proof of the theorem is similar to that of Theorem 8.4. □

For the associative algebra $\mathbb{E}_p(m^*,n)$ of Section 6, similar to the algebra $N(m^*,k,n)$, we define the non-associative algebra $N(\mathbb{E}_p(m^*,n))$ over a field of characteristic p as follows:

1. $N(\mathbb{E}_p(m^*,n))$ has the same standard basis

$$B_p=\{e^{a_1x_1^{i_1}}\cdots e^{a_mx_m^{i_m}}x_1^{j_1}\cdots x_n^{j_n}\partial_u|a_1,\dots,a_m,j_1,\dots,j_n\in\mathbb{Z}_p,1\le u\le n,m\le n\}$$

as $W(\mathbb{Z}_p(m^*,n))$ of Section 6, and

2. the multiplication of the algebra is defined as the multiplication (8.1) on the basis elements of B_p.

Then $N(\mathbb{E}_p(m^*,n))$ is a non-associative algebra as $N(m^*,k,n)$. So the antisymmetrized algebra $N(\mathbb{E}_p(m^*,n))^-$ of $N(\mathbb{E}_p(m^*,n))$ is the Witt-type Lie algebra $W(\mathbb{Z}_p(m^*,n))$.

THEOREM 8.6. *$N_p(\mathbb{E}_p(m^*,n))$ is simple*

PROOF. The proof of the theorem is standard and we omit it (see [**10**]). □

COROLLARY 8.7. *The subalgebra $N(\mathbb{E}(0,n))$ if $N(\mathbb{E}_p(m^*,n))$ is simple and the matrix ring $M_n(\mathbb{E}_p)$ is a subalgebra of $N(\mathbb{E}_p(0,n))$.*

PROOF. The proof of the corollary is straightforward by Lemma 8.2. □

9. Generalized Weyl-Type Modules and Related Topics

Let R be a ring and M be a R-module. A module is simple, it has only two submodules $\{0\}$ and itself. Let θ be a ring homomorphism from a ring R_1 to a ring R_2. Let M be an R_2-module. Then M is a R_1-module as follows:

$$r_1\cdot m=\theta(r_1)m$$

for $r_1\in R_1$ and $m\in M$. This module is called a twisted module and it is denoted by M_θ (see [3]) and we have the following diagram;

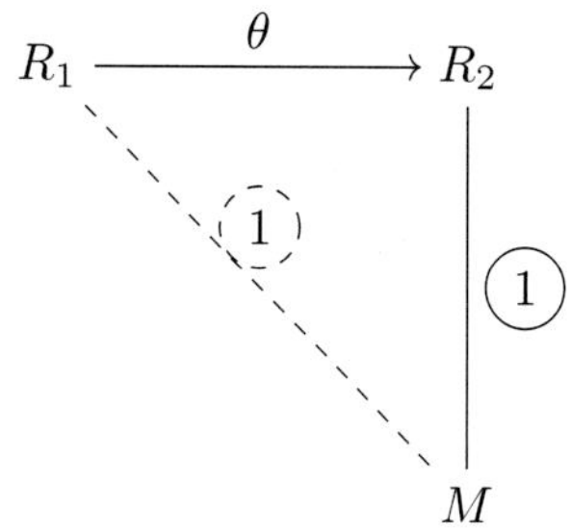

For any $f \in \mathbb{F}(m^*, k, n)$ and $\alpha \in A(\mathbb{F}(m^*, k, n))$, if we define $\alpha \cdot f = \alpha(f)$, then the $\mathbb{F}$-algebra $\mathbb{F}(m^*, k, n)$ is a left $A(\mathbb{F}(m^*, k, n))$-module. So for any automorphism θ of $A(\mathbb{F}(m^*, k, n))$, we define the twisted module $\mathbb{F}(m^*, k, n)_\theta$. For fixed non-zero algebraic numbers $\alpha_1, \dots, \alpha_m$, the $\mathbb{F}$-algebra $\mathbb{F}[e^{\pm\alpha_1 x_1}, \dots, e^{\pm\alpha_m x_m}, x_1, \dots, x_n]$ is a left $A(\mathbb{F}[e^{\pm\alpha_1 x_1}, \dots, e^{\pm\alpha_m x_m}, x_1, \dots, x_n])$-module for $m \le n$. From now on, a module means a left module and we omit $\cdot$ in $\alpha \cdot f$, i.e., we write αf for $\alpha \cdot f$.

THEOREM 9.1. *The $\mathbb{F}$-algebra $\mathbb{F}(m^*, k, n)$ is a simple $A(\mathbb{F}(m^*, k, n))$-module and for any automorphism θ of $A(\mathbb{F}(m^*, k, n))$, the twisted module $\mathbb{F}(m^*, k, n)_\theta$ is simple.*

PROOF. Since the algebra $A(\mathbb{F}(m^*, k, n))$ has the multiplicative identity 1, the proof of the theorem is similar to that of Theorem 3.5, and we omit. □

COROLLARY 9.2. *For a given non-zero algebraic number α, $\mathbb{F}[e^{\pm\alpha x_1}, \dots, e^{\pm\alpha x_m}, x_1, \dots, x_n]$ is a twisted left simple $A(\mathbb{F}[e^{\pm x_1}, \dots, e^{\pm x_m}, x_1, \dots, x_n])$-module.*

PROOF. Since by Lemma 6.7, there is an isomorphism ϕ from $A(\mathbb{F}[e^{\pm x_1}, \dots, e^{\pm x_n}, x_1, \dots, x_n])$ to $A(\mathbb{F}[e^{\pm\alpha x_1}, \dots, e^{\pm\alpha x_m}, x_1, \dots, x_n])$, we see that $\mathbb{F}[e^{\pm\alpha x_1}, \dots, e^{\pm\alpha x_m}, x_1, \dots, x_n]_\phi$ is a twisted left simple $A(\mathbb{F}[e^{\pm x_1}, \dots, e^{\pm x_m}, x_1, \dots, x_n])$-module. □

NOTE 2. For a given positive integer t and a non-zero scalar c, let $\theta_{c,t}$ be an $\mathbb{F}$-linear map from $A(\mathbb{F}(m^*, k, n))$ to itself such that $\theta_{c,t}(e^{x_j^{i_j}} x_j^s) = e^{x_j^{i_j}} x_j^s$ and $\theta_{c,t}(\partial_j) = \partial_j + c x_j^t$ for any $e^{x_j^{i_j}} x_j^s$, ∂_j, and x_j^t of $\mathbb{F}(m^*, k, n)_{\theta_{c,t}}$, $1 \le j \le n$. Then $\theta_{c,t}$ can be linearly extended to an automorphism of the algebra. So it induces an $A(\mathbb{F}(m^*, k, n))$ twisted module $\mathbb{F}(m^*, k, n)_{\theta_{c,t}}$. □

So we know that why we need to find the automorphisms of an algebra. Because of the automorphisms of $A(\mathbb{F}(m^*, k, n))$, we have the following results (see [3]).

THEOREM 9.3. *The twisted module $\mathbb{F}(m^*, k, n)_{\theta_{c,t}}$ is simple where $\theta_{c,t}$ is an automorphism of $A(\mathbb{F}(m^*, k, n))$ in Note 2 for $c \in \mathbb{F}^*$ and t is a positive integer.*

THEOREM 9.4. *Let t_1, t_2 be positive integers and c_1, c_2 be scalars of which at least one is non-zero. If $t_1 \ne t_2$ or $c_1 \ne c_2$, then the twisted modules $\mathbb{F}(m^*, k, n)_{\theta_{c_1,t_1}}$ and $\mathbb{F}(m^*, k, n)_{\theta_{c_2,t_2}}$ are not isomorphic.*

PROOF. Let us put $t_1 < t_2$ and assume that there is an isomorphism $\psi : \mathbb{F}(m^*, k, n)_{\theta_{c_1,t_1}} \to \mathbb{F}(n, n)_{\theta_{c_2,t_2}}$. Since 1 generates $\mathbb{F}(m^*, k, n)_{\theta_{c_1,t_1}}$, $\psi(1)$ generates $\mathbb{F}(m^*, k, n)_{\theta_{c_2,t_2}}$, we put $\psi(1) = f$ for $f \in \mathbb{F}(m^*, k, n)$. This implies that $\psi(\partial_i \cdot 1) = \partial_i \psi(1) = \partial_i(f) = \partial_i(f) + c_2 x_i^{t_2} f$ and on the other hand $\psi((\partial_i + c_1 x_i^{t_1})1) = \psi(c_1 x_i^{t_1} 1) = c_1 x_i^{t_1} f$ where $1 \le i \le n$. But $\partial_i(f) + c_2 x_i^{t_2} f = c_1 x_i^{t_1} f$ does not hold. This contradiction shows that the twisted modules are not isomorphic. □

THEOREM 9.5. *There are uncountably many pairwise non-isomorphic twisted modules* $\mathbb{F}(m^*, k, n)_\theta$ *on* $\mathbb{F}(m^*, k, n)$ *where* θ *is the automorphism of the Weyl-type algebra* $A(\mathbb{F}(m^*, k, n))$ *in Note 2.*

PROOF. Since there are uncountably many different transcendental numbers of $\mathbb{F}$, the proof of the theorem is straightforward by that of Theorem 9.4. □

COROLLARY 9.6. *The* $A(\mathbb{F}[(x^2+bx+1)^{\pm 1}, x^{\pm 1}])$*-module* $\mathbb{F}[(x^2+ax+1)^{\pm 1}, x^{\pm 1}]$ *is simple. For any automorphism* θ_f *of* $A(\mathbb{F}[(x^2+bx+1)^{\pm 1}, x^{\pm 1}])$, *the twisted Weyl-type module* $\mathbb{F}[(x^2+ax+1)^{\pm 1}, x^{\pm 1}]_{\theta_f}$ *is simple where* θ_f *is the map in Corollary 6.3. There are uncountably many pairwise non-isomorphic twisted modules* $A(\mathbb{F}[(x^2+bx+1)^{\pm 1}, x^{\pm 1}])_{\theta_f}$ *which is induced by the automorphism* θ_f *of Corollary 6.3.*

PROOF. The proof of the corollary is similar to that of the simplicity of the Weyl-type algebra $A(\mathbb{F}[(x^2+bx+1)^{\pm 1}, x^{\pm 1}])$. The rest proof of the corollary is straightforward. □

LEMMA 9.7. *For the function* $y = e^{e^{x^i}}$, *we have* $y^{(n)} = i^n e^{e^{x^i}} e^{nx^i} x^{ni-n} + *$ *where* $*$ *is the sum of the remaining terms of* $y^{(n)}$ *where* $n \in \mathbb{N}$, i *is a given positive integer.*

PROOF. For $y = e^{e^{x^i}}$, we have that $y' = ie^{e^{x^i}} e^{x^i} x^{i-1}$. So, by induction on n of $y^{(n)}$, we have that $y^{(n)} = i^n e^{e^{x^i}} e^{(n+1)x^i} x^{ni-n} + *$ where $*$ is the sum of the remaining term of $y^{(n)}$. These complete the proof of the lemma. □

PROPOSITION 9.8. *For a given positive integers* i, $y = e^{e^{x^i}}$ *is not the solution of a differential equation*

$$f_n(x)y^{(n)} + \cdots + f_1(x)y' + f_0(x)y + f_{-1}(x) = 0, \tag{9.1}$$

where $f_j(x) \in \mathbb{F}[x]$, $-1 \leq j \leq n \in \mathbb{N}$ *and* $n \in \mathbb{N}$ *(see* [**4**]*).*

PROOF. Let us use the method of undetermined polynomial coefficients for the proof of the proposition. Let us assume that the proposition is true, i.e., $y = e^{e^{x^i}}$ satisfies the differential equation (9.1). Since there are no non-zero polynomials $f_0(x)$ and $f_{-1}(x)$ such that $f_0(x)y + f_{-1}(x) = 0$, we can assume that $n > 1$. Let us assume that we can find polynomials $f_j(x) \in \mathbb{F}[x]$, $-1 \leq j \leq n$, that satisfy (9.1) where $f_n(x) \neq 0$. We have that

$$f_n(x)(e^{e^{x^i}})^{(n)} + \cdots + f_1(x)(e^{e^{x^i}})' + f_0(x)e^{e^{x^i}} + f_{-1}(x) = 0. \tag{9.2}$$

By Lemma 9.7 and $(e^{e^{x^i}})^{(n)}$ has only the non-zero term $i^n e^{e^{x^i}} e^{nx^i} x^{ni-n}$ of (9.2). Since the multiplication of two non-zero expolynomials is not zero, the equality (9.2) does not hold. We have a contradiction. So we have proven the proposition. □

PROPOSITION 9.9. *For non-zero real algebraic numbers* α *and* β, *if* $\beta \neq 1$, *then the real solution of* $e^{\alpha x^i} = \beta$ *is a transcendental number where* i *is a given positive integer* [**3**].

PROOF. Since an integer cannot be a solution of the equation of the proposition, the proof of the proposition is obvious by the Lindemann-Weierstrass Theorem. □

COROLLARY 9.10. *For non-zero rational numbers a and b, if $b \neq 1$, then the real solution of $e^{ax^i} = b$ is a transcendental number where i is a given positive integer* [**3**].

PROOF. Since an integer cannot be a solution of the equation of the proposition, the proof of the corollary is straightforward by Proposition 9.9. $\square$

It is easy to prove that for any prime p, the real solution of $x^x = p$ is a transcendental number (see [**16**]). It is well-known that the real solution of $x^{x^x} = 10$ is an irrational number using the Schanuel's conjecture [**23**]. We prove that the real soution of the equation $x^{x^x} = 10$ is an irrational number here.

PROPOSITION 9.11. *For a given non-zero rational number a and a positive algebraic number b, if b is not one and if the equation $x^{ax^x} = b$ has no integer solution, then the real solution of the equation is an irrational number.*

PROOF. Since the equation of the proposition does not have an integer solution, let us assume that there is a rational number solution $\frac{\alpha}{\beta}$ of the equation where α and β are relatively prime positive integers. We have that $(\frac{\alpha}{\beta})^{a(\frac{\alpha}{\beta})^{\frac{\alpha}{\beta}}} = b$. This implies that $(\frac{\alpha}{\beta})^{a\sqrt[\beta]{\frac{\alpha^\alpha}{\beta^\alpha}}} = b$. Since α and β are relatively prime positive integers, $a\sqrt[\beta]{\frac{\alpha^\alpha}{\beta^\alpha}}$ is a non-zero radical number. By Gelfond-Schneider Theorem, $(\frac{\alpha}{\beta})^{a\sqrt[\beta]{\frac{\alpha^\alpha}{\beta^\alpha}}}$ is a transcendental number, but b is an algebraic number, we have a contradiction. So the real solution of the proposition is an irrational number. $\square$

Note that we use the fact that $x^8 = x^{2^3}$, etc. In general it is very difficult to decide the irrationality or transcendence of a real solution of higher order ladder equations like

$$x^{x^{x^{\cdot^{\cdot^{\cdot^{x^x}}}}}} = 1000,$$

or more generally

$$x^{f_1(x)x^{f_2(x)x^{\cdot^{\cdot^{\cdot^{f_{k-1}(x)x^{f_k(x)x}}}}}}} = g(x),$$

depending on the appropriate rational polynomials $f_1(x), \ldots, f_k(x), g(x)$. By Proposition 9.11, we find irrational numbers using the equation $x^{ax^{bx}} = c$ by taking appropriate rational numbers a, b, and c.

PROPOSITION 9.12. *For a given positive integers n, $y = x^{x^x}$ is not the solution of a differential equation*

$$f_n(x)y^{(n)} + \cdots + f_1(x)y' + f_0(x)y + f_{-1}(x) = 0, \tag{9.3}$$

where $f_i(x) \in \mathbb{F}[x]$, $-1 \leq i \leq n \in \mathbb{N}$ (see [**4**]*).*

PROOF. The proof of the proposition is similar to the proof of Proposition 9.8, so let us omit it. $\square$

COROLLARY 9.13. *For a given positive integers n, $y = x^{x^x}$ is not the solution of a differential equation*

$$f_n(x)y^{(n)} + \cdots + f_1(x)y' + f_0(x)y + f_{-1}(x) = 0, \tag{9.4}$$

where $f_i(x) \in \mathbb{F}[x^{\pm 1}]$, $-1 \leq i \leq n \in \mathbb{N}$ (see [**4**]*).*

PROOF. The proof of the corollary is straight forward by that of Proposition 9.12, and we omit. □

10. Conclusion

The classical Cartan type Lie algebras have four types of the infinite dimensional Lie algebras, Witt, Special, Hamiltonian (Poission), and contact Lie algebras (see [**8**] and [**18**]). We define Witt-type Lie algebras which contain the Witt algebra in this work. So we can define other generalized Cartan types of Lie algebras on $\mathbb{F}(m^*, k, n)$ or its generalized algebra (see [**17**]). By Propositions 9.8 and 9.12, and Corollary 9.13, there are uncountably many transcendental differentiable functions which do not satisfy a differential equation whose coefficients are the polynomials of $\mathbb{C}[x]$ or the elements of the Laurent extension $\mathbb{C}[x^{\pm 1}]$ where $\mathbb{C}$ is the set of all complex numbers.

Acknowledgment

The authors thanks to the referee for the valuable comments for the manuscript.

References

[1] A. Belov-Kanel and M. Kontsevich, *Automorphisms of the Weyl algebra*, Lett. Math. Phys. **74** (2005), no. 2, 181–199, DOI 10.1007/s11005-005-0027-5. MR2191954

[2] J.-E. Björk, *Rings of differential operators*, North-Holland Publishing Co., Amsterdam, 1979.

[3] R. M. Chaphalkar, S.-G. Hwang, C. H. Lee, and K.-B. Nam, *New Gelfond-type transcendental numbers*, Preprint, `arXiv:2106.04055 [math.NT]`, 3 Jun 2021.

[4] S. C. Coutinho, *A primer of algebraic D-modules*, London Mathematical Society Student Texts, vol. 33, Cambridge University Press, Cambridge, 1995, DOI 10.1017/CBO9780511623653. MR1356713

[5] J. Dixmier, *Sur les algèbres de Weyl* (French), Bull. Soc. Math. France **96** (1968), 209–242. MR242897

[6] J. E. Humphreys, *Introduction to Lie algebras and representation theory*, Graduate Texts in Mathematics, Vol. 9, Springer-Verlag, New York-Berlin, 1972. MR323842

[7] N. Jacobson, *Lie algebras*, Dover Publications, Inc., New York, 1979. Republication of the 1962 original. MR559927

[8] V. G. Kac, *A description of the filtered Lie algebras with which graded Lie algebras of Cartan type are associated* (Russian), Izv. Akad. Nauk SSSR Ser. Mat. **38** (1974), 800–834. MR369452

[9] N. Kawamoto, A. Mitsukawa, K.-B. Nam, and M.-O. Wang, *The automorphisms of generalized Witt type Lie algebras*, J. Lie Theory **13** (2003), no. 2, 573–578. MR2003161

[10] J. Lee and K.-B. Nam, *Non-associative algebras containing the matrix ring*, Linear Algebra Appl. **429** (2008), no. 1, 72–78, DOI 10.1016/j.laa.2008.02.005. MR2419141

[11] L. Makar-Limanov, *On automorphisms of Weyl algebra* (English, with French summary), Bull. Soc. Math. France **112** (1984), no. 3, 359–363. MR794737

[12] K.-B. Nam, *Graded radical W type Lie algebras. I*, Int. J. Math. Math. Sci. **31** (2002), no. 6, 339–343, DOI 10.1155/S016117120210812X. MR1927827

[13] K.-B. Nam, S.-G. Hwang, K.-S. Lee, and J. Zhou, *Random insertion method based security protocols using transcendental number generators*, Proceedings of the Future of Information and Communication Conference (FICC) (Cham), Lecture Notes in Networks and Systems, vol. 921, Springer, 2024.

[14] K.-B. Nam and J. Pakianathan, *On generalized Witt algebras in one variable*, Turkish J. Math. **35** (2011), no. 3, 405–436. MR2867327

[15] H. G. Park, J. Lee, S. H. Choi, X. Chen, and K.-B. Nam, *Automorphism groups of some algebras*, Sci. China Ser. A **52** (2009), no. 2, 323–328, DOI 10.1007/s11425-009-0007-9. MR2491733

[16] R. S. Raja Durai, S.-G. Hwang, A. Kumar, and K.-B. Nam, *RSA & RIM method of cryptography*, Proceedings of the Future of Information and Communication Conference (FICC), Lecture Notes in Networks and Systems, Springer, 2025, to appear.

[17] R. S. Raja Durai, S. Sharma, and K.-B. Nam, *Structure and automorphisms of Weyl-type algebras over transcendental exponential rings*, preprint, 2024.

[18] A. N. Rudakov, *Automorphism groups of infinite-dimensional simple Lie algebras* (Russian), Izv. Akad. Nauk SSSR Ser. Mat. **33** (1969), 748–764. MR252462

[19] J.-P. Serre, *Complex semisimple Lie algebras*, Springer-Verlag, New York, 1987. Translated from the French by G. A. Jones, DOI 10.1007/978-1-4757-3910-7. MR914496

[20] S. Sinel'shchikov, *The Laurent extension of quantum plane: a complete list of $U_q(\mathfrak{sl}_2)$-symmetries*, SIGMA Symmetry Integrability Geom. Methods Appl. **15** (2019), Paper No. 38, 33, DOI 10.3842/SIGMA.2019.038. MR3948253

[21] E. H. Spanier, *Algebraic topology*, McGraw-Hill Book Co., New York-Toronto-London, 1966. MR210112

[22] H. Strade and R. L. Wilson, *Classification of simple Lie algebras over algebraically closed fields of prime characteristic*, Bull. Amer. Math. Soc. (N.S.) **24** (1991), no. 2, 357–362, DOI 10.1090/S0273-0979-1991-16033-7. MR1071032

[23] M. Waldschmidt, *Personal communication with tung nguyen*, 2024, personal communication.

Department of Mathematics, Jaypee University of Information Technology (JUIT), Waknaghat, Solan-173234, Himachal Pradesh, India

Email address: `Supriya15998@gmail.com`

Department of Mathematics, Jaypee University of Information Technology (JUIT), Waknaghat, Solan-173234, Himachal Pradesh, India

Email address: `rsraja.durai@juit.ac.in`

Department of Mathematics, University of Wisconsin-Whitewater, Whitewater, Wisconsin 53190

Email address: `namk@uww.edu`

Selected Published Titles in This Series

845 **Xueqing Chen and Yiqiang Li, Editors,** Cluster Algebras, Hall Algebras and Representation Theory, 2026

843 **José Ignacio Cogolludo-Augustín and Eriko Hironaka, Editors,** Algebraic and Topological Interplay of Algebraic Varieties, 2026

842 **Federico Binda, José Ignacio Burgos Gil, James Lewis, Deepam Patel, and Gregory Pearlstein, Editors,** Regulators V, 2026

841 **Gus Schrader and Boris Tsygan, Editors,** Homotopical Methods in Geometry and Physics, 2026

840 **John W. Jones, Jennifer Paulhus, Andrew V. Sutherland, and John Voight, Editors,** LuCaNT: Databases, Algorithms, and Computational Number Theory, 2026

839 **Sangita Jha, William Ott, and Mrinal Kanti Roychowdhury, Editors,** Current Trends in Dynamical Systems, Fractal Geometry, and Their Applications, 2026

838 **Julia E. Bergner, Joachim Kock, and Maru Sarazola, Editors,** Higher Segal Spaces and Applications, 2026

837 **Yiqiang Li and Changlong Zhong, Editors,** Representation Theory and Flag Varieties, 2026

835 **Mee Seong Im and Tony Shaska, Editors,** Artificial Intelligence and Mathematics Research, 2026

834 **Mee Seong Im, Hanna E. Makaruk, Bach Nguyen, and Robert Owczarek, Editors,** Inverse Problems, Geometric and Topological Aspects of Representation Theory, and Low-Dimensional Topology, 2026

832 **Nils Bruin, David Kohel, and Chloe Martindale, Editors,** Arithmetic, Geometry, Cryptography and Coding Theory, 2026

831 **Irena Swanson and Kei-ichi Watanabe, Editors,** Lectures on Tight Closure and Its Applications, 2025

830 **O. Bezushchak, Yu. Drozd, P. Etingof, V. Futorny, R. Grigorchuk, and E. Zelmanov, Editors,** Modern Algebra Volume 2, 2025

829 **O. Bezushchak, Yu. Drozd, P. Etingof, V. Futorny, R. Grigorchuk, and E. Zelmanov, Editors,** Modern Algebra Volume 1, 2025

828 **Hafedh Herichi, Maria Rosaria Lancia, Therese Basa Landry, Anna Rozanova-Pierrat, and Steffen Winter, Editors,** Fractal Geometry in Pure and Applied Mathematics, 2025

827 **Carmen Caprau, J. Scott Carter, Neslihan Gügümcü, and Sam Nelson, Editors,** Algebraic Structures in Knot Theory, 2025

826 **Ahmed Laghribi and André Leroy, Editors,** Algebraic Structures and Applications, 2025

825 **Saurabh Verma, María A. Navascués, and Amit Priyadarshi, Editors,** Advances in Dimension Theory, Fractal Functions and Measures, 2025

824 **Hanna E. Makaruk and Robert M. Owczarek, Editors,** Inverse Problems: In Memory of Professor Zbigniew Oziewicz, 2025

823 **Mikhail Bershtein, Anton Dzhamay, and Andrei Okounkov, Editors,** The Versatility of Integrability, 2025

822 **Ahmad Barhoumi, Roozbeh Gharakhloo, and Andrei Martinez-Finkelshtein, Editors,** Recent Developments in Orthogonal Polynomials, 2025

821 **Bogdan D. Suceavă, Adara M. Blaga, Marian Ioan Munteanu, Cezar Oniciuc, Shoo Seto, Gabriel-Eduard Vîlcu, and Lihan Wang, Editors,** Recent Advances in Differential Geometry and Related Areas, 2025

820 **Sudeb Mitra, Rakesh Kumar Parmar, Saminathan Ponnusamy, and Toshiyuki Sugawa, Editors,** Geometric Function Theory and Related Topics, 2025

Kolibri 360 GmbH
Ferdinand-Jühlke-Straße 7
99095 Erfurt, Deutschland
produktsicherheit@kolibri360.de